CAMBRIDGE:
PRINTED BY W. METCALFE AND SON, TRINITY STREET.

CONTENTS.

[Junior readers will find all essential parts of the theory of Analytical Geometry included in Chapters I., II., V., VI., X., XI., XII., omitting the articles marked with asterisks.]

CHAPTER I.

THE POINT.

CHAPTER II.

THE RIGHT LINE.

CHAPTER III.

EXAMPLES ON THE RIGHT LINE.

CHAPTER IV.

THE RIGHT LINE.—ABRIDGED NOTATION.

CHAPTER V.

RIGHT LINES.

CHAPTER VI.

THE CIRCLE.

CHAPTER VII.

EXAMPLES ON CIRCLE.

CHAPTER VIII.

PROPERTIES OF TWO OR MORE CIRCLES.

CHAPTER IX.

THE CIRCLE—ABRIDGED NOTATION.

CHAPTER X.

GENERAL EQUATION OF SECOND DEGREE.

CHAPTER XI.

CENTRAL EQUATIONS.

CHAPTER XII.

THE PARABOLA.

CHAPTER XIII.

EXAMPLES ON CONICS.

CHAPTER XIV.

ABRIDGED NOTATION.

CHAPTER XV.

RECIPROCAL POLARS.

CHAPTER XVI.

HARMONIC AND ANHARMONIC PROPERTIES.

CHAPTER XVII.

THE METHOD OF PROJECTION.

CHAPTER XVIII.

INVARIANTS AND COVARIANTS.

CHAPTER XIX.

THE METHOD OF INFINITESIMALS.

NOTES.

ANALYTIC GEOMETRY.

CHAPTER I.

THE POINT.

1. THE following method of determining the position of any point on a plane was introduced by Des Cartes in his *Géométrie*, 1637, and has been generally used by succeeding geometers.

We are supposed to be given the position of two fixed right lines XX', YY' intersecting in the point O. Now, if through any point P we draw PM, PN parallel to YY' and XX', it is plain that, if we knew the position of the point P, we should know the lengths of the parallels PM, PN; or, *vice versâ*, that if we knew the lengths of PM, PN, we should know the position of the point P.

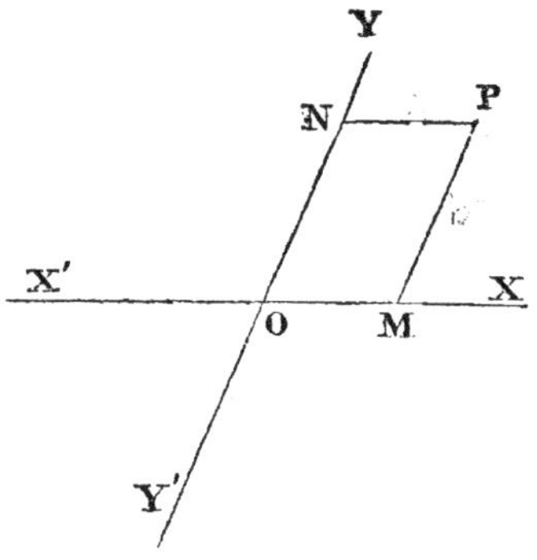

Suppose, for example, that we are given $PN = a$, $PM = b$, we need only measure $OM = a$ and $ON = b$, and draw the parallels PM, PN, which will intersect in the point required.

It is usual to denote PM parallel to OY by the letter y, and PN parallel to OX by the letter x, and the point P is said to be determined by the two equations $x = a$, $y = b$.

2. The parallels PM, PN are called the *coordinates* of the point P. PM is often called the *ordinate* of the point P; while PN, which is equal to OM the intercept *cut off* by the ordinate, is called the *abscissa*.

The fixed lines XX' and YY' are termed the *axes of co-ordinates*, and the point O, in which they intersect, is called the *origin*. The axes are said to be rectangular or oblique, according as the angle at which they intersect is a right angle or oblique.

It will readily be seen that the coordinates of the point M on the preceding figure are $x = a$, $y = 0$; that those of the point N are $x = 0$, $y = b$; and of the origin itself are $x = 0$, $y = 0$.

3. In order that the equations $x = a$, $y = b$ should only be satisfied by *one* point, it is necessary to pay attention, not only to the *magnitudes*, but also to the *signs* of the co-ordinates.

If we paid no attention to the signs of the coordinates, we might measure $OM = a$ and $ON = b$, on either side of the origin, and any of the four points P, P_1, P_2, P_3 would satisfy the equations $x = a$, $y = b$. It is possible, however, to distinguish algebraically between the lines OM, OM' (which are equal in magnitude, but opposite in direction) by giving them different signs. We lay down a rule that, if lines measured in one direction be considered as positive, lines measured in the opposite direction must be considered as negative. It is, of course, arbitrary in which direction we measure positive lines, but it is customary to consider OM (measured to the *right* hand) and ON (measured *upwards*) as positive, and OM', ON' (measured in the opposite directions) as negative lines.

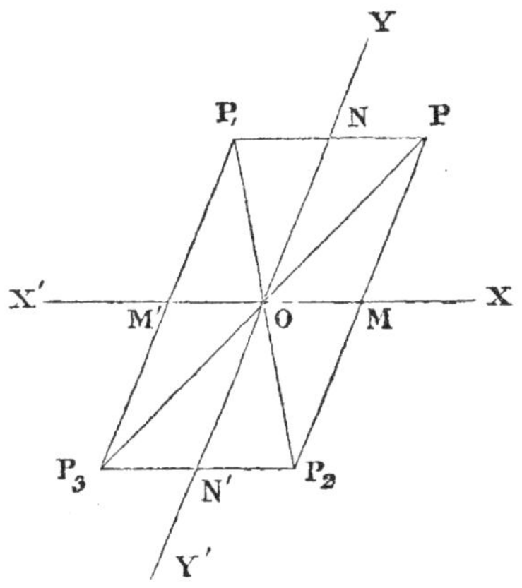

Introducing these conventions, the four points P, P_1, P_2, P_3 are easily distinguished. Their co-ordinates are, respectively,

$$\left.\begin{array}{l} x = +a \\ y = +b \end{array}\right\}, \quad \left.\begin{array}{l} x = -a \\ y = +b \end{array}\right\}, \quad \left.\begin{array}{l} x = +a \\ y = -b \end{array}\right\}, \quad \left.\begin{array}{l} x = -a \\ y = -b \end{array}\right\}.$$

These distinctions of sign can present no difficulty to the learner, who is supposed to be already acquainted with trigonometry.

N.B.—The points whose coordinates are $x=a$, $y=b$, or $x=x'$, $y=y'$, are generally briefly designated as the point (a, b), or the point $x'y'$.

It appears from what has been said, that the points $(+a, +b)$, $(-a, -b)$ lie on a right line passing through the origin; that they are equidistant from the origin, and on opposite sides of it.

4. *To express the distance between two points $x'y'$, $x''y''$, the axes of coordinates being supposed rectangular.*

By Euclid I. 47,

$$PQ^2 = PS^2 + SQ^2, \text{ but } PS = PM - QM' = y' - y'',$$

and $$QS = OM - OM' = x' - x'';$$

hence

$$\delta^2 = PQ^2 = (x' - x'')^2 + (y' - y'')^2.$$

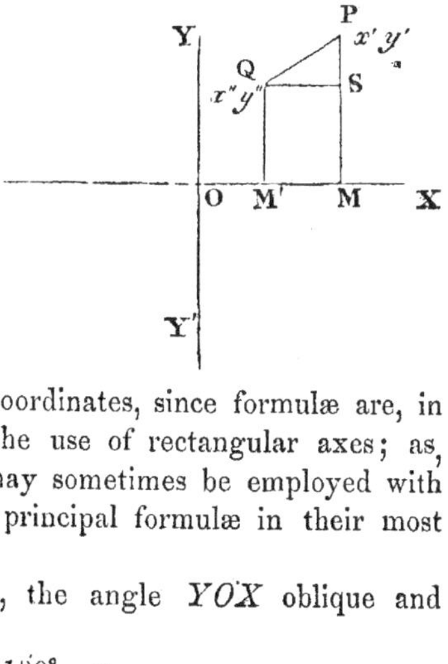

To express the distance of any point from the origin, we must make $x''=0$, $y''=0$ in the above, and we find

$$\delta^2 = x'^2 + y'^2.$$

5. In the following pages we shall but seldom have occasion to make use of oblique coordinates, since formulæ are, in general, much simplified by the use of rectangular axes; as, however, oblique coordinates may sometimes be employed with advantage, we shall give the principal formulæ in their most general form.

Suppose, in the last figure, the angle YOX oblique and $=\omega$, then

$$PSQ = 180° - \omega,$$

and $$PQ^2 = PS^2 + QS^2 - 2PS.\,QS.\cos PSQ,$$

or, $$PQ^2 = (y' - y'')^2 + (x' - x'')^2 + 2\,(y' - y'')\,(x' - x'')\cos\omega.$$

Similarly, the square of the distance of a point, $x'y'$, from the origin $= x'^2 + y'^2 + 2x'y' \cos\omega$.

In applying these formulæ, attention must be paid to the signs of the coordinates. If the point Q, for example, were in the angle XOY', the sign of y'' would be changed, and the line PS would be the *sum* and not the *difference* of y' and y''. The learner will find no difficulty, if, having written the coordinates with their proper signs, he is careful to take for PS and QS the *algebraic* difference of the corresponding pair of coordinates.

Ex. 1. Find the lengths of the sides of a triangle, the coordinates of whose vertices are $x' = 2, y' = 3$; $x'' = 4, y'' = -5$; $x''' = -3, y''' = -6$, the axes being rectangular. *Ans.* $\sqrt{68}$, $\sqrt{50}$, $\sqrt{106}$.

Ex. 2. Find the lengths of the sides of a triangle, the coordinates of whose vertices are the same as in the last example, the axes being inclined at an angle of 60°. *Ans.* $\sqrt{52}$, $\sqrt{57}$, $\sqrt{151}$.

Ex. 3. Express that the distance of the point xy from the point (2, 3) is equal to 4. *Ans.* $(x-2)^2 + (y-3)^2 = 16$

Ex. 4. Express that the point xy is equidistant from the points (2, 3), (4, 5). *Ans.* $(x-2)^2 + (y-3)^2 = (x-4)^2 + (y-5)^2$; or $x + y = 7$.

Ex. 5. Find the point equidistant from the points (2, 3), (4, 5), (6, 1). Here we have two equations to determine the two unknown quantities x, y.

Ans. $x = \frac{13}{3}$, $y = \frac{8}{3}$, and the common distance is $\frac{\sqrt{(50)}}{3}$.

6. The distance between two points, being expressed in the form of a square root, is necessarily susceptible of a double sign. If the distance PQ, measured from P to Q, be considered positive, then the distance QP, measured from Q to P, is considered negative. If indeed we are only concerned with the single distance between two points, it would be unmeaning to affix any sign to it, since by prefixing a sign we in fact direct that this distance shall be added to, or subtracted from, some other distance. But suppose we are given three points P, Q, R in a right line, and know the distances PQ, QR, we may infer $PR = PQ + QR$. And with the explanation now given, this equation remains true, even though the point R lie between P and Q. For, in that case, PQ and QR are measured in opposite directions, and PR, which is their arithmetical difference, is still their algebraical sum. Except in the case of lines parallel to one of the axes, no convention has been established as to which shall be considered the positive direction.

7. *To find the coordinates of the point cutting in a given ratio $m : n$, the line joining two given points $x'y'$, $x''y''$.*

Let x, y be the coordinates of the point R which we seek to determine, then

$m : n :: PR : RQ :: MS : SN$,

or

$m : n :: x' - x : x - x''$,

or $mx - mx'' = nx' - nx$,

hence

$$x = \frac{mx'' + nx'}{m + n}.$$

In like manner

$$y = \frac{my'' + ny'}{m + n}.$$

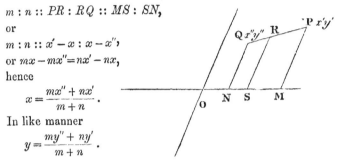

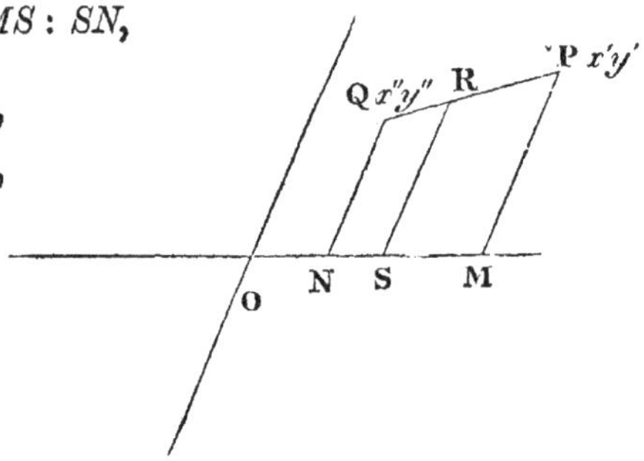

If the line were to be cut *externally* in the given ratio we should have

$$m : n :: x - x' : x - x'',$$

and therefore $$x = \frac{mx'' - nx'}{m - n}, \quad y = \frac{my'' - ny'}{m - n}.$$

It will be observed that the formulæ for external section are obtained from those for internal section by changing the sign of the ratio; that is, by changing $m : + n$ into $m : - n$. In fact, in the case of internal section, PR and RQ are measured in the same direction, and their ratio (Art. 6) is to be counted as positive. But in the case of external section PR and RQ are measured in opposite directions, and their ratio is negative.

Ex. 1. To find the coordinates of the middle point of the line joining the points $x'y'$, $x''y''$. *Ans.* $x = \frac{x' + x''}{2}$, $y = \frac{y' + y''}{2}$.

Ex. 2. To find the coordinates of the middle points of the sides of the triangle, the coordinates of whose vertices are (2, 3), (4, − 5), (− 3, − 6). *Ans.* $(\frac{1}{2}, -\frac{11}{2})$, $(-\frac{1}{2}, -\frac{3}{2})$, $(3, -1)$.

Ex. 3. The line joining the points (2, 3), (4, − 5) is trisected; to find the coordinates of the point of trisection nearest the former point. *Ans.* $x = \frac{8}{3}$, $y = \frac{1}{3}$.

Ex. 4. The coordinates of the vertices of a triangle being $x'y'$, $x''y''$, $x'''y'''$, to find the coordinates of the point of trisection (remote from the vertex) of the line joining any vertex to the middle point of the opposite side.

Ans. $x = \frac{1}{3}(x' + x'' + x''')$, $y = \frac{1}{3}(y' + y'' + y''')$.

Ex. 5. To find the coordinates of the intersection of the bisectors of sides of the triangle, the coordinates of whose vertices are given in Ex. 2. *Ans.* $x = 1$, $y = -\frac{8}{3}$.

Ex. 6. Any side of a triangle is cut in the ratio $m : n$, and the line joining this to the opposite vertex is cut in the ratio $m + n : l$; to find the coordinates of the point of section. *Ans.* $x = \dfrac{lx' + mx'' + nx'''}{l + m + n}$, $y = \dfrac{ly' + my'' + ny'''}{l + m + n}$.

TRANSFORMATION OF COORDINATES.*

8. When we know the coordinates of a point referred to one pair of axes, it is frequently necessary to find its coordinates referred to another pair of axes. This operation is called *the transformation of coordinates.*

We shall consider three cases separately; first, we shall suppose the origin changed, but the new axes parallel to the old; secondly, we shall suppose the directions of the axes changed, but the origin to remain unaltered; and thirdly, we shall suppose both origin and directions of axes to be altered.

First. Let the new axes be parallel to the old.

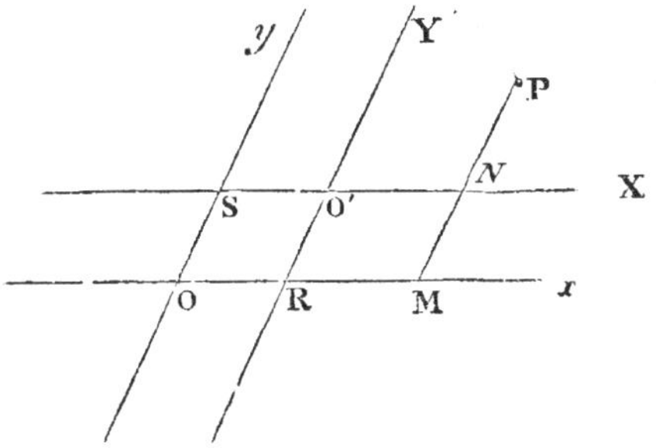

Let Ox, Oy be the old axes, $O'X$, $O'Y$ the new axes. Let the coordinates of the new origin referred to the old be x', y', or $O'S = x'$, $O'R = y'$. Let the old coordinates be x, y, the new X, Y, then we have

$$OM = OR + RM, \text{ and } PM = PN + NM,$$

that is $$x = x' + X, \text{ and } y = y' + Y.$$

These formulæ are, evidently, equally true, whether the axes be oblique or rectangular.

9. Secondly, let the directions of the axes be changed, while the origin is unaltered.

---

* The beginner may postpone the rest of this chapter till he has read to the end of Art. 41.

Let the original axes be Ox, Oy, so that we have $OQ=x$, $PQ=y$. Let the new axes be OX, OY, so that we have $ON=X$, $PN=Y$. Let OX, OY make angles respectively α, β, with the old axis of x, and angles α', β' with the old axis of y; and if the angle xOy between the old axes be ω, we have obviously $\alpha+\alpha'=\omega$, since $XOx+XOy=xOy$; and in like manner $\beta+\beta'=\omega$.

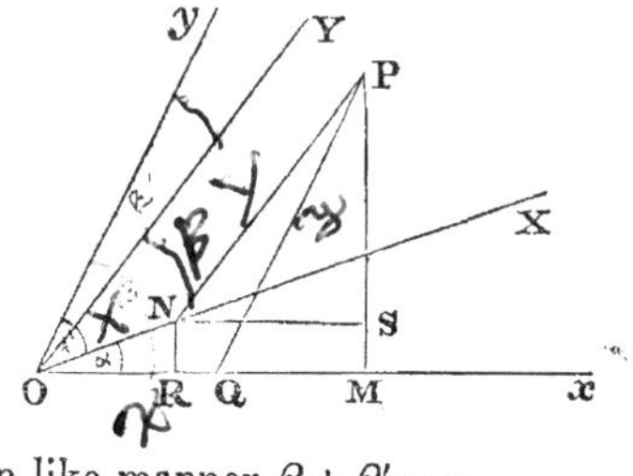

The formulæ of transformation are most easily obtained by expressing the perpendiculars from P on the original axes, in terms of the new coordinates and the old. Since

$$PM=PQ\sin PQM, \text{ we have } PM=y\sin\omega.$$

But also $PM=NR+PS=ON\sin NOR+PN\sin PNS$.

Hence $\quad y\sin\omega=X\sin\alpha+Y\sin\beta.$

In like manner

$$x\sin\omega=X\sin\alpha'+Y\sin\beta';$$

or $$x\sin\omega=X\sin(\omega-\alpha)+Y\sin(\omega-\beta).$$

In the figure the angles α, β, ω are all measured on the same side of Ox; and α', β', ω all on the same side of Oy. If any of these angles lie on the opposite side it must be given a negative sign. Thus, if OY lie to the left of Oy, the angle β is greater than ω, and β' $(=\omega-\beta)$ is negative, and therefore the coefficient of Y in the expression for $x\sin\omega$ is negative. This occurs in the following special case, to which, as the one which most frequently occurs in practice, we give a separate figure.

To transform from a system of rectangular axes to a new rectangular system making an angle θ with the old.

Here we have

$$\alpha=\theta,\ \beta=90+\theta,$$
$$\alpha'=90-\theta,\ \beta'=-\theta;$$

and the general formulæ become

$$y=X\sin\theta+Y\cos\theta,$$
$$x=X\cos\theta-Y\sin\theta;$$

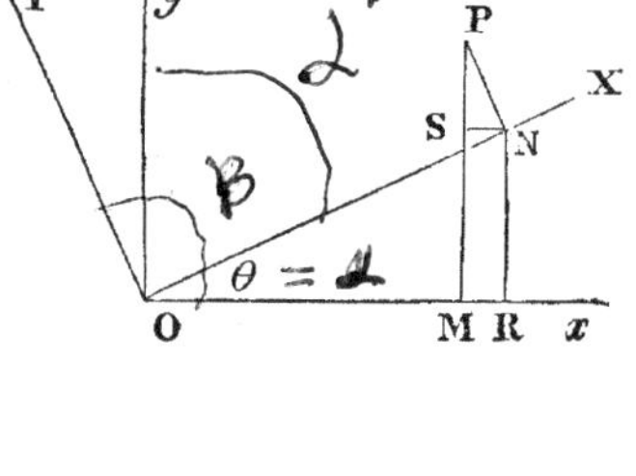

For Ex. 2, P8 cf. Ex., P75 of G. Hale Puckle

NB: Right lines refer to straight lines.

the truth of which may also be seen directly, since $y = PS + NR$, $x = OR - SN$, while

$PS = PN \cos\theta$, $NR = ON \sin\theta$; $OR = ON \cos\theta$, $SN = PN \sin\theta$.

There is only one other case of transformation which often occurs in practice.

To transform from oblique coordinates to rectangular, retaining the old axis of x.

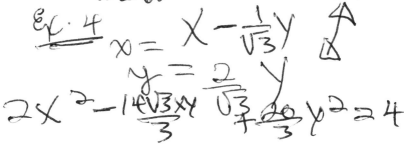

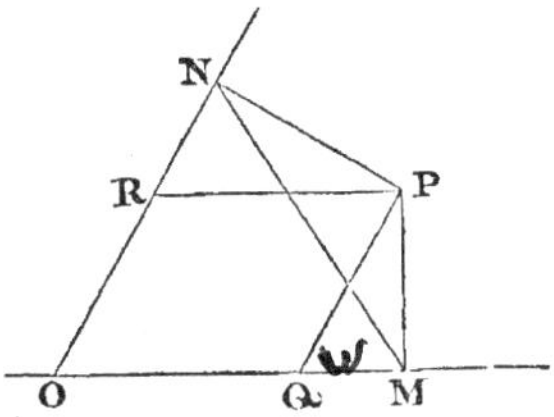

We may use the general formulæ making

$\alpha = 0$, $\beta = 90$, $\alpha' = \omega$, $\beta' = \omega - 90$.

But it is more simple to investigate the formulæ directly. We have OQ and PQ for the old x and y, OM and PM for the new; and, since $PQM = \omega$, we have

$$Y = y \sin\omega, \quad X = x + y \cos\omega;$$

while from these equations we get the expressions for the old coordinates in terms of the new

$$y \sin\omega = Y, \quad x \sin\omega = X \sin\omega - Y \cos\omega.$$

10. Thirdly, by combining the transformations of the two preceding articles, we can find the coordinates of a point referred to two new axes in any position whatever. We first find the coordinates (by Art. 8) referred to a pair of axes through the new origin parallel to the old axes, and then (by Art. 9) we can find the coordinates referred to the required axes.

The general expressions are obviously obtained by adding x' and y' to the values for x and y given in the last article.

Ex. 1. The coordinates of a point satisfy the relation

$$x^2 + y^2 - 4x - 6y = 18;$$

what will this become if the origin be transformed to the point (2, 3)?

Ans. $X^2 + Y^2 = 31$.

Ex. 2. The coordinates of a point to a set of rectangular axes satisfy the relation $y^2 - x^2 = 6$; what will this become if transformed to axes bisecting the angles between the given axes? *Ans.* $XY = 3$.

✓ Ex. 3. Transform the equation $2x^2 - 5xy + 2y^2 = 4$ from axes inclined to each other at an angle of 60° to the right lines which bisect the angles between the given axes. ?? $\omega = 60°$ *Ans.* $X^2 - 27Y^2 + 12 = 0$.

Ex. 4. Transform the same equation to rectangular axes, retaining the old axis of x. $\omega = 60°$ *Ans.* $3X^2 + 10Y^2 - 7XY\sqrt{3} = 6$.

Ex. 4

$x = X - \frac{1}{\sqrt{3}}Y$

$y = \frac{2}{\sqrt{3}}Y$

$2X^2 - \frac{14\sqrt{3}}{3}XY + \frac{20}{3}Y^2 = 4$

Ex. 5. It is evident that when we change from one set of rectangular axes to another, $x^2 + y^2$ must $= X^2 + Y^2$, since both express the square of the distance of a point from the origin. Verify this by squaring and adding the expressions for X and Y in Art. 9.

Ex. 6. Verify in like manner in general that

$$x^2 + y^2 + 2xy \cos xOy = X^2 + Y^2 + 2XY \cos XOY.$$

If we write $X \sin \alpha + Y \sin \beta = L$, $X \cos \alpha + Y \cos \beta = M$, the expressions in Art. 9 may be written $y \sin \omega = L$, $x \sin \omega = M \sin \omega - L \cos \omega$; whence

$$\sin^2\omega\,(x^2 + y^2 + 2xy \cos \omega) = (L^2 + M^2) \sin^2\omega.$$

But $\quad L^2 + M^2 = X^2 + Y^2 + 2XY \cos (\alpha - \beta)$, and $\alpha - \beta = XOY$.

11. *The degree of any equation between the coordinates is not altered by transformation of coordinates.*

Transformation cannot *increase* the degree of the equation; for if the highest terms in the given equation be x^m, y^m, &c., those in the transformed equation will be

$\{x' \sin \omega + x \sin (\omega - \alpha) + y \sin (\omega - \beta)\}^m$, $(y' \sin \omega + x \sin \alpha + y \sin \beta)^m$, &c., which evidently cannot contain powers of x or y above the mth degree. Neither can transformation *diminish* the degree of an equation, since by transforming the transformed equation back again to the old axes, we must fall back on the original equation, and if the first transformation had diminished the degree of the equation, the second should increase it, contrary to what has just been proved.

POLAR COORDINATES.

12. Another method of expressing the position of a point is often employed.

If we were given a fixed point O, and a fixed line through it OB, it is evident that we should know the position of any point P, if we knew the length OP, and also the angle POB. The line OP is called the *radius vector;* the fixed point is called the *pole;* and this method is called the method of *polar coordinates.*

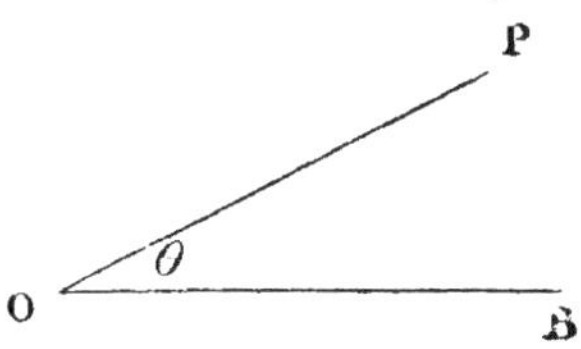

It is very easy, being given the x and y coordinates of a point, to find its polar ones, or *vice versâ.*

First, let the fixed line coincide with the axis of x, then we have

$$OP : PM :: \sin PMO : \sin POM;$$

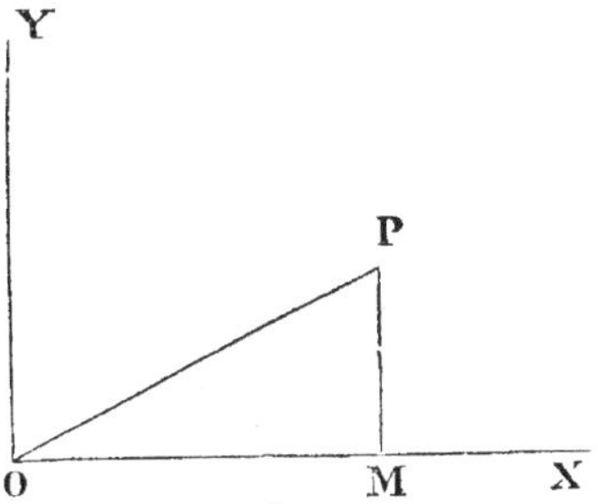

denoting OP by ρ, POM by θ, and YOX by ω, then

$$PM \text{ or } y = \frac{\rho \sin\theta}{\sin\omega};$$

and similarly, $$OM = x = \frac{\rho \sin(\omega - \theta)}{\sin\omega}.$$

For the more ordinary case of rectangular coordinates, $\omega = 90°$, and we have simply

$$x = \rho \cos\theta \text{ and } y = \rho \sin\theta.$$

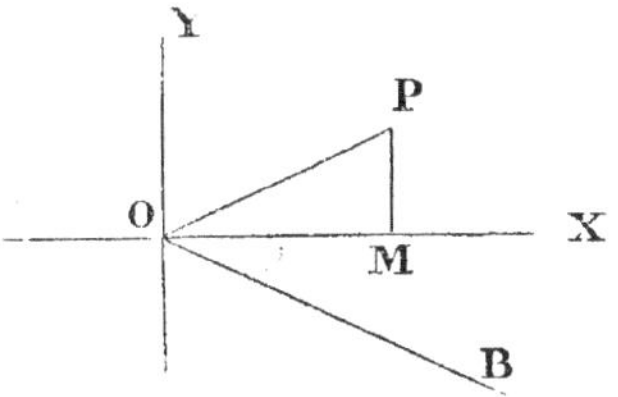

Secondly, let the fixed line OB not coincide with the axis of x, but make with it an angle $= \alpha$, then

$$POB = \theta \text{ and } POM = \theta - \alpha,$$

and we have only to substitute $\theta - \alpha$ for θ in the preceding formulæ.

For rectangular coordinates we have

$$x = \rho \cos(\theta - \alpha) \text{ and } y = \rho \sin(\theta - \alpha).$$

Ex. 1. Change to polar coordinates the following equations in rectangular coordinates:

$x^2 + y^2 = 5mx$. *Ans.* $\rho = 5m \cos\theta$.
$x^2 - y^2 = a^2$. *Ans.* $\rho^2 \cos 2\theta = a^2$.

Ex. 2. Change to rectangular coordinates the following equations in polar coordinates:

$\rho^2 \sin 2\theta = 2a^2$. *Ans.* $xy = a^2$.
$\rho^2 = a^2 \cos 2\theta$. *Ans.* $(x^2 + y^2)^2 = a^2 (x^2 - y^2)$.
$\rho^{\frac{1}{2}} \cos \frac{1}{2}\theta = a^{\frac{1}{2}}$. *Ans.* $x^2 + y^2 = (2a - x)^2$.
$\rho^{\frac{1}{2}} = a^{\frac{1}{2}} \cos \frac{1}{2}\theta$. *Ans.* $(2x^2 + 2y^2 - ax)^2 = a^2 (x^2 + y^2)$.

13. *To express the distance between two points, in terms of their polar coordinates.*

Let P and Q be the two points,

$$OP = \rho', \quad POB = \theta';$$
$$OQ = \rho'', \quad QOB = \theta'';$$

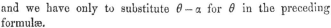

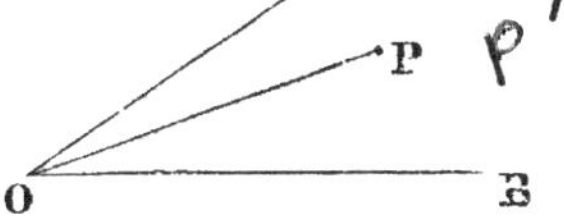

then $$PQ^2 = OP^2 + OQ^2 - 2OP.OQ.\cos POQ,$$

or $$\delta^2 = \rho'^2 + \rho''^2 - 2\rho'\rho'' \cos(\theta'' - \theta').$$

CHAPTER II.

THE RIGHT LINE.

14. *Any two equations between the coordinates represent geometrically one or more points.*

If the equations be both of the first degree (see Ex. 5, p. 4) they denote a single point. For solving the equations for x and y, we obtain a result of the form $x = a$, $y = b$, which, as was proved in the last chapter, represents a point.

If the equations be of higher degree, they represent more points than one. For, eliminating y between the equations, we obtain an equation containing x only; let its roots be α_1, α_2, α_3, &c. Now, if we substitute any of these values (α_1) for x in the original equations, we get two equations in y, which must have a common root (since the result of elimination between the equations is rendered $= 0$ by the supposition $x = \alpha_1$). Let this common root be $y = \beta_1$. Then the values $x = \alpha_1$, $y = \beta_1$, at once satisfy both the given equations, and denote a point which is represented by these equations. So, in like manner, is the point whose coordinates are $x = \alpha_2$, $y = \beta_2$, &c.

Ex. 1. What point is denoted by the equations $3x + 5y = 13$, $4x - y = 2$?

Ans. $x = 1$, $y = 2$

Ex. 2. What points are represented by the two equations $x^2 + y^2 = 5$, $xy = 2$? Eliminating y between the equations, we get $x^4 - 5x^2 + 4 = 0$. The roots of this equation are $x^2 = 1$ and $x^2 = 4$, and, therefore, the four values of x are

$$x = +1, \; x = -1, \; x = +2, \; x = -2.$$

Substituting these successively in the second equation, we obtain the corresponding values of y,

$$y = +2, \; y = -2, \; y = +1, \; y = -1.$$

The two given equations, therefore, represent the four points

$$(+1, +2), \; (-1, -2), \; (+2, +1), \; (-2, -1).$$

Ex. 3. What points are denoted by the equations

$$x - y = 1, \; x^2 + y^2 = 25?$$ *Ans.* (4, 3), (− 3, − 4).

Ex. 4. What points are denoted by the equations

$$x^2 - 5x + y + 3 = 0, \; x^2 + y^2 - 5x - 3y + 6 = 0?$$

Ans. (1, 1), (2, 3), (3, 3), (4, 1).

15. *A single equation between the coordinates denotes a geometrical locus.*

One equation evidently does not afford us conditions enough to determine the two unknown quantities x, y; and an indefinite number of systems of values of x and y can be found which will satisfy the given equation. And yet the coordinates of any point *taken at random* will not satisfy it. The assemblage then of points, whose coordinates *do* satisfy the equation, forms a *locus*, which is considered the geometrical signification of the given equation.

Thus, for example, we saw (Ex. 3, p. 4) that the equation

$$(x-2)^2+(y-3)^2=16$$

expresses that the distance of the point xy from the point $(2, 3)=4$. This equation then is satisfied by the coordinates of any point on the circle whose centre is the point (2, 3), and whose radius is 4; and by the coordinates of no other point. This circle then is the locus which the equation is said to represent.

We can illustrate by a still simpler example, that a single equation between the coordinates signifies a locus. Let us recall the construction by which (p. 1) we determined the position of a point from the two equations $x=a$, $y=b$. We took $OM=a$; we drew MK parallel to OY; and then, measuring $MP=b$, we found P, the point required. Had we been given a different value of y, $x=a$, $y=b'$, we should proceed as before, and we should find a point P' still situated on the line MK, but at a different distance from M. Lastly, if the value of y were left wholly indeterminate, and we were merely given the single equation $x=a$, we should know that the point P was situated *somewhere* on the line MK, but its position in that line would not be determined. Hence the line MK is the locus of all the points represented by the equation $x=a$,

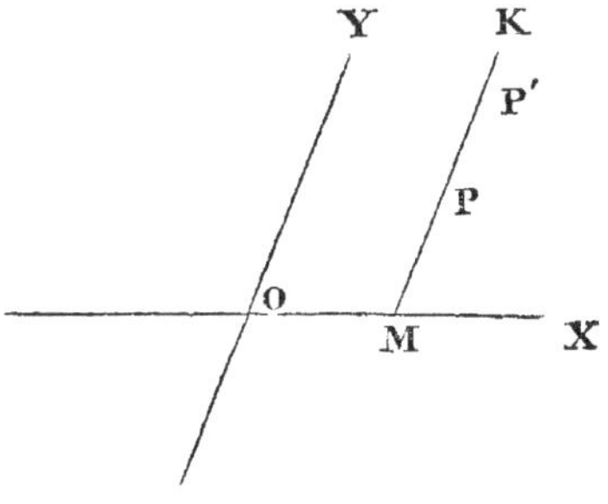

since, whatever point we take on the line MK, the x of that point will always $= a$.

16. In general, if we are given an equation of any degree between the coordinates, let us assume for x any value we please ($x = a$), and the equation will enable us to determine a finite number of values of y answering to this particular value of x; and, consequently, the equation will be satisfied for each of the points (p, q, r, &c.), whose x is the assumed value, and whose y is that found from the equation. Again, assume for x any other value ($x = a'$), and we find, in like manner, another series of points, p', q', r', whose coordinates satisfy the equation. So again, if we assume $x = a''$ or $x = a'''$, &c. Now, if x be supposed to take successively all possible values, the assemblage of points found as above will form a *locus*, every point of which satisfies the conditions of the equation, and which is, therefore, its geometrical signification.

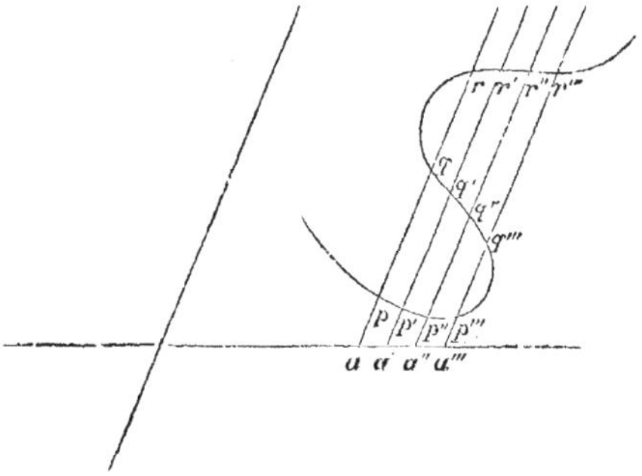

We can find in the manner just explained as many points of this locus as we please, until we have enough to represent its figure to the eye.

Ex. 1. Represent in a figure* a series of points which satisfy the equation $y = 2x + 3$.

Ans. Giving x the values -2, -1, 0, 1, 2, &c., we find for y, -1, 1, 3, 5, 7, &c., and the corresponding points will be seen all to lie on a right line.

Ex. 2. Represent the locus denoted by the equation $y = x^2 - 3x - 2$.

Ans. To the values for x, -1, $-\frac{1}{2}$, 0, $\frac{1}{2}$, 1, $\frac{3}{2}$, 2, $\frac{5}{2}$, 3, $\frac{7}{2}$, 4; correspond for y, 2, $-\frac{1}{4}$, -2, $-\frac{13}{4}$, -4, $-\frac{17}{4}$, -4, $-\frac{13}{4}$, -2, $-\frac{1}{4}$, 2. If the points thus denoted be laid down on paper, they will sufficiently exhibit the form of the curve, which may be continued indefinitely by giving x greater positive or negative values.

Ex. 3. Represent the curve $y = 3 \pm \sqrt{(20 - x - x^2)}$.

Here to each value of x correspond two values of y. No part of the curve lies to the right of the line $x = 4$, or to the left of the line $x = -5$, since by giving greater positive or negative values to x, the value of y becomes imaginary.

* The learner is recommended to use paper ruled into little squares, which is sold under the name of logarithm paper.

17. The whole science of Analytic Geometry is founded on the connexion which has been thus proved to exist between an equation and a locus. If a curve be defined by any geometrical property, it will be our business to deduce from that property an equation which must be satisfied by the coordinates of every point on the curve. Thus, if a circle be defined as the locus of a point (x, y), whose distance from a fixed point (a, b) is constant, and equal to r, then *the equation of the circle* in rectangular coordinates is (Art. 4),

$$(x-a)^2+(y-b)^2=r^2.$$

On the other hand, it will be our business when an equation is given, to find the figure of the curve represented, and to deduce its geometrical properties. In order to do this systematically, we make a classification of equations according to their degrees, and beginning with the simplest, examine the form and properties of the locus represented by the equation. The degree of an equation is estimated by the highest value of the sum of the indices of x and y in any term. Thus the equation $xy+2x+3y=4$ is of the second degree, because it contains the term xy. If this term were absent, it would be of the first degree. A curve is said to be of the nth degree when the equation which represents it is of that degree.

We commence with the equation of the first degree, and we shall prove that this always represents a *right line*, and, conversely, that the equation of a right line is always of the first degree.

18. We have already (Art. 15) interpreted the simplest case of an equation of the first degree, namely, the equation $x=a$. In like manner, the equation $y=b$ represents a line PN parallel to the axis OX, and meeting the axis OY at a distance from the origin $ON=b$. If we suppose b to be equal to nothing, we see that the equation $y=0$ denotes the axis OX; and in like manner that $x=0$ denotes the axis OY.

Let us now proceed to the case next in order of simplicity, and let us examine what relation subsists between the coordinates of points situated on a right line passing through the origin.

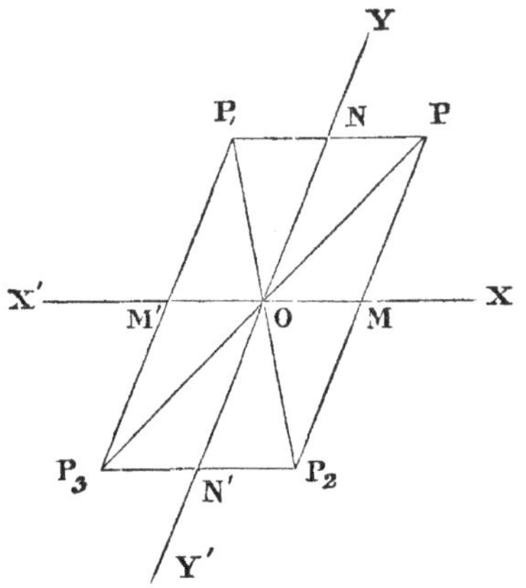

If we take any point P on such a line, we see that *both* the coordinates PM, OM, will vary in length, but that the *ratio* $PM : OM$ will be constant, being = to the ratio

$$\sin POM : \sin MPO.$$

Hence we see that the equation

$$y = \frac{\sin POM}{\sin MPO} x$$

will be satisfied for every point of the line OP, and therefore this equation is said to be the equation of the line OP.

Conversely, if we were asked what locus was represented by the equation

$$y = mx,$$

write the equation in the form $\frac{y}{x} = m$, and the question is: "To find the locus of a point P, such that, if we draw PM, PN parallel to two fixed lines, the ratio $PM : PN$ may be constant." Now this locus evidently is a right line OP, passing through O, the point of intersection of the two fixed lines, and dividing the angle between them in such a manner that

$$\sin POM = m \sin PON.$$

If the axes be rectangular, $\sin PON = \cos POM$; therefore, $m = \tan POM$, and the equation $y = mx$ represents a right line passing through the origin, and making an angle with the axis of x, whose tangent is m.

19. An equation of the form $y = +mx$ will denote a line OP, situated in the angles YOX, $Y'OX'$. For it appears, from the equation $y = +mx$, that whenever x is positive y will be positive, and whenever x is negative y will be negative. Points, therefore, represented by this equation must have their coordinates either both positive or both negative, and such points we saw (Art. 3) lie only in the angles YOX, $Y'OX'$.

On the contrary, in order to satisfy the equation $y = -mx$, if x be positive y must be negative, and if x be negative y must be positive. Points, therefore, satisfying this equation will have their coordinates of *different* signs; and the line represented by the equation, must, therefore (Art. 3), lie in the angles $Y'OX$, YOX'.

20. Let us now examine how to represent a right line PQ, situated in any manner with regard to the axes.

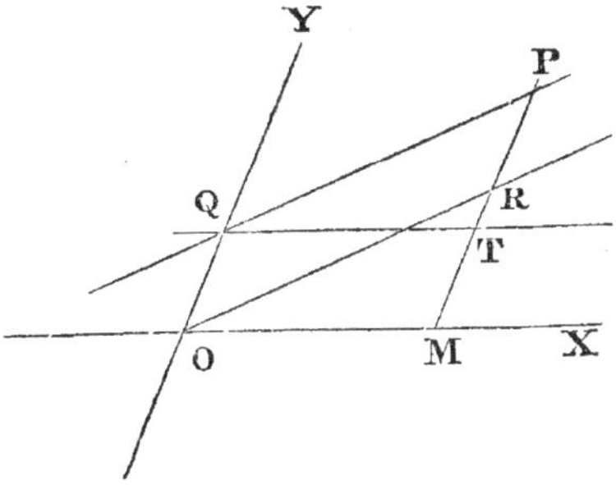

Draw OR through the origin parallel to PQ, and let the ordinate PM meet OR in R. Now it is plain (as in Art. 18), that the ratio $RM : OM$ will be always constant (RM always equal, suppose, to $m.OM$); but the ordinate PM differs from RM by the constant length $PR = OQ$, which we shall call b. Hence we may write down the equation

$$PM = RM + PR, \text{ or } PM = m.OM + PR,$$

that is $$y = mx + b.$$

The equation, therefore, $y = mx + b$, being satisfied by every point of the line PQ, is said to be the equation of that line.

It appears from the last Article, that m will be positive or negative according as OR, parallel to the right line PQ, lies in the angle YOX, or $Y'OX$. And, again, b will be positive or negative according as the point Q, in which the line meets OY, lies *above* or *below* the origin.

Conversely, the equation $y = mx + b$ will always denote a right line; for the equation can be put into the form

$$\frac{y-b}{x} = m.$$

Now, since if we draw the line QT parallel to OM, TM will be $= b$, and PT therefore $= y - b$, the question becomes: "To find the locus of a point, such that, if we draw PT parallel to OY to meet the fixed line QT, PT may be to QT in a constant ratio;" and this locus evidently is the right line PQ passing through Q.

The most general equation of the first degree, $Ax+By+C=0$, can obviously be reduced to the form $y=mx+b$, since it is equivalent to

$$y=-\frac{A}{B}x-\frac{C}{B};$$

this equation therefore *always* represents a right line.

21. From the last Articles we are able to ascertain the geometrical meaning of the constants in the equation of a right line. If the right line represented by the equation $y=mx+b$ make an angle $=\alpha$ with the axis of x, and $=\beta$ with the axis of y, then (Art. 18)

$$m=\frac{\sin\alpha}{\sin\beta};$$

and if the axes be rectangular, $m=\tan\alpha$.

We saw (Art. 20) that b is the intercept which the line cuts off on the axis of y.

If the equation be given in the general form $Ax+By+C=0$, we can reduce it, as in the last Article, to the form $y=mx+b$, and we find that

$$-\frac{A}{B}=\frac{\sin\alpha}{\sin\beta},$$

or if the axes be rectangular $=\tan\alpha$; and that $-\frac{C}{B}$ is the length of the intercept made by the line on the axis of y.

Cor. The lines $y=mx+b$, $y=m'x+b'$ will be parallel to each other if $m=m'$, since then they will both make the same angle with the axis. Similarly the lines $Ax+By+C=0$, $A'x+B'y+C'=0$, will be parallel if

$$\frac{A}{B}=\frac{A'}{B'}.$$

Beside the forms $Ax+By+C=0$ and $y=mx+b$, there are two other forms in which the equation of a right line is frequently used; these we next proceed to lay before the reader.

22. *To express the equation of a line MN in terms of the intercepts $OM = a$, $ON = b$ which it cuts off on the axes.*

We can derive this from the form already considered

$$Ax + By + C = 0, \text{ or } \frac{A}{C}x + \frac{B}{C}y + 1 = 0.$$

This equation must be satisfied by the coordinates of *every* point on MN, and therefore by those of M, which (see Art. 2) are $x = a$, $y = 0$. Hence we have

$$\frac{A}{C}a + 1 = 0, \quad \frac{A}{C} = -\frac{1}{a}.$$

In like manner, since the equation is satisfied by the coordinates of N, ($x = 0$, $y = b$), we have

$$\frac{B}{C} = -\frac{1}{b}.$$

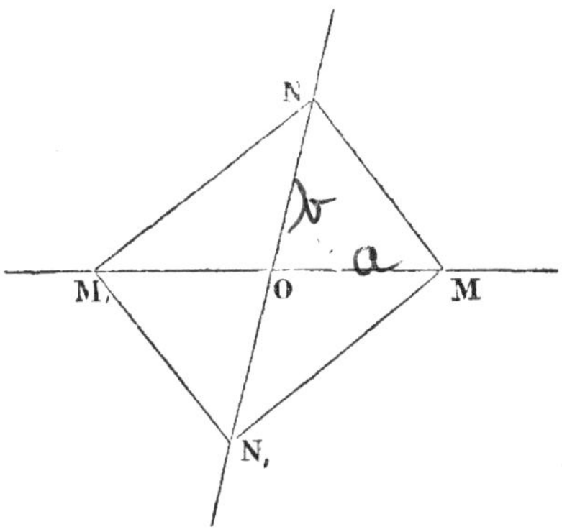

Substituting which values in the general form, it becomes

$$\frac{x}{a} + \frac{y}{b} = 1.$$

This equation holds whether the axes be oblique or rectangular.

It is plain that the position of the line will vary with the signs of the quantities a and b For example, the equation $\frac{x}{a} + \frac{y}{b} = 1$, which cuts off positive intercepts on both axes, represents the line MN on the preceding figure; $\frac{x}{a} - \frac{y}{b} = 1$, cutting off a positive intercept on the axis of x, and a negative intercept on the axis of y, represents MN'.

Similarly, $$-\frac{x}{a} + \frac{y}{b} = 1 \text{ represents } NM';$$

and $$-\frac{x}{a} - \frac{y}{b} = 1 \text{ represents } M'N'.$$

By dividing by the constant term, any equation of the first degree can evidently be reduced to some one of these four forms.

Ex. 1. Examine the position of the following lines, and find the intercepts they make on the axes:

$$2x - 3y = 7;\quad 3x + 4y + 9 = 0;$$

$$3x + 2y = 6;\quad 4y - 5x = 20.$$

Ex. 2. The sides of a triangle being taken for axes, form the equation of the line joining the points which cut off the mth part of each, and shew, by Art. 21, that it is parallel to the base.

Ans. $\frac{x}{a} + \frac{y}{b} = \frac{1}{m}$.

23. *To express the equation of a right line in terms of the length of the perpendicular on it from the origin, and of the angles which this perpendicular makes with the axes.*

Let the length of the perpendicular $OP = p$, the angle POM which it makes with the axis of $x = \alpha$, $PON = \beta$, $OM = a$, $ON = b$.

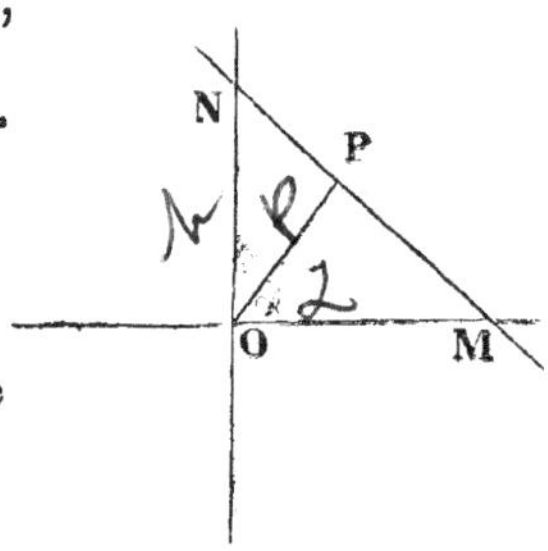

We saw (Art. 22) that the equation of the right line MN was

$$\frac{x}{a} + \frac{y}{b} = 1.$$

Multiply this equation by p, and we have

$$\frac{p}{a}x + \frac{p}{b}y = p.$$

But $\frac{p}{a} = \cos\alpha$, $\frac{p}{b} = \cos\beta$; therefore the equation of the line is

$$x \cos\alpha + y \cos\beta = p.$$

In rectangular coordinates, which we shall generally use, we have $\beta = 90° - \alpha$; and the equation becomes $x \cos\alpha + y \sin\alpha = p$. This equation will include the four cases of Art. 22, if we suppose that α may take any value from 0 to 360°. Thus, for the position NM', α is between 90° and 180°, and the coefficient of x is negative. For the position $M'N'$, α is between 180° and 270°, and has both sine and cosine negative. For MN', α is between 270° and 360°, and has a negative sine and positive cosine. In the last two cases, however, it is more convenient to write the formula $x \cos\alpha + y \sin\alpha = -p$, and consider α to denote the angle, ranging between 0 and 180°, made with the positive direction of the axis of x, by the perpendicular *produced*. In using, then, the formula $x \cos\alpha + y \sin\alpha = p$, we suppose p to be capable of a double sign, and α to denote the angle, not exceeding 180°, made with the axis of x either by the perpendicular or its production.

The general form $Ax+By+C=0$, can easily be reduced to the form $x \cos\alpha + y \sin\alpha = p$; for, dividing it by $\sqrt{(A^2+B^2)}$, we have

$$\frac{A}{\sqrt{(A^2+B^2)}}x + \frac{B}{\sqrt{(A^2+B^2)}}y + \frac{C}{\sqrt{(A^2+B^2)}} = 0.$$

But we may take

$$\frac{A}{\sqrt{(A^2+B^2)}} = \cos\alpha, \text{ and } \frac{B}{\sqrt{(A^2+B^2)}} = \sin\alpha,$$

since the sum of squares of these two quantities $=1$.

Hence we learn that $\frac{A}{\sqrt{(A^2+B^2)}}$ and $\frac{B}{\sqrt{(A^2+B^2)}}$ are respectively the cosine and sine of the angle which the perpendicular from the origin on the line $(Ax+By+C=0)$ makes with the axis of x, and that $\frac{C}{\sqrt{(A^2+B^2)}}$ is the length of this perpendicular.

*24. *To reduce the equation $Ax+By+C=0$ (referred to oblique coordinates) to the form $x \cos\alpha + y \cos\beta = p$.*

Let us suppose that the given equation when multiplied by a certain factor R is reduced to the required form, then $RA=\cos\alpha$, $RB=\cos\beta$. But it can easily be proved that, if α and β be any two angles whose sum is ω, we shall have

$$\cos^2\alpha + \cos^2\beta - 2\cos\alpha\cos\beta\cos\omega = \sin^2\omega.$$

Hence $$R^2(A^2+B^2-2AB\cos\omega) = \sin^2\omega,$$

and the equation reduced to the required form is

$$\frac{A\sin\omega}{\sqrt{(A^2+B^2-2AB\cos\omega)}}x + \frac{B\sin\omega}{\sqrt{(A^2+B^2-2AB\cos\omega)}}y + \frac{C\sin\omega}{\sqrt{(A^2+B^2-2AB\cos\omega)}} = 0.$$

And we learn that

$$\frac{A\sin\omega}{\sqrt{(A^2+B^2-2AB\cos\omega)}}, \quad \frac{B\sin\omega}{\sqrt{(A^2+B^2-2AB\cos\omega)}},$$

* Articles and Chapters marked with an asterisk may be omitted on a first reading.

are respectively the cosines of the angles that the perpendicular from the origin on the line $Ax + By + C = 0$ makes with the axes of x and y; and that $\frac{C \sin \omega}{\sqrt{(A^2 + B^2 - 2AB \cos \omega)}}$ is the length of this perpendicular. This length may be also easily calculated by dividing the double area of the triangle NOM, $(ON . OM \sin \omega)$ by the length of MN, expressions for which are easily found.

The square root in the denominators is, of course, susceptible of a double sign, since the equation may be reduced to either of the forms

$$x \cos \alpha + y \cos \beta - p = 0,\ x \cos (\alpha + 180^\circ) + y \cos (\beta + 180^\circ) + p = 0.$$

25. *To find the angle between two lines whose equations with regard to rectangular axes are given.*

The angle between the lines is manifestly equal to the angle between the perpendiculars on the lines from the origin; if therefore these perpendiculars make with the axis of x the angles α, α', we have (Art. 23)

$$\cos \alpha = \frac{A}{\sqrt{(A^2 + B^2)}};\ \sin \alpha = \frac{B}{\sqrt{(A^2 + B^2)}};$$

$$\cos \alpha' = \frac{A'}{\sqrt{(A'^2 + B'^2)}};\ \sin \alpha' = \frac{B'}{\sqrt{(A'^2 + B'^2)}}.$$

Hence
$$\sin (\alpha - \alpha') = \frac{BA' - AB'}{\sqrt{(A^2 + B^2)}\sqrt{(A'^2 + B'^2)}};$$

$$\cos (\alpha - \alpha') = \frac{AA' + BB'}{\sqrt{(A^2 + B^2)}\sqrt{(A'^2 + B'^2)}};$$

and therefore
$$\tan (\alpha - \alpha') = \frac{BA' - AB'}{AA' + BB'}.$$

Cor. 1. The two lines are parallel to each other when

$$BA' - AB' = 0 \quad \text{(Art 21)},$$

since then the angle between them vanishes.

Cor. 2. The two lines are perpendicular to each other when $AA' + BB' = 0$, since then the tangent of the angle between them becomes infinite.

If the equations of the lines had been given in the form

$$y = mx + b,\ y = m'x + b';$$

since the angle between the lines is the difference of the angles they make with the axis of x, and since (Art. 21) the tangents of these angles are m and m', it follows that the tangent of the required angle is $\frac{m - m'}{1 + mm'}$; that the lines are parallel if $m = m'$; and perpendicular to each other if $mm' + 1 = 0$.

*26. *To find the angle between two lines, the coordinates being oblique.*

We proceed as in the last article, using the expressions of Art. 24,

$$\cos\alpha = \frac{A \sin\omega}{\sqrt{(A^2 + B^2 - 2AB\cos\omega)}},$$

$$\cos\alpha' = \frac{A' \sin\omega}{\sqrt{(A'^2 + B'^2 - 2A'B'\cos\omega)}};$$

consequently,

$$\sin\alpha = \frac{B - A\cos\omega}{\sqrt{(A^2 + B^2 - 2AB\cos\omega)}};$$

$$\sin\alpha' = \frac{B' - A'\cos\omega}{\sqrt{(A'^2 + B'^2 - 2A'B'\cos\omega)}}.$$

Hence

$$\sin(\alpha - \alpha') = \frac{(BA' - AB')\sin\omega}{\sqrt{(A^2 + B^2 - 2AB\cos\omega)}\sqrt{(A'^2 + B'^2 - 2A'B'\cos\omega)}},$$

$$\cos(\alpha - \alpha') = \frac{BB' + AA' - (AB' + A'B)\cos\omega}{\sqrt{(A^2 + B^2 - 2AB\cos\omega)}\sqrt{(A'^2 + B'^2 - 2A'B'\cos\omega)}},$$

$$\tan(\alpha - \alpha') = \frac{(BA' - AB')\sin\omega}{AA' + BB' - (AB' + BA')\cos\omega}.$$

Cor. 1. The lines are parallel if $BA' = AB'$.

Cor. 2. The lines are perpendicular to each other if

$$AA' + BB' = (AB' + BA')\cos\omega.$$

27. *A right line can be found to satisfy any two conditions.*

Each of the forms that we have given of the general equation of a right line includes two constants. Thus the forms $y = mx + b$, $x\cos\alpha + y\sin\alpha = p$, involve the constants m and b, p and α. The only form which appears to contain more constants is $Ax + By + C = 0$; but in this case we are concerned not with the absolute magnitudes, but only with the mutual ratios of the quantities A, B, C. For if we multiply or divide the equation by any constant it will still represent the same line: we may divide therefore by C, when the equation will only contain the two constants $\frac{A}{C}$, $\frac{B}{C}$. Choosing, then, any of these forms, such as $y = mx + b$, to represent a line in general, we may consider m and b as two unknown quantities to be determined. And when any two conditions are given we are able to find the values of m and b, corresponding to the particular line which satisfies these conditions. This is sufficiently illustrated by the examples in Arts. 28, 29, 32, 33.

28. *To find the equation of a right line parallel to a given one, and passing through a given point $x'y'$.*

If the line $y = mx + b$ be parallel to a given one, the constant m is known (Cor., Art. 21). And if it pass through a fixed point, the equation, being true for *every* point on the line, is true for the point $x'y'$, and therefore we have $y' = mx' + b$, which determines b. The required equation then is

$$y = mx + y' - mx', \text{ or } y - y' = m(x - x').$$

If in this equation we consider m as indeterminate, we have the general equation of a right line passing through the point $x'y'$.

29. *To find the equation of a right line passing through two fixed points $x'y'$, $x''y''$.*

We found, in the last article, that the general equation of a right line passing through $x'y'$ is one which may be written in the form

$$\frac{y - y'}{x - x'} = m,$$

where m is indeterminate. But since the line must also pass through the point $x''y''$, this equation must be satisfied when the coordinates x'', y'', are substituted for x and y; hence

$$\frac{y'' - y'}{x'' - x'} = m.$$

Substituting this value of m, the equation of the line becomes

$$\frac{y-y'}{x-x'}=\frac{y''-y'}{x''-x'}.$$

In this form the equation can be easily remembered, but, clearing it of fractions, we obtain it in a form which is sometimes more convenient,

$$(y'-y'')\,x-(x'-x'')\,y+x'y''-y'x''=0.$$

The equation may also be written in the form

$$(x-x')\,(y-y'')=(x-x'')\,(y-y').$$

For this is the equation of a right line, since the terms xy, which appear on both sides, destroy each other; and it is satisfied either by making $x=x'$, $y=y'$, or $x=x''$, $y=y''$. Expanding it, we find the same result as before.

Cor. The equation of the line joining the point $x'y'$ *to the origin* is $y'x=x'y$.

Ex. 1. Form the equations of the sides of a triangle, the coordinates of whose vertices are $(2, 1)$, $(3, -2)$, $(-4, -1)$. *Ans.* $x+7y+11=0$, $3y-x=1$, $3x+y=7$.

Ex. 2. Form the equations of the sides of the triangle formed by $(2, 3)$, $(4, -5)$, $(-3, -6)$. *Ans.* $x-7y=39$, $9x-5y=3$, $4x+y=11$.

Ex. 3. Form the equation of the line joining the points

$$x'y' \text{ and } \frac{mx'+nx''}{m+n},\ \frac{my'+ny''}{m+n}.$$

Ans. $(y'-y'')\,x-(x'-x'')\,y+x'y''-y'x''=0$.

Ex. 4. Form the equation of the line joining

$$x'y' \text{ and } \frac{x''+x'''}{2},\ \frac{y''+y'''}{2}.$$

Ans. $(y''+y'''-2y')\,x-(x''+x'''-2x')\,y+x''y'-y''x'+x'''y'-y'''x'=0$.

Ex. 5. Form the equations of the bisectors of the sides of the triangle described in Ex. 2. *Ans.* $17x-3y=25$, $7x+9y+17=0$, $5x-6y=21$.

Ex. 6. Form the equation of the line joining

$$\frac{lx'-mx''}{l-m},\ \frac{ly'-my''}{l-m} \text{ to } \frac{lx'-nx'''}{l-n},\ \frac{ly'-ny'''}{l-n}.$$

Ans. $x\{l(m-n)\,y'+m\,(n-l)\,y''+n\,(l-m)\,y'''\}-y\,\{l\,(m-n)\,x'+m\,(n-l)\,x''+n\,(l-m)\,x'''\}$
$=lm\,(y'x''-x'y'')+mn\,(y''x'''-x''y''')+nl\,(y'''x'-y'x''')$.

30. *To find the condition that three points shall lie on one right line.*

We found (in Art. 29) the equation of the line joining two of them, and we have only to see if the coordinates of the third will satisfy this equation. The condition, therefore, is

$$(y_1-y_2)\,x_3-(x_1-x_2)\,y_3+(x_1y_2-x_2y_1)=0,$$

which can be put into the more symmetrical form

$$y_1(x_2 - x_3) + y_2(x_3 - x_1) + y_3(x_1 - x_2) = 0.*$$

31. *To find the coordinates of the point of intersection of two right lines whose equations are given.*

Each equation expresses a relation which must be satisfied by the coordinates of the point required; we find its coordinates, therefore, by solving for the two unknown quantities x and y, from the two given equations.

We said (Art. 14) that the position of a point was determined, being given two equations between its coordinates. The reader will now perceive that each equation represents a locus on which the point must lie, and that the point is the intersection of the two loci represented by the equations. Even the simplest equations to represent a point, viz. $x = a$, $y = b$, are the equations of two parallels to the axes of coordinates, the intersection of which is the required point. When the equations are both of the first degree they denote but one point; for each equation represents a right line, and two right lines can only intersect in one point. In the more general case, the loci represented by the equations are curves of higher dimensions, which will intersect each other in more points than one.

Ex. 1. To find the coordinates of the vertices of the triangle the equations of whose sides are $x + y = 2$; $x - 3y = 4$; $3x + 5y + 7 = 0$.

Ans. $(-\frac{1}{14}, -\frac{19}{14})$, $(\frac{17}{2}, -\frac{13}{2})$, $(\frac{5}{2}, -\frac{1}{2})$.

Ex. 2. To find the coordinates of the intersections of

$$3x + y - 2 = 0;\ x + 2y = 5;\ 2x - 3y + 7 = 0.$$

Ans. $(\frac{1}{7}, \frac{17}{7})$, $(-\frac{1}{11}, \frac{25}{11})$, $(-\frac{1}{5}, \frac{13}{5})$.

Ex. 3. Find the coordinates of the intersections of

$$2x + 3y = 13;\ 5x - y = 7;\ x - 4y + 10 = 0.$$

Ans. They meet in the point (2, 3).

Ex. 4. Find the coordinates of the vertices, and the equations of the diagonals, of the quadrilateral the equations of whose sides are

$$2y - 3x = 10,\ 2y + x = 6,\ 16x - 10y = 33,\ 12x + 14y + 29 = 0.$$

Ans. $(-1, \frac{7}{2})$, $(3, \frac{3}{2})$, $(\frac{1}{2}, -\frac{5}{2})$, $(-3, \frac{1}{2})$; $6y - x = 6$, $8x + 2y + 1 = 0$.

---

* In using this and other similar formulæ, which we shall afterwards have occasion to employ, the learner must be careful to take the coordinates in a fixed order (see engraving). For instance, in the second member of the formula just given, y_2 takes the place of y_1, x_3 of x_2, and x_1 of x_3. Then, in the third member, we advance from y_2 to y_3, from x_3 to x_1, and from x_1 to x_2, always proceeding in the order just indicated.

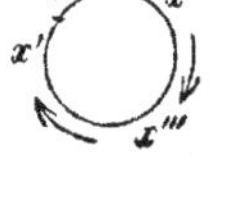

Ex. 5. Find the intersections of opposite sides of the same quadrilateral, and the equation of the line joining them. *Ans.* $(83, \frac{259}{2})$, $(-\frac{71}{5}, \frac{101}{10})$, $162y - 199x = 4462$.

Ex. 6. Find the diagonals of the parallelogram formed by

$$x = a,\ x = a',\ y = b,\ y = b'.$$

Ans. $(b - b')\,x - (a - a')\,y = a'b - ab'$; $(b - b')\,x + (a - a')\,y = ab - a'b'$.

Ex. 7. The axes of coordinates being the base of a triangle and the bisector of the base, form the equations of the two bisectors of sides, and find the coordinates of their intersection. Let the coordinates of the vertex be $0, y'$, those of the base angles $x', 0$; and $-x', 0$. *Ans.* $3x'y - y'x - x'y' = 0$; $3x'y + y'x - x'y' = 0$; $\left(0, \frac{y'}{3}\right)$.

Ex. 8. Two opposite sides of a quadrilateral are taken for axes, and the other two are

$$\frac{x}{2a} + \frac{y}{2b} = 1,\ \frac{x}{2a'} + \frac{y}{2b'} = 1;$$

find the coordinates of the middle points of diagonals. *Ans.* (a, b'), (a', b).

Ex. 9. In the same case find the coordinates of the middle point of the line joining the intersections of opposite sides.

Ans. $\dfrac{a'b \,.\, a - ab' \,.\, a'}{a'b - ab'}$, $\dfrac{a'b \,.\, b' - ab' \,.\, b}{a'b - ab'}$; and the form of the result shows (Art. 7) that this point divides externally, in the ratio $a'b : ab'$, the line joining the two middle points (a, b'), (a', b).

32. *To find the equation to rectangular axes of a right line passing through a given point, and perpendicular to a given line,* $y = mx + b$.

The condition that two lines should be perpendicular, being $mm' = -1$ (Art. 25), we have at once for the equation of the required perpendicular

$$y - y' = -\frac{1}{m}(x - x').$$

It is easy, from the above, to see that the equation of the perpendicular from the point $x'y'$ on the line $Ax + By + C = 0$ is

$$A\,(y - y') = B\,(x - x'),$$

that is to say, *we interchange the coefficients of x and y, and alter the sign of one of them.*

Ex. 1. To find the equations of the perpendiculars from each vertex on the opposite side of the triangle $(2, 1)$, $(3, -2)$, $(-4, -1)$.

The equations of the sides are (Art. 29, Ex. 1)

$$x + 7y + 11 = 0,\ 3y - x = 1,\ 3x + y = 7;$$

and the equations of the perpendiculars

$$7x - y = 13,\ 3x + y = 7,\ 3y - x = 1.$$

The triangle is consequently right-angled.

Ex. 2. To find the equations of the perpendiculars at the middle points of the side of the same triangle. The coordinates of the middle points being

$$(-\tfrac{1}{2}, -\tfrac{3}{2}),\ (-1, 0),\ (\tfrac{5}{2}, -\tfrac{1}{2}).$$

The perpendiculars are

$$7x - y + 2 = 0,\ 3x + y + 3 = 0,\ 3y - x + 4 = 0, \text{ intersecting in } (-\tfrac{1}{2}, -\tfrac{3}{2}).$$

Ex. 3. Find the equations of the perpendiculars from the vertices of the triangle (2, 3), (4, − 5), (− 3, − 6) (see Art. 29, Ex. 2).

Ans. $7x + y = 17,\ 5x + 9y + 25 = 0,\ x - 4y = 21$; intersecting in $(\tfrac{89}{29}, -\tfrac{130}{29})$.

Ex. 4. Find the equations of the perpendiculars at the middle points of the sides of the same triangle.

Ans. $7x + y + 2 = 0,\ 5x + 9y + 16 = 0,\ x - 4y = 7$; intersecting in $(-\tfrac{1}{29}, -\tfrac{51}{29})$.

Ex. 5. To find in general the equations of the perpendiculars from the vertices on the opposite sides of a triangle, the coordinates of whose vertices are given.

$$\textit{Ans.}\ (x'' - x''')\,x + (y'' - y''')\,y + (x'x''' + y'y''') - (x'x'' + y'y'') = 0,$$
$$(x''' - x')\,x + (y''' - y')\,y + (x''x' + y''y') - (x''x''' + y''y''') = 0,$$
$$(x' - x'')\,x + (y' - y'')\,y + (x'''x'' + y'''y'') - (x'''x' + y'''y') = 0.$$

Ex. 6. Find the equations of the perpendiculars at the middle points of the sides.

$$\textit{Ans.}\ (x'' - x''')\,x + (y'' - y''')\,y = \tfrac{1}{2}(x''^2 - x'''^2) + \tfrac{1}{2}(y''^2 - y'''^2),$$
$$(x''' - x')\,x + (y''' - y')\,y = \tfrac{1}{2}(x'''^2 - x'^2) + \tfrac{1}{2}(y'''^2 - y'^2),$$
$$(x' - x'')\,x + (y' - y'')\,y = \tfrac{1}{2}(x'^2 - x''^2) + \tfrac{1}{2}(y'^2 - y''^2).$$

Ex. 7. Taking for axes the base of a triangle and the perpendicular on it from the vertex, find the equations of the other two perpendiculars, and the coordinates of their intersection. The coordinates of the vertex are now $(0, y')$, and of the base angles $(x'', 0)$, $(-x''', 0)$.

$$\textit{Ans.}\ x'''(x - x'') + y'y = 0,\ x''(x + x''') - y'y = 0,\ \left(0, \frac{x''x'''}{y'}\right).$$

Ex. 8. Using the same axes, find the equations of the perpendiculars at the middle points of sides, and the coordinates of their intersection.

$$\textit{Ans.}\ 2(x'''x + y'y) = y'^2 - x'''^2,\ 2(x''x - y'y) = x''^2 - y'^2,\ 2x = x'' - x''',\ \left(\frac{x'' - x'''}{2}, \frac{y'^2 - x''x'''}{2y'}\right).$$

Ex. 9. Form the equation of the perpendicular from $x'y'$ on the line $x\cos\alpha + y\sin\alpha = p$; and find the coordinates of the intersection of this perpendicular with the given line.

Ans. $\{x' + \cos\alpha\,(p - x'\cos\alpha - y'\sin\alpha),\ y' + \sin\alpha\,(p - x'\cos\alpha - y'\sin\alpha)\}$.

Ex. 10. Find the distance between the latter point and $x'y'$.

Ans. $\pm(p - x'\cos\alpha - y'\sin\alpha)$.

33. *To find the equation of a line passing through a given point and making a given angle ϕ, with a given line $y = mx + b$ (the axes of coordinates being rectangular).*

Let the equation of the required line be

$$y - y' = m'(x - x'),$$

and the formula of Art. 25,

$$\tan\phi = \frac{m - m'}{1 + mm'},$$

enables us to determine

$$m' = \frac{m - \tan\phi}{1 + m\tan\phi}.$$

34. *To find the length of the perpendicular from any point $x'y'$ on the line whose equation is $x \cos\alpha + y \cos\beta - p = 0$.*

We have already indicated (Ex. 9 and 10, Art. 32) one way of solving this question, and we wish now to shew how the same result may be obtained geometrically. From the given point Q draw QR parallel to the given line, and QS perpendicular. Then $OK = x'$, and OT will be $= x' \cos\alpha$. Again, since $SQK = \beta$, and $QK = y'$,

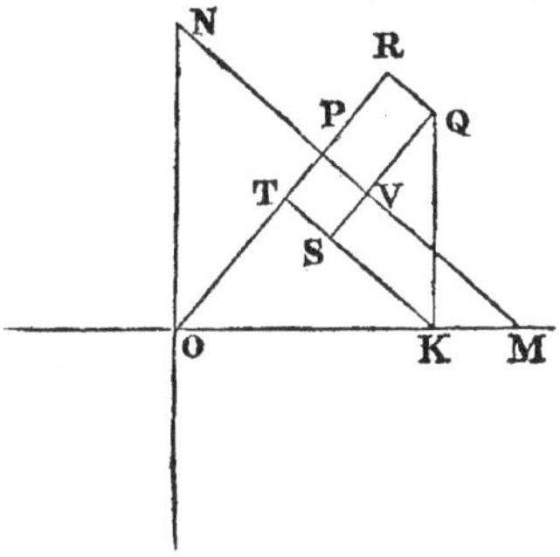

$$RT = QS = y' \cos\beta;$$

hence $$x' \cos\alpha + y' \cos\beta = OR.$$

Subtract OP, the perpendicular from the origin, and

$$x' \cos\alpha + y' \cos\beta - p = PR = \text{the perpendicular } QV.$$

But if in the figure the point Q had been taken on the side of the line next the origin, OR would have been less than OP, and we should have obtained for the perpendicular the expression $p - x' \cos\alpha - y' \cos\beta$; and we see that the perpendicular changes sign as we pass from one side of the line to the other. If we were only concerned with one perpendicular, we should only look to its absolute magnitude, and it would be unmeaning to prefix any sign. But if we were comparing the perpendiculars from two points, such as Q and S, it is evident (Art. 6) that the distances QV, SV, being measured in opposite directions, must be taken with opposite signs. We may then at pleasure choose for the expression for the length of the perpendicular either $\pm(p - x' \cos\alpha - y' \cos\beta)$. If we choose that form in which the absolute term is positive, this is equivalent to saying that the perpendiculars which fall on the side of the line next the origin are to be regarded as positive, and those on the other side as negative; and *vice versâ* if we choose the other form.

If the equation of the line had been given in the form $Ax + By + C = 0$, we have only (Art. 24) to reduce it to the form

$$x \cos\alpha + y \cos\beta - p = 0,$$

and the length of the perpendicular from any point $x'y'$

$$= \frac{Ax' + By' + C}{\sqrt{(A^2 + B^2)}}, \text{ or } \frac{(Ax' + By' + C) \sin \omega}{\sqrt{(A^2 + B^2 - 2AB \cos \omega)}},$$

according as the axes are rectangular or oblique. By comparing the expression for the perpendicular from $x'y'$ with that for the perpendicular from the origin, we see that $x'y'$ lies on the same side of the line as the origin when $Ax' + By' + C$ has the same sign as C, and *vice versâ*.

The condition that any point $x'y'$ should be on the right line $Ax + By + C = 0$, is, of course, that the coordinates $x'y'$ should satisfy the given equation, or

$$Ax' + By' + C = 0.$$

And the present Article shows that this condition is merely the algebraical statement of the fact, that the perpendicular from the point $x'y'$ on the given line is $= 0$.

Ex. 1. Find the length of the perpendicular from the origin on the line

$$3x + 4y + 20 = 0,$$

the axes being rectangular. *Ans.* 4.

Ex. 2. Find the length of the perpendicular from the point (2, 3) on $2x + y - 4 = 0$.

Ans. $\frac{3}{\sqrt{5}}$, and the given point is on the side remote from the origin.

Ex. 3. Find the lengths of the perpendiculars from each vertex on the opposite side of the triangle (2, 1), (3, − 2), (− 4, − 1).

Ans. $2\sqrt{(2)}$, $\sqrt{(10)}$, $2\sqrt{(10)}$, and the origin is within the triangle.

Ex. 4. Find the length of the perpendicular from (3, − 4) on $4x + 2y = 7$, the angle between the axes being 60°.

Ans. $\frac{3}{4}$, and the point is on the side next the origin.

Ex. 5. Find the length of the perpendicular from the origin on

$$a(x - a) + b(y - b) = 0. \qquad \textit{Ans. } \sqrt{(a^2 + b^2)}.$$

35. *To find the equation of a line bisecting the angle between two lines,* $x \cos\alpha + y \sin\alpha - p = 0$, $x \cos\beta + y \sin\beta - p' = 0$.

We find the equation of this line most simply by expressing algebraically the property that the perpendiculars let fall from any point xy of the bisector on the two lines are equal. This immediately gives us the equation

$$x \cos\alpha + y \sin\alpha - p = \pm (x \cos\beta + y \sin\beta - p'),$$

since each side of this equation denotes the length of one of those perpendiculars (Art. 34).

If the equations had been given in the form $Ax + By + C = 0$, $A'x + B'y + C' = 0$, the equation of a bisector would be

$$\frac{Ax + By + C}{\sqrt{(A^2 + B^2)}} = \pm \frac{A'x + B'y + C'}{\sqrt{(A'^2 + B'^2)}}.$$

It is evident from the double sign that there are two bisectors: one such that the perpendicular on what we agree to consider the positive side of one line is equal to the perpendicular on the negative side of the other; the other such that the equal perpendiculars are either both positive or both negative.

If we choose that sign which will make the two constant terms of the same sign, it follows, from Art. 34, that we shall have the bisector of that angle in which the origin lies; and if we give the constant terms opposite signs, we shall have the equation of the bisector of the supplemental angle.

Ex. 1. Reduce the equations of the bisectors of the angles between two lines to the form $x \cos \alpha + y \sin \alpha = p$.

Ans. $x \cos \{\frac{1}{2}(\alpha + \beta) + 90^\circ\} + y \sin \{\frac{1}{2}(\alpha + \beta) + 90^\circ\} = \dfrac{p - p'}{2 \sin \frac{1}{2}(\alpha - \beta)}$;

$x \cos \frac{1}{2}(\alpha + \beta) + y \sin \frac{1}{2}(\alpha + \beta) = \dfrac{p + p'}{2 \cos \frac{1}{2}(\alpha - \beta)}$.

Ex. 2. Find the equations of the bisectors of the angles between

$3x + 4y - 9 = 0$, $12x + 5y - 3 = 0$.

Ans. $7x - 9y + 34 = 0$, $9x + 7y = 12$.

36. *To find the area of the triangle formed by three points.*

If we multiply the length of the line joining two of the points, by the perpendicular on that line from the third point, we shall have double the area. Now the length of the perpendicular from x_3y_3 on the line joining x_1y_1, x_2y_2, the axes being rectangular, is (Arts. 29, 34)

$$\frac{(y_1 - y_2)x_3 - (x_1 - x_2)y_3 + x_1y_2 - x_2y_1}{\sqrt{\{(y_1 - y_2)^2 + (x_1 - x_2)^2\}}},$$

and the denominator of this fraction is the length of the line joining x_1y_1, x_2y_2, hence

$$y_1(x_2 - x_3) + y_2(x_3 - x_1) + y_3(x_1 - x_2)$$

represents double the area formed by the three points.

If the axes be oblique, it will be found, on repeating the investigation with the formulæ for oblique axes, that the only change that will occur is that the expression just given is to be multiplied by $\sin \omega$. Strictly speaking, we ought to prefix to these expressions the double sign implicitly involved in the square root used in finding them. If we are concerned with a single area we look only to its absolute magnitude without regard to sign. But if, for example, we are comparing two triangles whose vertices x_3y_3, x_4y_4, are on opposite sides of the line joining the base angles x_1y_1, x_2y_2, we must give their areas different signs; and the quadrilateral space included by the four points is the sum instead of the difference of the two triangles.

Cor. 1. Double the area of the triangle formed by the lines joining the points x_1y_1, x_2y_2 to the origin is $y_1x_2 - y_2x_1$, as appears by making $x_3 = 0$, $y_3 = 0$, in the preceding formula.

Cor. 2. The condition that three points should be on one right line, when interpreted geometrically, asserts that the area of the triangle formed by the three points becomes $= 0$ (Art. 30).

37. *To express the area of a polygon in terms of the co-ordinates of its angular points.*

Take any point xy within the polygon, and connect it with all the vertices x_1y_1, $x_2y_2 \ldots x_ny_n$; then evidently the area of the polygon is the sum of the areas of all the triangles into which the figure is thus divided. But by the last Article double these areas are respectively

$$x(y_1 - y_2) - y(x_1 - x_2) + x_1y_2 - x_2y_1,$$
$$x(y_2 - y_3) - y(x_2 - x_3) + x_2y_3 - x_3y_2,$$
$$x(y_3 - y_4) - y(x_3 - x_4) + x_3y_4 - x_4y_3,$$
$$\cdots\cdots\cdots\cdots\cdots\cdots\cdots\cdots$$
$$x(y_{n-1} - y_n) - y(x_{n-1} - x_n) + x_{n-1}y_n - x_ny_{n-1},$$
$$x(y_n - y_1) - y(x_n - x_1) + x_ny_1 - x_1y_n.$$

When we add these together, the parts which multiply x and y vanish, as they evidently ought to do, since the value of the total area must be independent of the manner in which we divide it into triangles; and we have for double the area

$$(x_1y_2 - x_2y_1) + (x_2y_3 - x_3y_2) + (x_3y_4 - x_4y_3) + \ldots (x_ny_1 - x_1y_n).$$

This may be otherwise written,

$$x_1(y_2 - y_n) + x_2(y_3 - y_1) + x_3(y_4 - y_2) + \ldots x_n(y_1 - y_{n-1}),$$

or else

$$y_1(x_n - x_2) + y_2(x_1 - x_3) + y_3(x_2 - x_4) + \ldots y_n(x_{n-1} - x_1).$$

Ex. 1. Find the area of the triangle (2, 1), (3, – 2), (– 4, – 1). *Ans.* 10.

Ex. 2. Find the area of the triangle (2, 3), (4, – 5), (– 3, – 6). *Ans.* 29.

Ex. 3. Find the area of the quadrilateral (1, 1), (2, 3), (3, 3), (4, 1). *Ans.* 4.

38. *To find the condition that three right lines shall meet in a point.*

Let their equations be

$$Ax + By + C = 0,\quad A'x + B'y + C' = 0,\quad A''x + B''y + C'' = 0.$$

If they intersect, the coordinates of the intersection of two of them must satisfy the third equation. But the coordinates of the intersection of the first two are $\frac{BC' - B'C}{AB' - A'B},\ \frac{CA' - C'A}{AB' - A'B}$.

Substituting in the third, we get, for the required condition,

$$A''(BC' - B'C) + B''(CA' - C'A) + C''(AB' - A'B) = 0,$$

which may be also written in either of the forms

$$A(B'C'' - B''C') + B(C'A'' - C''A') + C(A'B'' - A''B') = 0,$$

$$A(B'C'' - B''C') + A'(B''C - BC'') + A''(BC' - B'C) = 0.$$

*39. *To find the area of the triangle formed by the three lines* $Ax + By + C = 0,\ A'x + B'y + C' = 0,\ A''x + B''y + C'' = 0.$

By solving for x and y from each pair of equations in turn we obtain the coordinates of the vertices, and substituting them in the formula of Art. 36 we obtain for the double area the expression

$$\frac{BC' - B'C}{AB' - BA'}\left\{\frac{A'C'' - C'A''}{B'A'' - A'B''} - \frac{A''C - C''A}{B''A - A''B}\right\}$$
$$+ \frac{B'C'' - B''C'}{A'B'' - B'A''}\left\{\frac{A''C - C''A}{B''A - A''B} - \frac{AC' - CA'}{BA' - AB'}\right\}$$
$$+ \frac{B''C - BC''}{A''B - B''A}\left\{\frac{AC' - CA'}{BA' - AB'} - \frac{A'C'' - C'A''}{B'A'' - A'B''}\right\}.$$

But if we reduce to a common denominator, and observe that the numerator of the fraction between the first brackets is

$$\{A''(BC' - B'C) + A(B'C'' - B''C') + A'(B''C - C''B)\}$$

multiplied by A'', and that the numerators of the fractions between the second and third brackets are the same quantity multiplied respectively by A and A', we get for the double area the expression

$$\frac{\{A(B'C'' - B''C') + A'(B''C - BC'') + A''(BC' - B'C)\}^2}{(AB' - BA')(A'B'' - B'A'')(A''B - B''A)}.$$

If the three lines meet in a point, this expression for the area vanishes (Art. 38); if any two of them are parallel, it becomes infinite (Art. 25).

40. *Given the equations of two right lines, to find the equation of a third through their point of intersection.*

The method of solving this question, which will first occur to the reader, is to obtain the coordinates of the point of intersection by Art 31, and then to substitute these values for $x'y'$ in the equation of Art. 28, viz., $y - y' = m(x - x')$. The question, however, admits of an easier solution by the help of the following important principle: *If $S = 0$, $S' = 0$, be the equations of any two loci, then the locus represented by the equation $S + kS' = 0$ (where k is any constant) passes through every point common to the two given loci.* For it is plain that any coordinates which satisfy the equation $S = 0$, and also satisfy the equation $S' = 0$, must likewise satisfy the equation $S + kS' = 0$.

Thus, then, the equation

$$(Ax + By + C) + k(A'x + B'y + C') = 0,$$

which is obviously the equation of a right line, denotes one passing through the intersection of the right lines

$$Ax + By + C = 0, \quad A'x + B'y + C' = 0,$$

for if the coordinates of the point common to them both be substituted in the equation $(Ax + By + C) + k(A'x + B'y + C') = 0$, they will satisfy it, since they make each member of the equation separately $= 0$.

Ex. 1. To find the equation of the line joining to the origin the intersection of

$$Ax + By + C = 0, \quad A'x + B'y + C' = 0.$$

Multiply the first by C', the second by C, and subtract, and the equation of the required line is $(AC' - A'C)x + (BC' - CB')y = 0$; for it passes through the origin (Art. 18), and by the present article it passes through the intersection of the given lines.

Ex. 2. To find the equation of the line drawn through the intersection of the same lines, parallel to the axis of x. *Ans.* $(BA' - AB')y + CA' - AC' = 0$.

Ex. 3. To find the equation of the line joining the intersection of the same lines to the point $x'y'$. Writing down by this article the general equation of a line through the intersection of the given lines, we determine k from the consideration that it must be satisfied by the coordinates $x'y'$, and find for the required equation

$$(Ax + By + C)(A'x' + B'y' + C') = (Ax' + By' + C)(A'x + B'y + C').$$

Ex. 4. Find the equation of the line joining the point (2, 3) to the intersection of $2x + 3y + 1 = 0$, $3x - 4y = 5$.

Ans. $11(2x + 3y + 1) + 14(3x - 4y - 5) = 0$; or $64x - 23y = 59$.

41. The principle established in the last article gives us a test for three lines intersecting in the same point, often more convenient in practice than that given in Art 38. *Three right lines will pass through the same point if their equations being multiplied each by any constant quantity, and added together, the sum is identically* $= 0$; that is to say, if the following relation be true, no matter what x and y are:

$$l(Ax + By + C) + m(A'x + B'y + C') + n(A''x + B''y + C'') = 0.$$

For then those values of the coordinates which make the first two members severally $= 0$ must also make the third $= 0$.

Ex. 1. The three bisectors of the sides of a triangle meet in a point. Their equations are (Art. 29, Ex. 4)

$$(y'' + y''' - 2y')x - (x'' + x''' - 2x')y + (x''y' - y''x') + (x'''y' - y'''x') = 0,$$
$$(y''' + y' - 2y'')x - (x''' + x' - 2x'')y + (x'''y'' - y'''x'') + (x'y'' - y'x'') = 0,$$
$$(y' + y'' - 2y''')x - (x' + x'' - 2x''')y + (x'y''' - y'x''') + (x''y''' - y''x''') = 0.$$

And since the three equations when added together vanish identically, the lines represented by them meet in a point. Its coordinates are found, by solving between any two, to be $\frac{1}{3}(x' + x'' + x''')$, $\frac{1}{3}(y' + y'' + y''')$.

Ex. 2. Prove the same thing, taking for axes two sides of the triangle whose lengths are a and b.

Ans. $\frac{2x}{a} + \frac{y}{b} - 1 = 0$, $\frac{x}{a} + \frac{2y}{b} - 1 = 0$, $\frac{x}{a} - \frac{y}{b} = 0$.

Ex. 3. The three perpendiculars of a triangle, and the three perpendiculars at middle points of sides respectively meet in a point. For the equations of Ex. 5 and 6, Art. 32, when added together, vanish identically.

Ex. 4. The three bisectors of the angles of a triangle meet in a point. For their equations are

$$(x\cos\alpha + y\sin\alpha - p) - (x\cos\beta + y\sin\beta - p') = 0,$$
$$(x\cos\beta + y\sin\beta - p') - (x\cos\gamma + y\sin\gamma - p'') = 0.$$
$$(x\cos\gamma + y\sin\gamma - p'') - (x\cos\alpha + y\sin\alpha - p) = 0.$$

*42. *To find the coordinates of the intersection of the line joining the points $x'y'$, $x''y''$, with the right line $Ax + By + C = 0$.*

We give this example in order to illustrate a method (which we shall frequently have occasion to employ) of determining the point in which the line joining two given points is met by a given locus. We know (Art. 7) that the coordinates of any point on the line joining the given points must be of the form

$$x = \frac{mx'' + nx'}{m + n},\quad y = \frac{my'' + ny'}{m + n};$$

and we take as our unknown quantity $\frac{m}{n}$, the ratio, namely, in which the line joining the points is cut by the given locus; and we determine this unknown quantity from the condition, that the coordinates just written shall satisfy the equation of the locus. Thus, in the present example, we have

$$A\frac{mx''+nx'}{m+n}+B\frac{my''+ny'}{m+n}+C=0;$$

hence
$$\frac{m}{n}=-\frac{Ax'+By'+C}{Ax''+By''+C};$$

and consequently the coordinates of the required point are

$$x=\frac{(Ax'+By'+C)\,x''-(Ax''+By''+C)\,x'}{(Ax'+By'+C)-(Ax''+By''+C)},$$

with a similar expression for y. This value for the ratio $m:n$ might also have been deduced geometrically from the consideration that the ratio in which the line joining $x'y'$, $x''y''$ is cut, is equal to the ratio of the perpendiculars from these points upon the given line; but (Art. 34) these perpendiculars are

$$\frac{Ax'+By'+C}{\sqrt{(A^2+B^2)}}\text{ and }\frac{Ax''+By''+C}{\sqrt{(A^2+B^2)}}.$$

The negative sign in the preceding value arises from the fact that, in the case of *internal* section to which the positive sign of $m:n$ corresponds (Art. 7), the perpendiculars fall on opposite sides of the given line, and must, therefore, be understood as having different signs (Art. 34).

If a right line cut the sides of a triangle BC, CA, AB, in the points LMN, then

$$\frac{BL.CM.AN}{LC.MA.NB}=-1.$$

Let the coordinates of the vertices be $x'y'$, $x''y''$, $x'''y'''$, then

$$\frac{BL}{LC}=-\frac{Ax''+By''+C}{Ax'''+By'''+C},$$

$$\frac{CM}{MA}=-\frac{Ax'''+By'''+C}{Ax'+By'+C},$$

$$\frac{AN}{NB}=-\frac{Ax'+By'+C}{Ax''+By''+C},$$

and the truth of the theorem is manifest.

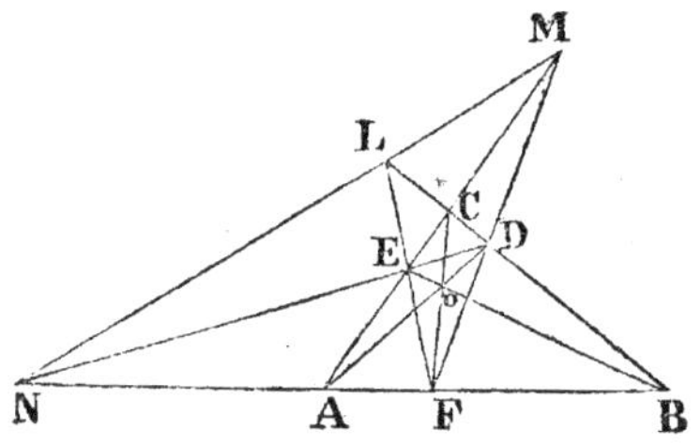

*43. *To find the ratio in which the line joining two points* x_1y_1, x_2y_2, *is cut by the line joining two other points* x_3y_3, x_4y_4.

The equation of this latter line is (Art. 29)

$$(y_3 - y_4)\,x - (x_3 - x_4)\,y + x_3y_4 - x_4y_3 = 0.$$

Therefore, by the last article,

$$\frac{m}{n} = -\frac{(y_3 - y_4)\,x_1 - (x_3 - x_4)\,y_1 + x_3y_4 - x_4y_3}{(y_3 - y_4)\,x_2 - (x_3 - x_4)\,y_2 + x_3y_4 - x_4y_3}.$$

It is plain (by Art. 36) that this is the ratio of the two triangles whose vertices are x_1y_1, x_3y_3, x_4y_4, and x_2y_2, x_3y_3, x_4y_4, as is also geometrically evident.

If the lines connecting any assumed point with the vertices of a triangle meet the opposite sides BC, CA, AB respectively, in D, E, F, then

$$\frac{BD.CE.AF}{DC.EA.FB} = +1.$$

Let the assumed point be x_4y_4, and the vertices x_1y_1, x_2y_2, x_3y_3, then

$$\frac{BD}{DC} = \frac{x_1(y_2 - y_4) + x_2(y_4 - y_1) + x_4(y_1 - y_2)}{x_1(y_4 - y_3) + x_4(y_3 - y_1) + x_3(y_1 - y_4)},$$

$$\frac{CE}{EA} = \frac{x_2(y_3 - y_4) + x_3(y_4 - y_2) + x_4(y_2 - y_3)}{x_1(y_2 - y_4) + x_2(y_4 - y_1) + x_4(y_1 - y_2)},$$

$$\frac{AF}{FB} = \frac{x_1(y_4 - y_3) + x_4(y_3 - y_1) + x_3(y_1 - y_4)}{x_2(y_3 - y_4) + x_3(y_4 - y_2) + x_4(y_2 - y_3)},$$

and the truth of the theorem is evident.

44. *To find the polar equation of a right line* (see Art. 12).

Suppose we take, as our fixed axis, OP the perpendicular on the given line, then let OR be any *radius vector* drawn from the pole to the given line

$$OR = \rho,\quad ROP = \theta;$$

but, plainly,

$$OR\cos\theta = OP,$$

hence the equation is

$$\rho\cos\theta = p.$$

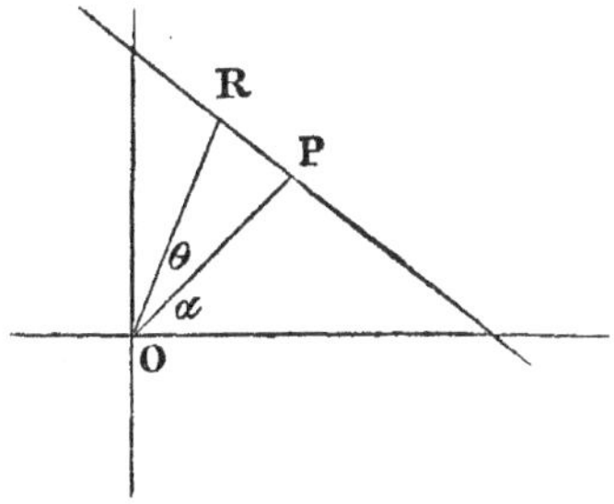

If the fixed axis be OA making an angle α with the perpendicular, then $ROA = \theta$, and the equation is

$$\rho \cos(\theta - \alpha) = p.$$

This equation may also be obtained by transforming the equation with regard to rectangular coordinates,

$$x \cos\alpha + y \sin\alpha = p.$$

Rectangular coordinates are transformed to polar by writing for x, $\rho \cos\theta$, and for y, $\rho \sin\theta$ (see Art. 12); hence the equation becomes

$$\rho(\cos\theta \cos\alpha + \sin\theta \sin\alpha) = p;$$

or, as we got before, $\rho \cos(\theta - \alpha) = p$.

An equation of the form

$$\rho(A \cos\theta + B \sin\theta) = C$$

can be (as in Art. 23) reduced to the form $\rho \cos(\theta - \alpha) = p$, by dividing by $\sqrt{(A^2 + B^2)}$; we shall then have

$$\cos\alpha = \frac{A}{\sqrt{(A^2 + B^2)}}, \sin\alpha = \frac{B}{\sqrt{(A^2 + B^2)}}, p = \frac{C}{\sqrt{(A^2 + B^2)}}.$$

Ex. 1. Reduce to rectangular coordinates the equation

$$\rho = 2a \sec\left(\theta + \frac{\pi}{6}\right).$$

Ex. 2. Find the polar coordinates of the intersection of the following lines, and also the angle between them: $\rho \cos\left(\theta - \frac{\pi}{2}\right) = 2a$, $\rho \cos\left(\theta - \frac{\pi}{6}\right) = a$.

Ans. $\rho = 2a$, $\theta = \frac{\pi}{2}$, angle $= \frac{\pi}{3}$.

Ex. 3. Find the polar equation of the line passing through the points whose polar coordinates are ρ', θ'; ρ', θ''.

Ans. $\rho'\rho'' \sin(\theta' - \theta'') + \rho''\rho \sin(\theta'' - \theta) + \rho\rho' \sin(\theta - \theta') = 0$.

CHAPTER III.

EXAMPLES ON THE RIGHT LINE.

45. Having in the last chapter laid down principles by which we are able to express algebraically the position of any point or right line, we proceed to give some further examples of the application of this method to the solution of geometrical problems. The learner should diligently exercise himself in working out such questions until he has acquired quickness and readiness in the use of this method. In working such examples our equations may generally be much simplified by a judicious choice of axes of coordinates; since, by choosing for axes two of the most remarkable lines on the figure, several of our expressions will often be much shortened. On the other hand, it will sometimes happen that by choosing axes unconnected with the figure, the equations will gain in symmetry more than an equivalent for what they lose in simplicity. The reader may compare the two solutions of the same question, given Ex. 1 and 2, Art. 41, where, though the first solution is the longest, it has the advantage that the equation of one bisector being formed, those of the others can be written down without further calculation.

Since expressions containing angles become more complicated by the use of oblique coordinates, it will be generally advisable to use rectangular axes in any question in which the consideration of angles is involved.

46. *Loci.*—Analytical geometry adapts itself with peculiar readiness to the investigation of loci. We have only to find what relation the conditions of the question assign between the coordinates of the point whose locus we seek, and then the statement of this relation in algebraical language gives us at once the *equation* of the required locus.

Ex. 1. Given base and difference of squares of sides of a triangle, to find the locus of vertex.

Let us take for axes the base and a perpendicular through its middle point. Let the half base = c, and let the coordinates of the vertex be x, y. Then

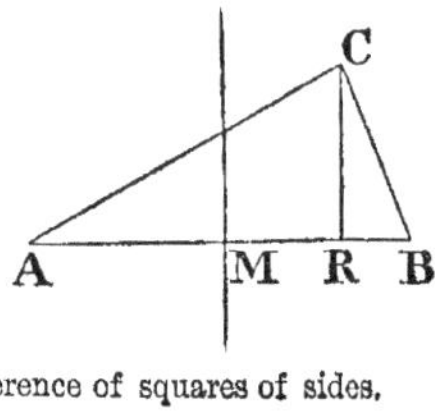

$$AC^2 = y^2 + (c + x)^2,^* \; BC^2 = y^2 + (c - x)^2,$$

$$AC^2 - BC^2 = 4cx,$$

and the equation of the locus is $4cx = m^2$. The locus is therefore a line perpendicular to the base at a distance from the middle point $x = \frac{m^2}{4c}$. It is easy to see that the difference of squares of segments of base = difference of squares of sides.

Ex. 2. Find locus of vertex, given base and $\cot A + m \cot B$.

It is evident, from the figure, that

$$\cot A = \frac{AR}{CR} = \frac{c + x}{y}; \; \cot B = \frac{c - x}{y},$$

and the required equation is $c + x + m(c - x) = py$, the equation of a right line.

Ex. 3. Given base and sum of sides of a triangle, if the perpendicular be produced beyond the vertex until its whole length is equal to one of the sides, to find the locus of the extremity of the perpendicular.

Take the same axes, and let us inquire what relation exists between the coordinates of the point whose locus we are seeking. The x of this point plainly is MR, and the y is, by hypothesis, $= AC$; and if m be the given sum of sides,

$$BC = m - y.$$

Now (Euclid II. 13) $\quad BC^2 = AB^2 + AC^2 - 2AB \, . \, AR$;

or $\quad (m - y)^2 = 4c^2 + y^2 - 4c(c + x).$

Reducing this equation we get

$$2my - 4cx = m^2,$$

the equation of a right line.

Ex. 4. Given two fixed lines, OA and OB, if any line AB be drawn to intersect them parallel to a third fixed line OC, to find the locus of the point P where AB is cut in a given ratio; viz. $PA = nAB$.

Let us take the lines OA, OC for axes, and let the equation of OB be $y = mx$. Then since the point B lies on the latter line, its ordinate is m times its abscissa; or $AB = mOA$. Therefore $PA = mnOA$; but PA and OA are the coordinates of the point P, whose locus is therefore a right line through the origin, having for its equation

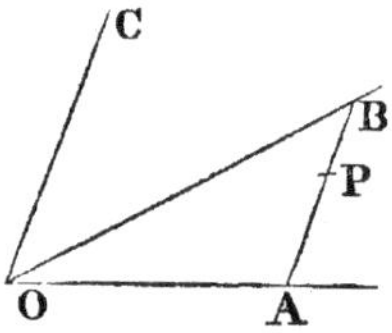

$$y = mnx.$$

---

* This is a particular case of Art. 4, and $c + x$ is the algebraic difference of the abscissæ of the points A and C (see remarks at top of p. 4). Beginners often reason that since the line AR consists of the parts $AM = -c$, and $MR = x$, its length is $-c + x$, and not $c + x$, and therefore that $AC^2 = y^2 + (x - c)^2$. It is to be observed that the sign given to a line depends not on the side of the origin on which it lies, but on the direction in which it is measured. We go from A to R by proceeding in the positive direction $AM = c$, and still further in the same direction $MR = x$, therefore the length $AR = c + x$; but we may proceed from R to B by first going in the negative direction $RM = -x$, and then in the opposite direction $MB = c$, hence the length RB is $c - x$.

Ex. 5. PA drawn parallel to OC, as before, meets any number of fixed lines in points B, B', B'', &c., and PA is taken proportional to the sum of all the ordinates BA, $B'A$, &c., find the locus of P.

Ans. If the equations of the lines be

$$y = mx,\ y = m'x + n',\ y = m''x + n'',\ \&c.,$$

the equation of the locus is

$$ky = mx + (m'x + n') + (m''x + n'') + \&c.$$

Ex. 6. Given bases and sum of areas of any number of triangles having a common vertex, to find its locus.

Let the equations of the bases be

$$x\cos\alpha + y\sin\alpha - p = 0,\ x\cos\beta + y\sin\beta - p_1 = 0,\ \&c.,$$

and their lengths, a, b, c, &c.; and let the given sum $= m^2$; then, since (Art. 34) $x\cos\alpha + y\sin\alpha - p$ denotes the perpendicular from the point xy on the first line, $a(x\cos\alpha + y\sin\alpha - p)$ will be double the area of the first triangle, &c., and the equation of the locus will be

$$a(x\cos\alpha + y\sin\alpha - p) + b(x\cos\beta + y\sin\beta - p_1) + c(x\cos\gamma + y\sin\gamma - p_2) + \&c. = 2m^2,$$

which, since it contains x and y only in the first degree, will represent a right line.

Ex. 7. Given vertical angle and sum of sides of a triangle, find the locus of the point where the base is cut in a given ratio.

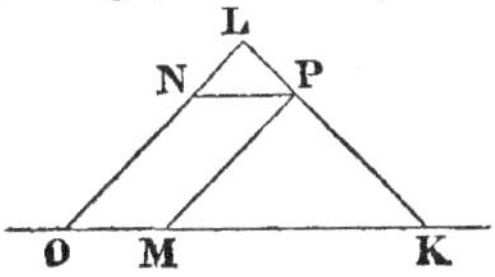

The sides of the triangle are taken for axes, and the ratio $PK : PL$ is given $= n : m$. Then by similar triangles,

$$OK = \frac{(m+n)\,x}{m},\ OL = \frac{(m+n)\,y}{n},$$

and the locus is a right line whose equation is $\frac{x}{m} + \frac{y}{n} = \frac{s}{m+n}$.

Ex. 8. Find the locus of P, if when perpendiculars PM, PN are let fall on two fixed lines, $OM + ON$ is given.

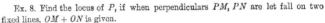

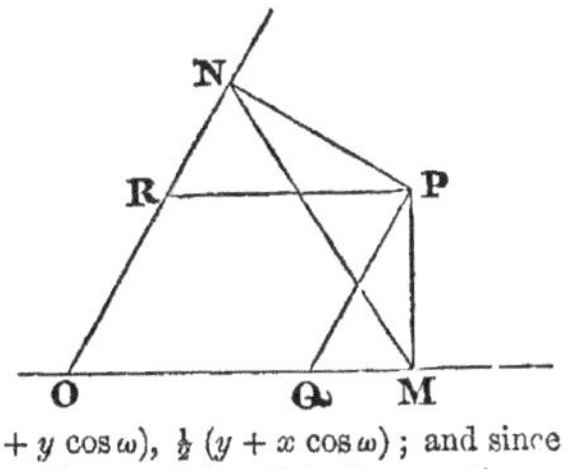

Taking the fixed lines for axes, it is evident that $OM = x + y\cos\omega$, $ON = y + x\cos\omega$, and the locus is $x + y =$ constant.

Ex. 9. Find the locus if MN be parallel to a fixed line.

Ans. $y + x\cos\omega = m(x + y\cos\omega)$.

Ex. 10. If MN be bisected [or cut in a given ratio] by a given line $y = mx + n$.

The coordinates of the middle point expressed in terms of the coordinates of P are $\frac{1}{2}(x + y\cos\omega)$, $\frac{1}{2}(y + x\cos\omega)$; and since these satisfy the equation of the given line, the coordinates of P satisfy the equation

$$y + x\cos\omega = m(x + y\cos\omega) + 2n.$$

Ex. 11. P moves along a given line $y = mx + n$, find the locus of the middle point of MN. If the coordinates of P be α, β, and those of the middle point x, y, it has just been proved that $2x = \alpha + \beta\cos\omega$, $2y = \beta + \alpha\cos\omega$. Whence solving for α, β,

$$\alpha\sin^2\omega = 2x - 2y\cos\omega,\ \beta\sin^2\omega = 2y - 2x\cos\omega.$$

But α, β are connected by the relation $\beta = m\alpha + n$, hence

$$2y - 2x\cos\omega = m(2x - 2y\cos\omega) + n\sin^2\omega.$$

47. It is customary to denote by x and y the coordinates of a variable point which describes a locus, and the coordinates of fixed points by accented letters. Accordingly in the preceding examples we have from the first denoted by x and y the coordinates of the point whose locus we seek. But frequently in finding a locus it is necessary to form the equations of lines connected with the figure; and there is danger of confusion between the x and y, which are the running coordinates of a point on one of these lines, and the x and y of the point whose locus we seek. In such cases it is convenient at first to denote the coordinates of the latter point by other letters such as α, β, until we have succeeded in obtaining a relation connecting these coordinates. Having thus found the equation of the locus, we may if we please replace α, β by x and y, so as to write the equation in the ordinary form in which the letters x and y are used to denote the coordinates of the point which describes the locus.

Ex. 1. Find the locus of the vertex of a triangle, given the base CD, and the ratio $AM : NB$ of the parts into which the sides divide a fixed line AB parallel to the base. Take AB and a perpendicular to it through A for axes, and it is necessary to express AM, NB in terms of the coordinates of P. Let these coordinates be $\alpha\beta$, and let the coordinates of C, D be $x'y'$, $x''y'$, the y' of both being the same since CD is parallel to AB. Then the equation of PC joining the points $\alpha\beta$, $x'y'$ is (Art. 29)

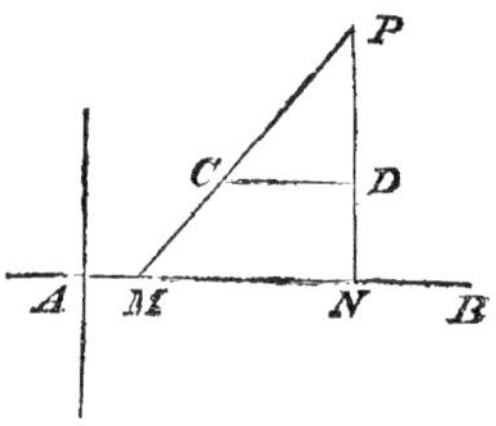

$$(\beta - y')\,x - (\alpha - x')\,y = \beta x' - \alpha y'.$$

This equation being satisfied by the x and y of every point on the line PC is satisfied by the point M, whose $y = 0$ and whose $x = AM$. Making then $y = 0$ in this equation we get

$$AM = \frac{\beta x' - \alpha y'}{\beta - y'}.$$

In like manner,

$$AN = \frac{\beta x'' - \alpha y'}{\beta - y'};$$

and if $AB = c$, the relation $AM = kBN$ gives

$$\frac{\beta x' - \alpha y'}{\beta - y'} = k\left(c - \frac{\beta x'' - \alpha y'}{\beta - y'}\right).$$

We have now expressed the conditions of the problem in terms of the coordinates of the point P; and now that there is no further danger of confusion, we may replace α, β, by x, y; when the equation of the locus, cleared of fractions, becomes

$$yx' - xy' = k\,\{c\,(y - y') - (yx'' - xy')\}.$$

Ex. 2. Two vertices of a triangle ABC move on fixed right lines LM, LN, and the three sides pass through three fixed points O, P, Q which lie on a right line find the locus of the third vertex.

Take for axis of x the right line OP, containing the three fixed points, and for axis of y the line OL joining the intersection of the two fixed lines to the point O through which the base passes. Let the coordinates of C be α, β, and let $OL = b$, $OM = a$, $ON = a'$, $OP = c$, $OQ = c'$. Then obviously the equations of LM, LN are

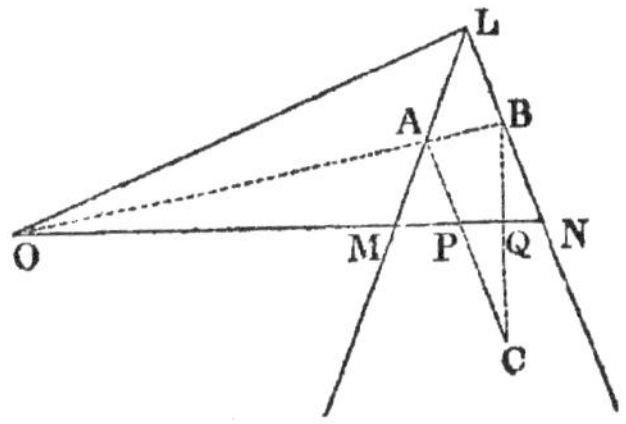

$$\frac{x}{a} + \frac{y}{b} = 1 \text{ and } \frac{x}{a'} + \frac{y}{b} = 1.$$

The equation of CP through $\alpha\beta$ and P ($y = 0$, $x = c$) is

$$(\alpha - c)\, y - \beta x + \beta c = 0.$$

The coordinates of A, the intersection of this line with

$$\frac{x}{a} + \frac{y}{b} = 1,$$

are

$$x_1 = \frac{ab\,(\alpha - c) + ac\beta}{b\,(\alpha - c) + a\beta};\ y_1 = \frac{b\,(a - c)\,\beta}{b\,(\alpha - c) + a\beta}.$$

The coordinates of B are found by simply accentuating the letters in the preceding:

$$x_2 = \frac{a'b\,(\alpha - c') + a'c'\beta}{b\,(\alpha - c') + a'\beta};\ y_2 = \frac{b\,(a' - c')\,\beta}{b\,(\alpha - c') + a'\beta}.$$

Now the condition that two points x_1y_1, x_2y_2 shall lie on a right line passing through the origin is (Art. 30) $\frac{y_1}{x_1} = \frac{y_2}{x_2}$.

Applying this condition we have

$$\frac{b\,(a - c)\,\beta}{ab\,(\alpha - c) + ac\beta} = \frac{b\,(a' - c')\,\beta}{a'b\,(\alpha - c') + a'c'\beta}.$$

We have now derived from the conditions of the problem a relation which must be satisfied by $\alpha\beta$ the coordinates of C; and if we replace α, β by x, y we have the equation of the locus written in its ordinary form. Clearing of fractions, we have

$$(a - c)\,[a'b\,(x - c') + a'c'y] = (a' - c')\,[ab\,(x - c) + acy],$$

or

$$\frac{(ac' - a'c)\,x}{cc'\,(a - a') - aa'\,(c - c')} + \frac{y}{b} = 1,$$

the equation of a right line through the point L.

Ex. 8. If in the last example the points P, Q lie on a right line passing not through O but through L, find the locus of vertex.

We shall first solve the general problem in which the points P, Q have any position. We take the fixed lines LM, LN for axes. Let the coordinates of P, Q, O, C be respectively $x'y'$, $x''y''$, $x'''y'''$, $\alpha\beta$; and the condition which we want to express is that if we join CP, CQ, and then join the points A, B, in which these lines meet the axes, the line AB shall pass through O. The equation of CP is $(\beta - y')\,x - (\alpha - x')\,y = \beta x' - \alpha y'$.

And the intercept which it makes on the axis of x is

$$LA = \frac{\beta x' - \alpha y'}{\beta - y'}.$$

In like manner the intercept which CQ makes on the axis of y is

$$LB = \frac{\alpha y'' - \beta x''}{\alpha - x''}.$$

The equation of AB is

$$\frac{x}{LA} + \frac{y}{LB} = 1, \text{ or } \frac{x\,(\beta - y')}{\beta x' - \alpha y'} + \frac{y\,(\alpha - x'')}{\alpha y'' - \beta x''} = 1.$$

And the condition of the problem is that this equation shall be satisfied by the coordinates $x'''y'''$. In order then that the point C may fulfil the conditions of the problem, its coordinates $\alpha\beta$ must be connected by the relation

$$\frac{x'''(\beta - y')}{\beta x' - \alpha y'} + \frac{y'''(\alpha - x'')}{\alpha y'' - \beta x''} = 1.$$

When this equation is cleared of fractions, it in general involves the coordinates $\alpha\beta$ in the second degree. But suppose that the points $x'y'$, $x''y''$ lie on the same line passing through the origin $y = mx$, so that we have $y' = mx'$, $y'' = mx''$, the equation may be written

$$\frac{x'''(\beta - y')}{x'(\beta - \alpha m)} + \frac{y'''(\alpha - x'')}{x''(\alpha m - \beta)} = 1.$$

Clearing of fractions and replacing α, β by x and y, the locus is a right line, viz.

$$x'''x''(y - y') - y'''x'(x - x'') = x'x''(mx - y).$$

48. It is often convenient, instead of expressing the conditions of the problem directly in terms of the coordinates of the point whose locus we are seeking, to express them in the first instance in terms of some other lines of the figure; we must then obtain as many relations as are necessary in order to eliminate the indeterminate quantities thus introduced, so as to have remaining a relation between the coordinates of the point whose locus is sought. The following Examples will sufficiently illustrate this method.

Ex. 1. To find the locus of the middle points of rectangles inscribed in a given triangle.

Let us take for axes CR and AB; let $CR = p$, $RB = s$, $AR = s'$. The equations of AC and BC are

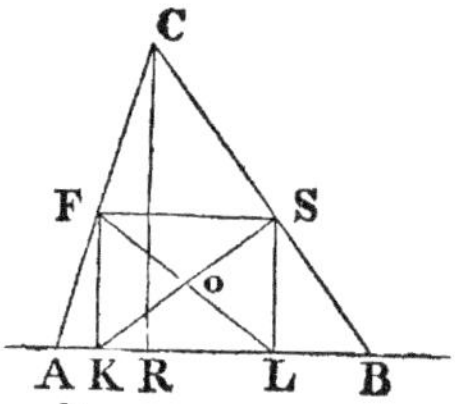

$$\frac{y}{p} - \frac{x}{s'} = 1 \text{ and } \frac{y}{p} + \frac{x}{s} = 1.$$

Now if we draw any line FS parallel to the base at a distance $FK = k$, we can find the abscissæ of the points F and S, in which the line FS meets AC and BC, by substituting in the equations of AC and BC the value $y = k$. Thus we get from the first equation

$$\frac{k}{p} - \frac{x}{s'} = 1; \therefore x \text{ or } RK = -s'\left(1 - \frac{k}{p}\right);$$

and from the second equation

$$\frac{k}{p} + \frac{x}{s} = 1; \therefore x \text{ or } RL = s\left(1 - \frac{k}{p}\right).$$

Having the abscissæ of F and S, we have (by Art. 7) the abscissa of the middle point of FS, viz. $x = \frac{s - s'}{2} \cdot \left(1 - \frac{k}{p}\right)$. This is evidently the abscissa of the middle point of the rectangle. But its ordinate is $y = \frac{1}{2}k$. Now we want to find a relation which will subsist between this ordinate and abscissa whatever k be. We have only then to eliminate k between these equations, by substituting in the first the value of k $(= 2y)$, derived from the second, when we have

$$2x = (s - s')\left(1 - \frac{2y}{p}\right),$$

or
$$\frac{2x}{s-s'}+\frac{2y}{p}=1.$$
This is the equation of the locus which we seek. It obviously represents a right line, and if we examine the intercepts which it cuts off on the axes, we shall find it to be the line joining the middle point of the perpendicular CR to the middle point of the base.

Ex. 2. A line is drawn parallel to the base of a triangle, and the points where it meets the sides joined to any two fixed points on the base; to find the locus of the point of intersection of the joining lines.

We shall preserve the same axes, &c., as in Ex. 1, and let the coordinates of the fixed points T and V, on the base, be for T $(m, 0)$, and for V $(n, 0)$.

The equation of FT will be found to be
$$\left\{s'\left(1-\frac{k}{p}\right)+m\right\}y+kx-km=0,$$
and that of SV to be
$$\left\{s\left(1-\frac{k}{p}\right)-n\right\}y-kx+kn=0.$$
Now since the point whose locus we are seeking lies on both the lines FT, SV, each of the equations just written expresses a relation which must be satisfied by its coordinates. Still, since these equations involve k, they express relations which are only true for that particular point of the locus which corresponds to the case where the parallel FS is drawn at a height k above the base. If, however, between the equations we eliminate the indeterminate k, we shall obtain a relation involving only the coordinates and known quantities, and which, since it must be satisfied whatever be the position of the parallel FS, will be the required equation of the locus.

In order, then, to eliminate k between the equations, put them into the form
$$FT \qquad (s'+m)\,y-k\left(\frac{s'}{p}\,y-x+m\right)=0,$$
and
$$SV \qquad (s-n)\,y-k\left(\frac{s}{p}\,y+x-n\right)=0;$$
and eliminating k we get for the equation of the locus
$$(s-n)\left(\frac{s'}{p}\,y-x+m\right)=(s'+m)\left(\frac{s}{p}\,y+x-n\right):$$
But this is the equation of a right line, since x and y are only in the first degree.

Ex. 3. A line is drawn parallel to the base of a triangle, and its extremities joined transversely to those of the base; to find the locus of the point of intersection of the joining lines.

This is a particular case of the foregoing, but admits of a simple solution by choosing for axes the sides of the triangle AC and CB. Let the lengths of those lines be a, b, and let the lengths of the proportional intercepts made by the parallel be μa, μb. Then the equations of the transversals will be
$$\frac{x}{a}+\frac{y}{\mu b}=1 \text{ and } \frac{x}{\mu a}+\frac{y}{b}=1.$$
Subtract one from the other, divide by the constant $1-\frac{1}{\mu}$, and we get for the equation of the locus
$$\frac{x}{a}-\frac{y}{b}=0,$$
which we have elsewhere found (see p. 34) to be the equation of the bisector of the base of the triangle.

Ex. 4. Given two fixed points A and B, one on each of the axes, if A' and B' be taken on the axes so that $OA'+OB'=OA+OB$: find the locus of the intersection of AB', $A'B$.

Let $OA = a$, $OB = b$, $OA' = a + k$, then, from the conditions of the problem, $OB' = b - k$. The equations of AB', $A'B$ are respectively

$$\frac{x}{a} + \frac{y}{b-k} = 1, \quad \frac{x}{a+k} + \frac{y}{b} = 1,$$

or

$$bx + ay - ab + k(a - x) = 0,$$
$$bx + ay - ab + k(y - b) = 0.$$

Subtracting, we eliminate k, and find for the equation of the locus

$$x + y = a + b.$$

Ex. 5. If on the base of a triangle we take any portion AT, and on the other side of the base another portion BS, in a fixed ratio to AT and draw ET and FS parallel to a fixed line CR; to find the locus of O, the point of intersection of EB and FA.

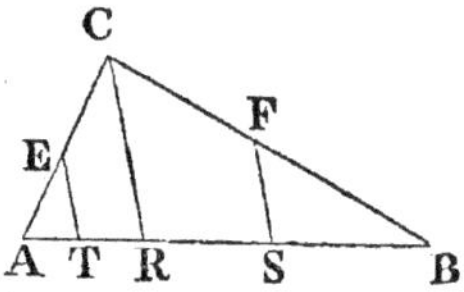

Take AB and CR for axes; let $AT = k$, $BR = s$, $AR = s'$, $CR = p$, let the fixed ratio be m, then BS will $= mk$; the coordinates of S will be $(s - mk, 0)$, and of T $\{-(s' - k), 0\}$.

The ordinates of E and F will be found by substituting these values of x in the equations of AC and BC. We get for

$$E, \quad x = -(s' - k), \; y = \frac{pk}{s'},$$

and for

$$F, \quad x = s - mk, \quad y = \frac{mpk}{s}.$$

Now form the equations of the transverse lines, and the equation of EB is

$$(s + s' - k)\, y + \frac{pk}{s'}\, x - \frac{pks}{s'} = 0,$$

and the equation of AF is

$$(s + s' - mk)\, y - \frac{mpk}{s}\, x - \frac{mpks'}{s} = 0.$$

To eliminate k, subtract one equation from the other, and the result, divided by k, will be

$$(m - 1)\, y + \left(\frac{mp}{s} + \frac{p}{s'}\right) x + \left(\frac{mps'}{s} - \frac{ps}{s'}\right) = 0,$$

which is the equation of a right line.

Ex. 6. PP' and QQ' are any two parallels to the sides of a parallelogram; to find the locus of the intersection of the lines PQ and $P'Q'$.

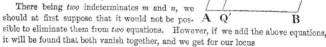

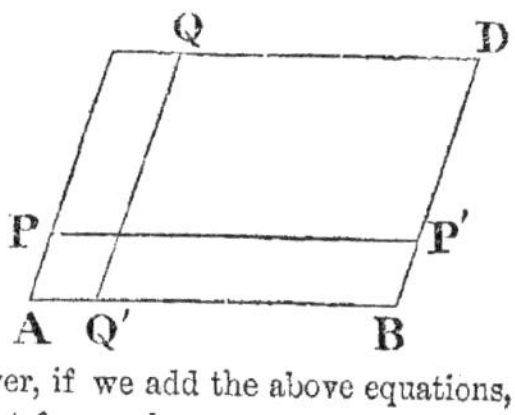

Let us take two of the sides for our axes, and let the lengths of the sides be a and b, and let $AQ' = m$, $AP = n$. Then the equation of PQ, joining P $(0, n)$ to Q (m, b) is

$$(b - n)\, x - my + mn = 0,$$

and the equation of $P'Q'$ joining P' (a, n) to Q' $(m, 0)$ is

$$nx - (a - m)\, y - mn = 0.$$

There being *two* indeterminates m and n, we should at first suppose that it would not be possible to eliminate them from *two* equations. However, if we add the above equations, it will be found that both vanish together, and we get for our locus

$$bx - ay = 0,$$

the equation of the diagonal of the parallelogram.

Ex. 7. Given a point and two fixed lines; draw any two lines through the fixed point, and join transversely the points where they meet the fixed lines; to find the locus of intersection of the transverse lines.

Take the fixed lines for axes, and let the equations of the lines through the fixed point be

$$\frac{x}{m}+\frac{y}{n}=1, \text{ and } \frac{x}{m'}+\frac{y}{n'}=1.$$

The conditions that these lines should pass through the fixed point $x'y'$ give us

$$\frac{x'}{m}+\frac{y'}{n}=1, \text{ and } \frac{x'}{m'}+\frac{y'}{n'}=1;$$

or, subtracting,

$$x'\left(\frac{1}{m}-\frac{1}{m'}\right)+y'\left(\frac{1}{n}-\frac{1}{n'}\right)=0.$$

Now the equations of the tranverse lines clearly are

$$\frac{x}{m}+\frac{y}{n'}=1, \text{ and } \frac{x}{m'}+\frac{y}{n}=1;$$

or, subtracting,

$$x\left(\frac{1}{m}-\frac{1}{m'}\right)-y\left(\frac{1}{n}-\frac{1}{n'}\right)=0.$$

Now from this and the equation just found we can eliminate

$$\left(\frac{1}{m}-\frac{1}{m'}\right) \text{ and } \left(\frac{1}{n}-\frac{1}{n'}\right),$$

and we have $$x'y+y'x=0,$$
the equation of a right line through the origin.

Ex. 8. At any point of the base of a triangle is drawn a line of given length, parallel to a given one, and so as to be cut in a given ratio by the base; find the locus of the intersection of the lines joining its extremities to those of the base.

49. The fundamental idea of Analytic Geometry is that every geometrical condition to be fulfilled by a point leads to an equation which must be satisfied by its coordinates. It is important that the beginner should quickly make himself expert in applying this idea, so as to be able to express by an equation any given geometrical condition. We add, therefore, for his further exercise, some examples of loci which lead to equations of degrees higher than the first. The interpretation of such equations will be the subject of future chapters, but the method of arriving at the equations, which is all with which we are here concerned, is precisely the same as when the locus is a right line. In fact, until the problem has been solved, we do not know what will be the degree of the resulting equation. The examples that follow are purposely chosen so as to admit of treatment similar to that pursued in former examples, according to the order of which they are arranged. In each of the answers given it is supposed that the same axes are chosen, and that the letters have the same meaning as in the corresponding previous example.

Ex. 1. Find the locus of vertex of a triangle, given base and sum of squares of sides. *Ans.* $x^2 + y^2 = \frac{1}{2}m^2 - c^2$.

Ex. 2. Given base and m squares of one side $\pm n$ squares of the other.
Ans. $(m \pm n)(x^2 + y^2) + 2(m \mp n) cx + (m \pm n) c^2 = p^2$.

Ex. 3. Given base and ratio of sides.

Ex. 4. Given base and product of tangents of base angles.
In this and the Examples next following, the learner will use the values of the tangents of the base angles given Ex. 2, Art. 46. *Ans.* $y^2 + m^2x^2 = m^2c^2$.

Ex. 5. Given base and vertical angle or, in other words, base and sum of base angles. *Ans.* $x^2 + y^2 - 2cy \cot C = c^2$.

Ex. 6. Given base and difference of base angles. *Ans.* $x^2 - y^2 + 2xy \cot D = c^2$.

Ex. 7. Given base, and that one base angle is double the other.
Ans. $3x^2 - y^2 + 2cx = c^2$.

Ex. 8. Given base, and $\tan C = m \tan B$. *Ans.* $m(x^2 + y^2 - c^2) = 2c(c - x)$.

Ex. 9. PA is drawn parallel to OC, as in Ex. 4, p. 39, meeting two fixed lines in points B, B'; and PA^2 is taken $= PB.PB'$, find the locus of P.
Ans. $mx(m'x + n') = y(mx + m'x + n')$.

Ex. 10. PA is taken the harmonic mean between AB and AB'.
Ans. $2mx(m'x + n') = y(mx + m'x + n')$.

Ex. 11. Given vertical angle of a triangle, find the locus of the point where the base is cut in a given ratio, if the area also is given. *Ans.* $xy =$ constant.

Ex. 12. If the base is given. *Ans.* $\dfrac{x^2}{m^2} + \dfrac{y^2}{n^2} - \dfrac{2xy \cos\omega}{mn} = \dfrac{b^2}{(m+n)^2}$.

Ex. 13. If the base pass through a fixed point. *Ans.* $\dfrac{mx'}{x} + \dfrac{ny'}{y} = m + n$.

Ex. 14. Find the locus of P [Ex. 8, p. 40] if MN is constant.
Ans. $x^2 + y^2 + 2xy \cos\omega =$ constant.

Ex. 15. If MN pass through a fixed point. *Ans.* $\dfrac{x'}{x + y\cos\omega} + \dfrac{y'}{y + x\cos\omega} = 1$.

Ex. 16. If MN pass through a fixed point, find the locus of the intersection of parallels to the axes through M and N. *Ans.* $\dfrac{x'}{x} + \dfrac{y'}{y} = 1$.

Ex. 17. Find the locus of P [*Ex.* 1, p. 41] if the line CD be not parallel to AB.

Ex. 18. Given base CD of a triangle, find the locus of vertex, if the intercept AB on a given line is constant.
Ans. $(x'y - y'x)(y - y'') - (x''y - y''x)(y - y') = c(y - y')(y - y'')$.

50. *Problems where it is required to prove that a moveable right line passes through a fixed point.*

We have seen (Art. 40) that the line

$$Ax + By + C + k(A'x + B'y + C') = 0;$$

or, what is the same thing,

$$(A + kA')x + (B + kB')y + C + kC' = 0,$$

where k is indeterminate, always passes through a fixed point, namely, the intersection of the lines

$$Ax + By + C = 0, \text{ and } A'x + B'y + C' = 0.$$

Hence, *if the equation of a right line contain an indeterminate quantity in the first degree, the right line will always pass through a fixed point.*

Ex. 1. Given vertical angle of a triangle and the sum of the reciprocals of the sides, the base will always pass through a fixed point.

Take the sides for axes; the equation of the base is $\frac{x}{a} + \frac{y}{b} = 1$, and we are given the condition

$$\frac{1}{a} + \frac{1}{b} = \frac{1}{m}, \text{ or } \frac{1}{b} = \frac{1}{m} - \frac{1}{a},$$

therefore, equation of base is

$$\frac{x}{a} + \frac{y}{m} - \frac{y}{a} = 1,$$

where m is constant and a indeterminate, that is

$$\frac{1}{a}(x - y) + \frac{y}{m} - 1 = 0,$$

where $\frac{1}{a}$ is indeterminate. Hence the base must always pass through the intersection of the two lines $x - y = 0$, and $y = m$.

Ex. 2. Given three fixed lines OA, OB, OC, meeting in a point, if the three vertices of a triangle move one on each of these lines, and two sides of the triangle pass through fixed points, to prove that the remaining side passes through a fixed point.

Take for axes the fixed lines OA, OB on which the base angles move, then the line OC on which the vertex moves will have an equation of the form $y = mx$, and let the fixed points be $x'y'$, $x''y''$. Now, in any position of the vertex, let its coordinates be $x = a$, and consequently $y = ma$; then the equation of AC is

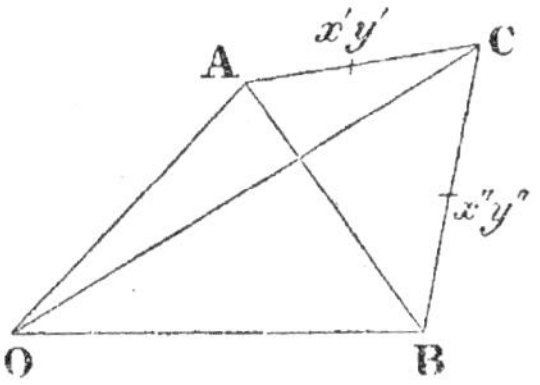

$$(x' - a)\,y - (y' - ma)\,x + a\,(y' - mx') = 0.$$

Similarly, the equation of BC is

$$(x'' - a)\,y - (y'' - ma)\,x + a\,(y'' - mx'') = 0.$$

Now the length of the intercept OA is found by making $x = 0$ in equation AC, or

$$y = -\frac{a\,(y' - mx')}{x' - a}.$$

Similarly, OB is found by making $y = 0$ in BC, or

$$x = \frac{a\,(y'' - mx'')}{y'' - ma}.$$

Hence, from these intercepts, equation of AB is

$$x\,\frac{y'' - ma}{y'' - mx''} - y\,\frac{x' - a}{y' - mx'} = a.$$

But since a is indeterminate, and only in the first degree, this line always passes through a fixed point. The particular point is found by arranging the equation in the form

$$\frac{y''}{y'' - mx''}\,x - \frac{x'}{y' - mx'}\,y - a\left(\frac{mx}{y'' - mx''} - \frac{y}{y' - mx'} + 1\right) = 0.$$

Hence the line passes through the intersection of the two lines

$$\frac{y''}{y''-mx''}\,x - \frac{x'}{y'-mx'}\,y = 0,$$

and

$$\frac{mx}{y''-mx''} - \frac{y}{y'-mx'} + 1 = 0.$$

Ex. 3. If in the last example the line on which the vertex C moves do not pass through O, to determine whether in any case the base will pass through a fixed point.

We retain the same axes and notation as before, with the only difference that the equation of the line on which C moves will be $y = mx + n$, and the coordinates of the vertex in any position will be a, and $ma + n$. Then the equation of AC is

$$(x'-a)\,y - (y'-ma-n)\,x + a\,(y'-mx') - nx' = 0.$$

The equation of BC is

$$(x''-a)\,y - (y''-ma-n)\,x + a\,(y''-mx'') - nx'' = 0,$$

$$OA = -\frac{a\,(y'-mx')-nx'}{x'-a}\,;\quad OB = \frac{a\,(y''-mx'')-nx''}{y''-ma-n}.$$

The equation of AB is therefore

$$x\,\frac{y''-ma-n}{a\,(y''-mx'')-nx''} - y\,\frac{x'-a}{a\,(y'-mx')-nx'} = 1.$$

Now when this is cleared of fractions, it will in general contain a in the second degree, and therefore the base will in general not pass through a fixed point; *if, however, the points $x'y'$, $x''y''$ lie in a right line ($y = kx$) passing through O*, we may substitute in the denominators $y'' = kx''$, and $y' = kx'$, and the equation becomes

$$x\,\frac{y''-ma-n}{x''} - y\,\frac{x'-a}{x'} = a\,(k-m) - n,$$

which contains a *in the first degree* only, and therefore denotes a right line passing through a fixed point.

Ex. 4. If a line be such that the sum of the perpendiculars let fall on it from a number of fixed points, each multiplied by a constant, may $= 0$, it will pass through a fixed point.

Let the equation of the line be

$$x\cos\alpha + y\sin\alpha - p = 0,$$

then the perpendicular on it from $x'y'$ is

$$x'\cos\alpha + y'\sin\alpha - p,$$

and the conditions of the problem give us

$$m'\,(x'\cos\alpha + y'\sin\alpha - p) + m''\,(x''\cos\alpha + y''\sin\alpha - p)$$
$$+ m'''\,(x'''\cos\alpha + y'''\sin\alpha - p) + \&c. = 0.$$

Or, using the abbreviations $\Sigma\,(mx')$ for the sum* of the mx, that is,

$$m'x' + m''x'' + m'''x''' + \&c.,$$

and in like manner $\Sigma\,(my')$ for

$$m'y' + m''y'' + m'''y''' + \&c.,$$

and $\Sigma\,(m)$ for the sum of the m's or

$$m' + m'' + m''' + \&c.,$$

---

* By sum we mean the *algebraic* sum, for any of the quantities m', m'', &c. may be negative.

we may write the preceding equation

$$\Sigma\,(mx')\cos\alpha+\Sigma\,(my')\sin\alpha-p\Sigma\,(m)=0.$$

Substituting in the original equation the value of p hence obtained, we get for the equation of the moveable line

$$x\Sigma\,(m)\cos\alpha+y\Sigma\,(m)\sin\alpha-\Sigma\,(mx')\cos\alpha-\Sigma\,(my')\sin\alpha=0,$$

or

$$x\Sigma\,(m)-\Sigma\,(mx')+\{y\Sigma\,(m)-\Sigma\,(my')\}\tan\alpha=0.$$

Now as this equation involves the indeterminate $\tan\alpha$ in the first degree, the line passes through the fixed point determined by the equations

$$x\Sigma\,(m)-\Sigma\,(mx')=0,\ \text{and}\ y\Sigma\,(m)-\Sigma\,(my')=0,$$

or, writing at full length,

$$x=\frac{m'x'+m''x''+m'''x'''+\&c.}{m'+m''+m'''+\&c.},\quad y=\frac{m'y'+m''y''+m'''y'''+\&c.}{m'+m''+m'''+\&c.}.$$

This point has sometimes been called the *centre of mean position* of the given points.

51. If the equation of any line involve the coordinates of a certain point $x'y'$ in the first degree, thus,

$$(Ax'+By'+C)\,x+(A'x'+B'y'+C')\,y+(A''x'+B''y'+C'')=0;$$

then if the point $x'y'$ move along a right line, the line whose equation has just been written will always pass through a fixed point. For, suppose the point always to lie on the line

$$Lx'+My'+N=0,$$

then if, by the help of this relation, we eliminate x' from the given equation, the indeterminate y' will remain in it of the first degree, therefore the line will pass through a fixed point.

Or, again, *if the coefficients in the equation* $Ax+By+C=0$ *be connected by the relation* $aA+bB+cC=0$ (*where* a, b, c *are constant and* A, B, C *may vary*), *the line represented by this equation will always pass through a fixed point.*

For by the help of the given relation we can eliminate C, and write the equation

$$(cx-a)\,A+(cy-b)\,B=0,$$

a right line passing through the point $\left(x=\frac{a}{c},\ y=\frac{b}{c}\right)$.

52. *Polar Coordinates.*—It is, in general, convenient to use this method, if the question be to find the locus of the extremities of lines drawn through a fixed point according to any given law.

Ex. 1. A and B are two fixed points; draw through B any line, and let fall on a perpendicular from A, AP; produce AP so that the rectangle $AP.AQ$ may be constant; to find the locus of the point Q.

Take A for the pole, and AB for the fixed axis, then AQ is our radius vector, designated by ρ, and the angle $QAB = \theta$, and our object is to find the relation existing between ρ and θ. Let us call the constant length $AB = c$, and from the right-angled triangle APB we have $AP = c \cos\theta$, but $AP \cdot AQ = \text{const.} = k^2$: therefore

$$\rho c \cos\theta = k^2, \text{ or } \rho \cos\theta = \frac{k}{c};$$

but we have seen (Art. 44) that this is the equation of a right line perpendicular to AB, and at a distance from $A = \frac{k^2}{c}$

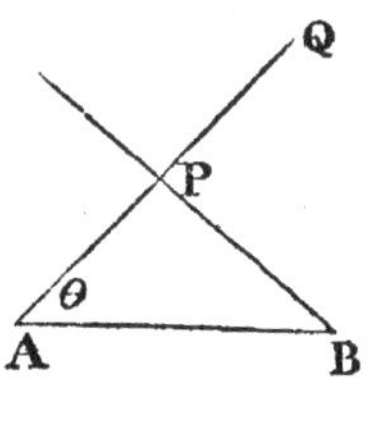

Ex. 2. Given the angles of a triangle; one vertex A is fixed, another B moves along a fixed right line: to find the locus of the third.

Take the fixed vertex A for pole, and AP perpendicular to the fixed line for axis, then $AC = \rho$, $CAP = \theta$. Now since the angles of ABC are given, AB is in a fixed ratio to $AC (= mAC)$ and $BAP = \theta - \alpha$; but $AP = AB \cos BAP$; therefore, if we call AP, a, we have

$$m\rho \cos(\theta - \alpha) = a,$$

which (Art. 44) is the equation of a right line, making an angle α with the given line, and at a distance from $A = \frac{a}{m}$.

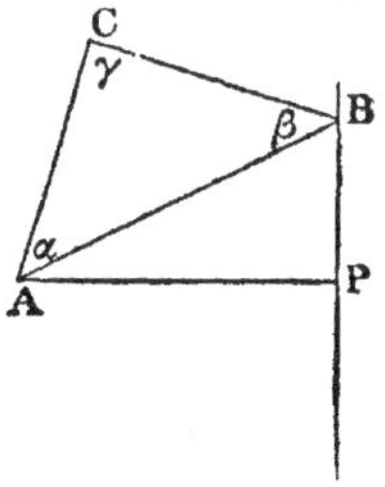

Ex. 3. Given base and sum of sides of a triangle, if at either extremity of the base B a perpendicular be erected to the conterminous side BC; to find the locus of P the point where it meets CP the external bisector of vertical angle.

Let us take the point B for our pole, then BP will be our radius vector ρ; and let us take the base produced for our fixed axis, then $PBD = \theta$, and our object is to express ρ in terms of θ. Let us designate the sides and opposite angles of the triangle a, b, c, A, B, C, then it is easy to see that the angle $BCP = 90° - \frac{1}{2}C$, and from the triangle PCB that $a = \rho \tan \frac{1}{2}C$. Hence it is evident that if we could express a and $\tan\frac{1}{2}C$ in terms of θ, we could express ρ in terms of θ. Now from the triangle ABC we have

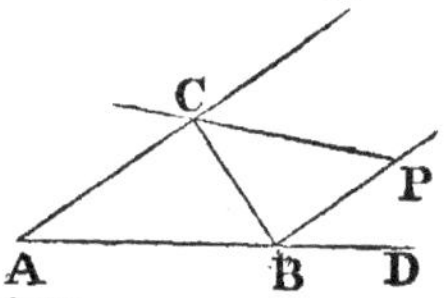

$$b^2 = a^2 + c^2 - 2ac \cos B,$$

but if the given sum of sides be m, we may substitute for b, $m - a$; and $\cos B$ plainly $= \sin\theta$; hence

$$m^2 - 2am + a^2 = a^2 + c^2 - 2ac \sin\theta,$$

and

$$a = \frac{m^2 - c^2}{2(m - c \sin\theta)}.$$

Thus we have expressed a in terms of θ and constants, and it only remains to find an expression for $\tan\frac{1}{2}C$.

Now

$$\tan\tfrac{1}{2}C = \frac{b \sin C}{b(1 + \cos C)},$$

but $b \sin C = c \sin B = c \cos\theta$, and $b \cos C = a - c \cos B = a - c \sin\theta$;

hence

$$\tan\tfrac{1}{2}C = \frac{c \cos\theta}{m - c \sin\theta}.$$

We are now able to express ρ in terms of θ, for, substitute in the equation $a = \rho \tan \frac{1}{2}C$, the values we have found for a and $\tan \frac{1}{2}C$, and we get

$$\frac{m^2 - c^2}{2(m - c\sin\theta)} = \frac{\rho c \cos\theta}{(m - c\sin\theta)}, \text{ or } \rho\cos\theta = \frac{m^2 - c^2}{2c}.$$

Hence the locus is a line perpendicular to the base of the triangle at a distance from $B = \frac{m^2 - c^2}{2c}$.

The student may exercise himself with the corresponding locus, if CP had been the *internal* bisector, and if the *difference* of sides had been given.

Ex. 4. Given n fixed right lines and a fixed point O; if through this point any radius vector be drawn meeting the right lines in the points $r_1, r_2, r_3 \ldots r_n$, and on this a point R be taken such that $\frac{n}{OR} = \frac{1}{Or_1} + \frac{1}{Or_2} + \frac{1}{Or_3} + \ldots \frac{1}{Or_n}$, to find the locus of R.

Let the equations of the right lines be

$$\rho\cos(\theta - \alpha) = p_1;\ \rho\cos(\theta - \beta) = p_2, \text{ \&c.}$$

Then it is easy to see that the equation of the locus is

$$\frac{n}{\rho} = \frac{\cos(\theta - \alpha)}{p_1} + \frac{\cos(\theta - \beta)}{p_2} + \text{\&c.}$$

the equation of a right line (Art. 44). This theorem is only a particular case of a general one, which we shall prove afterwards.

We add, as in Art. 49, a few examples leading to equations of higher degree.

Ex. 5. BP is a fixed line whose equation is $\rho\cos\theta = m$, and on each radius vector is taken a constant length PQ; to find the locus of Q [see fig., Ex. 1].

AP is by hypothesis $= \frac{m}{\cos\theta}$; therefore $AQ = \rho = \frac{m}{\cos\theta} + d$, which, transformed to rectangular coordinates, is $(x - m)^2 (x^2 + y^2) = d^2x^2$.

Ex. 6. Find the locus of Q, if P describe any locus whose polar equation is given, $\rho = \phi(\theta)$. We are by hypothesis given AP in terms of θ, but AP is the ρ of the locus $- d$; we have therefore only to substitute in the given equation $\rho - d$ for ρ.

Ans. $\rho - d = \phi(\theta)$.

Ex. 7. If AQ be produced so that AQ may be double AP, then AP is half the ρ of the locus, and we must substitute half ρ for ρ in the given equation.

Ex. 8. If the angle PAB were bisected, and on the bisector a portion AP' be taken so that $AP'^2 = mAP$, find the locus of P' when P describes the right line $\rho\cos\theta = m$. PAB is now twice the θ of the locus, and therefore $AP = \frac{m}{\cos 2\theta}$, and the equation of the locus is $\rho^2 \cos 2\theta = m^2$.

*CHAPTER IV.

APPLICATION OF ABRIDGED NOTATION TO THE EQUATION OF THE RIGHT LINE.

53. We have seen (Art. 40) that the line

$$(x \cos\alpha + y \sin\alpha - p) - k(x \cos\beta + y \sin\beta - p') = 0$$

denotes a line passing through the intersection of the lines

$$x \cos\alpha + y \sin\alpha - p = 0, \quad x \cos\beta + y \sin\beta - p' = 0.$$

We shall often find it convenient to use abbreviations for these quantities. Let us call

$$x \cos\alpha + y \sin\alpha - p, \ \alpha; \quad x \cos\beta + y \sin\beta - p', \ \beta.$$

Then the theorem just stated may be more briefly expressed; the equation $\alpha - k\beta = 0$ denotes a line passing through the intersection of the two lines denoted by $\alpha = 0$, $\beta = 0$. We shall for brevity call these the lines α, β, and their point of intersection the point $\alpha\beta$. We shall, too, have occasion often to use abbreviations for the equations of lines in the form $Ax + By + C = 0$. We shall in these cases make use of Roman letters, reserving the letters of the Greek alphabet to intimate that the equation is in the form

$$x \cos\alpha + y \sin\alpha - p = 0.$$

54. We proceed to examine the meaning of the coefficient k in the equation $\alpha - k\beta = 0$. We saw (Art. 34) that the quantity α (that is, $x \cos\alpha + y \sin\alpha - p$) denotes the length of the perpendicular PA let fall from any point xy on the line OA (which we suppose represented by α). Similarly, that β is the length of the perpendicular PB from the point xy on the line OB, represented by β. Hence the equation $\alpha - k\beta = 0$ asserts that if, from any point of the locus represented by it, perpendiculars be let fall on the lines OA, OB, the ratio of these perpendiculars (that is, $PA : PB$) will be constant and $= k$. Hence

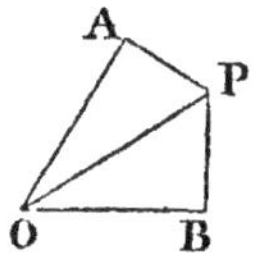

the locus represented by $\alpha - k\beta = 0$ is a right line through O, and

$$k = \frac{PA}{PB}, \text{ or } = \frac{\sin POA}{\sin POB}.$$

It follows from the conventions concerning signs (Art. 34) that $\alpha + k\beta = 0$ denotes a right line dividing *externally* the angle AOB into parts such that $\frac{\sin POA}{\sin POB} = k$. It is, of course, assumed in what we have said that the perpendiculars PA, PB are those which we agree to consider positive; those on the opposite sides of α, β being regarded as negative.

Ex. 1. To express in this notation the proof that the three bisectors of the angles of a triangle meet in a point.

The equations of the three bisectors are obviously (see Arts. 35, 54) $\alpha - \beta = 0$, $\beta - \gamma = 0$, $\gamma - \alpha = 0$, which, added together, vanish identically.

Ex. 2. Any two of the external bisectors of the angles of a triangle meet on the third internal bisector.

Attending to the convention about signs, it is easy to see that the equations of two external bisectors are $\alpha + \beta = 0$, $\alpha + \gamma = 0$, and subtracting one from the other we get $\beta - \gamma = 0$, the equation of the third internal bisector.

Ex. 3. The three perpendiculars of a triangle meet in a point.

Let the angles opposite to the sides α, β, γ be A, B, C respectively. Then since the perpendicular divides any angle of the triangle into parts, which are the complements of the remaining two angles, therefore (by Art. 54) the equations of the perpendiculars are

$$\alpha \cos A - \beta \cos B = 0, \ \beta \cos B - \gamma \cos C = 0, \ \gamma \cos C - \alpha \cos A = 0,$$

which obviously meet in a point.

Ex. 4. The three bisectors of the sides of a triangle meet in a point.

The ratio of the perpendiculars on the sides from the point where the bisector meets the base plainly is $\sin A : \sin B$. Hence the equations of the three bisectors are

$$\alpha \sin A - \beta \sin B = 0, \ \beta \sin B - \gamma \sin C = 0, \ \gamma \sin C - \alpha \sin A = 0.$$

Ex. 5. The lengths of the sides of a quadrilateral are a, b, c, d; find the equation of the line joining middle points of diagonals.

Ans. $a\alpha - b\beta + c\gamma - d\delta = 0$; for this line evidently passes through the intersection of $a\alpha - b\beta$, and $c\gamma - d\delta$; but, by the last example, these are the bisectors of the base of two triangles having one diagonal for their common base. In like manner $a\alpha - d\delta$, $b\beta - c\gamma$ intersect in the middle point of the other diagonal.

Ex. 6 To form the equation of a perpendicular to the base of a triangle at its extremity. *Ans.* $\alpha + \gamma \cos B = 0$.

Ex. 7. If there be two triangles such that the perpendiculars from the vertices of one on the sides of the other meet in a point, then, *vice versâ*, the perpendiculars from the vertices of the second on the sides of the first will meet in a point.

Let the sides be α, β, γ, α', β', γ', and let us denote by $(\alpha\beta)$ the angle between α and β. Then the equation of the perpendicular

from $\alpha\beta$ on γ' is $\alpha \cos(\beta\gamma') - \beta \cos(\alpha\gamma') = 0$,
from $\beta\gamma$ on α' is $\beta \cos(\gamma\alpha') - \gamma \cos(\beta\alpha') = 0$,
from $\gamma\alpha$ on β' is $\gamma \cos(\alpha\beta') - \alpha \cos(\gamma\beta') = 0$.

The condition that these should meet in a point is found by eliminating β between the first two, and examining whether the resulting equation coincides with the third. It is

$$\cos(\alpha\beta')\cos(\beta\gamma')\cos(\gamma\alpha') = \cos(\alpha'\beta)\cos(\beta'\gamma)\cos(\gamma'\alpha).$$

But the symmetry of this equation shews that this is also the condition that the perpendiculars from the vertices of the second triangle on the sides of the first should meet in a point.

55. The lines $\alpha - k\beta = 0$, and $k\alpha - \beta = 0$, are plainly such that one makes the same angle with the line α which the other makes with the line β, and are therefore equally inclined to the bisector $\alpha - \beta$.

Ex. If through the vertices of a triangle there be drawn any three lines meeting in a point, the three lines drawn through the same angles, equally inclined to the bisectors of the angles, will also meet in a point.

Let the sides of the triangle be α, β, γ, and let the equations of the first three lines be

$$l\alpha - m\beta = 0,\ m\beta - n\gamma = 0,\ n\gamma - l\alpha = 0,$$

which, by the principle of Art. 41, are the equations of three lines meeting in a point, and which obviously pass through the points $\alpha\beta$, $\beta\gamma$, and $\gamma\alpha$. Now, from this Article, the equations of the second three lines will be

$$\frac{\alpha}{l} - \frac{\beta}{m} = 0,\ \frac{\beta}{m} - \frac{\gamma}{n} = 0,\ \text{and } \frac{\gamma}{n} - \frac{\alpha}{l} = 0,$$

which (by Art. 41) must also meet in a point.

56. The reader is probably already acquainted with the following fundamental geometrical theorem :—"*If a pencil of four right lines meeting in a point O be intersected by any transverse right line in the four points A, P, P', B, then the ratio* $\frac{AP.P'B}{AP'.PB}$ *is constant, no matter how the transverse line be drawn.*" This ratio is called the *anharmonic ratio* of the pencil. In fact, let the perpendicular from O on the transverse line $= p$; then $p.AP = OA.OP.\sin AOP$ (both being double the area of the triangle AOP); $p.P'B = OP'.OB \sin P'OB$; $p.AP' = OA.OP' \sin AOP'$; $p.PB = OP.OB.\sin POB$; hence

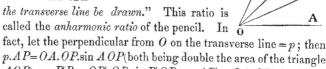

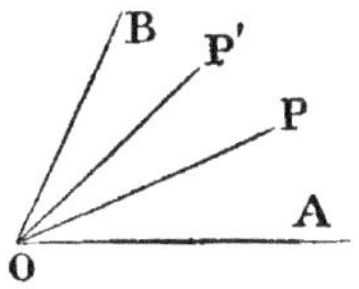

$$p^2.AP.P'B = OA.OP.OP'.OB.\sin AOP.\sin P'OB;$$
$$p^2.AP'.PB = OA.OP'.OP.OB.\sin AOP'.\sin POB;$$
$$\frac{AP.P'B}{AP'.PB} = \frac{\sin AOP.\sin P'OB}{\sin AOP'.\sin POB};$$

but the latter is a constant quantity, independent of the position of the transverse line.

57. If $\alpha - k\beta = 0$, $\alpha - k'\beta = 0$, be the equations of two lines, then $\frac{k}{k'}$ will be the *anharmonic ratio* of the pencil formed by the four lines α, β, $\alpha - k\beta$, $\alpha - k'\beta$, for (Art. 54)

$$k = \frac{\sin AOP}{\sin POB}, \quad k' = \frac{\sin AOP'}{\sin P'OB},$$

therefore

$$\frac{k}{k'} = \frac{\sin AOP . \sin P'OB}{\sin AOP' . \sin POB},$$

but this is the anharmonic ratio of the pencil.

The pencil is a *harmonic* pencil when $\frac{k}{k'} = -1$, for then the angle AOB is divided internally and externally into parts whose sines are in the same ratio. Hence we have the important theorem, *two lines whose equations are* $\alpha - k\beta = 0$, $\alpha + k\beta = 0$, *form with* α, β *a harmonic pencil.*

58. In general the anharmonic ratio of four lines $\alpha - k\beta$, $\alpha - l\beta$, $\alpha - m\beta$, $\alpha - n\beta$ is $\frac{(n-l)(m-k)}{(n-m)(l-k)}$. For let the pencil be cut by any parallel to β in the four points K, L, M, N, and the ratio is $\frac{NL.MK}{NM.LK}$. But since β has the same value for each of these four points, the perpendiculars from these points on α are (by virtue of the equations of the lines) proportional to k, l, m, n; and AK, AL, AM, AN are evidently proportional to these perpendiculars; hence NL is proportional to $n - l$; MK to $m - k$; NM to $n - m$; and LK to $l - k$.

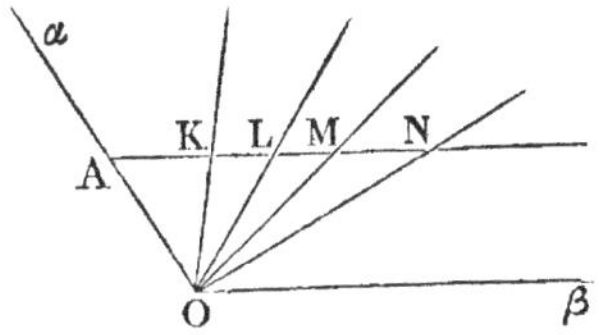

59. The theorems of the last two articles are true of lines represented in the form $P - kP'$, $P - lP'$, &c., where P, P' denote $ax + by + c$, $a'x + b'y + c'$, &c. For we can bring P to the form $x \cos \alpha + y \sin \alpha - p$ by dividing by a certain factor. The equations therefore $P - kP' = 0$, $P - lP' = 0$, &c., are equivalent to equations of the form $\alpha - k\rho\beta = 0$, $\alpha - l\rho\beta = 0$, &c., where ρ is the ratio of the factors by which P and P' must be divided in order to bring them to the forms α, β. But the expressions

for anharmonic ratio are unaltered when we substitute for k, l, m, n; $k\rho$, $l\rho$, $m\rho$, $n\rho$.

It is worthy of remark, that since the expressions for anharmonic ratio only involve the coefficients k, l, m, n, it follows that if we have a system of any number of lines passing through a point, $P-kP'$, $P-lP'$, &c.; and a second system of lines passing through another point, $Q-kQ'$, $Q-lQ'$, &c., the line $P-kP'$ being said to correspond to the line $Q-kQ'$, &c.; then the anharmonic ratio of any four lines of the one system is equal to that of the four corresponding lines of the other system. We shall hereafter often have occasion to speak of such systems of lines, which are called *homographic* systems.

60. *Given three lines* α, β, γ, *forming a triangle*;* *the equation of any right line*, $ax+by+c=0$, *can be thrown into the form*

$$l\alpha+m\beta+n\gamma=0.$$

Write at full length for α, β, γ the quantities which they represent, and $l\alpha+m\beta+n\gamma$ becomes

$$(l\cos\alpha+m\cos\beta+n\cos\gamma)\,x+(l\sin\alpha+m\sin\beta+n\sin\gamma)\,y$$
$$-(lp+mp'+np'')=0.$$

This will be identical with the equation of the given line, if we have

$$l\cos\alpha+m\cos\beta+n\cos\gamma=a,\quad l\sin\alpha+m\sin\beta+n\sin\gamma=b,$$
$$lp+mp'+np''=-c,$$

and we can evidently determine l, m, n, so as to satisfy these three equations.

The following examples will illustrate the principle that it is possible to express the equations of all the lines of any figure in terms of any three, $\alpha=0$, $\beta=0$, $\gamma=0$.

Ex. 1. To deduce analytically the harmonic properties of a complete quadrilateral. (See figure, next page).

Let the equation of AC be $\alpha=0$; of AB, $\beta=0$; of BD, $\gamma=0$; of AD $l\alpha-m\beta=0$; and of BC, $m\beta-n\gamma=0$. Then we are able to express in terms of these quantities the equations of all the other lines of the figure.

---

* We say "forming a triangle," for if the lines α β, γ meet in a point, $l\alpha+m\beta+n\gamma$ must always denote a line passing through the same point, since any values of the coordinates which make α, β, γ separately $=0$, must make $l\alpha+m\beta+n\gamma=0$.

For instance, the equation of CD is

$$l\alpha - m\beta + n\gamma = 0,$$

for it is the equation of a right line passing through the intersection of $l\alpha - m\beta$ and γ, that is, the point D, and of α and $m\beta - n\gamma$, that is, the point C. Again, $l\alpha - n\gamma = 0$ is the equation of OE, for it passes through $\alpha\gamma$ or E, and it also passes through the intersection of AD and BC, since it is $= (l\alpha - m\beta) + (m\beta - n\gamma)$.

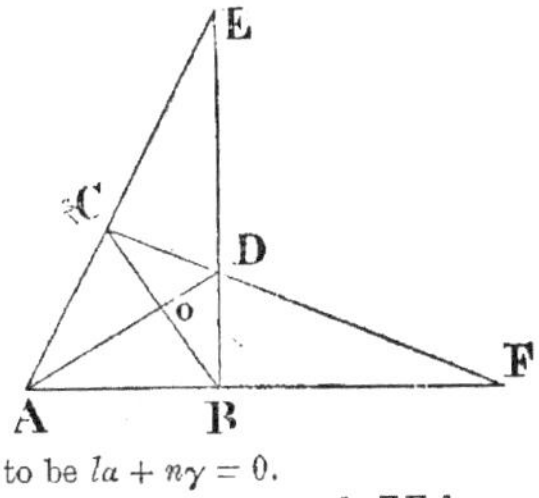

EF joins the point $\alpha\gamma$ to the point $(l\alpha - m\beta + n\gamma,\ \beta)$, and its equation will be found to be $l\alpha + n\gamma = 0$.

From Art. 57 it appears that the four lines EA, EO, EB, and EF form a harmonic pencil, for their equations have been shown to be

$$\alpha = 0,\ \ \gamma = 0,\ \text{ and } l\alpha \pm n\gamma = 0.$$

Again, the equation of FO, which joins the points $(l\alpha + n\gamma,\ \beta)$ and $(l\alpha - m\beta,\ m\beta - n\gamma)$ is

$$l\alpha - 2m\beta + n\gamma = 0.$$

Hence (Art. 57) the four lines FE, FC, FO, and FB are a harmonic pencil, for their equations are

$$l\alpha - m\beta + n\gamma = 0,\ \beta = 0,\ \text{ and } l\alpha - m\beta + n\gamma \pm m\beta = 0.$$

Again, OC, OE, OD, OF are a harmonic pencil, for their equations are

$$l\alpha - m\beta = 0,\ \ m\beta - n\gamma = 0,\ \text{ and } l\alpha - m\beta \pm (m\beta - n\gamma) = 0.$$

Ex. 2. To discuss the properties of the system of lines formed by drawing through the angles of a triangle three lines meeting in a point.

Let the equation of AB be $\gamma = 0$; of AC, $\beta = 0$; of BC, $\alpha = 0$; and let the lines OA, OB, OC, meeting in a point, be $m\beta - n\gamma$, $n\gamma - l\alpha$, $l\alpha - m\beta$ (see Art. 55).

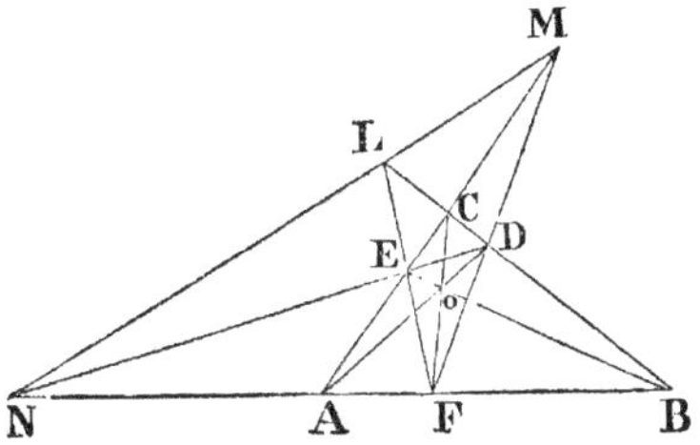

Now we can form the equations of all the other lines in the figure.

For example, the equation of EF is

$$m\beta + n\gamma - l\alpha = 0,$$

since it passes through the points $(\beta,\ n\gamma - l\alpha)$ or E, and $(\gamma,\ m\beta - l\alpha)$ or F.

In like manner, the equation of DF is

$$l\alpha - m\beta + n\gamma = 0,$$

and of DE

$$l\alpha + m\beta - n\gamma = 0.$$

Now we can prove that the three points L, M, N are all in one right line, whose equation is

$$l\alpha + m\beta + n\gamma = 0,$$

for this line passes through the points $(l\alpha + m\beta - n\gamma,\ \gamma)$ or N; $(l\alpha - m\beta + n\gamma,\ \beta)$ or M; and $(m\beta + n\gamma - l\alpha,\ \alpha)$ or L.

The equation of CN is

$$l\alpha + m\beta = 0,$$

for this is evidently a line through (α, β) or C, and it also passes through N, since it $= (l\alpha + m\beta + n\gamma) - n\gamma$.

Hence BN is cut harmonically, for the equations of the four lines CN, CA, CF, CB are

$$\alpha = 0,\ \beta = 0,\ l\alpha - m\beta = 0,\ l\alpha + m\beta = 0.$$

The equations of this example can be applied to many particular cases of frequent occurrence. Thus (see Ex. 3, p. 54) the equation of the line joining the feet of two perpendiculars of a triangle is $\alpha \cos A + \beta \cos B - \gamma \cos C = 0$; while $\alpha \cos A + \beta \cos B + \gamma \cos C$ passes through the intersections with the opposite sides of the triangle, of the lines joining the feet of the perpendiculars. In like manner $\alpha \sin A + \beta \sin B - \gamma \sin C$ represents the line joining the middle points of two sides, &c.

Ex. 3. Two triangles are said to be *homologous*, when the intersections of the corresponding sides lie on the same right line called the *axis of homology*; prove that the lines joining the corresponding vertices meet in a point [called the *centre of homology*].

Let the sides of the first triangle be α, β, γ; and let the line on which the corresponding sides meet be $l\alpha + m\beta + n\gamma$; then the equation of a line through the intersection of this with α must be of the form $l'\alpha + m\beta + n\gamma = 0$, and similarly those of the other two sides of the second triangle are

$$l\alpha + m'\beta + n\gamma = 0,\ l\alpha + m\beta + n'\gamma = 0.$$

But subtracting successively each of the last three equations from another, we get for the equations of the lines joining corresponding vertices

$$(l - l')\,\alpha = (m - m')\,\beta,\ (m - m')\,\beta = (n - n')\,\gamma,\ (n - n')\,\gamma = (l - l')\,\alpha,$$

which obviously meet in a point.

61. *To find the condition that two lines* $l\alpha + m\beta + n\gamma$, $l'\alpha + m'\beta + n'\gamma$ *may be mutually perpendicular.*

Write the equations at full length as in Art. 60, and apply the criterion of Art. 25, Cor. 2 ($AA' + BB' = 0$), when we find

$$ll' + mm' + nn' + (mn' + m'n)\cos(\beta - \gamma) + (nl' + n'l)\cos(\gamma - \alpha)$$
$$+ (lm' + l'm)\cos(\alpha - \beta) = 0.$$

Now since β and γ are the angles made with the axis of x by the perpendiculars on the lines β, γ, $\beta - \gamma$ is the angle between those perpendiculars, which again is equal or supplemental to the angle between the lines themselves. If we suppose the origin to be within the triangle, and A, B, C to be the angles of the triangle, $\beta - \gamma$ is the supplement of A. The condition for perpendicularity therefore is

$$ll' + mm' + nn' - (mn' + m'n)\cos A - (nl' + n'l)\cos B - (lm' + l'm)\cos C = 0.$$

As a particular case of the above, the condition that $l\alpha + m\beta + n\gamma$ may be perpendicular to γ is

$$n = m\cos A + l\cos B.$$

In like manner we find the length of the perpendicular from $x'y'$

on $l\alpha + m\beta + n\gamma$. Write the equation at full length and apply the formula of Art. 34, when, if we write $x' \cos\alpha + y' \sin\alpha - p = \alpha'$, &c., the result is

$$\frac{l\alpha' + m\beta' + n\gamma'}{\sqrt{(l^2 + m^2 + n^2 - 2mn\cos A - 2nl\cos B - 2lm\cos C)}}.$$

Ex. 1. To find the equation of a perpendicular to γ through its extremity. The equation is of the form $l\alpha + n\gamma = 0$. And the condition of this article gives $n = l\cos B$, as in Ex. 6, p. 54.

Ex. 2. To find the equation of a perpendicular to γ through its middle point. The middle point being the intersection of γ with $\alpha\sin A - \beta\sin B$, the equation of any line through it is of the form $\alpha\sin A - \beta\sin B + n\gamma = 0$, and the condition of this article gives $n = \sin(A - B)$.

Ex. 3. The three perpendiculars at middle points of sides meet in a point. For eliminating α, β, γ in turn between

$$\alpha\sin A - \beta\sin B + \gamma\sin(A - B) = 0,\quad \beta\sin B - \gamma\sin C + \alpha\sin(B - C) = 0,$$

we get for the lines joining to the three vertices the intersection of two perpendiculars $\frac{\alpha}{\cos A} = \frac{\beta}{\cos B} = \frac{\gamma}{\cos C}$; and the symmetry of the equations proves that the third perpendicular passes through the same point. The equations of the perpendiculars vanish when multiplied by $\sin^2 C$, $\sin^2 A$, $\sin^2 B$, and added together.

Ex. 4. Find, by Art. 25, expressions for the sine, cosine, and tangent of the angle between $l\alpha + m\beta + n\gamma$, $l'\alpha + m'\beta + n'\gamma$.

Ex. 5. Prove that $\alpha\cos A + \beta\cos B + \gamma\cos C$ is perpendicular to

$$\alpha\sin A\cos A\sin(B - C) + \beta\sin B\cos B\sin(C - A) + \gamma\sin C\cos C\sin(A - B).$$

Ex. 6. Find the equation of a line through the point $\alpha'\beta'\gamma'$ perpendicular to the line γ. *Ans.* $\alpha(\beta' + \gamma'\cos A) - \beta(\alpha' + \gamma'\cos B) + \gamma(\beta'\cos B - \alpha'\cos A)$.

62. We have seen that we can express the equation of any right line in the form $l\alpha + m\beta + n\gamma = 0$, and so solve any problem by a set of equations expressed in terms of α, β, γ, without any direct mention of x and y. This suggests a new way of looking at the principle laid down in Art. 60. Instead of regarding α as a mere abbreviation for the quantity $x\cos\alpha + y\sin\alpha - p$, we may look upon it as simply denoting the length of the perpendicular from a point on the line α. We may imagine a system of *trilinear coordinates* in which the position of a point is defined by its distances from three fixed lines, and in which the position of any right line is defined by a homogeneous equation between these distances, of the form

$$l\alpha + m\beta + n\gamma = 0.$$

The advantage of trilinear coordinates is, that whereas in

Cartesian (or x and y) coordinates the utmost simplification we can introduce is by choosing *two* of the most remarkable lines in the figure for axes of coordinates, we can in trilinear coordinates obtain still more simple expressions by choosing *three* of the most remarkable lines for the lines of reference α, β, γ. The reader will compare the brevity of the expressions in Art. 54 with those corresponding in Chap. II.

63. The perpendiculars from any point O on α, β, γ are connected by the relation $a\alpha + b\beta + c\gamma = M$, where a, b, c, are the sides, and M double the area, of the triangle of reference. For evidently $a\alpha$, $b\beta$, $c\gamma$ are respectively double the areas of the triangles OBC, OCA, OAB. The reader may suppose that this is only true if the point O be taken *within* the triangle; but he is to remember that if the point O were on the other side of any of the lines of reference (α), we must give a negative sign to that perpendicular, and the quantity $a\alpha + b\beta + c\gamma$ would then be double $OCA + OAB - OBC$, that is, still = double the area of the triangle. Since $\sin A$ is proportional to a, it is plain that $\alpha \sin A + \beta \sin B + \gamma \sin C$ is also constant, a theorem which may otherwise be proved by writing α, β, γ at full length, as in Art. 60, multiplying by $\sin(\beta - \gamma)$, $\sin(\gamma - \alpha)$, $\sin(\alpha - \beta)$, respectively, and adding, when the coefficients of x and y vanish, and the sum is therefore constant.

The theorem of this article enables us always to use *homogeneous* equations in α, β, γ, for if we are given such an equation as $\alpha = 3$, we can throw it into the homogeneous form

$$M\alpha = 3\,(a\alpha + b\beta + c\gamma).$$

64. *To express in trilinear coordinates the equation of the parallel to a given line* $l\alpha + m\beta + n\gamma$.

In Cartesian coordinates two lines $Ax + By + C$, $Ax + By + C'$, are parallel if their equations differ only by a constant. It follows then that

$$l\alpha + m\beta + n\gamma + k\,(\alpha \sin A + \beta \sin B + \gamma \sin C) = 0$$

denotes a line parallel to $l\alpha + m\beta + n\gamma$, since the two equations differ only by a quantity which has been just proved to be constant.

In the same case $Ax + By + C + (Ax + By + C')$ denotes a line also parallel to the two given lines and half-way between them; hence if two equations $P = 0$, $P' = 0$ are so connected that $P - P' =$ constant, then $P + P'$ denotes a parallel to P and P' half-way between them.

Ex. 1. To find the equation of a parallel to the base of a triangle drawn through the vertex. *Ans.* $\alpha \sin A + \beta \sin B = 0$.

For this, obviously, is a line through $\alpha\beta$; and writing the equation in the form

$$\gamma \sin C - (\alpha \sin A + \beta \sin B + \gamma \sin C) = 0,$$

it appears that it differs only by a constant from $\gamma = 0$.

We see, also, that the parallel $\alpha \sin A + \beta \sin B$, and the bisector of the base $\alpha \sin A - \beta \sin B$, form a harmonic pencil with α, β, (Art. 57).

Ex. 2. The line joining the middle points of sides of a triangle is parallel to the base. Its equation (see Ex. 2, p. 58) is

$$\alpha \sin A + \beta \sin B - \gamma \sin C = 0, \text{ or } 2\gamma \sin C = \alpha \sin A + \beta \sin B + \gamma \sin C.$$

Ex. 3. The line $a\alpha - b\beta + c\gamma - d\delta$ (see Ex. 5, Art. 54) passes through the middle point of the line joining $\alpha\gamma$, $\beta\delta$. For $(a\alpha + c\gamma) + (b\beta + d\delta)$ is constant, being twice the area of the quadrilateral; hence $a\alpha + c\gamma$, $b\beta + d\delta$ are parallel, and $(a\alpha + c\gamma) - (b\beta + d\delta)$ is also parallel and half-way between them. It therefore bisects the line joining $(\alpha\gamma)$, which is a point on the first line, to $(\beta\delta)$ which is a point on the second.

65. *To write in the form $l\alpha + m\beta + n\gamma = 0$ the equation of the line joining two given points $x'y'$, $x''y''$.*

Let α', as before, denote the quantity $x' \cos \alpha + y' \sin \alpha - p$. Then the condition that the coordinates $x'y'$ shall satisfy the equation $l\alpha + m\beta + n\gamma = 0$ may be written

$$l\alpha' + m\beta' + n\gamma' = 0.$$

Similarly we have $$l\alpha'' + m\beta'' + n\gamma'' = 0.$$

Solving for $\frac{l}{n}$, $\frac{m}{n}$, from these two equations, and substituting in the given form, we obtain for the equation of the line joining the two points

$$\alpha (\beta'\gamma'' - \gamma'\beta'') + \beta (\gamma'\alpha'' - \gamma''\alpha') + \gamma (\alpha'\beta'' - \alpha''\beta') = 0.$$

It is to be observed that the equations in trilinear coordinates being homogeneous, we are not concerned with the actual lengths of the perpendiculars from any point on the lines of reference, but only with their mutual ratios. Thus the preceding equation is not altered if we write $\rho\alpha'$, $\rho\beta'$, $\rho\gamma'$, for α', β', γ'. Accordingly, if a point be given as the intersection of the lines $\frac{\alpha}{l} = \frac{\beta}{m} = \frac{\gamma}{n}$, we may take l, m, n as the trilinear coordinates of that point. For let ρ be the common value of these fractions, and the actual lengths of the perpendiculars on α, β, γ are $l\rho$, $m\rho$, $n\rho$, where ρ is given by the equation $al\rho + bm\rho + cn\rho = M$, but, as has been just proved, we do not need to determine ρ. Thus, in applying the equation of this article, we may take for the coordinates of intersection of bisectors of sides, $\sin B \sin C$, $\sin C \sin A$, $\sin A \sin B$; of intersection of perpendiculars, $\cos B \cos C$, $\cos C \cos A$, $\cos A \cos B$; of centre of inscribed circle 1, 1, 1; of centre of circumscribing circle $\cos A$, $\cos B$, $\cos C$, &c.

Ex. 1. Find the equation of the line joining intersections of perpendiculars, and of bisectors of sides (see Art. 61, Ex. 5).

Ans. $\alpha \sin A \cos A \sin (B - C) + \beta \sin B \cos B \sin (C - A) + \gamma \sin C \cos C \sin (A - B) = 0.$

Ex. 2. Find equation of line joining centres of inscribed and circumscribing circles.

Ans. $\alpha (\cos B - \cos C) + \beta (\cos C - \cos A) + \gamma (\cos A - \cos B) = 0.$

66. It is proved, as in Art. 7, that the length of the perpendicular on α from the point which divides in the ratio $l : m$ the line joining two points whose perpendiculars are α', α'' is $\frac{l\alpha' + m\alpha''}{l + m}$. Consequently the coordinates of the point dividing in the ratio $l : m$ the line joining $\alpha'\beta'\gamma'$, $\alpha''\beta''\gamma''$ are $l\alpha' + m\alpha''$, $l\beta' + m\beta''$, $l\gamma' + m\gamma''$. It is otherwise evident that this point lies on the line joining the given points, for if $\alpha'\beta'\gamma'$, $\alpha''\beta''\gamma''$ both satisfy the equation of a line $A\alpha + B\beta + C\gamma = 0$, so will also $l\alpha' + m\alpha''$, &c. It follows hence, without difficulty, that $l\alpha' - m\alpha''$, &c., is the fourth harmonic to $l\alpha' + m\alpha''$, α', α''; that the anharmonic ratio of $\alpha' - k\alpha''$, $\alpha' - l\alpha''$, $\alpha' - m\alpha''$, $\alpha' - n\alpha''$ is $\frac{(n - l)(m - k)}{(n - m)(l - k)}$; and also that, given two systems of points on two right lines $\alpha' - k\alpha''$, $\alpha' - l\alpha''$, &c., $\alpha''' - k\alpha''''$, $\alpha''' - l\alpha''''$, &c.; these systems are *homographic*, the anharmonic ratio of any four points on one line being equal to that of the four corresponding points on the other.

Ex. The intersection of perpendiculars, of bisectors of sides, and the centre of circumscribing circle lie on a right line. For the coordinates of these points are $\cos B \cos C$, &c., $\sin B \sin C$, &c., and $\cos A$, &c. But the last set of coordinates may be written $\sin B \sin C - \cos B \cos C$, &c.

The point whose coordinates are $\cos (B - C)$, $\cos (C - A)$, $\cos (A - B)$ evidently lies on the same right line and is a fourth harmonic to the three preceding. It will be found hereafter that this is the centre of the circle through the middle points of the sides.

67. *To examine what line is denoted by the equation*

$$\alpha \sin A + \beta \sin B + \gamma \sin C = 0.$$

This equation is included in the general form of an equation of a right line, but we have seen (Art. 63) that the left-hand member is constant, and never $= 0$. Let us return, however, to the general equation of the right line $Ax + By + C = 0$. We saw that the intercepts cut off on the axes are $-\frac{C}{A}$, $-\frac{C}{B}$; consequently, the smaller A and B become the greater will be the intercepts on the axes, and therefore the more remote the line represented. Let A and B be both $= 0$, then the intercepts become infinite, and the line is altogether situated at an infinite distance from the origin. Now it was proved (Art. 63) that the equation under consideration is equivalent to $0x + 0y + C = 0$, and though it cannot be satisfied by any finite values of the coordinates, it may by infinite values, since the product of nothing by infinity may be finite. It appears then that $\alpha \sin A + \beta \sin \beta + \gamma \sin C$ *denotes a right line situated altogether at an infinite distance from the origin;* and that the equation of an infinitely distant right line, in Cartesian coordinates, is $0 \cdot x + 0 \cdot y + C = 0$. We shall, for shortness, commonly cite the latter equation in the less accurate form $C = 0$.

68. We saw (Art. 64) that a line parallel to the line $\alpha = 0$ has an equation of the form $\alpha + C = 0$. Now the last Article shows that this is only an additional illustration of the principle of Art. 40. For a parallel to α may be considered as intersecting it at an infinite distance, but (Art. 40) an equation of the form $\alpha + C = 0$ represents a line through the intersection of the lines $\alpha = 0$, $C = 0$, or (Art. 67) through the intersection of the line α with the line at infinity.

69. We have to add that Cartesian coordinates are only a particular case of trilinear. There appears, at first sight, to be an essential difference between them, since trilinear equations are always homogeneous, while we are accustomed to speak of Cartesian equations as containing an absolute term, terms of the first degree, terms of the second degree, &c. A little reflection, however, will show that this difference is only apparent, and

that Cartesian equations must be equally homogeneous in reality, though not in form. The equation $x = 3$, for example, must mean that the line x is equal to three feet or three inches, or, in short, to three times some linear unit; the equation $xy = 9$ must mean that the rectangle xy is equal to nine *square* feet or *square* inches, or to nine *squares* of some linear unit; and so on.

If we wish to have our equations homogeneous in form as well as in reality, we may denote our linear unit by z, and write the equation of the right line

$$Ax + By + Cz = 0.$$

Comparing this with the equation

$$A\alpha + B\beta + C\gamma = 0,$$

and remembering (Art. 67) that when a line is at an infinite distance its equation takes the form $z = 0$, we learn that *equations in Cartesian coordinates are only the particular form assumed by trilinear equations when two of the lines of reference are what are called the coordinate axes, while the third is at an infinite distance.*

70. We wish in conclusion to give a brief account of what is meant by systems of *tangential coordinates*, in which the position of a right line is expressed by coordinates, and that of a point by an equation. In this volume we limit ourselves to what is not so much a new system of coordinates as a new way of speaking of the equations already in use. If the equation (Cartesian or trilinear) of any line be $\lambda x + \mu y + \nu z = 0$, then evidently, if λ, μ, ν be known, the position of the line is known; and we may call these three quantities (or rather their mutual ratios with which only we are concerned) the coordinates of the right line. If the line pass through a fixed point $x'y'z'$, the relation must be fulfilled $x'\lambda + y'\mu + z'\nu = 0$; if therefore we are given any equation connecting the coordinates of a line, of the form $a\lambda + b\mu + c\nu = 0$, this denotes that the line passes through the fixed point (a, b, c), (see Art. 51), and the given equation may be called the equation of that point. Further, we may use abbreviations for the equations of points, and may denote by α, β the quantities $x'\lambda + y'\mu + z'\nu$, $x''\lambda + y''\mu + z''\nu$; then it is evident that $l\alpha + m\beta = 0$ is the equation of a point dividing in a given ratio the line joining the points α, β; that $l\alpha = m\beta$, $m\beta = n\gamma$, $n\gamma = l\alpha$ are the equations of three points which lie on a right line; that $\alpha + k\beta$, $\alpha - k\beta$ denote two points harmonically conjugate with regard to α, β, &c. We content ourselves here with indicating analogies which we shall hereafter develope more fully; for we shall have occasion to show that theorems concerning points are so connected with theorems concerning lines, that when either is known the other can be inferred, and often that the same equations differently interpreted will prove either theorem. Theorems so connected are called *reciprocal* theorems.

Ex. Interpret in tangential coordinates the equations used in Art. 60, Ex. 2.

Let α, β, γ denote the points A, B, C; $m\beta - n\gamma$, $n\gamma - l\alpha$, $l\alpha - m\beta$, the points L, M, N; then $m\beta + n\gamma - l\alpha$, $n\gamma + l\alpha - m\beta$, $l\alpha + m\beta - n\gamma$ denote the vertices of the triangle formed by LA, MB, NC; and $l\alpha + m\beta + n\gamma$ denotes a point O in which meet the lines joining the vertices of this new triangle to the corresponding vertices of the original: $m\beta + n\gamma$, $n\gamma + l\alpha$, $l\alpha + m\beta$ denote D, E, F. It is easy hence to see the points in the figure, which are harmonically conjugate.

CHAPTER V.

EQUATIONS ABOVE THE FIRST DEGREE REPRESENTING RIGHT LINES.

71. BEFORE proceeding to speak of the *curves* represented by equations above the first degree, we shall examine some cases where these equations represent *right lines.*

If we take any number of equations $L = 0$, $M = 0$, $N = 0$, &c., and multiply them together, the compound equation LMN &c. $= 0$ will represent the aggregate of all the lines represented by its factors; for it will be satisfied by the values of the coordinates which make any of its factors $= 0$. Conversely, *if an equation of any degree can be resolved into others of lower degrees, it will represent the aggregate of all the loci represented by its different factors.* If, then, an equation of the nth degree can be resolved into n factors of the first degree, it will represent n right lines.

72. *A homogeneous equation of the nth degree in x and y denotes n right lines passing through the origin.*

Let the equation be

$$x^n - px^{n-1}y + qx^{n-2}y^2 - \&\text{c.} \ldots + ty^n = 0.$$

Divide by y^n, and we get

$$\left(\frac{x}{y}\right)^n - p\left(\frac{x}{y}\right)^{n-1} + q\left(\frac{x}{y}\right)^{n-2} - \&\text{c.} = 0.$$

Let a, b, c, &c., be the n roots of this equation, then it is resolvable into the factors

$$\left(\frac{x}{y} - a\right)\left(\frac{x}{y} - b\right)\left(\frac{x}{y} - c\right) \&\text{c.} = 0,$$

and the original equation is therefore resolvable into the factors

$$(x - ay)(x - by)(x - cy) \&\text{c.} = 0.$$

It accordingly represents the n right lines $x - ay = 0$, &c., all of which pass through the origin. Thus, then, in particular, the homogeneous equation

$$x^2 - pxy + qy^2 = 0$$

represents the two right lines $x - ay = 0$, $x - by = 0$, where a and b are the two roots of the quadratic

$$\left(\frac{x}{y}\right)^2 - p\left(\frac{x}{y}\right) + q = 0.$$

It is proved, in like manner, that the equation

$$(x-a)^n - p(x-a)^{n-1}(y-b) + q(x-a)^{n-2}(y-b)^2 \ldots + t(y-b)^n = 0$$

denotes n right lines passing through the point (a, b).

Ex. 1. What locus is represented by the equation $xy = 0$?

Ans. The two axes; since the equation is satisfied by either of the suppositions $x = 0$, $y = 0$.

Ex. 2. What locus is represented by $x^2 - y^2 = 0$?

Ans. The bisectors of the angles between the axes, $x \pm y = 0$ (see Art. 35).

Ex. 3. What locus is represented by $x^2 - 5xy + 6y^2 = 0$? *Ans.* $x - 2y = 0$, $x - 3y = 0$.

Ex. 4. What locus is represented by $x^2 - 2xy \sec\theta + y^2 = 0$?

Ans. $x = y \tan(45^\circ \pm \frac{1}{2}\theta)$.

Ex. 5. What lines are represented by $x^2 - 2xy \tan\theta - y^2 = 0$?

Ex. 6. What lines are represented by $x^3 - 6x^2y + 11xy^2 - 6y^3 = 0$?

73. Let us examine more minutely the three cases of the solution of the equation $x^2 - pxy + qy^2 = 0$, according as its roots are real and unequal, real and equal, or both imaginary.

The first case presents no difficulty: a and b are the tangents of the angles which the lines make with the axis of y (the axes being supposed rectangular), p is therefore the sum of those tangents, and q their product.

In the second case, when $a = b$, it was once usual among geometers to say that the equation represented but one right line $(x - ay = 0)$. We shall find, however, many advantages in making the language of geometry correspond exactly to that of algebra, and as we do not say that the equation above has *only one* root, but that it has *two equal* roots, so we shall not say that it represents *only one* line, but that it represents *two coincident* right lines.

Thirdly, let the roots be both imaginary. In this case no real coordinates can be found to satisfy the equation, except the coordinates of the origin $x = 0$, $y = 0$; hence it was usual to say that in this case the equation did not represent right lines, but was the equation of the origin. Now this language appears to us very objectionable, for we saw (Art. 14) that *two* equations are required to determine any point, hence we are unwilling to acknowledge *any single* equation as the equation of a point. Moreover, we have been hitherto accustomed to find that two *different* equations always had *different* geometrical significations, but here we should have innumerable equations, all purporting to be the equation of the same point; for it is obviously immaterial what the values of p and q are, provided only that they give imaginary values for the roots, that is to say, provided that p^2 be less than $4q$. We think it, therefore, much preferable to make our language correspond exactly to the language of algebra; and as we do not say that the equation above has *no* roots when p^2 is less than $4q$, but that it has two *imaginary* roots, so we shall not say that, in this case, it represents *no* right lines, but that it represents two *imaginary* right lines. In short, the equation $x^2 - pxy + qy^2 = 0$ being *always* reducible to the form $(x - ay)(x - by) = 0$, we shall always say that it represents two right lines drawn through the origin; but when a and b are real, we shall say that these lines are real; when a and b are equal, that the lines coincide; and when a and b are imaginary, that the lines are imaginary. It may seem to the student a matter of indifference which mode of speaking we adopt; we shall find, however, as we proceed, that we should lose sight of many important analogies by refusing to adopt the language here recommended.

Similar remarks apply to the equation

$$Ax^2 + Bxy + Cy^2 = 0,$$

which can be reduced to the form $x^2 - pxy + qy^2 = 0$, by dividing by the coefficient of x^2. This equation will always represent two right lines through the origin; these lines will be real if $B^2 - 4AC$ be positive, as at once appears from solving the equation; they will coincide if $B^2 - 4AC = 0$; and they will be imaginary if $B^2 - 4AC$ be negative. So, again, the same language is used if we meet with equal or imaginary roots in the solution of the general homogeneous equation of the nth degree.

74. *To find the angle contained by the lines represented by the equation* $x^2 - pxy + qy^2 = 0$.

Let this equation be equivalent to $(x - ay)(x - by) = 0$, then the tangent of the angle between the lines is (Art. 25) $\dfrac{a - b}{1 + ab}$

but the product of the roots of the given equation $= q$, and their difference $= \sqrt{(p^2 - 4q)}$. Hence

$$\tan\phi = \frac{\sqrt{(p^2 - 4q)}}{1 + q}.$$

If the equation had been given in the form

$$Ax^2 + Bxy + Cy^2 = 0,$$

it would have been found that

$$\tan\phi = \frac{\sqrt{(B^2 - 4AC)}}{A + C}.$$

COR. The lines will cut at right angles, or $\tan\phi$ will become infinite, if $q = -1$ in the first case, or if $A + C = 0$ in the second.

Ex. Find the angle between the lines

$x^2 + xy - 6y^2 = 0.$ *Ans.* 45°

$x^2 - 2xy\sec\theta + y^2 = 0.$ *Ans.* θ.

*If the axes be oblique we find, in like manner,

$$\tan\phi = \frac{\sin\omega\sqrt{(B^2 - 4AC)}}{A + C - B\cos\omega}.$$

75. *To find the equation which will represent the lines bisecting the angles between the lines represented by the equation*

$$Ax^2 + Bxy + Cy^2 = 0.$$

Let these lines be $x - ay = 0$, $x - by = 0$; let the equation of the bisector be $x - \mu y = 0$, and we seek to determine μ. Now (Art. 18) μ is the tangent of the angle made by this bisector with the axis of y, and it is plain that this angle is half the sum of the angles made with this axis by the lines themselves. Equating, therefore, tangent of twice this angle to tangent of sum, we get

$$\frac{2\mu}{1 - \mu^2} = \frac{a + b}{1 - ab};$$

but, from the theory of equations,

$$a + b = -\frac{B}{A}, \quad ab = \frac{C}{A};$$

therefore

$$\frac{2\mu}{1 - \mu^2} = -\frac{B}{A - C},$$

or

$$\mu^2 - 2\frac{A - C}{B}\mu - 1 = 0.$$

This gives us a quadratic to determine μ, one of whose roots will be the tangent of the angle made with the axis of y by the *internal* bisector of the angle between the lines, and the other the tangent of the angle made by the *external* bisector. We can find the combined equation of both lines by substituting in the last quadratic for μ its value $= \frac{x}{y}$, and we get

$$x^2 - 2\frac{A-C}{B}xy - y^2 = 0,*$$

and the form of this equation shows that the bisectors cut each other at right angles (Art. 74).

The student may also obtain this equation by forming (Art. 35) the equations of the internal and external bisectors of the angle between the lines $x - ay = 0$, $x - by = 0$, and multiplying them together, when he will have

$$\frac{(x-ay)^2}{1+a^2} = \frac{(x-by)^2}{1+b^2},$$

and then clearing of fractions, and substituting for $a + b$, and ab their values in terms of A, B, C, the equation already found is obtained.

76. We have seen that an equation of the second degree *may* represent two right lines; but such an equation in general cannot be resolved into the product of two factors of the first degree, unless its coefficients fulfil a certain relation, which can be most easily found as follows. Let the general equation of the second degree be written

$$ax^2 + 2hxy + by^2 + 2gx + 2fy + c = 0,\dagger$$

or

$$ax^2 + 2(hy + g)x + by^2 + 2fy + c = 0.$$

---

* It is remarkable that the roots of this last equation will always be real, even the roots of the equation $Ax^2 + Bxy + Cy^2 = 0$ be imaginary, which leads to the curious result, that a pair of imaginary lines has a pair of real lines bisecting the angle between them. It is the existence of such relations between real and imaginary lines which makes the consideration of the latter profitable.

† It might seem more natural to write this equation

$$ax^2 + bxy + cy^2 + dx + ey + f = 0,$$

but as it is desirable that the equation should be written with the same letters all through the book, I have decided on using, from the first, the form which will hereafter be found most convenient and symmetrical. It will appear hereafter

Solving this equation for x we get

$$ax = -(hy+g) \pm \sqrt{\{(h^2-ab)\,y^2 + 2\,(hg-af)\,y + (g^2-ac)\}}.$$

In order that this may be capable of being reduced to the form $x = my + n$, it is necessary that the quantity under the radical should be a perfect square, in which case the equation would denote two right lines according to the different signs we give the radical. But the condition that the radical should be a perfect square is

$$(h^2-ab)\,(g^2-ac) = (hg-af)^2.$$

Expanding, and dividing by a, we obtain the required condition, viz.

$$abc + 2fgh - af^2 - bg^2 - ch^2 = 0.*$$

1. Verify that the following equation represents right lines, and find the lines:

$$x^2 - 5xy + 4y^2 + x + 2y - 2 = 0.$$

Ans. Solving for x as in the text, the lines are found to be

$$x - y - 1 = 0,\quad x - 4y + 2 = 0.$$

Ex. 2. Verify that the following equation represents right lines:

$$(\alpha x + \beta y - r^2)^2 = (\alpha^2 + \beta^2 - r^2)\,(x^2 + y^2 - r^2).$$

Ex. 3. What lines are represented by the equation

$$x^2 - xy + y^2 - x - y + 1 = 0\,?$$

Ans. The imaginary lines $x + \theta y + \theta^2 = 0$, $x + \theta^2 y + \theta = 0$, where θ is one of the imaginary cube roots of 1.

Ex. 4. Determine h, so that the following equation may represent right lines:

$$x^2 + 2hxy + y^2 - 5x - 7y + 6 = 0.$$

Ans. Substituting these values of the coefficients in the general condition, we get for h the quadratic $12h^2 - 35h + 25 = 0$, whose roots are $\frac{5}{3}$ and $\frac{5}{4}$.

*77. The method used in the preceding Article, though the most simple in the case of the equation of the second degree, is not applicable to equations of higher degrees; we therefore give another solution of the same problem. It is required to ascertain

---

that this equation is intimately connected with the homogeneous equation in three variables, which may be most symmetrically written

$$ax^2 + by^2 + cz^2 + 2fyz + 2gzx + 2hxy = 0.$$

The form in the text is derived from this by making $z = 1$. The coefficient 2 is affixed to certain terms, because formulæ connected with the equation, which we shall have occasion to use, thus become simpler and more easy to be remembered.

* If the coefficients f, g, h in the equation had been written without numerical multipliers, this condition would have been

$$4abc + fgh - af^2 - bg^2 - ch^2 = 0.$$

whether the given equation of the second degree can be identical with the product of the equations of two right lines

$$(\alpha x + \beta y - 1)(\alpha' x + \beta' y - 1) = 0.$$

Multiply out this product, and equate the coefficient of each term to the corresponding coefficient in the general equation of the second degree, having previously divided the latter by c, so as to make the absolute term in each equation $= 1$. We thus obtain five equations, viz.

$$\alpha\alpha' = \frac{a}{c},\quad \alpha + \alpha' = -\frac{2g}{c},\quad \beta\beta' = \frac{b}{c},\quad \beta + \beta' = -\frac{2f}{c},\quad \alpha\beta' + \alpha'\beta = \frac{2h}{c};$$

from which eliminating the four unknown quantities α, α', β, β', we obtain the required condition. The first four of the equations at once give us two quadratics for determining α, α'; β, β'; which indeed might have been also obtained from the consideration that these quantities are the reciprocals of the intercepts made by the lines on the axes; and that the intercepts made by the locus on the axes are found (by making alternately $x = 0$, $y = 0$, in the general equation) from the equations

$$ax^2 + 2gx + c = 0,\quad by^2 + 2fy + c = 0.$$

We can now complete the elimination by solving the quadratics, substituting in the fifth equation and clearing of radicals; or we may proceed more simply as follows: Since nothing shews whether the root α of the first quadratic is to be combined with the root β or β' of the second, it is plain that $\frac{2h}{c}$ may have either of the values $\alpha\beta' + \alpha'\beta$ or $\alpha\beta + \alpha'\beta'$. This is also evident geometrically, since if the locus meet the axes in the points L, L'; M, M'; it is plain that if it represent right lines at all, these must be either the pair LM, $L'M'$, or else LM', $L'M$, whose equations are

$$(\alpha x + \beta y - 1)(\alpha' x + \beta' y - 1) = 0, \text{ or } (\alpha x + \beta' y - 1)(\alpha' x + \beta y - 1) = 0.$$

The sum then of the two quantities $\alpha\beta' + \alpha'\beta$, $\alpha\beta + \alpha'\beta'$

$$= (\alpha + \alpha')(\beta + \beta') = \frac{4fg}{c^2},$$

and their product

$$= \alpha\alpha'(\beta^2 + \beta'^2) + \beta\beta'(\alpha^2 + \alpha'^2) = \frac{a}{c}\frac{(4f^2 - 2bc)}{c^2} + \frac{b}{c}\frac{(4g^2 - 2ac)}{c^2}.$$

Hence $\frac{h}{c}$ is given by the quadratic

$$\frac{h^2}{c^2} - \frac{fg}{c^2} \cdot \frac{2h}{c} + \frac{af^2 + bg^2 - abc}{c^3} = 0,$$

which, cleared of fractions, is the condition already obtained.

Ex. To determine h so that $x^2 + 2hxy + y^2 - 5x - 7y + 6 = 0$ may represent right lines (see Ex 4, p. 72).

The intercepts on the axes are given by the equations

$$x^2 - 5x + 6 = 0, \; y^2 - 7y + 6 = 0,$$

whose roots are $x = 2$, $x = 3$; $y = 1$, $y = 6$. Forming, then, the equation of the lines joining the points so found, we see that if the equation represent right lines, it must be of one or other of the forms

$$(x + 2y - 2)\,(2x + y - 6) = 0, \; (x + 3y - 3)\,(3x + y - 6) = 0,$$

whence, multiplying out, h is determined.

*78. *To find how many conditions must be satisfied in order that the general equation of the nth degree may represent right lines.*

We proceed as in the last Article; we compare the general equation, having first by division made the absolute term $= 1$, with the product of the n right lines

$$(\alpha x + \beta y - 1)\,(\alpha' x + \beta' y - 1)\,(\alpha'' x + \beta'' y - 1) \text{ \&c.} = 0.$$

Let the number of terms in the general equation be N; then from a comparison of coefficients we obtain $N - 1$ equations (the absolute term being already the same in both); $2n$ of these equations are employed in determining the $2n$ unknown quantities α, α', &c., whose values being substituted in the remaining equations afford $N - 1 - 2n$ conditions. Now if we write the general equation

$$\begin{array}{c} A \\ + Bx + Cy \\ + Dx^2 + Exy + Fy^2 \\ + Gx^3 + Hx^2y + Kxy^2 + Ly^3 \\ + \text{\&c.} = 0, \end{array}$$

it is plain that the number of terms is the sum of the arithmetic series

$$N = 1 + 2 + 3 + \ldots (n + 1) = \frac{(n+1)(n+2)}{1.2};$$

hence $$N - 1 = \frac{n(n+3)}{1.2}; \; N - 1 - 2n = \frac{n(n-1)}{1.2}.$$

CHAPTER VI.

THE CIRCLE.

79. BEFORE proceeding to the discussion of the general equation of the second degree, it seems desirable that we should shew, in the simple case of the circle, how all the properties of a curve may be deduced from its equation, without assuming any previous acquaintance with the geometrical theory.

The equation, to rectangular axes, of the circle whose centre is the point $(\alpha\beta)$ and radius is r, has already (Art. 17) been found to be

$$(x-\alpha)^2+(y-\beta)^2=r^2.$$

Two particular cases of this equation deserve attention, as occurring frequently in practice. Let the centre be the origin, then $\alpha=0$, $\beta=0$, and the equation is

$$x^2+y^2=r^2.$$

Let the axis of x be a diameter, and the axis of y a perpendicular at its extremity, then $\alpha=r$, $\beta=0$, and the equation becomes

$$x^2+y^2=2rx.$$

80. It will be observed that the equation of the circle, to rectangular axes, does not contain the term xy, and that the coefficients of x^2 and y^2 are equal. The general equation therefore

$$ax^2+2hxy+by^2+2gx+2fy+c=0$$

cannot represent a circle, unless we have $h=0$ and $a=b$. Any equation of the second degree which fulfils these two conditions may be reduced to the form $(x-\alpha)^2+(y-\beta)^2=r^2$, by a process corresponding to that used in the solution of quadratic equations. If the common coefficient of x^2 and y^2 be not already unity, by division make it so; then having put the terms containing x and y on the left-hand side of the equation, and the constant term on the right, complete the squares by adding to both sides the sum of the squares of half the coefficients of x and y.

Ex. Reduce to the form $(x-\alpha)^2+(y-\beta)^2=r^2$, the equations

$$x^2+y^2-2x-4y=20\,;\; 3x^2+3y^2-5x-7y+1=0.$$

Ans. $(x-1)^2+(y-2)^2=25$; $(x-\frac{5}{6})^2+(y-\frac{7}{6})^2=\frac{62}{36}$; and the coordinates of the centre and the radius are (1, 2), and 5 in the first case; $(\frac{5}{6}, \frac{7}{6})$ and $\frac{1}{6}\sqrt{(62)}$ in the second.

If we treat in like manner the equation

$$a(x^2+y^2)+2gx+2fy+c=0,$$

we get
$$\left(x+\frac{g}{a}\right)^2+\left(y+\frac{f}{a}\right)^2=\frac{g^2+f^2-ac}{a^2},$$

then the coordinates of the centre are $\frac{-g}{a}$, $\frac{-f}{a}$, and the radius is $\frac{1}{a}\sqrt{(g^2+f^2-ac)}$.

If g^2+f^2-ac is negative, the radius of the circle is imaginary, and the equation being equivalent to $(x-\alpha)^2+(y-\beta)^2+r^2=0$ cannot be satisfied by any real values of x and y.

If $g^2+f^2=ac$, the radius is nothing, and the equation being equivalent to $(x-\alpha)^2+(y-\beta)^2=0$, can be satisfied by no coordinates save those of the point $(\alpha\beta)$. In this case then the equation used to be called the equation of that point, but for the reason stated (Art. 73) we prefer to call it the equation of *an infinitely small circle* having that point for centre. We have seen (Art. 73) that it may also be considered as the equation of the two imaginary lines $(x-\alpha)\pm(y-\beta)\sqrt{(-1)}$ passing through the point $(\alpha\beta)$. So in like manner the equation $x^2+y^2=0$ may be regarded as the equation of an infinitely small circle having the origin for centre, or else of the two imaginary lines $x\pm y\sqrt{(-1)}$.

81. The equation of the circle to oblique axes is not often used. It is found by expressing (Art. 5) that the distance of any point from the centre is equal to the radius, and is

$$(x-\alpha)^2+2(x-\alpha)(y-\beta)\cos\omega+(y-\beta)^2=r^2.$$

If we compare this with the general equation, we see that the latter cannot represent a circle unless $a=b$ and $h=a\cos\omega$. When these conditions are fulfilled we find by comparison of coefficients that the coordinates of the centre and the radius are given by the equations

$$\alpha+\beta\cos\omega=-\frac{g}{a},\; \beta+\alpha\cos\omega=-\frac{f}{a},\; \alpha^2+\beta^2+2\alpha\beta\cos\omega-r^2=\frac{c}{a}.$$

Since α, β are determined from the first two equations, which do not contain c, we learn that *two circles will be concentric if their equations differ only in the constant term.*

Again, if $c = 0$, the origin is on the curve. For then the equation is satisfied by the coordinates of the origin $x = 0$, $y = 0$. The same argument proves that *if an equation of any degree want the absolute term, the curve represented passes through the origin.*

82. *To find the coordinates of the points in which a given right line* $x \cos\alpha + y \sin\alpha = p$ *meets a given circle* $x^2 + y^2 = r^2$.

Equating to each other the values of y found from the two equations we get, for determining x, the equation

$$\frac{p - x\cos\alpha}{\sin\alpha} = \sqrt{(r^2 - x^2)},$$

or, reducing $\quad x^2 - 2px\cos\alpha + p^2 - r^2\sin^2\alpha = 0;$

hence, $\quad x = p\cos\alpha \pm \sin\alpha\sqrt{(r^2 - p^2)},$

and, in like manner,

$$y = p\sin\alpha \mp \cos\alpha\sqrt{(r^2 - p^2)}.$$

(The reader may satisfy himself, by substituting these values in the given equations, that the $-$ in the value of y corresponds to the $+$ in the value of x, and *vice versâ*).

Since we obtained a quadratic to determine x, and since every quadratic has two roots, real or imaginary, we must, in order to make our language conform to the language of algebra, assert that *every* line meets a circle in two points, real or imaginary. Thus, when p is greater than r, that is to say, when the distance of the line from the centre is greater than the radius, the line, geometrically considered, does not meet the circle; yet we have seen that analysis furnishes definite imaginary values for the coordinates of intersection. Instead then of saying that the line meets the circle in no points, we shall say that it meets it in two imaginary points, just as we do not say that the corresponding quadratic has no roots, but that it has two imaginary roots. By an imaginary point we mean nothing more than a point, one or both of whose coordinates are imaginary. It is a purely analytical conception, which we do not attempt to represent geometrically; just as when we find imaginary values for roots of an equation, we do not try to attach an arithmetical meaning to our result. But attention to these imaginary points is necessary to preserve generality in our reasonings, for we shall presently meet with many cases in which the line joining two imaginary points is real, and enjoys all the geometrical properties of the corresponding line in the case where the points are real.

83. When $p=r$ it is evident, geometrically, that the line touches the circle, and our analysis points to the same conclusion, since the two values of x in this case become *equal*, as do likewise the two values of y. Consequently the points answering to these two values, which are in general different, will in this case coincide. We shall, therefore, not say that the tangent meets the circle in only one point, but rather that it meets it in two coincident points; just as we do not say that the corresponding quadratic has only one root, but rather that it has two equal roots. And in general *we define the tangent to any curve as the line joining two indefinitely near points on that curve.*

We can in like manner find a quadratic to determine the points where the line $Ax + By + C$ meets a circle given by the general equation. When this quadratic has equal roots the line is a tangent.

Ex. 1. Find the coordinates of the intersections of $x^2 + y^2 = 65$; $3x + y = 25$.
Ans. (7, 4) and (8, 1)

Ex. 2. Find intersections of $(x - c)^2 + (y - 2c)^2 = 25c^2$; $4x + 3y = 35c$.
Ans. The line touches at the point $(5c, 5c)$.

Ex. 3. When will $y = mx + b$ touch $x^2 + y^2 = r^2$? *Ans.* When $b^2 = r^2 (1 + m^2)$.

Ex. 4. When will a line through the origin, $y = mx$, touch

$$a (x^2 + 2xy \cos \omega + y^2) + 2gx + 2fy + c = 0?$$

The points of meeting are given by the equation

$$a (1 + 2m \cos \omega + m^2) x^2 + 2 (g + fm) x + c = 0,$$

which will have equal roots when

$$(g + fm)^2 = ac (1 + 2m \cos \omega + m^2).$$

We have thus a quadratic for determining m.

Ex. 5. Find the tangents from the origin to $x^2 + y^2 - 6x - 2y + 8 = 0$.
Ans. $x - y = 0$, $x + 7y = 0$.

84. When seeking to determine the position of a circle represented by a given equation, it is often as convenient to do so by finding the intercepts which it makes on the axes, as by finding its centre and radius. For a circle is known when

three points on it are known; the determination, therefore, of the four points where the circle meets the axes serves completely to fix its position. By making alternately $y=0$, $x=0$ in the general equation of the circle, we find that the points in which it meets the axes are determined by the quadratics

$$ax^2+2gx+c=0,\quad ay^2+2fy+c=0.$$

The axis of x will be a tangent when the first quadratic has equal roots, that is, when $g^2=ac$, and the axis of y when $f^2=ac$. Conversely, if it be required to find the equation of a circle making intercepts λ, λ' on the axis of x, we may take $a=1$, and we must have $2g=-(\lambda+\lambda')$, $c=\lambda\lambda'$. If it make intercepts μ, μ' on the axis of y, we must have $2f=-(\mu+\mu')$, $c=\mu\mu'$. Thus we see that we must have $\lambda\lambda'=\mu\mu'$ (Euc. III. 36).

Ex 1. Find the points where the axes are cut by $x^2+y^2-5x-7y+6=0$.

Ans. $x=3$, $x=2$; $y=6$, $y=1$.

Ex. 2. What is the equation of the circle which touches the axes at distances from the origin $=a$?

Ans. $x^2+y^2-2ax-2ay+a^2=0$.

Ex. 3. Find the equation of a circle, the axes being a tangent and any line through the point of contact. Here we have λ, λ', μ all $=0$; and it is easy to see from the figure that $\mu'=2r\sin\omega$, the equation therefore is

$$x^2+2xy\cos\omega+y^2-2ry\sin\omega=0.$$

85. *To find the equation of the tangent at the point $x'y'$ to a given circle.*

The tangent having been defined (Art. 83) as the line joining two indefinitely near points on the curve, its equation will be found by first forming the equation of the line joining any two points $(x'y',\ x''y'')$ on the curve, and then making $x'=x''$ and $y'=y''$ in that equation.

To apply this to the circle: first, let the centre be the origin, and, therefore, the equation of the circle $x^2+y^2=r^2$.

The equation of the line joining any two points $(x'y')$ and $(x''y'')$ is (Art. 29)

$$\frac{y-y'}{x-x'}=\frac{y'-y''}{x'-x''};$$

now if we were to make in this equation $y'=y''$ and $x'=x''$, the right-hand member would become indeterminate. The cause of this is, that we have not yet introduced the condition that the two points $(x'y',\ x''y'')$ are *on the circle*. By the help of this condition we shall be able to write the equation in a form which will not become indeterminate when the two points are made to coincide. For, since

$r^2 = x'^2 + y'^2 = x''^2 + y''^2$, we have $x'^2 - x''^2 = y''^2 - y'^2$,

and therefore $$\frac{y' - y''}{x' - x''} = -\frac{x' + x''}{y' + y''}.$$

Hence the equation of the chord becomes

$$\frac{y - y'}{x - x'} = -\frac{x' + x''}{y' + y''}.$$

And if we *now* make $x' = x''$ and $y' = y''$, we find for the equation of the tangent

$$\frac{y - y'}{x - x'} = -\frac{x'}{y'},$$

or, reducing, and remembering that $x'^2 + y'^2 = r^2$, we get finally

$$xx' + yy' = r^2.$$

Otherwise thus:* The equation of the chord joining two points on a circle may be written

$$(x - x')(x - x'') + (y - y')(y - y'') = x^2 + y^2 - r^2.$$

For this is the equation of a right line, since the terms $x^2 + y^2$ on each side destroy each other; and if we make $x = x'$, $y = y'$, the left-hand side vanishes identically, and the right-hand side vanishes, since the point $x'y'$ is on the circle. In like manner the equation is satisfied by the coordinates $x''y''$. This then is the equation of a chord; and the equation of the tangent got by making $x' = x''$, $y' = y''$, is

$$(x - x')^2 + (y - y')^2 = x^2 + y^2 - r^2,$$

which reduced, gives, as before, $xx' + yy' = r^2$.

If we were now to transform the equations to a new origin, so that the coordinates of the centre should become α, β, we must substitute (Art. 8) $x - \alpha$, $x' - \alpha$, $y - \beta$, $y' - \beta$, for x, x', y, y', respectively; the equation of the circle would become

$$(x - \alpha)^2 + (y - \beta)^2 = r^2,$$

and that of the tangent

$$(x - \alpha)(x' - \alpha) + (y - \beta)(y' - \beta) = r^2;$$

a form easily remembered from its similarity to the equation of the circle.

---

* This method is due to Mr. Burnside.

COR. The tangent is perpendicular to the radius, for the equation of the radius, the centre being origin, is easily seen to be $x'y - y'x = 0$; but this (Art. 32) is perpendicular to $xx' + yy' = r^2$.

86. The method used in the last article may be applied to the general equation*

$$ax^2 + 2hxy + by^2 + 2gx + 2fy + c = 0.$$

The equation of the chord joining two points on the curve may be written

$$a(x - x')(x - x'') + 2h(x - x')(y - y'') + b(y - y')(y - y'')$$
$$= ax^2 + 2hxy + by^2 + 2gx + 2fy + c.$$

For the equation represents a right line, the terms above the first degree destroying each other; and, as before, it is evidently satisfied by the two points on the curve $x'y'$, $x''y''$. Putting $x'' = x'$, $y'' = y'$, we get the equation of the tangent

$$a(x-x')^2 + 2h(x-x')(y-y') + b(y-y')^2 = ax^2 + 2hxy + by^2 + 2gx + 2fy + c;$$

or, expanding,

$$2ax'x + 2h(x'y + y'x) + 2by'y + 2gx + 2fy + c = ax'^2 + 2hx'y' + by'^2.$$

Add to both sides $2gx' + 2fy' + c$, and the right-hand side will vanish, because $x'y'$ satisfies the equation of the curve. Thus the equation of the tangent becomes

$$ax'x + h(x'y + y'x) + by'y + g(x + x') + f(y + y') + c = 0.$$

This equation will be more easily remembered if we compare it with the equation of the curve, when we see that it is derived from it by writing $x'x$ and $y'y$ for x^2 and y^2, $x'y + y'x$ for $2xy$, and $x' + x$, $y' + y$ for $2x$ and $2y$.

Ex. 1. Find the equations of the tangents to the curves $xy = c^2$ and $y^2 = px$.

Ans. $x'y + y'x = 2c^2$ and $2yy' = p(x + x')$.

Ex. 2. Find the tangent at the point (5, 4) to $(x - 2)^2 + (y - 3)^2 = 10$.

Ans. $3x + y = 19$.

Ex. 3. What is the equation of the chord joining the points $x'y'$, $x''y''$ on the circle $x^2 + y^2 = r^2$? *Ans.* $(x' + x'')x + (y' + y'')y = r^2 + x'x'' + y'y''$.

Ex. 4. Find the condition that $Ax + By + C = 0$ should touch

$$(x - \alpha)^2 + (y - \beta)^2 = r^2.$$

Ans. $\dfrac{A\alpha + B\beta + C}{\sqrt{(A^2 + B^2)}} = r$; since the perpendicular on the line from $\alpha\beta$ is equal to r.

---

* Of course when this equation represents a circle we must have $b = a$, $h = a \cos \omega$; but since the process is the same, whether or not b or h have these particular values, we prefer in this and one or two similar cases to obtain at once formulæ which will afterwards be required in our discussion of the general equation of the second degree.

87. *To draw a tangent to the circle* $x^2+y^2=r^2$ *from any point* $x'y'$. Let the point of contact be $x''y''$, then since, by hypothesis, the coordinates $x'y'$ satisfy the equation of the tangent at $x''y''$, we have the condition $x'x''+y'y''=r^2$.

And since $x''y''$ is on the circle, we have also

$$x''^2+y''^2=r^2.$$

These two conditions are sufficient to determine the coordinates x'', y''. Solving the equations we get

$$x''=\frac{r^2x'\pm ry'\sqrt{(x'^2+y'^2-r^2)}}{x'^2+y'^2},\quad y''=\frac{r^2y'\mp rx'\sqrt{(x'^2+y'^2-r^2)}}{x'^2+y'^2}.$$

Hence, from every point may be drawn *two* tangents to a circle. These tangents will be real when $x'^2+y'^2$ is $>r^2$, or the point outside the circle; they will be imaginary when $x'^2+y'^2$ is $<r^2$, or the point inside the circle; and they will coincide when $x'^2+y'^2=r^2$, or the point on the circle.

88. We have seen that the coordinates of the points of contact are found by solving for x and y from the equations

$$xx'+yy'=r^2;\quad x^2+y^2=r^2.$$

Now the geometrical meaning of these equations evidently is, that these points are the intersections of the circle $x^2+y^2=r^2$ with the right line $xx'+yy'=r^2$. This, last, then is the equation of the right line joining the points of contact of tangents from the point $x'y'$; as may also be verified by forming the equation of the line joining the two points whose coordinates were found in the last article.*

We see, then, that whether the tangents from $x'y'$ be real or imaginary, the line joining their points of contact will be the real line $xx'+yy'=r^2$, which we shall call the *polar* of $x'y'$ with regard to the circle. This line is evidently perpendicular to the

* In general the equation of the tangent to any curve expresses a relation connecting the coordinates of any point on the tangent, with the coordinates of the point of contact. If we are given a point on the tangent and required to find the point of contact, we have only to accentuate the coordinates of the point which is supposed to be known, and remove the accents from those of the point of contact, when we have the equation of a curve on which that point must lie, and whose intersection with the given curve determines the point of contact. Thus, if the equation of the tangent to a curve at any point $x'y'$ be $xx'^2+yy'^2=r^3$, the points of contact of tangents drawn from any point $x'y'$ must lie on the curve $x'x^2+y'y^2=r^3$. It is only in the case of curves of the second degree that the equation which determines the points of contact is similar in form to the equation of the tangent.

line $(x'y - y'x = 0)$, which joins $x'y'$ to the centre; and its distance from the centre (Art. 23) is $\frac{r^2}{\sqrt{(x'^2 + y'^2)}}$. Hence, the polar of any point P is constructed geometrically by joining it to the centre C, taking on the joining line a point M, such that $CM.CP = r^2$, and erecting a perpendicular to CP at M. We see, also, that the equation of the polar is similar in form to that of the tangent, only that in the former case the point $x'y'$ is not supposed to be necessarily on the circle; if, however, $x'y'$ be on the circle, then its polar is the tangent at that point.

89. *To find the equation of the polar of $x'y'$ with regard to the curve* $ax^2 + 2hxy + by^2 + 2gx + 2fy + c = 0$.

We have seen (Art. 86) that the equation of the tangent is

$$ax'x + h(x'y + y'x) + by'y + g(x + x') + f(y + y') + c = 0.$$

This expresses a relation between the coordinates xy of any point on the tangent, and those of the point of contact $x'y'$. We indicate that the former coordinates are known and the latter unknown, by accentuating the former, and removing the accents from the latter coordinates. But the equation, being symmetrical with respect to the coordinates xy, $x'y'$, is unchanged by this operation. The equation then written above (which when $x'y'$ is a point on the curve, represents the tangent at that point), when $x'y'$ is not on the curve, represents a line on which lie the points of contact of tangents real or imaginary from $x'y'$.

If we substitute $x'y'$ for xy in the equation of the polar we get the same result as if we made the same substitution in the equation of the curve. This result then vanishes when $x'y'$ is on the curve. Hence the polar of a point passes through that point only when the point is on the curve, in which case the polar is the tangent.

COR. The polar *of the origin* is $gx + fy + c = 0$.

Ex. 1. Find the polar of (4, 4) with regard to $(x-1)^2+(y-2)^2=13$. *Ans.* $3x+2y=20$.

Ex. 2. Find the polar of (4, 5) with regard to $x^2+y^2-3x-4y=8$. *Ans.* $5x+6y=48$.

Ex. 3. Find the pole of $Ax + By + C = 0$ with regard to $x^2 + y^2 = r^2$.

Ans. $\left(-\frac{Ar^2}{C}, -\frac{Br^2}{C}\right)$, as appears from comparing the given equation with

$$xx' + yy' = r^2.$$

Ex. 4. Find the pole of $3x + 4y = 7$ with regard to $x^2 + y^2 = 14$. *Ans.* (6, 8).

Ex. 5. Find the pole of $2x + 3y = 6$ with regard to $(x - 1)^2 + (y - 2)^2 = 12$.

Ans. (– 11, – 16).

90. *To find the length of the tangent drawn from any point to the circle* $(x-\alpha)^2+(y-\beta)^2-r^2=0$.

The square of the distance of any point from the centre

$$=(x-\alpha)^2+(y-\beta)^2;$$

and since this square exceeds the square of the tangent by the square of the radius, the square of the tangent from any point is found by substituting the coordinates of that point for x and y in the first member of the equation of the circle

$$(x-\alpha)^2+(y-\beta)^2-r^2=0.$$

Since the general equation to rectangular coordinates

$$a(x^2+y^2)+2gx+2fy+c=0,$$

when divided by a, is (Art. 80) equivalent to one of the form

$$(x-\alpha)^2+(y-\beta)^2-r^2=0,$$

we learn that the square of the tangent to a circle whose equation is given in its most general form is found by dividing by the coefficient of x^2, and then substituting in the equation the coordinates of the given point.

The square of the tangent from the origin is found by making x and $y=0$, and is, therefore, = the absolute term in the equation of the circle, divided by a.

The same reasoning is applicable if the axes be oblique.

*91. *To find the ratio in which the line joining two given points* $x'y'$, $x''y''$, *is cut by a given circle.*

We proceed precisely as in Art. 42. The coordinates of any point on the line must (Art. 7) be of the form

$$\frac{lx''+mx'}{l+m},\ \frac{ly''+my'}{l+m}.$$

Substituting these values in the equation of the circle

$$x^2+y^2-r^2=0,$$

and arranging, we have, to determine the ratio $l:m$, the quadratic

$$l^2(x''^2+y''^2-r^2)+2lm(x'x''+y'y''-r^2)+m^2(x'^2+y'^2-r^2)=0.$$

The values of $l:m$ being determined from this equation, we have at once the coordinates of the points where the right line meets the circle. The symmetry of the equation makes this method sometimes more convenient than that used (Art. 82).

If $x''y''$ lie on the polar of $x'y'$, we have $x'x''+y'y''-r^2=0$ (Art. 88), and the factors of the preceding equation must be of the form $l+\mu m$, $l-\mu m$; the line joining $x'y'$, $x''y''$ is therefore cut internally and externally in the same ratio, and we deduce the well-known theorem, *any line drawn through a point is cut harmonically by the point, the circle, and the polar of the point.*

*92. *To find the equation of the tangents from a given point to a given circle.*

We have already (Art. 87) found the coordinates of the points of contact; substituting, therefore, these values in the equation $xx''+yy''-r^2=0$, we have for the equation of one tangent

$$r(xx'+yy'-x'^2-y'^2)+(xy'-yx')\sqrt{(x'^2+y'^2-r^2)}=0,$$

and for that of the other

$$r(xx'+yy'-x'^2-y'^2)-(xy'-yx')\sqrt{(x'^2+y'^2-r^2)}=0.$$

These two equations multiplied together give the equation of the pair of tangents in a form free from radicals. The preceding article enables us, however, to obtain this equation in a still more simple form. For the equation which determines $l:m$ will have equal roots if the line joining $x'y'$, $x''y''$ touch the given circle; if then $x''y''$ be any point on either of the tangents through $x'y'$, its coordinates must satisfy the condition

$$(x'^2+y'^2-r^2)(x^2+y^2-r^2)=(xx'+yy'-r^2)^2.$$

This, therefore, is the equation of the pair of tangents through the point $x'y'$. It is not difficult to prove that this equation is identical with that obtained by the method first indicated.

The process used in this and the preceding article is equally applicable to the general equation. We find in precisely the same way that $l:m$ is determined from the quadratic

$$l^2(ax''^2+2hx''y''+by''^2+2gx''+2fy''+c)$$
$$+2lm\{ax'x''+h(x'y''+x''y')+by'y''+g(x'+x'')+f(y'+y'')+c\}$$
$$+m^2(ax'^2+2hx'y'+by'^2+2gx'+2fy'+c)=0;$$

from which we infer, as before, that when $x''y''$ lies on the polar of $x'y'$ the line joining these points is cut harmonically; and also that the equation of the pair of tangents from $x'y'$ is

$$(ax'^2+2hx'y'+by'^2+2gx'+2fy'+c)(ax^2+2hxy+by^2+2gx+2fy+c)$$
$$=\{ax'x+h(x'y+xy')+byy'+g(x+x')+f(y+y')+c\}^2.$$

93. *To find the equation of a circle passing through three given points.*

We have only to write down the general equation

$$x^2 + y^2 + 2gx + 2fy + c = 0,$$

and then substituting in it, successively, the coordinates of each of the given points, we have three equations to determine the three unknown quantities g, f, c. We might also obtain the equation by determining the coordinates of the centre and the radius, as in Ex. 5, p. 4.

Ex. 1. Find the circle through (2, 3), (4, 5), (6, 1).

Ans. $(x - \frac{13}{3})^2 + (y - \frac{8}{3})^2 = \frac{50}{9}$ (see p. 4).

Ex. 2. Find the circle through the origin and through (2, 3) and (3, 4).

Here $c = 0$, and we have $13 + 4g + 6f = 0$, $25 + 6g + 8f = 0$, whence $2g = -23$, $2f = 11$.

Ex. 3. Taking the same axes as in Art. 48, Ex. 1, find the equation of the circle through the origin and through the middle points of sides; and shew that it also passes through the middle point of base.

Ans. $2p(x^2 + y^2) - p(s - s')x - (p^2 + ss')y = 0.$

*94. *To express the equation of the circle through three points $x'y'$, $x''y''$, $x'''y'''$ in terms of the coordinates of those points.*

We have to substitute in

$$x^2 + y^2 + 2gx + 2fy + c = 0,$$

the values of g, f, c derived from

$$(x'^2 + y'^2) + 2gx' + 2fy' + c = 0,$$
$$(x''^2 + y''^2) + 2gx'' + 2fy'' + c = 0,$$
$$(x'''^2 + y'''^2) + 2gx''' + 2fy''' + c = 0.$$

The result of thus eliminating g, f, c between these four equations will be found to be*

$$\begin{aligned}&(x^2 + y^2)\{x'(y'' - y''') + x''(y''' - y') + x'''(y' - y'')\}\\ &- (x'^2 + y'^2)\{x''(y''' - y) + x'''(y - y'') + x(y'' - y''')\}\\ &+ (x''^2 + y''^2)\{x'''(y - y') + x(y' - y''') + x'(y''' - y)\}\\ &- (x'''^2 + y'''^2)\{x(y' - y'') + x'(y'' - y) + x''(y - y')\} = 0,\end{aligned}$$

as may be seen by multiplying each of the four equations by the quantities which multiply $(x^2 + y^2)$ &c. in the last written equation, and adding them together, when the quantities multiplying g, f, c will be found to vanish identically.

---

* The reader who is acquainted with the determinant notation will at once see how the equation of the circle may be written in the form of a determinant.

If it were required to find the condition that four points should lie on a circle, we have only to write x_4, y_4 for x and y in the last equation. It is easy to see that the following is the geometrical interpretation of the resulting condition. If A, B, C, D be any four points on a circle, and O any fifth point taken arbitrarily, and if we denote by BCD the area of the triangle BCD, &c., then

$$OA^2.BCD+OC^2.ABD=OB^2.ACD+OD^2.ABC.$$

95. We shall conclude this chapter by showing how to find the *polar equation* of a circle.

We may either obtain it by substituting for x, $\rho\cos\theta$, and for y, $\rho\sin\theta$ (Art. 12), in either of the equations of the circle already given,

$$a(x^2+y^2)+2gx+2fy+c=0, \text{ or } (x-\alpha)^2+(y-\beta)^2=r^2,$$

or else we may find it independently, from the definition of the circle, as follows:

Let O be the pole, C the centre of the circle, and OC the fixed axis; let the distance $OC=d$, and let OP be any radius vector, and, therefore, $=\rho$, and the angle $POC=\theta$, then we have

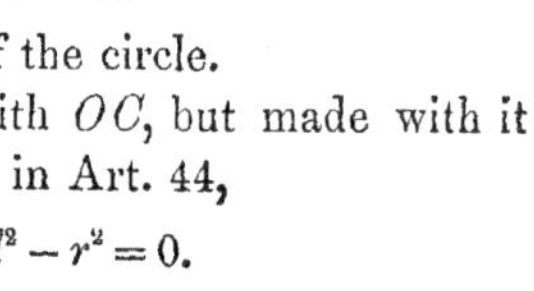

$$PC^2=OP^2+OC^2-2OP.OC\cos POC,$$

that is, $$r^2=\rho^2+d^2-2\rho d\cos\theta,$$

or $$\rho^2-2d\rho\cos\theta+d^2-r^2=0.$$

This, therefore, is the polar equation of the circle.

If the fixed axis did not coincide with OC, but made with it any angle α, the equation would be, as in Art. 44,

$$\rho^2-2d\rho\cos(\theta-\alpha)+d^2-r^2=0.$$

If we suppose the pole on the circle, the equation will take a simpler form, for then $r=d$, and the equation will be reduced to

$$\rho=2r\cos\theta,$$

a result which we might have also obtained at once geometrically from the property that the angle in a semicircle is right; or else by substituting for x and y their polar values in the equation (Art. 79) $$x^2+y^2=2rx.$$

CHAPTER VII.

THEOREMS AND EXAMPLES ON THE CIRCLE.

96. HAVING in the last chapter shown how to form the equations of the circle, and of the most remarkable lines related to it, we proceed in this chapter to illustrate these equations by examples, and to apply them to the establishment of some of the principal properties of the circle. We recommend the reader first to refer to the answers to the examples of Art. 49, to examine in each case whether the equation represents a circle, and if so to determine its position either (Art. 80) by finding the coordinates of the centre and the radius, or (Art. 84) by finding the points where the circle meets the axes. We add a few more examples of circular loci.

Ex. 1. Given base and vertical angle, find the locus of vertex, the axes having any position.

Let the coordinates of the extremities of base be $x'y'$, $x''y''$. Let the equation of one side be

$$y - y' = m(x - x'),$$

then the equation of the other side, making with this the angle C, will be (Art. 33)

$$(1 + m \tan C)(y - y'') = (m - \tan C)(x - x'').$$

Eliminating m, the equation of the locus is

$$\tan C\{(y - y')(y - y'') + (x - x')(x - x'')\} + x(y' - y'') - y(x' - x'') + x'y'' - y'x'' = 0.$$

If C be a right angle, the equations of the sides are

$$y - y' = m(x - x');\quad m(y - y'') + (x - x'') = 0,$$

and that of the locus

$$(y - y')(y - y'') + (x - x')(x - x'') = 0.$$

Ex. 2. Given base and vertical angle, find the locus of the intersection of perpendiculars of the triangle.

The equations of the perpendiculars to the sides are

$$m(y - y'') + (x - x'') = 0,\quad (m - \tan C)(y - y') + (1 + m \tan C)(x - x') = 0.$$

Eliminating m, the equation of the locus is

$$\tan C\{(y - y')(y - y'') + (x - x')(x - x'')\} = x(y' - y'') - y(x' - x'') + x'y'' - y'x'';$$

an equation which only differs from that of the last article by the sign of $\tan C$, and which is therefore the locus we should have found for the vertex had we been given the same base and a vertical angle equal to the supplement of the given one.

Ex. 3. Given any number of points, to find locus of a point such that m' times square of its distance from the first $+ m''$ times square of its distance from the second $+$ &c. $=$ a constant; or (adopting the notation used in Ex. 4, p. 49) such that $\Sigma(mr^2)$ may be constant.

The square of the distance of any point xy from $x'y'$ is $(x-x')^2+(y-y')^2$. Multiply this by m', and add it to the corresponding terms found by expressing the distance of the point xy from the other points $x''y''$, &c. If we adopt the notation of p. 49, we may write for the equation of the locus

$$\Sigma(m)x^2+\Sigma(m)y^2-2\Sigma(mx')x-2\Sigma(my')y+\Sigma(mx'^2)+\Sigma(my'^2)=C.$$

Hence the locus will be a circle, the coordinates of whose centre will be

$$x=\frac{\Sigma(mx')}{\Sigma(m)},\quad y=\frac{\Sigma(my')}{\Sigma(m)},$$

that is to say, the centre will be the point which, in p. 50, was called the centre of mean position of the given points.

If we investigate the value of the radius of this circle we shall find

$$R^2\Sigma(m)=\Sigma(mr^2)-\Sigma(m\rho^2),$$

where $\Sigma(mr^2)=C=$ sum of m times square of distance of each of the given points from any point on the circle, and $\Sigma(m\rho^2)=$ sum of m times square of distance of each point from the centre of mean position.

Ex. 4. Find the locus of a point O, such that if parallels be drawn through it to the three sides of a triangle, meeting them in points B, C; C', A'; A'', B''; the sum may be given of the three rectangles

$$BO.OC+C'O.OA'+A''O.OB''.$$

Taking two sides for axes, the equation of the locus is

$$x\left(a-x-\frac{a}{b}y\right)+y\left(b-y-\frac{b}{a}x\right)+\frac{c^2xy}{ab}=m^2,$$

or $$x^2+y^2+2xy\cos C-ax-by+m^2=0.$$

This represents a circle, which, as is easily seen, is concentric with the circumscribing circle, the coordinates of the centre in both cases being given by the equations $2(\alpha+\beta\cos C)=a$, $2(\beta+\alpha\cos C)=b$. These last two equations enable us to solve the problem to find the locus of the centre of circumscribing circle, when two sides of a triangle are given in position, and any relation connecting their lengths is given.

Ex. 5. Find the locus of a point O, if the line joining it to a fixed point makes the same intercept on the axis of x as is made on the axis of y by a perpendicular through O to the joining line.

Ex. 6. Find the locus of a point such that if it be joined to the vertices of a triangle, and perpendiculars to the joining lines erected at the vertices, these perpendiculars meet in a point.

97. We shall next give one or two examples involving the problem of Art. 82, to find the coordinates of the points where a given line meets a given circle.

Ex. 1. To find the locus of the middle points of chords of a given circle drawn parallel to a given line.

Let the equation of any of the parallel chords be

$$x\cos\alpha+y\sin\alpha-p=0,$$

where α is, by hypothesis, given, and p is indeterminate; the abscissæ of the points where this line meets the circle are (Art. 82) found from the equation

$$x^2-2px\cos\alpha+p^2-r^2\sin^2\alpha=0.$$

Now, if the roots of this equation be x' and x'', the x of the middle point of the chord will (Art. 7) be $\frac{1}{2}(x' + x'')$, or, from the theory of equations, will $= p\cos\alpha$. In like manner, the y of the middle point will equal $p\sin\alpha$. Hence the equation of the locus is $y = x\tan\alpha$, that is, a right line drawn through the centre perpendicular to the system of parallel chords, since α is the angle made with the axis of x by a perpendicular to any of the chords.

Ex. 2. To find the condition that the intercept made by the circle on the line

$$x\cos\alpha + y\sin\alpha = p$$

should subtend a right angle at the point $x'y'$.

We found (Art. 96, Ex. 1) the condition that the lines joining the points $x''y''$, $x'''y'''$ to xy should be at right angles to each other; viz.

$$(x - x'')(x - x''') + (y - y'')(y - y''') = 0.$$

Let $x''y''$, $x'''y'''$ be the points where the line meets the circle, then, by the last example,

$$x'' + x''' = 2p\cos\alpha,\quad x''x''' = p^2 - r^2\sin^2\alpha,\quad y'' + y''' = 2p\sin\alpha,\quad y''y''' = p^2 - r^2\cos^2\alpha.$$

Putting in these values, the required condition is

$$x'^2 + y'^2 - 2px'\cos\alpha - 2py'\sin\alpha + 2p^2 - r^2 = 0.$$

Ex. 3. To find the locus of the middle point of a chord which subtends a right angle at a given point.

If x and y be the coordinates of the middle point, we have, by Ex. 1,

$$p\cos\alpha = x,\quad p\sin\alpha = y,\quad p^2 = x^2 + y^2,$$

and, substituting these values, the condition found in the last example becomes

$$(x - x')^2 + (y - y')^2 + x^2 + y^2 = r^2.$$

Ex. 4. Given a line and a circle, to find a point such that if any chord be drawn through it, and perpendiculars let fall from its extremities on the given line, the rectangle under these perpendiculars may be constant.

Take the given line for axis of x, and let the axis of y be the perpendicular on it from the centre of the given circle, whose equation will then be

$$x^2 + (y - \beta)^2 = r^2.$$

Let the coordinates of the sought point be $x'y'$, then the equation of any line through it will be $y - y' = m(x - x')$. Eliminate x between these two equations and we get a quadratic for y, the product of whose roots will be found to be

$$\frac{(y' - mx')^2 + m^2(\beta^2 - r^2)}{1 + m^2}.$$

This will not be independent of m unless the numerator be divisible by $1 + m^2$, and it will be found that this cannot be the case unless $x' = 0$, $y'^2 = \beta^2 - r^2$.

Ex. 5. To find the condition that the intercept made on $x\cos\alpha + y\sin\alpha - p$ by the circle

$$x^2 + y^2 + 2gx + 2fy + c = 0$$

may subtend a right angle at the origin. The equation of the pair of lines joining the extremities of the chord to the origin may be written down at once. For if we multiply the terms of the second degree in the equation of the circle by p^2, those of the first degree by $p(x\cos\alpha + y\sin\alpha)$, and the absolute term by $(x\cos\alpha + y\sin\alpha)^2$, we get an equation homogeneous in x and y, which therefore represents right lines drawn through the origin; and it is satisfied by those points on the circle for which $x\cos\alpha + y\sin\alpha = p$. The equation expanded and arranged is

$$(p^2 + 2gp\cos\alpha + c\cos^2\alpha)x^2 + 2(gp\sin\alpha + fp\cos\alpha + c\sin\alpha\cos\alpha)xy$$
$$+ (p^2 + 2fp\sin\alpha + c\sin^2\alpha)y^2 = 0.$$

These two lines cut at right angles (Art. 74) if

$$2p^2 + 2p(g \cos \alpha + f \sin \alpha) + c = 0.$$

Ex. 6. To find the locus of the foot of the perpendicular from the origin on a chord which subtends a right angle at the origin. The polar coordinates of the locus are p and α in the equation last found; and the equation of the locus is therefore

$$2(x^2 + y^2) + 2gx + 2fy + c = 0.$$

It will be found on examination that this is the same circle as in Ex. 3.

Ex. 7. If any chord be drawn through a fixed point on a diameter of a circle and its extremities joined to either end of the diameter, the joining lines cut off on the tangent at the other end portions whose rectangle is constant.

Find, as in Ex. 5, the equation of the lines joining to the origin the intersections of $x^2 + y^2 - 2rx$ with the chord $y = m(x - x')$ which passes through the fixed point $(x', 0)$. The intercepts on the tangent are found by putting $x = 2r$ in this equation and seeking the corresponding values of y. The product of these values will be found to be independent of m, viz. $4r^2 \frac{x' - 2r}{x'}$.

98. We shall next obtain from the equations (Art. 88) a few of the properties of poles and polars.

If a point A lie on the polar of B, then B lies on the polar of A. For the condition that $x'y'$ should lie on the polar of $x''y''$ is $x'x'' + y'y'' = r^2$; but this is also the condition that the point $x''y''$ should lie on the polar of $x'y'$. It is equally true if we use the general equation (Art. 89) that the result of substituting the coordinates $x''y''$ in the equation of the polar of $x'y'$ is the same as that of substituting the coordinates of $x'y'$ in the polar of $x''y''$. This theorem then, and those which follow, are true of all curves of the second degree. It may be otherwise stated thus: *if the polar of B pass through a fixed point A, the locus of B is the polar of A.*

99. Given a circle and a triangle ABC, if we take the polars with respect to the circle of A, B, C, we form a new triangle $A'B'C'$ called the *conjugate* triangle, A' being the pole of BC, B' of CA, and C' of AB. In the particular case where the polars of A, B, C respectively are BC, CA, AB, the second triangle coincides with the first, and the triangle is called a *self-conjugate* triangle.

The lines AA', BB', CC', joining the corresponding vertices of a triangle and of its conjugate, meet in a point.

The equation of the line joining the point $x'y'$ to the intersection of the two lines $xx''+yy''-r^2=0$ and $xx'''+yy'''-r^2=0$ is (Art. 40, Ex. 3)

$$AA',\ (x'x'''+y'y'''-r^2)(xx''+yy''-r^2) - (x'x''+y'y''-r^2)(xx'''+yy'''-r^2)=0.$$

In like manner

$$BB',\ (x'x''+y'y''-r^2)(xx'''+yy'''-r^2) - (x''x'''+y''y'''-r^2)(xx'+yy'-r^2)=0$$

and $CC',\ (x''x'''+y''y'''-r^2)(xx'+yy'-r^2)$

$$-(x'x'''+y'y'''-r^2)(xx''+yy''-r^2)=0;$$

and by Art. 41 these lines must pass through the same point.

The following is a particular case of the theorem just proved: *If a circle be inscribed in a triangle, and each vertex of the triangle joined to the point of contact of the circle with the opposite side, the three joining lines will meet in a point.*

The proof just given applies equally if we use the general equation. If we write for shortness $P_1=0$ for the equation of the polar of $x'y'$, $(ax'x+\&c.=0)$; and in like manner P_2, P_3 for the polars of $x''y''$, $x'''y'''$; and if we write $[1, 2]$ for the result of substituting the coordinates $x''y''$ in the polar of $x'y'$, $(ax'x''+\&c.)$, then the equations are easily seen to be

$$AA' \quad [1, 3]P_2=[1, 2]P_3,$$
$$BB' \quad [1, 2]P_3=[2, 3]P_1,$$
$$CC' \quad [2, 3]P_1=[1, 3]P_2,$$

which denote three lines meeting in a point. It follows (Art. 60, Ex. 3) that the intersections of corresponding sides of a triangle and its conjugate lie in one right line.

100. *Given any point O, and any two lines through it; join both directly and transversely the points in which these lines meet a circle; then, if the direct lines intersect each other in P and the transverse in Q, the line PQ will be the polar of the point O with regard to the circle.*

Take the two fixed lines for axes, and let the intercepts made on them by the circle be λ and λ', μ and μ'. Then

$$\frac{x}{\lambda}+\frac{y}{\mu}-1=0,\ \frac{x}{\lambda'}+\frac{y}{\mu'}-1=0$$

will be the equations of the direct lines; and

$$\frac{x}{\lambda'} + \frac{y}{\mu} - 1 = 0, \quad \frac{x}{\lambda} + \frac{y}{\mu'} - 1 = 0,$$

the equations of the transverse lines. Now, the equation of the line PQ will be

$$\frac{x}{\lambda} + \frac{x}{\lambda'} + \frac{y}{\mu} + \frac{y}{\mu'} - 2 = 0,$$

for (see Art. 40) this line passes through the intersection of

$$\frac{x}{\lambda} + \frac{y}{\mu} - 1, \quad \frac{x}{\lambda'} + \frac{y}{\mu'} - 1,$$

and also of
$$\frac{x}{\lambda} + \frac{y}{\mu'} - 1, \quad \frac{x}{\lambda'} + \frac{y}{\mu} - 1.$$

If the equation of the curve be

$$ax^2 + 2hxy + by^2 + 2gx + 2fy + c = 0,$$

λ and λ' are determined from the equation $ax^2 + 2gx + c = 0$ (Art. 84), therefore,

$$\frac{1}{\lambda} + \frac{1}{\lambda'} = -\frac{2g}{c}, \text{ and } \frac{1}{\mu} + \frac{1}{\mu'} = -\frac{2f}{c}.$$

Hence, equation of PQ is

$$gx + fy + c = 0;$$

but we saw (Art. 89) that this was the equation of the polar of the origin O. Hence it appears that if the point O were given, and the two lines through it were *not* fixed, the *locus* of the points P and Q would be the polar of the point O.

101. *Given any two points A and B, and their polars with respect to a circle whose centre is O; let fall a perpendicular AP from A on the polar of B, and a perpendicular BQ from B on the polar of A, then* $\dfrac{OA}{AP} = \dfrac{OB}{BQ}$.

The equation of the polar of A $(x'y')$ is $xx' + yy' - r^2 = 0$; and BQ, the perpendicular on this line from B $(x''y'')$, is (Art. 34)

$$\frac{x'x'' + y'y'' - r^2}{\sqrt{(x'^2 + y'^2)}}.$$

Hence, since $\sqrt{(x'^2 + y'^2)} = OA$, we find

$$OA . BQ = x'x'' + y'y'' - r^2;$$

and, for the same reason,

$$OB.AP = x'x'' + y'y'' - r^2.$$

Hence
$$\frac{OA}{AP} = \frac{OB}{BQ}.$$

102. In working out questions on the circle it is often convenient, instead of denoting the position of a point on the curve by its *two* coordinates $x'y'$, to express both these in terms of a single independent variable. Thus, let θ' be the angle which the radius to $x'y'$ makes with the axis of x, then $x' = r\cos\theta'$, $y' = r\sin\theta'$, and on substituting these values our formulæ will generally become simplified.

The equation of the tangent at the point $x'y'$ will by this substitution become

$$x\cos\theta' + y\sin\theta' = r;$$

and the equation of the chord joining $x'y'$, $x''y''$, which (Art. 86, Ex. 3) is

$$x(x' + x'') + y(y' + y'') = r^2 + x'x'' + y'y'',$$

will, by a similar substitution, become

$$x\cos\tfrac{1}{2}(\theta' + \theta'') + y\sin\tfrac{1}{2}(\theta' + \theta'') = r\cos\tfrac{1}{2}(\theta' - \theta''),$$

θ' and θ'' being the angles which radii drawn to the extremities of the chord make with the axis of x.

This equation might also have been obtained directly from the general equation of a right line (Art. 23) $x\cos\alpha + y\sin\alpha = p$, for the angle which the perpendicular on the chord makes with the axis is plainly half the sum of the angles made with the axis by radii to its extremities, and the perpendicular on the chord

$$= r\cos\tfrac{1}{2}(\theta' - \theta'').$$

Ex. 1. To find the coordinates of the intersection of tangents at two given points on the circle The tangents being

$$x\cos\theta' + y\sin\theta' = r, \quad x\cos\theta'' + y\sin\theta'' = r,$$

the coordinates of their intersection are

$$x = r\frac{\cos\frac{1}{2}(\theta' + \theta'')}{\cos\frac{1}{2}(\theta' - \theta'')}, \quad y = r\frac{\sin\frac{1}{2}(\theta' + \theta'')}{\cos\frac{1}{2}(\theta' - \theta'')}.$$

Ex. 2. To find the locus of the intersection of tangents at the extremities of a chord whose length is constant.

Making the substitution of this article in

$$(x' - x'')^2 + (y' - y'')^2 = \text{constant},$$

it reduces to $\cos(\theta' - \theta'') = \text{constant}$, or $\theta' - \theta'' = \text{constant}$. If the given length of the chord be $2r \sin\delta$, then $\theta' - \theta'' = 2\delta$. The coordinates therefore found in the last example fulfil the condition

$$(x^2 + y^2) \cos^2\delta = r^2.$$

Ex. 3. What is the locus of a point where a chord of a constant length is cut in a given ratio?

Writing down (Art. 7) the coordinates of the point where the chord is cut in a given ratio, it will be found that they satisfy the condition $x^2 + y^2 =$ constant.

103. We have seen that the tangent to any circle $x^2 + y^2 = r^2$ has an equation of the form

$$x \cos\theta + y \sin\theta = r;$$

and it can be proved, in like manner, that the equation of the tangent to $(x-\alpha)^2 + (y-\beta)^2 = r^2$ may be written

$$(x-\alpha)\cos\theta + (y-\beta)\sin\theta = r.$$

Conversely, then, if the equation of any right line contain an indeterminate θ in the form

$$(x-\alpha)\cos\theta + (y-\beta)\sin\theta = r,$$

that line will touch the circle $(x-\alpha)^2 + (y-\beta)^2 = r^2$.

Ex. 1. If a chord of a constant length be inscribed in a circle, it will always touch another circle. For, in the equation of the chord

$$x \cos\tfrac{1}{2}(\theta' + \theta'') + y \sin\tfrac{1}{2}(\theta' + \theta'') = r \cos\tfrac{1}{2}(\theta' - \theta'');$$

by the last article, $\theta' - \theta''$ is known, and $\theta' + \theta''$ indeterminate; the chord, therefore, always touches the circle

$$x^2 + y^2 = r^2 \cos^2\delta.$$

Ex. 2. Given any number of points, if a right line be such that m' times the perpendicular on it from the first point $+ m''$ times the perpendicular from the second + &c. be constant, the line will always touch a circle.

This only differs from Ex. 4, p. 49, in that the sum, in place of being $= 0$, is constant. Adopting then the notation of that Article, instead of the equation there found,

$$\{x\Sigma(m) - \Sigma(mx')\}\cos\alpha + \{y\Sigma(m) - \Sigma(my')\}\sin\alpha = 0,$$

we have only to write

$$\{x\Sigma m - \Sigma(mx')\}\cos\alpha + \{y\Sigma(m) - \Sigma(my')\}\sin\alpha = \text{constant}.$$

Hence this line must always touch the circle

$$\left\{x - \frac{\Sigma(mx')}{\Sigma(m)}\right\}^2 + \left\{y - \frac{\Sigma(my')}{\Sigma(m)}\right\}^2 = \text{constant},$$

whose centre is the centre of mean position of the given points.

104. We shall conclude this chapter with some examples of the use of polar coordinates.

Ex. 1. If through a fixed point any chord of a circle be drawn, the rectangle under its segments will be constant (*Euclid* III. 35, 36).

Take the fixed point for the pole, and the polar equation is (Art. 95)

$$\rho^2 - 2\rho d \cos\theta + d^2 - r^2 = 0;$$

the roots of which equation in ρ are evidently OP, OP', the values of the radius vector answering to any given value of θ or POC.

Now, by the theory of equations, $OP \cdot OP'$, the product of these roots will $= d^2 - r^2$, a quantity *independent* of θ, and therefore constant, whatever be the direction in which the line OP is drawn. If the point O be outside the circle, it is plain that $d^2 - r^2$ must be = the square of the tangent.

Ex. 2. If through a fixed point O any chord of a circle be drawn, and OQ taken an arithmetic mean between the segments OP, OP', to find the locus of Q.

We have $OP + OP'$, or the sum of the roots of the quadratic in the last example, $= 2d \cos\theta$; but $OP + OP' = 2OQ$, therefore

$$OQ = d \cos\theta.$$

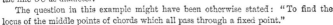

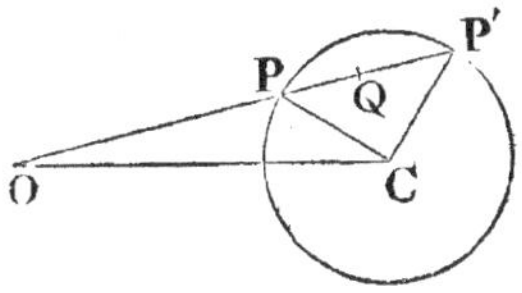

Hence the polar equation of the locus is

$$\rho = d \cos\theta.$$

Now it appears from the final equation (Art. 95) that this is the equation of a circle described on the line OC as diameter.

The question in this example might have been otherwise stated: "To find the locus of the middle points of chords which all pass through a fixed point."

Ex. 3. If the line OQ had been taken a *harmonic* mean between OP and OP' to find the locus of Q.

That is to say, $OQ = \dfrac{2OP.OP'}{OP+OP'}$, but $OP.OP' = d^2 - r^2$, and $OP + OP' = 2d\cos\theta$; therefore the polar equation of the locus is

$$\rho = \frac{d^2 - r^2}{d \cos\theta}, \text{ or } \rho \cos\theta = \frac{d^2 - r^2}{d}.$$

This is the equation of a right line (Art. 44) perpendicular to OC, and at a distance from $O = d - \dfrac{r^2}{d}$, and, therefore, at a distance from $C = \dfrac{r^2}{d}$. Hence (Art. 88) the locus is the *polar* of the point O.

We can, in like manner, solve this and similar questions when the equation is given in the form

$$a(x^2 + y^2) + 2gx + 2fy + c = 0,$$

for, transforming to polar coordinates, the equation becomes

$$\rho^2 + 2\left(\frac{g}{a}\cos\theta + \frac{f}{a}\sin\theta\right)\rho + \frac{c}{a} = 0,$$

and, proceeding precisely as in this example, we find, for the locus of harmonic means,

$$\rho = -\frac{c}{g\cos\theta + f\sin\theta},$$

and, returning to rectangular coordinates, the equation of the locus is

$$gx + fy + c = 0,$$

the same as the equation of the polar obtained already (Art. 89).

Ex. 4. Given a point and a right line or circle; if on OP the radius vector to the line or circle a part OQ be taken inversely as OP, find the locus of Q.

Ex. 5. Given vertex and vertical angle of a triangle and rectangle under sides, if one extremity of the base describe a right line or a circle, find the locus described by the other extremity.

Take the vertex for pole; let the lengths of the sides be ρ and ρ', and the angles they make with the axis θ and θ', then we have $\rho\rho' = k^2$ and $\theta \quad \theta' = C$.

The student must write down the polar equation of the locus which one base angle is said to describe; this will give him a relation between ρ and θ; then, writing for ρ, $\frac{k^2}{\rho'}$, and for θ, $C + \theta'$, he will find a relation between ρ' and θ', which will be the polar equation of the locus described by the other base angle.

This example might be solved in like manner, if the *ratio* of the sides, instead of their rectangle, had been given.

Ex. 6. Through the intersection of two circles a right line is drawn; find the locus of the middle point of the portion intercepted between the circles.

The equations of the circles will be of the form

$$\rho = 2r \cos(\theta - \alpha); \quad \rho = 2r' \cos(\theta - \alpha');$$

and the equation of the locus will be

$$\rho = r \cos(\theta - \alpha) + r' \cos(\theta - \alpha');$$

which also represents a circle.

Ex. 7. If through any point O, on the circumference of a circle, any three chords be drawn, and on each, as diameter, a circle be described, these three circles (which, of course, all pass through O) will intersect in three other points, which lie in one right line (See *Cambridge Mathematical Journal*, vol. I. p. 169).

Take the fixed point O for pole, then if d be the diameter of the original circle, its polar equation will be (Art. 95)

$$\rho = d \cos \theta.$$

In like manner, if the diameter of one of the other circles make an angle α with the fixed axis, its length will be $= d \cos \alpha$, and the equation of this circle will be

$$\rho = d \cos \alpha \cos(\theta - \alpha).$$

The equation of another circle will, in like manner, be

$$\rho = d \cos \beta \cos(\theta - \beta).$$

To find the polar coordinates of the point of intersection of these two, we should seek what value of θ would render

$$\cos \alpha \cos(\theta - \alpha) = \cos \beta \cos(\theta - \beta),$$

and it is easy to find that θ must $= \alpha + \beta$, and the corresponding value of $\rho = d \cos \alpha \cos \beta$.

Similarly, the polar coordinates of the intersection of the first and third circles are

$$\theta = \alpha + \gamma, \text{ and } \rho = d \cos \alpha \cos \gamma.$$

Now, to find the polar equation of the line joining these two points, take the general equation of a right line, $\rho \cos(k - \theta) = p$ (Art. 44), and substitute in it successively these values of θ and ρ, and we shall get two equations to determine p and k. We shall get

$$p = d \cos \alpha \cos \beta \cos\{k - (\alpha + \beta)\} = d \cos \alpha \cos \gamma \cos\{k - (\alpha + \gamma)\}.$$

Hence $$k = \alpha + \beta + \gamma, \text{ and } p = d \cos \alpha \cos \beta \cos \gamma.$$

The symmetry of these values shows that it is the same right line which joins the intersections of the first and second, and of the second and third circles, and, therefore, that the three points are in a right line.

CHAPTER VIII.

PROPERTIES OF A SYSTEM OF TWO OR MORE CIRCLES.

105. *To find the equation of the chord of intersection of two circles.*

If $S=0$, $S'=0$ be the equations of two circles, then any equation of the form $S+kS'=0$ will be the equation of a figure passing through their points of intersection (Art. 40).

Let us write down the equations

$$S=(x-\alpha)^2+(y-\beta)^2-r^2=0,$$
$$S'=(x-\alpha')^2+(y-\beta')^2-r'^2=0,$$

and it is evident that the equation $S+kS'=0$ will in general represent a circle, since the coefficient of $xy=0$, and that of $x^2=$ that of y^2. There is one case, however, where it will represent a right line, namely, when $k=-1$. The terms of the second degree then vanish, and the equation becomes

$$S-S'=2(\alpha'-\alpha)x+2(\beta'-\beta)y+r'^2-r^2+\alpha^2-\alpha'^2+\beta^2-\beta'^2=0.$$

This is, therefore, the equation of the right line passing through the points of intersection of the two circles.

What has been proved in this article may be stated as in Art. 50. If the equation of a circle be of the form $S+kS'=0$ involving an indeterminate k in the first degree, the circle passes through two fixed points, namely, the two points common to the circles S and S'.

106. The points common to the circles S and S' are found by seeking, as in Art. 82, the points in which the line $S-S'$ meets either of the given circles. These points will be real, coincident, or imaginary, according to the nature of the roots of the resulting equation; but it is remarkable that, whether the circles meet in real or imaginary points, the equation of the chord of intersection, $S-S'=0$, always represents a real line, having important geometrical properties in relation to the two circles. This is in conformity with our assertion (Art. 82), that

the line joining two points may preserve its existence and its properties when these points have become imaginary.

In order to avoid the harshness of calling the line $S - S'$, the chord of intersection in the case where the circles do not *geometrically* appear to intersect, it has been called* the *radical axis* of the two circles.

107. We saw (Art. 90) that if the coordinates of any point xy be substituted in S, it represents the square of the tangent drawn to the circle S from the point xy. So also S' is the square of the tangent drawn to the circle S'; hence the equation $S - S' = 0$ asserts, that *if from any point on the radical axis tangents be drawn to the two circles, these tangents will be equal.*

The line $(S - S')$ possesses this property whether the circles meet in real points or not. When the circles do not meet in real points, the position of the radical axis is determined geometrically by cutting the line joining their centres, so that the difference of the squares of the parts may = the difference of the squares of the radii, and erecting a perpendicular at this point; as is evident, since the tangents from this point must be equal to each other.

If it were required to find the locus of a point whence tangents to two circles have *a given ratio*, it appears, from Art. 90, that the equation of the locus will be $S - k^2S' = 0$, which (Art. 105) represents a circle passing through the real or imaginary points of intersection of S and S'. When the circles S and S' do not intersect in real points, we may express the relation which they bear to the circle $S - k^2S'$, by saying that the three circles have a common radical axis.

Ex. Find the coordinates of the centre, and the radius of $kS + lS'$.

Ans. Coordinates are $\frac{k\alpha + l\alpha'}{k + l}$, $\frac{k\beta + l\beta'}{k + l}$; that is to say, the line joining the centres of S, S' is divided in the ratio $k : l$. Radius is given by the equation

$$(k + l)^2 r''^2 = (k + l)(kr^2 + lr'^2) - klD^2,$$

where D is the distance between the centres of S and S'.

108. *Given any three circles, if we take the radical axis of each pair of circles, these three lines will meet in a point, which is called the* radical centre *of the three circles.*

---

* By M. Gaultier, of Tours (*Journal de l'École Polytechnique*, Cahier XVI. 1813).

For the equations of the three radical axes are

$$S - S' = 0, \quad S' - S'' = 0, \quad S'' - S = 0,$$

which, by Art. 41, meet in a point.

From this theorem we immediately derive the following:

If several circles pass through two fixed points, their chords of intersection with a fixed circle will pass through a fixed point.

For, imagine one circle through the two given points to be fixed, then its chord of intersection with the given circle will be fixed; and its chord of intersection with any variable circle drawn through the given points will plainly be the fixed line joining the two given points. These two lines determine by their intersection a fixed point through which the chord of intersection of the variable circle with the first given circle must pass.

Ex. 1. Find the radical axis of

$$x^2 + y^2 - 4x - 5y + 7 = 0; \quad x^2 + y^2 + 6x + 8y - 9 = 0.$$

Ans. $10x + 13y = 16$.

Ex. 2. Find the radical centre of

$$(x-1)^2 + (y-2)^2 = 7; \quad (x-3)^2 + y^2 = 5; \quad (x+4)^2 + (y+1)^2 = 9.$$

Ans. $(-\frac{1}{16}, -\frac{25}{16})$.

*109. A system of circles having a common radical axis possesses many remarkable properties, which are more easily investigated by taking the radical axis for the axis of y, and the line joining the centres for the axis of x. Then the equation of any circle will be

$$x^2 + y^2 - 2kx \pm \delta^2 = 0,$$

where δ^2 is the same for all the circles of the system, and the equations of the different circles are obtained by giving different values to k. For it is evident (Art. 80) that the centre is on the axis of x, at the variable distance k; and if we make $x = 0$ in the equation, we see that no matter what the value of k may be, the circle passes through the fixed points on the axis of y, $y^2 \pm \delta^2 = 0$. These points are imaginary when we give δ^2 the sign +, and real when we give it the sign −.

*110. *The polars of a given point, with regard to a system of circles having a common radical axis, always pass through a fixed point.*

The equation of the polar of $x'y'$ with regard to

$$x^2 + y^2 - 2kx + \delta^2 = 0,$$

is (Art. 89) $$xx' + yy' - k(x + x') + \delta^2 = 0;$$
therefore, since this involves the indeterminate k in the first degree, the line will always pass through the intersection of $xx' + yy' + \delta^2 = 0$, and $x + x' = 0$.

*111. *There can always be found two points, however, such that their polars, with regard to any of the circles, will not only pass through a fixed point, but will be altogether fixed.*

This will happen when $xx' + yy' + \delta^2 = 0$ and $x + x' = 0$ represent the same right line, for this right line will then be the polar whatever the value of k. But that this should be the case we must have

$$y' = 0 \text{ and } x'^2 = \delta^2, \text{ or } x' = \pm\delta.$$

The two points whose coordinates have been just found have many remarkable properties in the theory of these circles, and are such that the polar of either of them, with regard to any of the circles, is a line drawn through the other, perpendicular to the line of centres. These points are real when the circles of the system have common two imaginary points, and imaginary when they have real points common.

The equation of the circle may be written in the form

$$y^2 + (x - k)^2 = k^2 - \delta^2,$$

which evidently cannot represent a real circle if k^2 be less than δ^2; and if $k^2 = \delta^2$, then the equation (Art. 80) will represent a circle of infinitely small radius, the coordinates of whose centre are $y = 0$, $x = \pm\delta$. Hence the points just found may themselves be considered as circles of the system, and have, accordingly, been termed by Poncelet* the *limiting* points of the system of circles.

*112. If from any point on the radical axis we draw tangents to all these circles, the locus of the point of contact must be a circle, since we proved (Art. 107) that all these tangents were equal. It is evident, also, that this circle cuts any of the given system at right angles, since its radii are tangents to the given system. The equation of this circle can be readily found.

---

* *Traité des Propriétés Projectives*, p. 41.

The square of the tangent from any point ($x=0$, $y=h$) to the circle

$$x^2+y^2-2kx+\delta^2=0,$$

being found by substituting these coordinates in this equation is $h^2+\delta^2$; and the circle whose centre is the point ($x=0$, $y=h$), and whose radius squared $=h^2+\delta^2$, must have for its equation

$$x^2+(y-h)^2=h^2+\delta^2,$$

or

$$x^2+y^2-2hy=\delta^2.$$

Hence, whatever be the point taken on the radical axis (*i.e.* whatever the value of h may be), still this circle will always pass through the fixed points ($y=0$, $x=\pm\delta$) found in the last Article. And we infer that *all circles which cut the given system at right angles pass through the limiting points of the system.*

Ex. 1. Find the condition that two circles

$$x^2+y^2+2gx+2fy+c=0,\quad x^2+y^2+2g'x+2f'y+c'=0$$

should cut at right angles. Expressing that the square of the distance between the centres is equal to the sum of the squares of the radii, we have

$$(g-g')^2+(f-f')^2=g^2+f^2-c+g'^2+f'^2-c',$$

or, reducing,

$$2gg'+2ff'=c+c'.$$

Ex. 2. Find the circle cutting three circles orthogonally. We have three equations of the first degree to determine the three unknown quantities g, f, c; and the problem is solved as in Art. 94. Or the problem may be solved otherwise, since it is evident from this article that the centre of the required circle is the radical centre of the three circles, and the length of its radius equal to that of the tangent from the radical centre to any of the circles.

Ex. 3. Find the circle cutting orthogonally the three circles, Art. 108, Ex. 2.

Ans. $(x+\frac{1}{16})^2+(y+\frac{25}{16})^2=\frac{1746}{256}$.

Ex. 4. If a circle cut orthogonally three circles S', S'', S''', it cuts orthogonally any circle $kS'+lS''+mS'''=0$. Writing down the condition

$$2g(kg'+lg''+mg''')+2f(kf'+lf''+mf''')=(k+l+m)c+(kc'+lc''+mc'''),$$

we see that the coefficients of k, l, m vanish separately by hypothesis.

Similarly, a circle cutting S', S'' orthogonally, also cuts orthogonally $kS'+lS''$.

Ex. 5. A system of circles which cuts orthogonally two given circles S', S'' has a common radical axis. This, which has been proved in Art. 112, may be proved otherwise as follows: The two conditions

$$2gg'+2ff'=c+c',\quad 2gg''+2ff''=c+c'',$$

enable us to determine g and f linearly in terms of c. Substituting the values so found in

$$x^2+y^2+2gx+2fy+c=0,$$

the equation retains a single indeterminate c in the first degree, and therefore (Art. 105) denotes a system having a common radical axis.

Ex. 6. If AB be a diameter of a circle, the polar of A with respect to any circle which cuts the first orthogonally will pass through B.

Ex. 7. The square of the tangent from any point of one circle to another is proportional to the perpendicular from that point upon their radical axis.

Ex. 8. To find the angle (α) at which two circles intersect.

Let the radii of the circles be R, r, and let D be the distance between their centres, then

$$D^2 = R^2 + r^2 - 2Rr \cos\alpha,$$

since the angle at which the circles intersect is equal to that between the radii to the point of intersection.

When the circles are given by the general equations, this expression becomes

$$2Rr \cos\alpha = 2Gg + 2Ff - C - c.$$

If $S = 0$ be the equation of the circle whose radius is r, the coordinates of the centre of the other circle must fulfil the condition $R^2 - 2Rr\cos\alpha = S$, as is evident from Art. 90, since $D^2 - r^2$ is the square of the tangent to S from the centre of the other circle.

Ex. 9. If we are given the angles α, β at which a circle cuts two fixed circles S, S', the circle is not determined, since we have only two conditions; but we can determine the angle at which it cuts any circle of the system $kS + lS'$. For we have

$$R^2 - 2Rr\cos\alpha = S, \quad R^2 - 2Rr'\cos\beta = S',$$

whence
$$R^2 - 2R\frac{kr\cos\alpha + lr'\cos\beta}{k+l} = \frac{kS + lS'}{k+l},$$

which is the condition that the moveable circle should cut $kS + lS'$ at the constant angle γ; where $(k+l)\,r''\cos\gamma = kr\cos\alpha + lr'\cos\beta$, r'' being the radius of the circle $kS + lS'$.

Ex. 10. A circle which cuts two fixed circles at constant angles will also touch two fixed circles. For we can determine the ratio $k : l$, so that γ shall $= 0$, or $\cos\gamma = 1$. We have (Art. 107, Ex.)

$$(k+l)^2 r''^2 = (k+l)(kr^2 + lr'^2) - klD^2,$$

Substituting this value for r'' in the equation of the last example, we get a quadratic to determine $k : l$.

113. *To draw a common tangent to two circles.*

Let their equations be

$$(x-\alpha)^2 + (y-\beta)^2 = r^2 \qquad (S),$$

and
$$(x-\alpha')^2 + (y-\beta')^2 = r'^2 \qquad (S').$$

We saw (Art. 85) that the equation of a tangent to (S) was

$$(x-\alpha)(x'-\alpha) + (y-\beta)(y'-\beta) = r^2;$$

or, as in Art. 102, writing

$$\frac{x'-\alpha}{r} = \cos\theta, \quad \frac{y'-\beta}{r} = \sin\theta,$$

$$(x-\alpha)\cos\theta + (y-\beta)\sin\theta = r.$$

In like manner, any tangent to (S') is

$$(x-\alpha')\cos\theta' + (y-\beta')\sin\theta' = r'.$$

Now if we seek the conditions necessary that these two equations should represent *the same* right line; first, from comparing the ratio of the coefficients of x and y, we get $\tan\theta = \tan\theta'$, whence θ' either $= \theta$, or $= 180° + \theta$. If either of these conditions be fulfilled, we must equate the absolute terms, and we find, in the first case,

$$(\alpha - \alpha') \cos\theta + (\beta - \beta') \sin\theta + r - r' = 0,$$

and in the second case,

$$(\alpha - \alpha') \cos\theta + (\beta - \beta') \sin\theta + r + r' = 0.$$

Either of these equations would give us a quadratic to determine θ. The two roots of the first equation would correspond

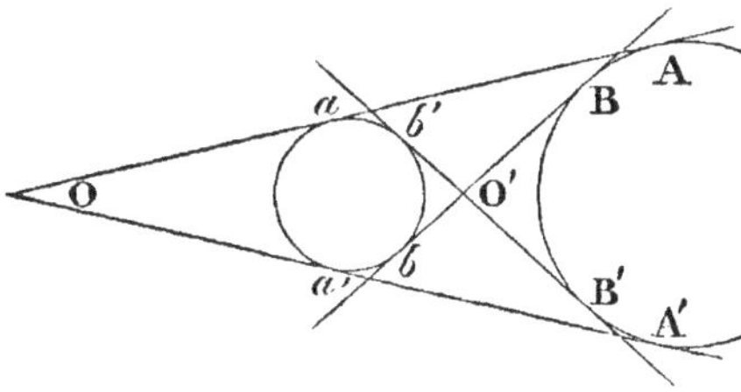

to the direct or exterior common tangents, Aa, $A'a'$; the roots of the second equation would correspond to the transverse or interior tangents, Bb, $B'b'$.

If we wished to find the coordinates of the point of contact of the common tangent with the circle (S), we must substitute, in the equation just found, for $\cos\theta$, its value, $\frac{x'-\alpha}{r}$, and for $\sin\theta$, $\frac{y'-\beta}{r}$, and we find

$$(\alpha - \alpha')(x' - \alpha) + (\beta - \beta')(y' - \beta) + r(r - r') = 0;$$

or else, $(\alpha - \alpha')(x' - \alpha) + (\beta - \beta')(y' - \beta) + r(r + r') = 0.$

The first of these equations, combined with the equation (S) of the circle, will give a quadratic, whose roots will be the coordinates of the points A and A', in which the direct common tangents touch the circle (S); and it will appear, as in Art. 88, that

$$(\alpha' - \alpha)(x - \alpha) + (\beta' - \beta)(y - \beta) = r(r - r')$$

is the equation of AA', the chord of contact of direct common tangents. So, likewise,

$$(\alpha' - \alpha)(x - \alpha) + (\beta' - \beta)(y - \beta) = r(r + r')$$

is the equation of the chord of contact of transverse common tangents. If the origin be the centre of the circle (S), then α and $\beta = 0$; and we find, for the equations of the chords of contact,

$$\alpha' x + \beta' y = r\,(r \mp r').$$

Ex. Find the common tangents to the circles

$$x^2 + y^2 - 4x - 2y + 4 = 0,\quad x^2 + y^2 + 4x + 2y - 4 = 0.$$

The chords of contact of common tangents with the first circle are

$$2x + y = 6,\quad 2x + y = 3.$$

The first chord meets the circle in the points $(2, 2)$, $(\frac{14}{5}, \frac{2}{5})$, the tangents at which are

$$y = 2,\quad 4x - 3y = 10,$$

and the second chord meets the circle in the points $(1, 1)$, $(\frac{7}{5}, \frac{1}{5})$, the tangents at which are

$$x = 1,\quad 3x + 4y = 5.$$

114. The points O and O', in which the direct or transverse tangents intersect, are (for a reason explained in the next Article) called the *centres of similitude* of the two circles.

Their coordinates are easily found, for O is the pole, with regard to circle (S), of the chord AA', whose equation is

$$\frac{(\alpha' - \alpha)\,r}{r - r'}(x - \alpha) + \frac{(\beta' - \beta)\,r}{r - r'}(y - \beta) = r^2.$$

Comparing this equation with the equation of the polar of the point $x'y'$,

$$(x' - \alpha)(x - \alpha) + (y' - \beta)(y - \beta) = r^2,$$

we get $$x' - \alpha = \frac{(\alpha' - \alpha)\,r}{r - r'},\ \text{ or }\ x' = \frac{\alpha' r - \alpha r'}{r - r'},$$

$$y' - \beta = \frac{(\beta' - \beta)\,r}{r - r'},\ \text{ or }\ y' = \frac{\beta' r - \beta r'}{r - r'}.$$

So, likewise, the coordinates of O' are found to be

$$x = \frac{\alpha' r + \alpha r'}{r + r'},\ \text{ and }\ y = \frac{\beta' r + \beta r'}{r + r'}.$$

These values of the coordinates indicate (see Art. 7) that the centres of similitude are the points where the line joining the centres is cut externally and internally in the ratio of the radii.

Ex. Find the common tangents to the circles

$$x^2 + y^2 - 6x - 8y = 0,\quad x^2 + y^2 - 4x - 6y = 3.$$

The equation of the pair of tangents through $x'y'$ to

$$(x - \alpha)^2 + (y - \beta)^2 = r^2$$

is found (Art. 92) to be

$$\{(x' - \alpha)^2 + (y' - \beta)^2 - r^2\}\{(x - \alpha)^2 + (y - \beta)^2 - r^2\} = \{(x - \alpha)(x' - \alpha) + (y - \beta)(y' - \beta) - r^2\}^2$$

Now the coordinates of the exterior centre of similitude are found to be $(-2, -1)$ and hence the pair of tangents through it is

$25(x^2 + y^2 - 6x - 8y) = (5x + 5y - 10)^2$; or $xy + x + 2y + 2 = 0$; or $(x + 2)(y + 1) = 0$.

As the given circles intersect in real points, the other two common tangents become imaginary; but their equation is found, by calculating the pair of tangents through the other centre of similitude $(\frac{23}{9}, \frac{31}{9})$, to be

$$40x^2 + xy + 40y^2 - 199x - 278y + 722 = 0.$$

115. *Every right line drawn through the intersection of common tangents is cut similarly by the two circles.*

It is evident that if on the radius vector to any point P there be taken a point Q, such that $OP = m$ times OQ, then the x and y of the point P will be respectively m times the x and y of the point Q; and that, therefore, if P describe any curve, the locus of Q is found by substituting mx, my for x and y in the equation of the curve described by P.

Now, if the common tangents be taken for axes, and if we denote Oa by a, OA by a', the equations of the two circles are (Art. 84, Ex. 2)

$$x^2 + y^2 + 2xy\cos\omega - 2ax - 2ay + a^2 = 0,$$

$$x^2 + y^2 + 2xy\cos\omega - 2a'x - 2a'y + a'^2 = 0.$$

But the second equation is what we should have found if we had substituted $\frac{ax}{a'}$, $\frac{ay}{a'}$ for x, y in the first equation; and it therefore represents the locus formed by producing each radius vector to the first circle in the ratio $a : a'$.

COR. Since the rectangle $O\rho . O\rho'$ is constant (see fig. next page), and since we have proved OR to be in a constant ratio to $O\rho$, it follows that the rectangle $OR . O\rho' = OR' . O\rho$ is constant, however the line be drawn through O.

116. *If through a centre of similitude we draw any two lines meeting the first circle in the points R, R', S, S', and the second in the points ρ, ρ', σ, σ', then the chords RS, $\rho\sigma$; $R'S'$, $\rho'\sigma'$ will be parallel, and the chords RS, $\rho'\sigma'$; $R'S'$, $\rho\sigma$ will meet on the radical axis of the two circles.*

Take OR, OS for axes, then we saw (Art. 115) that $OR = m\,O\rho$, $OS = m\,O\sigma$, and that if the equation of the circle $\rho\sigma\rho'\sigma'$ be

$$a(x^2 + 2xy\cos\omega + y^2) + 2gx + 2fy + c = 0,$$

that of the other will be

$$a(x^2 + 2xy\cos\omega + y^2)$$
$$+ 2m(gx + fy) + m^2c = 0,$$

and, therefore, the equation of the radical axis will be (Art. 105)

$$2(gx + fy) + (m + 1)c = 0.$$

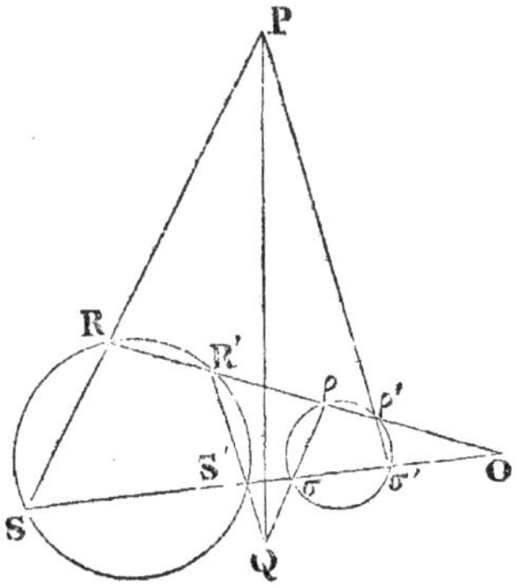

Now let the equations of $\rho\sigma$ and of $\rho'\sigma'$ be

$$\frac{x}{a} + \frac{y}{b} = 1, \quad \frac{x}{a'} + \frac{y}{b'} = 1,$$

then the equations of RS and $R'S'$ must be

$$\frac{x}{ma} + \frac{y}{mb} = 1, \quad \frac{x}{ma'} + \frac{y}{mb'} = 1.$$

It is evident, from the form of the equations, that RS is parallel to $\rho\sigma$; and RS and $\rho'\sigma'$ must intersect on the line

$$x\left(\frac{1}{a} + \frac{1}{a'}\right) + y\left(\frac{1}{b} + \frac{1}{b'}\right) = 1 + m,$$

or, as in Art. 100, on

$$2(gx + fy) + (m + 1)c = 0,$$

the radical axis of the two circles.

A particular case of this theorem is, that the tangents at R and ρ are parallel, and that those at R and ρ' meet on the radical axis.

117. *Given three circles S, S', S''; the line joining a centre of similitude of S and S' to a centre of similitude of S and S'' will pass through a centre of similitude of S' and S''.*

Form the equation of the line joining the first two of the points

$$\left(\frac{r\alpha' - \alpha r'}{r - r'}, \frac{r\beta' - \beta r'}{r - r'}\right), \left(\frac{r\alpha'' - \alpha r''}{r - r''}, \frac{r\beta'' - \beta r''}{r - r''}\right), \left(\frac{r'\alpha'' - r''\alpha'}{r' - r''}, \frac{r'\beta'' - r''\beta'}{r' - r''}\right),$$

(Art. 114), and we get (see Ex. 6, p. 24),

$$\{r(\beta' - \beta'') + r'(\beta'' - \beta) + r''(\beta - \beta')\}x$$
$$- \{r(\alpha' - \alpha'') + r'(\alpha'' - \alpha) + r''(\alpha - \alpha')\}y$$
$$= r(\beta'\alpha'' - \beta''\alpha') + r'(\beta''\alpha - \beta\alpha'') + r''(\beta\alpha' - \beta'\alpha).$$

Now the symmetry of this equation sufficiently shows, that the line it represents must pass through the third centre of similitude. This line is called an *axis of similitude* of the three circles.

Since for each pair of circles there are two centres of similitude, there will be in all *six* for the three circles, and these will be distributed along *four* axes of similitude, as represented in the figure. The equations of the other three will be found by changing the signs of either r, or r', or r'', in the equation just given.

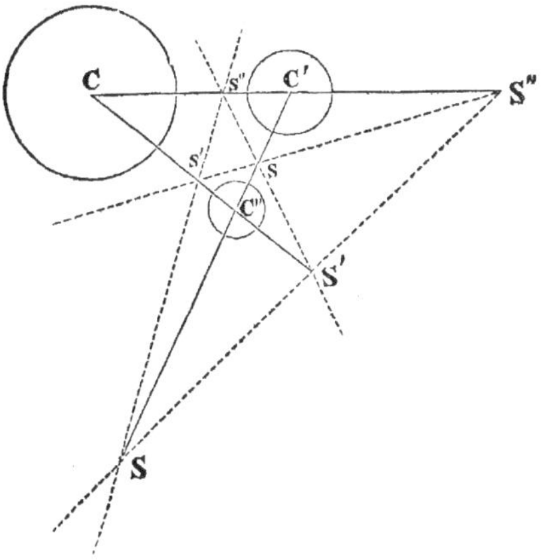

COR. *If a circle* (Σ) *touch two others* (S *and* S'), *the line joining the points of contact will pass through a centre of similitude of* S *and* S'. For when two circles touch, one of their centres of similitude will coincide with the point of contact.

If Σ touch S and S', either both externally or both internally, the line joining the points of contact will pass through the *external* centre of similitude of S and S'. If Σ touch one externally and the other internally, the line joining the points of contact will pass through the *internal* centre of similitude.

*118. *To find the locus of the centre of a circle cutting three given circles at equal angles.*

If a circle whose radius is R, cut at an angle α the three circles S, S', S'', then (Art. 112, Ex. 8) the coordinates of its centre fulfil the three conditions

$$S = R^2 - 2Rr\cos\alpha,\quad S' = R^2 - 2Rr'\cos\alpha,\quad S'' = R^2 - 2Rr''\cos\alpha.$$

From these conditions we can at once eliminate R^2 and $R\cos\alpha$. Thus, by subtraction,

$$S - S' = 2R(r' - r)\cos\alpha,\quad S - S'' = 2R(r'' - r)\cos\alpha,$$

whence $$(S - S')(r - r'') = (S - S'')(r - r'),$$

the equation of a line on which the centre must lie. It obviously passes through the radical centre (Art. 108); and if we write for $S-S'$, $S-S''$, their values (Art. 105), the coefficient of x in the equation is found to be

$$-2\{\alpha(r'-r'')+\alpha'(r''-r)+\alpha''(r-r')\},$$

while that of y is

$$-2\{\beta(r'-r'')+\beta'(r''-r)+\beta''(r-r')\}.$$

Now if we compare these values with the coefficients in the equation of the axis of similitude (Art. 117), we infer (Art. 32), that the locus is a perpendicular let fall from the radical centre on an axis of similitude.

It is of course optional which of two supplemental angles we consider to be the angle at which two circles intersect. The formula (Art. 112) which we have used assumes that the angle at which two circles cut is measured by the angle which the distance between their centres subtends at the point of meeting; and with this convention, the locus under consideration is a perpendicular on the external axis of similitude. If this limitation be removed, the formula we have used becomes $S=R^2\pm 2Rr\cos\alpha$; or, in other words, we may change the sign of either r, r', or r'' in the preceding formulæ, and therefore (Art. 117) the locus is a perpendicular on any of the four axes of similitude.*

When two circles touch internally, their angle of intersection vanishes, since the radii to the point of meeting coincide. But if they touch externally, their angle of intersection according to the preceding convention is 180°, one radius to the point of meeting being a continuation of the other. It follows, from

* In fact, all circles cutting three circles at equal angles have one of the axes of similitude for a common radical axis. Let Σ, Σ', Σ'' be three circles, all cutting the given circles at the same angles α, β, γ respectively. Then the coordinates of the centre of each of the circles S, S', S'' must fulfil the conditions

$$\Sigma=r^2-2rR\cos\alpha,\quad \Sigma'=r^2-2rR'\cos\beta,\quad \Sigma''=r^2-2rR''\cos\gamma;$$

whence $\quad (R\cos\alpha-R''\cos\gamma)(\Sigma-\Sigma')=(R\cos\alpha-R'\cos\beta)(\Sigma-\Sigma'').$

This which appears to be the equation of a right line is satisfied by the coordinates of the centre of S, of S', and of S'', three points which are not supposed to be on a right line. Now the only way in which what seems an equation of the first degree, such as $ax+by+c=a'x+b'y+c'$ can be satisfied by the coordinates of three points which are not on a right line, is if the equation is in truth an identical one, $a=a'$, $b=b'$, $c=c'$. The equation, therefore, written above denotes an identical relation of the form $\Sigma=k\Sigma'+l\Sigma''$, shewing that the three circles have a common radical axis.

what has been just proved, that the perpendicular on the external axis of similitude contains the centre of a circle touching three given circles, either all externally, or all internally. If we change the sign of r, the equation of the locus which we found denotes a perpendicular on one of the other axes of similitude which will contain the centre of the circle touching S externally, and the other two internally, or *vice versâ*. Eight circles in all can be drawn to touch three given circles, and their centres lie, a pair on each of the perpendiculars let fall from the radical centre on the four axes of similitude.

*119. *To describe a circle touching three given circles.* We have found one locus on which the centre must lie, and we could find another by eliminating R between the two conditions

$$S = R^2 + 2Rr, \quad S' = R^2 + 2Rr'.$$

The result, however, would not represent a circle, and the solution will therefore be more elementary, if instead of seeking the coordinates of the centre of the touching circle, we look for those of its point of contact with one of the given circles. We have already one relation connecting these coordinates, since the point lies on a given circle, therefore another relation between them will suffice completely to determine the point.*

Let us for simplicity take for origin the centre of the circle, the point of contact with which we are seeking, that is to say, let us take $\alpha = 0$, $\beta = 0$, then if A and B be the coordinates of the centre of Σ, the sought circle, we have seen that they fulfil the relations

$$S - S' = 2R(r - r'), \quad S - S'' = 2R(r - r'').$$

But if x and y be the coordinates of the point of contact of Σ with S, we have from similar triangles

$$A = \frac{x(R + r)}{r}, \quad B = \frac{y(R + r)}{r}.$$

Now if in the equation of any right line we substitute mx, my for x and y, the result will evidently be the same as if we multiply the whole equation by m, and subtract $(m - 1)$ times the absolute term. Hence, remembering that the absolute term in $S - S'$ is

* This solution is by M. Gergonne, *Annales des Mathématiques*, vol. VII. p. 289.

(Art. 105) $r'^2 - r^2 - \alpha'^2 - \beta'^2$, the result of making the above substitutions for A and B in $(S - S') = 2R(r - r')$ is

$$\frac{R+r}{r}(S - S') + \frac{R}{r}(\alpha'^2 + \beta'^2 + r^2 - r'^2) = 2R(r - r'),$$

or $$(R + r)(S - S') = R\{(r - r')^2 - \alpha'^2 - \beta'^2\}.$$

Similarly $$(R + r)(S - S'') = R\{(r - r'')^2 - \alpha''^2 - \beta''^2\}.$$

Eliminating R, the point of contact is determined as one of the intersections of the circle S with the right line

$$\frac{S - S'}{\alpha'^2 + \beta'^2 - (r - r')^2} = \frac{S - S''}{\alpha''^2 + \beta''^2 - (r - r'')^2}.$$

120. To complete the geometrical solution of the problem, it is necessary to show how to construct the line whose equation has been just found. It obviously passes through the radical centre of the circles; and a second point on it is found as follows: Write at full length for $S - S'$ (Art. 105), and the equation is

$$\frac{2\alpha'x + 2\beta'y + r'^2 - r^2 - \alpha'^2 - \beta'^2}{\alpha'^2 + \beta'^2 - (r - r')^2} = \frac{2\alpha''x + 2\beta''y + r''^2 - r^2 - \alpha''^2 - \beta''^2}{\alpha''^2 + \beta''^2 - (r - r'')^2}.$$

Add 1 to both sides of the equation, and we have

$$\frac{\alpha'x + \beta'y + (r' - r)r}{\alpha'^2 + \beta'^2 - (r - r')^2} = \frac{\alpha''x + \beta''y + (r'' - r)r}{\alpha''^2 + \beta''^2 - (r - r'')^2}.$$

showing that the above line passes through the intersection of

$$\alpha'x + \beta'y + (r' - r)r = 0, \quad \alpha''x + \beta''y + (r'' - r)r = 0.$$

But the first of these lines (Art. 113) is the chord of common tangents of the circles S and S'; or, in other words (Art. 114), is the polar with regard to S of the centre of similitude of these circles. And, in like manner, the second line is the polar of the centre of similitude of S and S''; therefore (since the intersection of any two lines is the pole of the line joining their poles) the intersection of the lines

$$\alpha'x + \beta'y + (r' - r)r = 0, \quad \alpha''x + \beta''y + (r'' - r)r = 0$$

is the pole of the axis of similitude of the three circles, with regard to the circle S.

Hence we obtain the following construction:

Drawing any of the four axes of similitude of the three circles, take its pole with respect to each circle, and join the points so found (P, P', P''') with the radical centre; then, if the joining lines meet the circles in the points

$$(a, b; a', b'; a'', b''),$$

the circle through a, a', a'' will be one of the touching circles, and that through b, b', b'' will be another. Repeating this process with the other three axes of similitude, we can determine the other six touching circles.

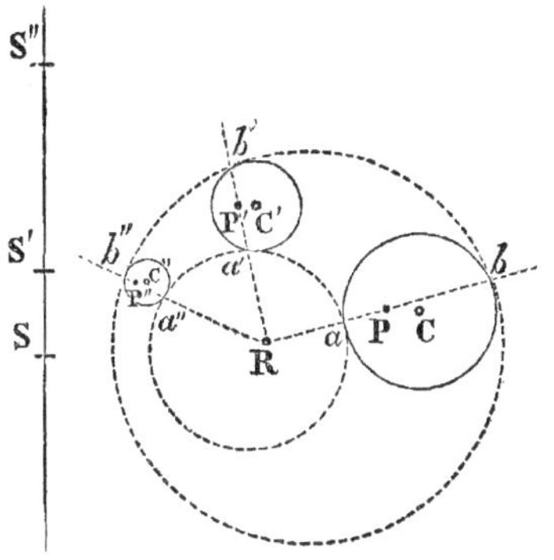

121. It is useful to show how the preceding results may be derived without algebraical calculations.

(1) By Cor., Art. 117, the lines ab, $a'b'$, $a''b''$ meet in a point, viz., the centre of similitude of the circles $aa'a''$, $bb'b''$.

(2) In like manner $a'a''$, $b'b''$ intersect in S, the centre of similitude of C', C''.

(3) Hence (Art. 116) the transverse lines $a'b'$, $a''b''$ intersect on the radical axis of C', C''. So again $a''b''$, ab intersect on the radical axis of C'', C. Therefore the point R (the centre of similitude of $aa'a''$, $bb'b''$) must be the radical centre of the circles C, C', C''.

(4) In like manner, since $a'b'$, $a''b''$ pass through a centre of similitude of $aa'a''$, $bb'b''$; therefore (Art. 116) $a'a''$, $b'b''$ meet on the radical axis of these two circles. So again the points S' and S'' must lie on the same radical axis; *therefore $SS'S''$, the axis of similitude of the circles C, C', C'', is the radical axis of the circles $aa'a''$, $bb'b''$.*

(5) Since $a''b''$ passes through the centre of similitude of $aa'a''$, $bb'b''$, therefore (Art. 116) the tangents to these circles where it meets them intersect on the radical axis $SS'S''$. But this point of intersection must plainly be the pole of $a''b''$ with regard to the circle C''. Now since the pole of $a''b''$ lies on $SS'S''$, therefore (Art. 98) the pole of $SS'S''$ with regard to C'' lies on $a''b''$. Hence $a''b''$ is constructed by joining the radical centre to the pole of $SS'S''$ with regard to C''.

(6) Since the centre of similitude of two circles is on the line joining their centres, and the radical axis is perpendicular to that line, we learn (as in Art. 118) that the line joining the centres of $aa'a''$, $bb'b''$ passes through R, and is perpendicular to $SS'S''$.

121 (*a*).* Dr. Casey has given a solution of the problem we are considering, depending on the following principle due to him: If four circles be all touched by the same fifth circle, the lengths of their common tangents are connected by the following relation, $\overline{12}.\overline{34} \pm \overline{14}.\overline{23} \pm \overline{13}.\overline{24} = 0$, where $\overline{12}$ denotes the length of a common tangent to the first and second circles, &c. This may be proved by expressing each common tangent in terms of the length of the line joining the points where the circles touch the common touching circle. Let R be the radius of the latter circle whose centre is O, r and r' of the circles whose centres are A and B, then, from the isosceles triangle aOb, we have

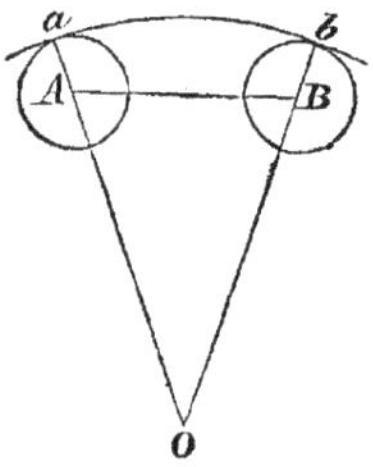

$$ab = 2R \sin \tfrac{1}{2} aOb.$$

But from the triangle AOB, whose base is D, and sides $R - r$, $R - r'$, we have $\sin^2 \frac{1}{2} aOb = \dfrac{D^2 - (r - r')^2}{4(R - r)(R - r')}$. Now the numerator of this fraction is the square of the common tangent $\overline{12}$, hence

$$ab = \frac{R.\overline{12}}{\sqrt{(R - r)(R - r')}}.$$

But since the four points of contact form a quadrilateral inscribed in a circle, its sides and diagonals are connected by the relation $ab.cd + ad.bc = ac.bd$. Substitute in this equation the expression just given for each chord in terms of the corresponding common tangent, and suppress the numerator R^2 and the denominator $\sqrt{(R - r)(R - r')(R - r'')(R - r''')}$ which are common to every term, and there remains the relation which we are required to prove.

121 (*b*). Let now the fourth circle reduce itself to a point, this will be a point on the circle touching the other three, and

* In order to avoid confusion in the references, I retain the numbering of the articles in the fourth edition, and mark separately those articles which have been since added.

$\overline{41}$, $\overline{42}$, $\overline{43}$ will denote the lengths of the tangents from that point to these three circles. But the lengths of these tangents are (Art. 90) the square roots of the results of substituting the coordinates of that point in the equations of the circles. We see then that the coordinates of any point on the circle which touches three others must fulfil the relation

$$\overline{23}\sqrt{(S)} \pm \overline{31}\sqrt{(S')} \pm \overline{12}\sqrt{(S'')} = 0.$$

If this equation be cleared of radicals it will be found to be one of the fourth degree, and when $\overline{23}$, $\overline{31}$, $\overline{12}$ are the direct common tangents, it will be the product of the equations of the two circles (see fig., p. 112) which touch either all externally or all internally.

121 (*c*). The principle just used may also be established without assuming the relation connecting the sides and diagonals of an inscribed quadrilateral. If on each radius vector OP to a curve we take, as in Ex. 4, p. 96, a part OQ inversely proportional to OP, the locus of Q is a curve which is called the *inverse* of the given curve. It is found without difficulty that the equation of the inverse of the circle $x^2 + y^2 + 2gx + 2fy + c$ is

$$c(x^2 + y^2) + 2gx + 2fy + 1 = 0,$$

which denotes a circle, except when $c = 0$ (that is to say, when the point O is on the circle), in which case the inverse is a right line. Conversely, the inverse of a right line is a circle passing through the point O. Now Dr. Casey has noticed that if we are given a pair of circles, and form the inverse pair with regard to any point, then the ratio of the square of a common tangent to the product of the radii is the same for each pair of circles.* For if in $g^2 + f^2 - c$, which (Art. 80) is r^2, we substitute for g, f, c; $\frac{g}{c}, \frac{f}{c}, \frac{1}{c}$, we find that the radius of the inverse circle is r divided by c; and if we make a similar substitution in $c + c' - 2gg' - 2ff'$, which (Ex. 1, p. 102) is $D^2 - r^2 - r'^2$, we get the same quantity divided by cc'. Hence the ratio of $D^2 - r^2 - r'^2$ to rr' is the same for a pair of circles

* This is equivalent (see Ex. 8, p. 103) to saying that the angle of intersection is the same for each pair, as may easily be proved geometrically.

and for the inverse pair; and, therefore, so is also the ratio to rr' of $D^2 - (r \pm r')^2$.

Consider now four circles touching the same right line in four points. Now the mutual distances of four points on a right line are connected by the relation $\overline{12}.\overline{34} + \overline{14}.\overline{32} = \overline{13}.\overline{24}$; as may easily be proved by the identical equation

$$(b-a)(d-c) + (d-a)(c-b) = (c-a)(d-b),$$

where a, b, c, d denote the distances of the points from any origin on the line. Thus then the common tangents of four circles which touch the same right line are connected by the relation which is to be proved. But if we take the inverse of the system with regard to any point, we get four circles touched by the same circle, and the relation subsists still; for if the equation be divided by the square root of the products of all the radii, it consists of members $\frac{\overline{12}}{\sqrt{(rr')}}$, $\frac{\overline{34}}{\sqrt{(r''r''')}}$, &c., which are unchanged by the process of inversion.

The relation between the common tangents being proved in this way,* we have only to suppose the four circles to become four points, when we deduce as a particular case the relation connecting the sides and diagonals of an inscribed quadrilateral. This method also shews that, in the case of two circles which touch the same side of the enveloping circle, we are to use the direct common tangent; but the transverse common tangent when one touches the concavity, and the other the convexity of that circle. Thus then we get the equation of the four pairs of circles which touch three given circles,

$$\overline{23}\sqrt{(S)} \pm \overline{31}\sqrt{(S')} \pm \overline{12}\sqrt{(S'')} = 0.$$

When $\overline{12}$, $\overline{23}$, $\overline{31}$ denote the lengths of the direct common tangents, this equation represents the pair of circles having the given circles either all inside or all outside. If $\overline{23}$ denotes a direct common tangent, and $\overline{31}$, $\overline{12}$ transverse, we get a pair of circles each having the first circle on one side, and the other two on the other. And, similarly, we get the other pairs of circles by taking in turn $\overline{31}$, $\overline{12}$ as direct common tangents, and the other common tangents transverse.

* Another proof will be given in the appendix to the next chapter.

*CHAPTER IX.

APPLICATION OF ABRIDGED NOTATION TO THE EQUATION OF THE CIRCLE.

122. If we have an equation of the second degree expressed in the abridged notation explained in Chap. IV., and if we desire to know whether it represents a circle, we have only to transform to x and y coordinates, by substituting for each abbreviation (α) its equivalent ($x\cos\alpha + y\sin\alpha - p$); and then to examine whether the coefficient of xy in the transformed equation vanishes, and whether the coefficients of x^2 and of y^2 are equal. This is sufficiently illustrated in the examples which follow.

When will the locus of a point be a circle if the product of perpendiculars from it on two opposite sides of a quadrilateral be in a given ratio to the product of perpendiculars from it on the other two sides?

Let α, β, γ, δ be the four sides of the quadrilateral, then the equation of the locus is at once written down $\alpha\gamma = k\beta\delta$, which represents a curve of the second degree passing through the angles of the quadrilateral, since it is satisfied by any of the four suppositions,

$$\alpha = 0,\ \beta = 0;\ \alpha = 0,\ \delta = 0;\ \beta = 0,\ \gamma = 0;\ \gamma = 0,\ \delta = 0.$$

Now, in order to ascertain whether this equation represents a circle, write it at full length

$$(x\cos\alpha + y\sin\alpha - p)(x\cos\gamma + y\sin\gamma - p'')$$
$$= k(x\cos\beta + y\sin\beta - p')(x\cos\delta + y\sin\delta - p''').$$

Multiplying out, equating the coefficient of x^2 to that of y^2, and putting that of $xy = 0$, we obtain the conditions

$$\cos(\alpha + \gamma) = k\cos(\beta + \delta);\ \sin(\alpha + \gamma) = k\sin(\beta + \delta).$$

Squaring these equations, and adding them, we find $k = \pm 1$; and if this condition be fulfilled, we must have

$$\alpha + \gamma = \beta + \delta, \text{ or else } = 180^\circ + \beta + \delta;$$

whence $\qquad \alpha - \beta = \delta - \gamma, \text{ or } 180 + \delta - \gamma.$

Recollecting (Art. 61) that $\alpha - \beta$ is the supplement of that angle between α and β, in which the origin lies, we see that this condition will be fulfilled if the quadrilateral formed by $\alpha\beta\gamma\delta$ be inscribable in a circle (Euc. III. 22). And it will be seen on examination that when the origin is within the quadrilateral we are to take $k = -1$, and that the angle (in which the origin lies) between α and β is supplemental to that between γ and δ; but that we are to take $k = +1$, when the origin is without the quadrilateral, and that the opposite angles are equal.

123. *When will the locus of a point be a circle, if the square of its distance from the base of a triangle be in a constant ratio to the product of its distances from the sides?*

Let the sides of the triangle be α, β, γ, and the equation of the locus is $\alpha\beta = k\gamma^2$. If now we look for the points where the line α meets this locus, by making in it $\alpha = 0$, we obtain the perfect square $\gamma^2 = 0$. Hence α meets the locus in two coincident points, that is to say (Art. 83), it touches the locus at the point $\alpha\gamma$. Similarly, β touches the locus at the point $\beta\gamma$. Hence α and β are both tangents, and γ their chord of contact. Now, to ascertain whether the locus is a circle, writing at full length as in the last article, and applying the tests of Art. 80, we obtain the conditions

$$\cos(\alpha + \beta) = k \cos 2\gamma; \quad \sin(\alpha + \beta) = k \sin 2\gamma;$$

whence (as in the last article) we get $k = 1$, $\alpha - \gamma = \gamma - \beta$, or the triangle is isosceles. Hence we may infer that *if from any point of a circle perpendiculars be let fall on any two tangents and on their chord of contact, the square of the last will be equal to the rectangle under the other two.*

Ex. When will the locus of a point be a circle if the sum of the squares of the perpendiculars from it on the sides of any triangle be constant?

The locus is $\alpha^2 + \beta^2 + \gamma^2 = c^2$; and the conditions that this should represent a circle are

$$\cos 2\alpha + \cos 2\beta + \cos 2\gamma = 0; \qquad \sin 2\alpha + \sin 2\beta + \sin 2\gamma = 0.$$

$$\cos 2\alpha = -2\cos(\beta + \gamma)\cos(\beta - \gamma); \quad \sin 2\alpha = -2\sin(\beta + \gamma)\cos(\beta - \gamma).$$

Squaring and adding,

$$1 = 4\cos^2(\beta - \gamma); \quad \beta - \gamma = 60^\circ.$$

And so, in like manner, each of the other two angles of the triangle is proved to be 60°, or the triangle must be equilateral.

124. *To obtain the equation of the circle circumscribing the triangle formed by the lines* $\alpha = 0$, $\beta = 0$, $\gamma = 0$.

Any equation of the form

$$l\beta\gamma + m\gamma\alpha + n\alpha\beta = 0$$

denotes a curve of the second degree circumscribing the given triangle, since it is satisfied by any of the suppositions

$$\alpha = 0,\ \beta = 0;\ \beta = 0,\ \gamma = 0;\ \gamma = 0,\ \alpha = 0.$$

The conditions that it should represent a circle are found, by the same process as in Art. 122, to be

$$l \cos(\beta + \gamma) + m \cos(\gamma + \alpha) + n \cos(\alpha + \beta) = 0,$$

$$l \sin(\beta + \gamma) + m \sin(\gamma + \alpha) + n \sin(\alpha + \beta) = 0.$$

Now we have seen (Art. 65) that when we are given a pair of equations of the form

$$l\alpha' + m\beta' + n\gamma' = 0,\ \ l\alpha'' + m\beta'' + n\gamma'' = 0,$$

l, m, n must be respectively proportional to $\beta'\gamma'' - \beta''\gamma'$, $\gamma'\alpha'' - \gamma''\alpha'$, $\alpha'\beta'' - \alpha''\beta'$. In the present case then l, m, n must be proportional to $\sin(\beta - \gamma)$, $\sin(\gamma - \alpha)$, $\sin(\alpha - \beta)$, or (Art. 61) to $\sin A$, $\sin B$, $\sin C$. Hence the equation of the circle circumscribing a triangle is

$$\beta\gamma \sin A + \gamma\alpha \sin B + \alpha\beta \sin C = 0.$$

125. The geometrical interpretation of the equation just found deserves attention. If from any point O we let fall perpendiculars OP, OQ, on the lines α, β, then (Art. 54) α, β are the lengths of these perpendiculars; and since the angle between them is the supplement of C, the quantity $\alpha\beta \sin C$ is double the area of the triangle OPQ. In like manner, $\gamma\alpha \sin B$ and $\beta\gamma \sin A$ are double the triangles OPR, OQR. Hence the quantity

$$\beta\gamma \sin A + \gamma\alpha \sin B + \alpha\beta \sin C$$

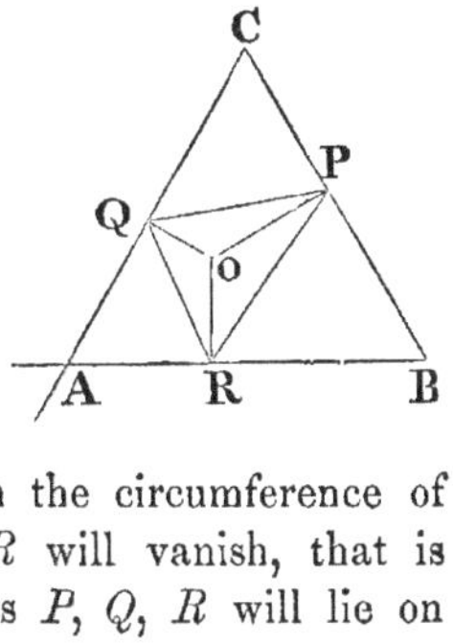

is double the area of the triangle PQR, and the equation found in the last article asserts that if the point O be taken on the circumference of the circumscribing circle, the area PQR will vanish, that is to say (Art. 36, Cor. 2), the three points P, Q, R will lie on one right line.

If it were required to find the locus of a point from which, if we let fall perpendiculars on the sides of a triangle, and join their feet, the triangle PQR so formed should have a constant magnitude, the equation of the locus would be

$$\beta\gamma \sin A + \gamma\alpha \sin B + \alpha\beta \sin C = \text{constant},$$

and, since this only differs from the equation of the circumscribing circle in the constant part, it is (Art. 81) the equation of a circle concentric with the circumscribing circle.*

126. The following inferences may be drawn from the equation $l\beta\gamma + m\gamma\alpha + n\alpha\beta = 0$, whether or not l, m, n have the values $\sin A$, $\sin B$, $\sin C$, and therefore lead to theorems true not only of the circle but of any curve of the second degree circumscribing the triangle. Write the equation in the form

$$\gamma\,(l\beta + m\alpha) + n\alpha\beta = 0;$$

and we saw in Art. 124 that γ meets the curve in the two points where it meets the lines α and β; since if we make $\gamma = 0$ in the equation, it reduces to $\alpha\beta = 0$. Now, for the same reason, the two points in which $l\beta + m\alpha$ meets the curve are the two points where *it* meets the lines α and β. But these two points coincide, since $l\beta + m\alpha$ passes through the point $\alpha\beta$. Hence the line $l\beta + m\alpha$, which meets the curve in two coincident points, is (Art. 83) the tangent at the point $\alpha\beta$.

In the case of the circle the tangent is $\alpha \sin B + \beta \sin A$. Now we saw (Art. 64) that $\alpha \sin A + \beta \sin B$ denotes a parallel to the base γ drawn through the vertex. Hence (Art. 55) the tangent makes the same angle with one side that the base makes with the other (Euc. III. 32).

---

* Consider a quadrilateral inscribed in a circle of which α, β, γ, δ are sides and ε a diagonal; then the equation of the circle may be written in either of the forms

$$\frac{\sin A}{\alpha} + \frac{\sin B}{\beta} + \frac{\sin E}{\varepsilon} = 0, \quad \frac{\sin C}{\gamma} + \frac{\sin D}{\delta} - \frac{\sin E}{\varepsilon} = 0,$$

where A is the angle in the segment subtended by α, &c., and we have written ε with a negative side in the second equation, because opposite sides of the line are considered in the two triangles. Hence, every point on the circle satisfies also the equation $\frac{\sin A}{\alpha} + \frac{\sin B}{\beta} + \frac{\sin C}{\gamma} + \frac{\sin D}{\delta} = 0$.

This equation when cleared of fractions is of the third degree, and represents, together with the circle, the line joining the intersections of $\alpha\gamma$, $\beta\delta$. In the same manner, if we have an inscribed polygon of any number of sides, Dr. Casey has shewn that an equation of similar form will be satisfied for any point of the circle.

Writing the equations of the tangents at the three vertices in the form

$$\frac{\beta}{m}+\frac{\gamma}{n}=0,\ \frac{\gamma}{n}+\frac{\alpha}{l}=0,\ \frac{\alpha}{l}+\frac{\beta}{m}=0,$$

we see that the three points in which each intersects the opposite side are in one right line, whose equation is

$$\frac{\alpha}{l}+\frac{\beta}{m}+\frac{\gamma}{n}=0.$$

Subtracting, one from another, the equations of the three tangents, we get the equations of the lines joining the vertices of the original triangle to the corresponding vertices of the triangle formed by the three tangents, viz.,

$$\frac{\beta}{m}-\frac{\gamma}{n}=0,\ \frac{\gamma}{n}-\frac{\alpha}{l}=0,\ \frac{\alpha}{l}-\frac{\beta}{m}=0,$$

three lines which meet in a point (Art. 40).*

127. If $\alpha'\beta'\gamma'$, $\alpha''\beta''\gamma''$ be the coordinates of any two points on the curve, the equation of the line joining them is

$$\frac{l\alpha}{\alpha'\alpha''}+\frac{m\beta}{\beta'\beta''}+\frac{n\gamma}{\gamma'\gamma''}=0;$$

for if we substitute in this equation $\alpha'\beta'\gamma'$ for $\alpha\beta\gamma$, the equation is satisfied, since $\alpha''\beta''\gamma''$ satisfy the equation of the curve, which may be written

$$\frac{l}{\alpha}+\frac{m}{\beta}+\frac{n}{\gamma}=0.$$

In like manner the equation is satisfied by the coordinates $\alpha''\beta''\gamma''$. It follows that the equation of the tangent at any point $\alpha'\beta'\gamma'$ may be written

$$\frac{l\alpha}{\alpha'^2}+\frac{m\beta}{\beta'^2}+\frac{n\gamma}{\gamma'^2}=0;$$

and conversely, that if $\lambda\alpha+\mu\beta+\nu\gamma=0$ is the equation of a tangent, the coordinates of the point of contact $\alpha'\beta'\gamma'$ are given by the equations

$$\frac{l}{\alpha'^2}=\lambda,\ \frac{m}{\beta'^2}=\mu,\ \frac{n}{\gamma'^2}=\nu.$$

---

* The theorems of this article are by M. Bobillier (*Annales des Mathématiques*, vol. XVIII. p. 320). The first equation of the next article is by M. Hermes.

Solving for α', β', γ' from these equations, and substituting in the equation of the curve, which must be satisfied by the point $\alpha'\beta'\gamma'$, we get

$$\sqrt{(l\lambda)} + \sqrt{(m\mu)} + \sqrt{(n\nu)} = 0.$$

This is *the condition that the line* $\lambda\alpha + \mu\beta + \nu\gamma$ *may touch* $\beta\gamma + m\gamma\alpha + n\alpha\beta$; or it may be called (see Art. 70) the *tangential equation* of the curve. The tangential equation might also be obtained by eliminating γ between the equation of the line and that of the curve, and forming the condition that the resulting equation in $\alpha : \beta$ may have equal roots.

128. *To find the conditions that the general equation of the second degree in* α, β, γ,

$$a\alpha^2 + b\beta^2 + c\gamma^2 + 2f\beta\gamma + 2g\gamma\alpha + 2h\alpha\beta = 0,$$

may represent a circle. [Dublin Exam. Papers, Jan. 1857].

It is convenient to avail ourselves of the result of Art. 124. Since the terms of the second degree, $x^2 + y^2$, are the same in the equations of all circles, the equations of two circles can only differ in the linear part; and if S represent a circle, an equation of the form $S + lx + my + n = 0$ may represent any circle whatever. In like manner, in trilinear coordinates, if we have found one equation which represents a circle, we have only to add to it terms $l\alpha + m\beta + n\gamma$ (which in order that the equation may be homogeneous we multiply by the constant $\alpha \sin A + \beta \sin B + \gamma \sin C$), and we shall have an equation which may represent any circle whatever. Thus then (Art. 124) the equation of any circle may be thrown into the form

$$(l\alpha + m\beta + n\gamma)(\alpha \sin A + \beta \sin B + \gamma \sin C)$$
$$+ k(\beta\gamma \sin A + \gamma\alpha \sin B + \alpha\beta \sin C) = 0.$$

If now we compare the coefficients of α^2, β^2, γ^2 in this form with those in the general equation, we see that, if the latter represent a circle, it must be reducible to the form

$$\left(\frac{a}{\sin A}\alpha + \frac{b}{\sin B}\beta + \frac{c}{\sin C}\gamma\right)(\alpha \sin A + \beta \sin B + \gamma \sin C)$$
$$+ k(\beta\gamma \sin A + \gamma\alpha \sin B + \alpha\beta \sin C) = 0$$

and a comparison of the remaining coefficients gives

$$2f \sin B \sin C = c \sin^2 B + b \sin^2 C + k \sin A \sin B \sin C,$$

$$2g \sin C \sin A = a \sin^2 C + c \sin^2 A + k \sin A \sin B \sin C,$$

$$2h \sin A \sin B = b \sin^2 A + a \sin^2 B + k \sin A \sin B \sin C,$$

whence eliminating k, we have the required conditions, viz.

$$b \sin^2 C + c \sin^2 B - 2f \sin B \sin C = c \sin^2 A + a \sin^2 C - 2g \sin C \sin A$$
$$= a \sin^2 B + b \sin^2 A - 2h \sin A \sin B.$$

If we have the equations of two circles written in the form

$$(l\alpha + m\beta + n\gamma)(\alpha \sin A + \beta \sin B + \gamma \sin C)$$
$$+ k(\beta\gamma \sin A + \gamma\alpha \sin B + \alpha\beta \sin C) = 0,$$

$$(l'\alpha + m'\beta + n'\gamma)(\alpha \sin A + \beta \sin B + \gamma \sin C)$$
$$+ k(\beta\gamma \sin A + \gamma\alpha \sin B + \alpha\beta \sin C) = 0,$$

it is evident that their radical axis is

$$l\alpha + m\beta + n\gamma - (l'\alpha + m'\beta + n'\gamma),$$

and that $l\alpha + m\beta + n\gamma$ is the radical axis of the first with the circumscribing circle.

Ex. 1. Verify that $\alpha\beta - \gamma^2$ represents a circle if $A = B$ (Art. 123).

The equation may be written

$$\alpha\beta \sin C + \beta\gamma \sin A + \gamma\alpha \sin B - \gamma(\alpha \sin A + \beta \sin B + \gamma \sin C) = 0.$$

Ex. 2. When will $a\alpha^2 + b\beta^2 + c\gamma^2$ represent a circle?

Ex. 3. The three middle points of sides, and the three feet of perpendiculars lie on a circle. The equation

$$\alpha^2 \sin A \cos A + \beta^2 \sin B \cos B + \gamma^2 \sin C \cos C - (\beta\gamma \sin A + \gamma\alpha \sin B + \alpha\beta \sin C) = 0,$$

represents a curve of the second degree passing through the points in question. For if we make $\gamma = 0$, we get

$$\alpha^2 \sin A \cos A + \beta^2 \sin B \cos B - \alpha\beta (\sin A \cos B + \sin B \cos A) = 0,$$

the factors of which are $\alpha \sin A - \beta \sin B$ and $\alpha \cos A - \beta \cos B$. Now the curve is a circle, for it may be written

$$(\alpha \cos A + \beta \cos B + \gamma \cos C)(\alpha \sin A + \beta \sin B + \gamma \sin C)$$
$$- 2(\beta\gamma \sin A + \gamma\alpha \sin B + \alpha\beta \sin C) = 0$$

Thus the radical axis of the circumscribing circle and of the circle through the middle points of sides is $\alpha \cos A + \beta \cos B + \gamma \cos C$, that is, the axis of homology of the given triangle with the triangle formed by joining the feet of perpendiculars.

129. We shall next show how to form the equations of the circles which touch the three sides of the triangle α, β, γ. The

general equation of a curve of the second degree touching the three sides is

$$l^2\alpha^2 + m^2\beta^2 + n^2\gamma^2 - 2mn\beta\gamma - 2nl\gamma\alpha - 2lm\alpha\beta = 0.*$$

Thus γ is a tangent, or meets the curve in two coincident points, since, if we make $\gamma = 0$ in the equation, we get the perfect square $l^2\alpha^2 + m^2\beta^2 - 2lm\alpha\beta = 0$. The equation may also be written in a convenient form

$$\sqrt{(l\alpha)} + \sqrt{(m\beta)} + \sqrt{(n\gamma)} = 0;$$

for, if we clear this equation of radicals, we shall find it to be identical with that just written.

Before determining the values of l, m, n, for which the equation represents a circle, we shall draw from it some inferences which apply to all curves of the second degree inscribed in the triangle. Writing the equation in the form

$$n\gamma\,(n\gamma - 2l\alpha - 2m\beta) + (l\alpha - m\beta)^2 = 0,$$

we see that the line $(l\alpha - m\beta)$, which obviously passes through the point $\alpha\beta$, passes also through the point where γ meets the curve. The three lines, then, which join the points of contact of the sides with the opposite angles of the circumscribing triangle are

$$l\alpha - m\beta = 0, \quad m\beta - n\gamma = 0, \quad n\gamma - l\alpha = 0,$$

and these obviously meet in a point.

The very same proof which showed that γ touches the curve shows also that $n\gamma - 2l\alpha - 2m\beta$ touches the curve, for when this quantity is put $= 0$, we have the perfect square $(l\alpha - m\beta)^2 = 0$; hence this line meets the curve in two coincident points, that is, touches the curve, and $l\alpha - m\beta$ passes through the point of contact. Hence, if the vertices of the triangle be joined to the

* Strictly speaking, the double rectangles in this equation ought to be written with the ambiguous sign $\pm$, and the argument in the text would apply equally. If, however, we give all the rectangles positive signs, or if we give one of them a positive sign, and the other two negative, the equation does not denote a proper curve of the second degree, but the square of some one of the lines $l\alpha \pm m\beta \pm n\gamma$. And the form in the text may be considered to include the case where one of the rectanglesi is negative and the other two positive, if we suppose that l, m, or n may denote a negative as well as a positive quantity.

points of contact of opposite sides, and at the points where the joining lines meet the circle again tangents be drawn, their equations are

$$2l\alpha + 2m\beta - n\gamma = 0, \quad 2m\beta + 2n\gamma - l\alpha = 0, \quad 2n\gamma + 2l\alpha - m\beta = 0.$$

Hence we infer that the three points, where each of these tangents meets the opposite side, lie in one right line,

$$l\alpha + m\beta + n\gamma = 0,$$

for this line passes through the intersection of the first line with γ, of the second with α, and of the third with β.

130. The equation of the chord joining two points $\alpha'\beta'\gamma'$, $\alpha''\beta''\gamma''$, on the curve is

$$\alpha\sqrt{(l)}\,\{\sqrt{(\beta'\gamma'')} + \sqrt{(\beta''\gamma')}\} + \beta\sqrt{(m)}\,\{\sqrt{(\gamma'\alpha'')} + \sqrt{(\gamma''\alpha')}\}$$
$$+ \gamma\sqrt{(n)}\,\{\sqrt{(\alpha'\beta'')} + \sqrt{(\alpha''\beta')}\} = 0.\text{*}$$

For substitute α', β', γ' for α, β, γ, and it will be found that the quantity on the left-hand side may be written

$$\{\sqrt{(\alpha'\beta'\gamma'')} + \sqrt{(\beta'\gamma'\alpha'')} + \sqrt{(\gamma'\alpha'\beta'')}\}\,\{\sqrt{(l\alpha')} + \sqrt{(m\beta')} + \sqrt{(n\gamma')}\}$$
$$- \sqrt{(\alpha'\beta'\gamma')}\,\{\sqrt{(l\alpha'')} + \sqrt{(m\beta'')} + \sqrt{(n\gamma'')}\},$$

which vanishes, since the points are on the curve. The equation of the tangent is found by putting α'', β'', $\gamma'' = \alpha'$, β', γ' in the above. Dividing by $2\sqrt{(\alpha'\beta'\gamma')}$, it becomes

$$\alpha\sqrt{\left(\frac{l}{\alpha'}\right)} + \beta\sqrt{\left(\frac{m}{\beta'}\right)} + \gamma\sqrt{\left(\frac{n}{\gamma'}\right)} = 0.$$

Conversely, if $\lambda\alpha + \mu\beta + \nu\gamma$ is a tangent, the coordinates of the point of contact are given by the equations

$$\sqrt{\left(\frac{l}{\alpha'}\right)} = \lambda, \quad \sqrt{\left(\frac{m}{\beta'}\right)} = \mu, \quad \sqrt{\left(\frac{n}{\gamma'}\right)} = \nu.$$

Solving for $\alpha'\beta'\gamma'$, and substituting in the equation of the curve, we get

$$\frac{l}{\lambda} + \frac{m}{\mu} + \frac{n}{\nu} = 0,$$

which is the condition that $\lambda\alpha + \mu\beta + \nu\gamma$ may be a tangent; that is to say, is the tangential equation of the curve.

---

* This equation is Dr. Hart's.

The reciprocity of tangential and ordinary equations will be better seen if we solve the converse problem, viz. to find the equation of the curve, the tangents to which fulfil the condition

$$\frac{l}{\lambda}+\frac{m}{\mu}+\frac{n}{\nu}=0,$$

We follow the steps of Art. 127. Let $\lambda'\alpha+\mu'\beta+\nu'\gamma$, $\lambda''\alpha+\mu''\beta+\nu''\gamma$ be any two lines, such that $\lambda'\mu'\nu'$, $\lambda''\mu''\nu''$ satisfy the above condition, and which therefore are tangents to the curve whose equation we are seeking; then

$$\frac{l\lambda}{\lambda'\lambda''}+\frac{m\mu}{\mu'\mu''}+\frac{n\nu}{\nu'\nu''}=0,$$

is the tangential equation of their point of intersection. For (Art. 70) any equation of the form $A\lambda+B\mu+C\nu=0$ is the condition that the line $\lambda\alpha+\mu\beta+\nu\gamma$ should pass through a certain point, or, in other words, is the tangential equation of a point; and the equation we have written being satisfied by the tangential coordinates of the two lines is the equation of their point of intersection. Making $\lambda', \mu', \nu'=\lambda'', \mu'', \nu''$ we learn that if there be two consecutive tangents to the curve, the equation of their point of intersection, or, in other words, of their point of contact, is

$$\frac{l\lambda}{\lambda'^2}+\frac{m\mu}{\mu'^2}+\frac{n\nu}{\nu'^2}=0.$$

The coordinates then of the point of contact are

$$\alpha=\frac{l}{\lambda'^2},\quad \beta=\frac{m}{\mu'^2},\quad \gamma=\frac{n}{\nu'^2}.$$

Solving for λ', μ', ν' from these equations, and substituting in the relation, which by hypothesis $\lambda'\mu'\nu'$ satisfy, we get the required equation of the curve

$$\sqrt{(l\alpha)}+\sqrt{(m\beta)}+\sqrt{(n\gamma)}=0.$$

131. The conditions that the equation of Art. 129 should represent a circle are (Art. 128)

$$m^2\sin^2 C+n^2\sin^2 B+2mn\sin B\sin C=n^2\sin^2 A+l^2\sin^2 C$$
$$+2nl\sin C\sin A=l^2\sin^2 B+m^2\sin^2 A+2lm\sin A\sin B,$$

or $m\sin C+n\sin B=\pm(n\sin A+l\sin C)=\pm(l\sin B+m\sin A)$.

Four circles then may be described to touch the sides of the given triangle, since, by varying the sign, these equations may be written in four different ways. If we choose in both cases the + sign, the equations are

$$l \sin C - m \sin C + n(\sin A - \sin B) = 0;$$

$$l \sin B + m(\sin A - \sin C) - n \sin B = 0.$$

The solution of which gives (see Art. 124)

$$l = \sin A(\sin B + \sin C - \sin A), \quad m = \sin B(\sin C + \sin A - \sin B),$$
$$n = \sin C(\sin A + \sin B - \sin C).$$

But since in a plane triangle

$$\sin B + \sin C - \sin A = 4 \cos \tfrac{1}{2}A \sin \tfrac{1}{2}B \sin \tfrac{1}{2}C,$$

these values for l, m, n are respectively proportional to $\cos^2 \frac{1}{2}A$, $\cos^2 \frac{1}{2}B$, $\cos^2 \frac{1}{2}C$, and the equation of the corresponding circle, which is the *inscribed* circle, is

$$\cos \tfrac{1}{2}A \sqrt{(\alpha)} + \cos \tfrac{1}{2}B \sqrt{(\beta)} + \cos \tfrac{1}{2}C \sqrt{(\gamma)} = 0,*$$

or $$\alpha^2 \cos^4 \tfrac{1}{2}A + \beta^2 \cos^4 \tfrac{1}{2}B + \gamma^2 \cos^4 \tfrac{1}{2}C - 2\beta\gamma \cos^2 \tfrac{1}{2}B \cos^2 \tfrac{1}{2}C$$
$$- 2\gamma\alpha \cos^2 \tfrac{1}{2}C \cos^2 \tfrac{1}{2}A - 2\alpha\beta \cos^2 \tfrac{1}{2}A \cos^2 \tfrac{1}{2}B = 0.$$

We may verify that this equation represents a circle by writing it in the form

$$\left(\frac{\alpha \cos^4 \frac{1}{2}A}{\sin A} + \frac{\beta \cos^4 \frac{1}{2}B}{\sin B} + \frac{\gamma \cos^4 \frac{1}{2}C}{\sin C}\right)(\alpha \sin A + \beta \sin B + \gamma \sin C)$$
$$- \frac{4 \cos^2 \frac{1}{2}A \cos^2 \frac{1}{2}B \cos^2 \frac{1}{2}C}{\sin A \sin B \sin C}(\beta\gamma \sin A + \gamma\alpha \sin B + \alpha\beta \sin C) = 0.$$

---

* Dr. Hart derives this equation from that of the circumscribing circle as follows Let the equations of the sides of the triangle formed by joining the points of contact of the inscribed circle be $\alpha' = 0$, $\beta' = 0$, $\gamma' = 0$, and let its angles be A', B', C'; then (Art. 124) the equation of the circle is

$$\beta'\gamma' \sin A' + \gamma'\alpha' \sin B' + \alpha'\beta' \sin C' = 0.$$

But (Art. 123) for every point of the circle we have $\alpha'^2 = \beta\gamma$, $\beta'^2 = \gamma\alpha$, $\gamma'^2 = \alpha\beta$, and it is easy to see that $A' = 90 - \frac{1}{2}A$, &c. Substituting these values, the equation of the circle becomes, as before,

$$\cos \tfrac{1}{2}A \sqrt{(\alpha)} + \cos \tfrac{1}{2}B \sqrt{(\beta)} + \cos \tfrac{1}{2}C \sqrt{(\gamma)} = 0.$$

If the equation of the note, p. 119, be treated similarly, we find that every point of the circle, of which α, β, γ, δ are tangents, satisfies the equation,

$$\frac{\cos \frac{1}{2}(12)}{\sqrt{(\alpha\beta)}} + \frac{\cos \frac{1}{2}(23)}{\sqrt{(\beta\gamma)}} + \frac{\cos \frac{1}{2}(34)}{\sqrt{(\gamma\delta)}} + \frac{\cos \frac{1}{2}(41)}{\sqrt{(\delta\alpha)}} = 0,$$

where (12) denotes the angle between $\alpha\beta$, &c. Similarly for any number of tangents.

In the same way, the equation of one of the exscribed circles is found to be

$$\alpha^2 \cos^4 \tfrac{1}{2}A + \beta^2 \sin^4 \tfrac{1}{2}B + \gamma^2 \sin^4 \tfrac{1}{2}C - 2\beta\gamma \sin^2 \tfrac{1}{2}B \sin^2 \tfrac{1}{2}C$$
$$+ 2\gamma\alpha \sin^2 \tfrac{1}{2}C \cos^2 \tfrac{1}{2}A + 2\alpha\beta \sin^2 \tfrac{1}{2}B \cos^2 \tfrac{1}{2}A = 0,$$

or $$\cos \tfrac{1}{2}A \sqrt{(-\alpha)} + \sin \tfrac{1}{2}B \sqrt{(\beta)} + \sin \tfrac{1}{2}C \sqrt{(\gamma)} = 0.$$

The negative sign given to α is in accordance with the fact, that this circle and the inscribed circle lie on opposite sides of the line α.

Ex. Find the radical axis of the inscribed circle and the circle through the middle points of sides.

The equation formed by the method of Art. 128 is

$$2 \cos^2 \tfrac{1}{2}A \cos^2 \tfrac{1}{2}B \cos^2 \tfrac{1}{2}C \{\alpha \cos A + \beta \cos B + \gamma \cos C\}$$
$$= \sin A \sin B \sin C \left(\alpha \frac{\cos^4 \frac{1}{2}A}{\sin A} + \beta \frac{\cos^4 \frac{1}{2}B}{\sin B} + \gamma \frac{\cos^4 \frac{1}{2}C}{\sin C}\right).$$

Divide by $2 \cos \frac{1}{2}A \cos \frac{1}{2}B \cos \frac{1}{2}C$, and the coefficient of α in this equation is

$$\cos \tfrac{1}{2}A \{2 \cos^2 \tfrac{1}{2}A \sin \tfrac{1}{2}B \sin \tfrac{1}{2}C - \cos A \cos \tfrac{1}{2}B \cos \tfrac{1}{2}C\},$$

or $$\cos \tfrac{1}{2}A \sin \tfrac{1}{2}(A - B) \sin \tfrac{1}{2}(A - C).$$

The equation of the radical axis then may be written

$$\frac{\alpha \cos \frac{1}{2}A}{\sin \frac{1}{2}(B - C)} + \frac{\beta \cos \frac{1}{2}B}{\sin \frac{1}{2}(C - A)} + \frac{\gamma \cos \frac{1}{2}C}{\sin \frac{1}{2}(A - B)} = 0;$$

and it appears from the condition of Art. 130 that this line touches the inscribed circle, the coordinates of the point of contact being $\sin^2 \frac{1}{2}(B-C)$, $\sin^2 \frac{1}{2}(C-A)$, $\sin^2 \frac{1}{2}(A-B)$. These values shew (Art. 66) that the point of contact lies on the line joining the two centres whose coordinates are 1, 1, 1, and $\cos(B - C)$, $\cos(C - A)$, $\cos(A - B)$.

In the same way it can be proved that the circle through the middle points of sides touches all the circles which touch the sides. This theorem is due to Feuerbach.*

---

* Dr. Casey has given a proof of Feuerbach's theorem, which will equally prove Dr. Hart's extension of it, viz. that the circles which touch three given circles can be distributed into sets of four, all touched by the same circle. The signs in the following correspond to a triangle whose sides are in order of magnitude a, b, c. The exscribed circles are numbered 1, 2, 3, and the inscribed 4; the lengths of the direct and transverse common tangents to the first two circles are written (12), (12)′. Then because the side a is touched by the circle 1 on one side, and by the other three circles on the other, we have (see p. 115)

$$(13)'(24) = (12)'(34) + (14)'(23).$$

Similarly $$(12)'(34) + (24)'(13) = (23)'(14),$$
$$(23)'(14) = (13)'(24) + (34)'(12),$$

whence, adding, we have $$(24)'(13) = (14)'(23) + (34)'(12);$$

showing that the four circles are also touched by a circle, having the circle 4 on one side and the other three on the other.

132. If the equation of a circle in trilinear coordinates is equivalent to an equation in rectangular coordinates, in which the coefficient of $x^2 + y^2$ is m, then the result of substituting in the equation the coordinates of any point is m times the square of the tangent from that point. This constant m is easily determined in practice if there be any point, the square of the tangent from which is known by geometrical considerations; and then the length of the tangent from any other point may be inferred. Also, if we have determined this constant m for two circles, and if we subtract, one from the other, the equations divided respectively by m and m', the difference which must represent the radical axis will always be divisible by $\alpha \sin A + \beta \sin B + \gamma \sin C$.

Ex. 1. Find the value of the constant m for the circle through the middle points of the sides

$$\alpha^2 \sin A \cos A + \beta^2 \sin B \cos B + \gamma^2 \sin C \cos C - \beta\gamma \sin A - \gamma\alpha \sin B - \alpha\beta \sin C = 0.$$

Since the circle cuts any side γ at points whose distances from the vertex A are $\frac{1}{2}c$ and $b \cos A$, the square of the tangent from A is $\frac{1}{2}bc \cos A$. But since for A we have $\beta = 0$, $\gamma = 0$, the result of substituting in the equation the coordinates of A is $\alpha'^2 \sin A \cos A$ (where α' is the perpendicular from A on the opposite side), or is $bc \sin A \sin B \sin C \cos A$. It follows that the constant m is $2 \sin A \sin B \sin C$.

Ex. 2. Find the constant m for the circle $\beta\gamma \sin A + \gamma\alpha \sin B + \alpha\beta \sin C$. If from the preceding equation we subtract the linear terms

$$(\alpha \cos A + \beta \cos B + \gamma \cos C)\ (\alpha \sin A + \beta \sin B + \gamma \sin C),$$

the coefficient of $x^2 + y^2$ is unaltered. The constant therefore for $\beta\gamma \sin A$ &c., is $-\sin A \sin B \sin C$. It follows that for an equation written in the form at the end of Art. 128 the constant is $-k \sin A \sin B \sin C$.

Ex. 3. To find the distance between the centres of the inscribed and circumscribing circle. We find $D^2 - R^2$, the square of the tangent from the centre of the inscribed to the circumscribing circle, by substituting $\alpha = \beta = \gamma = r$, to be $-\dfrac{r^2 (\sin A + \sin B + \sin C)}{\sin A \sin B \sin C}$ or, by a well-known formula, $= -2Rr$. Hence $D^2 = R^2 - 2Rr$.

Ex. 4. Find the distance between the centres of the inscribed circle and that through the middle points of sides. If the radius of the latter be ρ, making use of the formula,

$$\sin A \cos A + \sin B \cos B + \sin C \cos C = 2 \sin A \sin B \sin C,$$

we have $$D^2 - \rho^2 = r^2 - rR.$$

Assuming then that we otherwise know $R = 2\rho$, we have $D = r - \rho$; or the circles touch.

Ex. 5. Find the constant m for the equation of the inscribed circle given above.

Ans. $4 \cos^2 \frac{1}{2}A \cos^2 \frac{1}{2}B \cos^2 \frac{1}{2}C$

Ex. 6. Find the tangential equation of a circle whose centre is $\alpha'\beta'\gamma'$ and radius r. This is investigated as in Art. 86, Ex. 4; attending to the formula of Art. 61; and is found to be

$$(\lambda\alpha' + \mu\beta' + \nu\gamma')^2 = r^2 (\lambda^2 + \mu^2 + \nu^2 - 2\mu\nu \cos A - 2\nu\lambda \cos B - 2\lambda\mu \cos C).$$

The corresponding equation in α, β, γ is deduced from this by the method afterwards explained, Art. 285, and is

$$r^2(\alpha\sin A+\beta\sin B+\gamma\sin C)^2=(\beta\gamma'-\beta'\gamma)^2+(\gamma\alpha'-\gamma'\alpha)^2+(\alpha\beta'-\alpha'\beta)^2$$
$$-2(\gamma\alpha'-\gamma'\alpha)(\alpha\beta'-\alpha'\beta)\cos A-2(\alpha\beta'-\alpha'\beta)(\beta\gamma'-\beta'\gamma)\cos B-2(\beta\gamma'-\beta'\gamma)(\gamma\alpha'-\gamma'\alpha)\cos C.$$

This equation also gives an expression for the distance between any two points.

Ex. 7. The feet of the perpendiculars on the sides of the triangle of reference from the points α', β', γ'; $\frac{1}{\alpha'}, \frac{1}{\beta'}, \frac{1}{\gamma'}$; (see Art. 55) lie on the same circle. By the help of Ex. 6, p. 60, its equation is found to be

$$(\beta\gamma\sin A+\gamma\alpha\sin B+\alpha\beta\sin C)(\alpha'\sin A+\beta'\sin B+\gamma'\sin C)(\beta'\gamma'\sin A+\gamma'\alpha'\sin B+\alpha'\beta'\sin C)$$
$$=\sin A\sin B\sin C\,(\alpha\sin A+\beta\sin B+\gamma\sin C)$$
$$\left\{\frac{\alpha\alpha'(\beta'+\gamma'\cos A)(\gamma'+\beta'\cos A)}{\sin A}+\frac{\beta\beta'(\gamma'+\alpha'\cos B)(\alpha'+\gamma'\cos B)}{\sin B}+\frac{\gamma\gamma'(\alpha'+\beta'\cos C)(\beta'+\alpha'\cos C)}{\sin C}\right\}.$$

Ex. 8. It will appear afterwards that the centre of a circle is the pole of the line at infinity $\alpha\sin A+\beta\sin B+\gamma\sin C$; and it is evident that if we substitute the coordinates of the centre in the equation of a circle, for which the coefficient of x^2+y^2 has been made unity, we get the negative square of the radius. By these principles we establish the following expressions of Mr. Cathcart. The coordinates of the centre of the circle (Art. 128)

$$(l\alpha+m\beta+n\gamma)(\alpha\sin A+\&c.)+k(\beta\gamma\sin A+\&c.),$$

are $\frac{R}{k}(k\cos A+l-m\cos C-n\cos B)$, $\frac{R}{k}(k\cos B-l\cos C+m-n\cos A)$,

$$\frac{R}{k}(k\cos C-l\cos B-m\cos A+n),$$

where R is the radius of the circumscribing circle. The radius ρ is given by the equation

$$k^2\rho^2=R^2\{k^2+2k(l\cos A+m\cos B+n\cos C)$$
$$+l^2+m^2+n^2-2mn\cos A-2nl\cos B-2lm\cos C\},$$

and the angle of intersection of two circles is given by

$$\frac{\rho\rho'\cos\theta}{R^2}=1+\frac{l\cos A+m\cos B+n\cos C}{k}+\frac{l'\cos A+m'\cos B+n'\cos C}{k'}$$
$$+\frac{ll'+mm'+nn'-(mn'+m'n)\cos A-(nl'+n'l)\cos B-(lm'+l'm)\cos C}{kk'}.$$

DETERMINANT NOTATION.

132(*a*). In the earlier editions of this book I did not venture to introduce the determinant notation, and in the preceding pages I have not supposed the reader to be acquainted with it. But the knowledge of determinants has become so much more common now than it was, that there seems no reason for excluding the notation, at least from the less elementary chapters of the book. Thus the equation of the line joining two points (Art. 29), the double area of a triangle (Art. 36) and the condition (Art. 38), that three lines should meet in a point, may be written respectively

$$\begin{vmatrix} x, & y, & 1 \\ x', & y', & 1 \\ x'', & y'', & 1 \end{vmatrix} = 0, \quad \begin{vmatrix} x_1, & y_1, & 1 \\ x_2, & y_2, & 1 \\ x_3, & y_3, & 1 \end{vmatrix}, \quad \begin{vmatrix} A, & B, & C \\ A', & B', & C' \\ A'', & B'', & C'' \end{vmatrix} = 0.$$

Ex. 1. Find the area of the triangle contained by the three lines $l\alpha + m\beta + n\gamma$, $l'\alpha$ + &c., &c., (J. J. Walker).

Ans. If a, b, c be the sides and Δ the area of the triangle of reference

$$\frac{\Delta abc \begin{vmatrix} l, & m, & n \\ l', & m', & n' \\ l'', & m'', & n'' \end{vmatrix}^2}{\begin{vmatrix} a, & b, & c \\ l, & m, & n \\ l', & m', & n' \end{vmatrix} \begin{vmatrix} a, & b, & c \\ l', & m', & n' \\ l'', & m'', & n'' \end{vmatrix} \begin{vmatrix} a, & b, & c \\ l'', & m'', & n'' \\ l, & m, & n \end{vmatrix}}.$$

Ex. 2. The equation of the perpendicular from $\alpha'\beta'\gamma'$ on $l\alpha + m\beta + n\gamma = 0$, may be written

$$\begin{vmatrix} \alpha, & \alpha', & l - m\cos C - n\cos B \\ \beta, & \beta', & m - n\cos A - l\cos C \\ \gamma, & \gamma', & n - l\cos B - m\cos A \end{vmatrix} = 0.$$

132 (*b*). The equations of the circle through three points (Art 94), and of the circle cutting three at right angles (Ex. 2, p. 102), may be written respectively

$$\begin{vmatrix} x^2 + y^2, & x, & y, & 1 \\ x'^2 + y'^2, & x', & y', & 1 \\ x''^2 + y''^2, & x'', & y'', & 1 \\ x'''^2 + y'''^2, & x''', & y''', & 1 \end{vmatrix} = 0, \quad \begin{vmatrix} x^2 + y^2, & -x, & -y, & 1 \\ c', & g', & f', & 1 \\ c'', & g'', & f'', & 1 \\ c''', & g''', & f''', & 1 \end{vmatrix} = 0.$$

The equation of the latter circle may also be formed by the help of the principle (Ex. 6, p. 102), as the locus of the point whose polars with respect to three given circles meet in a point, in the form

$$\begin{vmatrix} x + g', & y + f', & g'x + f'y + c' \\ x + g'', & y + f'', & g''x + f''y + c'' \\ x + g''', & y + f''', & g'''x + f'''y + c''' \end{vmatrix} = 0.$$

The corresponding equation for any three curves of the second degree will be discussed hereafter.

132 (*c*). If the radius of a circle vanishes, $(x-\alpha)^2 + (y-\beta)^2 = 0$ the polar of any point $x'y'$, $(x' - \alpha)(x - \alpha) + (y' - \beta)(y - \beta) = 0$

evidently passes through the point $\alpha\beta$. It is in fact the perpendicular through that point to the line joining $\alpha\beta$, $x'y'$, as is evident geometrically. Hence then if the circle

$$x^2 + y^2 + 2gx + 2fy + c = 0$$

reduce to a point, that point which, as being the centre, is given by the equations $x + g = 0$, $y + f = 0$, also satisfies the equation of the polar of the origin $gx + fy + c = 0$.

If given three circles S', S'', S''' we examine in what cases $lS' + mS'' + nS'''$ can represent a point, we see that the coordinates of such a point must satisfy the three equations

$$l(x + g') + m(x + g'') + n(x + g''') = 0,$$
$$l(y + f') + m(y + f'') + n(y + f''') = 0,$$
$$l(g'x + f'y + c') + m(g''x + f''y + c'') + n(g'''x + f'''y + c''') = 0,$$

from which if we eliminate l, m, n, we get the same determinant as in the last article; showing that the orthogonal circle is the locus of all the points that can be represented by $lS' + mS'' + nS'''$.

The expression (Ex. 8, p. 103) for the angle at which two circles intersect may be written $2rr' \cos\theta = 2gg' + 2ff' - c - c'$. If now we calculate by the formula of p. 76 the radius of the circle $lS' + mS'' + nS'''$, and reduce the result by the formula just given, we find

$$(l + m + n)^2 r^2 = l^2 r'^2 + m^2 r''^2 + n^2 r'''^2$$
$$+ 2mnr''r''' \cos\theta' + 2nlr'''r' \cos\theta'' + 2lmr'r'' \cos\theta''',$$

where θ', θ'', θ''' are the angles at which the circles respectively intersect. And since the coordinates of the centre of $lS' + mS'' + nS'''$ are $\dfrac{lg' + mg'' + ng'''}{l + m + n}$, $\dfrac{lf' + mf'' + nf'''}{l + m + n}$, we see that these coordinates will represent a point on the orthogonal circle if l, m, n are connected by the relation $l^2 r'^2 + m^2 r''^2 + \&c. = 0$. If the three given circles be mutually orthogonal this relation reduces itself to its three first terms.*

132 (*d*). The condition that four circles may have a common orthogonal circle is found by eliminating C, F, G from the four conditions

$$2Gg + 2Ff - C - c = 0, \&c.,$$

* Casey, *Phil. Trans.*, 1871, p. 586.

and is

$$\begin{vmatrix} c\ , g\ , f\ , 1 \\ c', g', f', 1 \\ c'', g'', f'', 1 \\ c''', g''', f''', 1 \end{vmatrix} = 0.$$

Since c denotes the square of the tangent from the origin to the first circle, and since the origin may be any point, this condition, geometrically interpreted, expresses (see Art. 94) that the tangents from any point to four circles having a common orthogonal circle are connected by the relation

$$OA^2.BCD + OC^2.ABD = OB^2.ACD + OD^2.ABC.^*$$

132 (*e*). If a circle

$$x^2 + y^2 + 2Gx + 2Fy + C = 0,$$

cut three others at the same angle θ, we have, besides the equation first given, three others of the form

$$c' + 2Rr' \cos\theta - 2Gg' - 2Ff' + C = 0;$$

from which, eliminating G, F, C, we have

$$\begin{vmatrix} x^2 + y^2 , & -x, & -y, & 1 \\ c' + 2Rr' \cos\theta, & g', & f', & 1 \\ c'' + 2Rr'' \cos\theta, & g'', & f'', & 1 \\ c''' + 2Rr''' \cos\theta, & g''', & f''', & 1 \end{vmatrix}^{\dagger} = 0,$$

Now if we write $2R\cos\theta = \lambda$, the determinant just written is resolvable into

$$\begin{vmatrix} x^2+y^2, & -x, & -y, & 1 \\ c', & g', & f', & 1 \\ c'', & g'', & f'', & 1 \\ c''', & g''', & f''', & 1 \end{vmatrix} + \lambda \begin{vmatrix} 0, & -x, & -y, & 1 \\ r', & g', & f', & 1 \\ r'', & g'', & f'', & 1 \\ r''', & g''', & f''', & 1 \end{vmatrix} = 0.$$

The first determinant equated to zero is, as has just been pointed out, the equation of the orthogonal circle, and the second when expanded will be found to be the equation of the axis of similitude (Art. 117). Thus we have the theorem (Note, p. 109) that all circles cutting three circles at the same angle have a

---

* This theorem is Mr. R. J. Harvey's (Casey, *Trans. Royal Irish Acad*, XXIV. 458).

† Since this only differs from the equation of the orthogonal circle by writing $c' + \lambda r'$ for c', &c. we obtain another form for this determinant by making the same change in the last determinant of Art. 132 (*b*). I owe this form to Mr. Cathcart.

common radical axis, viz., the axis of similitude. If in the second determinant we change the sign either of r', r'', or r''', we get the equations of the other three axes of similitude. Now it has been stated (Art. 118) that it is optional which of two supplemental angles we consider to be the angle at which two circles intersect; and if in any line of the first determinant of this article we substitute for θ its supplement, this is equivalent to changing the sign of the corresponding r. Hence it is evident that we may have four systems of circles cutting the given three at equal angles, each system having a different one of the axes of similitude for radical axis; calculating by the usual formula the radius of the circle whose equation has been written above, we get R in terms of λ, and then from the equation $2R\cos\theta = \lambda$ we get a quadratic to determine the value of λ corresponding to any value of θ.

Ex. 1. To find the condition for the co-existence of the equations

$$ax + by + c = a'x + b'y + c' = a''x + b''y + c'' = a'''x + b'''y + c'''.$$

Let the common value of these quantities be λ; then eliminating x, y, λ from the four equations of the form $ax + by + c = \lambda$, we have the result in the form of a determinant

$$\begin{vmatrix} 1, & 1, & 1, & 1 \\ a, & a', & a'', & a''' \\ b, & b', & b'', & b''' \\ c, & c', & c'', & c''' \end{vmatrix} = 0,$$

or $A + C = B + D$, where A, B, C, D are the four minors got by erasing in turn each column, and the top row in this determinant.

To find the condition that four lines should touch the same circle, is the same as to find the condition for the co-existence of the equations $\alpha = \beta = \gamma = \delta$. In this case the determinants A, B, C, D geometrically represent the product of each side of the quadrilateral formed by the four lines, by the sines of the two adjacent angles.

Ex. 2. The expression, p. 129, for the distance between two points may be written

$$r^2 (\alpha \sin A + \beta \sin B + \gamma \sin C)^2 = \begin{vmatrix} 0, & 0, & \alpha, & \beta, & \gamma \\ 0, & 0, & \alpha', & \beta', & \gamma' \\ \alpha, & \alpha', & 1, & -\cos C, & -\cos B \\ \beta, & \beta', & -\cos C, & 1, & -\cos A \\ \gamma, & \gamma', & -\cos B, & -\cos A, & 1 \end{vmatrix},$$

and this determinant may be resolved into the product

$$\begin{vmatrix} \alpha, & \alpha', & -1 \\ \beta & \beta', & e^{iC} \\ \gamma, & \gamma', & e^{-iB} \end{vmatrix} \cdot \begin{vmatrix} \alpha, & \alpha', & -1 \\ \beta, & \beta', & e^{-iC} \\ \gamma, & \gamma', & e^{iB} \end{vmatrix},$$

or analogous factors arising from $A + B + C = \pi$.

Ex. 3. To find the relation connecting the mutual distances of four points on a circle. The investigation is Prof. Cayley's (see *Lessons on Higher Algebra*, p. 23). Multiply together according to the ordinary rule the determinants

$$\begin{vmatrix} x_1^2+y_1^2, & -2x_1, & -2y_1, & 1 \\ x_2^2+y_2^2, & -2x_2, & -2y_2, & 1 \\ x_3^2+y_3^2, & -2x_3, & -2y_3, & 1 \\ x_4^2+y_4^2, & -2x_4, & -2y_4, & 1 \end{vmatrix} \times \begin{vmatrix} 1, & x_1, & y_1, & x_1^2+y_1^2 \\ 1, & x_2, & y_2, & x_2^2+y_2^2 \\ 1, & x_3, & y_3, & x_3^2+y_3^2 \\ 1, & x_4, & y_4, & x_4^2+y_4^2 \end{vmatrix},$$

which are only different ways of writing the condition of Art. 94; and we get the required relation

$$\begin{vmatrix} 0, & (12)^2, & (13)^2, & (14)^2 \\ (12)^2, & 0, & (23)^2, & (24)^2 \\ (13)^2, & (23)^2, & 0, & (34)^2 \\ (14)^2, & (24)^2, & (34)^2, & 0 \end{vmatrix} = 0,$$

where $(12)^2$ is the square of the distance between two points. This determinant expanded is equivalent to $(12)(34) \pm (13)(42) \pm (14)(23) = 0$.

Ex. 4. To find the relation connecting the mutual distances of any four points in a plane. This investigation is also Prof. Cayley's (*Lessons on Higher Algebra*, p. 24). Prefix a unit and cyphers to each of the determinants in the last example; thus

$$\begin{vmatrix} 1, & 0, & 0, & 0 \\ x_1^2+y_1^2, & -2x_1, & -2y_1, & 1 \\ & \&c. & & \end{vmatrix} \times \begin{vmatrix} 0, & 0, & 0, & 1 \\ 1, & x_1, & y_1, & x_1^2+y_1^2 \\ & & \&c. & \end{vmatrix}.$$

We have then five rows and four columns, the determinant formed from which, according to the rules of multiplication, must vanish identically. But this is

$$\begin{vmatrix} 0, & 1, & 1, & 1, & 1 \\ 1, & 0, & (12)^2, & (13)^2, & (14)^2 \\ 1, & (12)^2, & 0, & (23)^2, & (24)^2 \\ 1, & (13)^2, & (23)^2, & 0, & (34)^2 \\ 1, & (14)^2, & (24)^2, & (34)^2, & 0 \end{vmatrix} = 0;$$

which, expanded, is

$$\begin{aligned} &(12)^2(34)^2\{(12)^2+(34)^2-(13)^2-(14)^2-(23)^2-(24)^2\} \\ +&(13)^2(24)^2\{(13)^2+(24)^2-(12)^2-(14)^2-(23)^2-(34)^2\} \\ +&(14)^2(23)^2\{(14)^2+(23)^2-(12)^2-(13)^2-(24)^2-(34)^2\} \\ +&(23)^2(34)^2(42)^2+(31)^2(14)^2(43)^2+(12)^2(24)^2(41)^2+(23)^2(31)^2(12)^2=0. \end{aligned}$$

If we write in the above a, b, c for 23, 31, 12; and $R+r$, $R+r'$, $R+r''$ for 14, 24, 34, we get a quadratic in R, whose roots are the lengths of the radii of the circles touching either all externally or internally three circles, whose radii are r, r', r'', and whose centres form a triangle whose sides are a, b, c.

Ex. 5. A relation connecting the lengths of the common tangents of any five circles may be obtained precisely as in the last example. Write down the two matrices

$$\begin{vmatrix} 1, & 0, & 0, & 0, & 0 \\ x'^2+y'^2-r'^2, & -2x', & -2y', & 2r', & 1 \\ x''^2+y''^2-r''^2, & -2x'', & -2y'', & 2r'', & 1 \\ & \&c. & & & \end{vmatrix}, \quad \begin{vmatrix} 0, & 0, & 0, & 0, & 1 \\ 1, & x', & y', & r', & x'^2+y'^2-r'^2 \\ 1, & x'', & y'', & r'', & x''^2+y''^2-r''^2 \\ & & \&c. & & \end{vmatrix},$$

where there are six rows and five columns, and the determinant formed according to the rules of multiplication must vanish. But this is

$$\begin{vmatrix} 0, & 1, & 1, & 1, & 1, & 1 \\ 1, & 0, & (12)^2, & (13)^2, & (14)^2, & (15)^2 \\ 1, & (12)^2, & 0, & (23)^2, & (24)^2, & (25)^2 \\ 1, & (13)^2, & (23)^2, & 0, & (34)^2, & (35)^2 \\ 1, & (14)^2, & (24)^2, & (34)^2, & 0, & (45)^2 \\ 1, & (15)^2, & (25)^2, & (35)^2, & (45)^2, & 0 \end{vmatrix} = 0,$$

where (12), &c. denote the lengths of the common tangents to each pair of circles. If we suppose the circle 5 to touch all the others, then (15), (25), (35), (45), all vanish, and we get, as a particular case of the above, Dr. Casey's relation between the common tangents of four circles touched by a fifth, in the form

$$\begin{vmatrix} 0, & (12)^2, & (13)^2, & (14)^2 \\ (12)^2, & 0, & (23)^2, & (24)^2 \\ (13)^2, & (23)^2, & 0, & (34)^2 \\ (14)^2, & (24)^2, & (34)^2, & 0 \end{vmatrix} = 0.$$

Ex. 6. Relation between the angles at which four circles whose radii are r, r', r'', r''' intersect. If the circle r have its centre at the point 1 in Ex. 4, r' at 2, &c. we may put for $\overline{12}^2 = r^2 + r'^2 - 2rr' \cos\overline{12}$, &c. in the determinant of that example which becomes then

$$\begin{vmatrix} 0, & 1, & 1, & 1, & 1 \\ 1, & 0, & r'^2+r^2-2r'r\cos\overline{21}, & r''^2+r^2-2r''r\cos\overline{31}, & r'''^2+r^2-2r'''r\cos\overline{41} \\ 1, & r^2+r'^2-2rr'\cos\overline{12}, & 0, & r''^2+r'^2-2r''r'\cos\overline{32}, & r'''^2+r'^2-2r'''r'\cos\overline{42} \\ 1, & r^2+r''^2-2rr''\cos\overline{13}, & r'^2+r''^2-2r'r''\cos\overline{23}, & 1, & r'''^2+r''^2-2r'''r''\cos\overline{43} \\ 1, & r^2+r'''^2-2rr'''\cos\overline{14}, & r'^2+r'''^2-2r'r'''\cos\overline{24}, & r''^2+r'''^2-2r''r'''\cos\overline{34}, & 0 \end{vmatrix} = 0,$$

subtracting from each row and column the first multiplied by corresponding square of radius and writing ρ for $\frac{1}{r}$, ρ' for $\frac{1}{r'}$, &c. this reduces to

$$\begin{vmatrix} 0, & \rho, & \rho', & \rho'', & \rho''' \\ \rho, & 1, & \cos\overline{21}, & \cos\overline{31}, & \cos\overline{41} \\ \rho', & \cos\overline{12}, & 1, & \cos\overline{32}, & \cos\overline{42} \\ \rho'', & \cos\overline{13}, & \cos\overline{23}, & 1, & \cos\overline{43} \\ \rho''', & \cos\overline{14}, & \cos\overline{24}, & \cos\overline{34}, & 1 \end{vmatrix} = 0.$$

If in this we let $\cos\overline{21} = \cos\overline{31} = \cos\overline{41} = \cos\theta$, we have the quadratic in λ mentioned at the end of Art. 132 *e*.

CHAPTER X.

PROPERTIES COMMON TO ALL CURVES OF THE SECOND DEGREE, DEDUCED FROM THE GENERAL EQUATION.

133. THE most general form of the equation of the second degree is

$$ax^2 + 2hxy + by^2 + 2gx + 2fy + c = 0,$$

where a, b, c, f, g, h are all constants.

It is our object in this chapter to classify the different curves which can be represented by equations of the general form just written, and to obtain some of the properties which are common to them all.*

Five relations between the coefficients are sufficient to determine a curve of the second degree. For though the general equation contains *six* constants, the nature of the curve depends not on the *absolute magnitude*, but on the *mutual ratios* of these coefficients; since, if we multiply or divide the equation by any constant, it will still represent the same curve. We may, therefore, divide the equation by c, so as to make the absolute term $= 1$, and there will then remain but five constants to be determined.

Thus, for example, a conic section can be described *through five points*. Substituting in the equation (as in Art. 93) the coordinates of each point $(x'y')$ through which the curve must pass, we obtain five relations between the coefficients, which will enable us to determine the five quantities, $\frac{a}{c}$, &c.

134. We shall in this chapter often have occasion to use the method of transformation of coordinates; and it will be useful

* We shall prove hereafter, that the section made by any plane in a cone standing on a circular base is a curve of the second degree, and, conversely, that there is no curve of the second degree which may not be considered as a *conic section*. It was in this point of view that these curves were first examined by geometers. We mention the property here, because we shall often find it convenient to use the terms "conic section," or "conic," instead of the longer appellation, "curve of the second degree."

to find what the general equation becomes when transformed to parallel axes through a new origin $(x'y')$. We form the new equation by substituting $x + x'$ for x, and $y + y'$ for y (Art. 8), and we get

$$a(x+x')^2+2h(x+x')(y+y')+b(y+y')^2+2g(x+x')+2f(y+y')+c=0.$$

Arranging this equation according to the powers of the variables, we find that the coefficients of x^2, xy, and y^2, will be, as before, a, $2h$, b; that

the new g, $\quad g' = ax' + hy' + g$;

the new f, $\quad f' = hx' + by' + f$;

the new c, $\quad c' = ax'^2 + 2hx'y' + by'^2 + 2gx' + 2fy' + c$.

Hence, *if the equation of a curve of the second degree be transformed to parallel axes through a new origin, the coefficients of the highest powers of the variables will remain unchanged, while the new absolute term will be the result of substituting in the original equation the coordinates of the new origin.**

135. *Every right line meets a curve of the second degree in two real, coincident, or imaginary points.*

This is inferred, as in Art. 82, from the fact that we get a quadratic equation to determine the points where any line $y = mx + n$ meets the curve. Thus, substituting this value of y in the equation of the second degree, we get a quadratic to determine the x of the points of intersection. In particular (see Art. 84) the points where the curve meets the axes are determined by the quadratics

$$ax^2 + 2gx + c = 0, \quad by^2 + 2fy + c = 0.$$

An apparent exception, however, may arise which does not present itself in the case of the circle. The quadratic may reduce to a simple equation in consequence of the vanishing of the coefficient which multiplies the square of the variable. Thus

$$xy + 2y^2 + x + 5y + 3 = 0$$

is an equation of the second degree; but if we make $y = 0$, we get only a simple equation to determine the point of meeting of the axis of x with the locus represented. Suppose, however, that in any quadratic $Ax^2 + 2Bx + C = 0$, the coefficient C

* This is equally true for equations of any degree, as can be proved in like manner.

vanishes, we do not say that the quadratic reduces to a simple equation; but we regard it still as a quadratic, one of whose roots is $x=0$, and the other $x=-\frac{2B}{A}$. Now this quadratic may be also written

$$C\left(\frac{1}{x}\right)^2+2B\left(\frac{1}{x}\right)+A=0;$$

and we see by parity of reasoning that, if A vanishes, we ought to regard this still as a quadratic equation, one of whose roots is $\frac{1}{x}=0$, or $x=\infty$; and the other $\frac{1}{x}=-\frac{2B}{C}$, or $x=-\frac{C}{2B}$. The same thing follows from the general solution of the quadratic, which may be written in either of the forms

$$x=\frac{-B\pm\sqrt{(B^2-AC)}}{A}=\frac{C}{-B\mp\sqrt{(B^2-AC)}};$$

the latter being the form got by solving the equation for the reciprocal of x, and the equivalence of the two forms is easily verified by multiplying across. Now the smaller A is, the more nearly does the radical become $=\pm B$; and therefore the last form of the solution shows that the smaller A is, the larger is one of the roots of the equation; and that when A vanishes we are to regard one of the roots as infinite. When, therefore, we apparently get a simple equation to determine the points in which any line meets the curve, we are to regard it as the limiting case of a quadratic of the form $0.x^2+2Bx+C=0$, one of whose roots is infinite; and we are to regard this as indicating that one of the points where the line meets the curve is infinitely distant. Thus the equation, selected as an example, which may be written $(y+1)(x+2y+3)=0$, represents two right lines, one of which meets the axis of x in a finite point, and the other being parallel to it meets it in an infinitely distant point.

In like manner, if in the equation $Ax^2+2Bx+C=0$, both B and C vanish, we say that it is a quadratic equation, both of whose roots are $x=0$; so if both B and A vanish we are to say that it is a quadratic equation, both of whose roots are $x=\infty$. With the explanation here given, and taking account of infinitely distant as well as of imaginary points, we can assert that *every* right line meets a curve of the second degree in two points.

136. The equation of the second degree transformed to polar coordinates* is

$$(a \cos^2\theta + 2h \cos\theta \sin\theta + b \sin^2\theta)\,\rho^2 + 2\,(g \cos\theta + f \sin\theta)\,\rho + c = 0;$$

and the roots of this quadratic are the two values of the length of the radius vector corresponding to any assigned value of θ. Now we have seen in the last article that one of these values will be infinite, (that is to say, the radius vector will meet the curve in an infinitely distant point,) when the coefficient of ρ^2 vanishes. But this condition will be satisfied for two values of θ, namely those given by the quadratic

$$a + 2h \tan\theta + b \tan^2\theta = 0.$$

Hence, *there can be drawn through the origin two real, coincident, or imaginary lines, which will meet the curve at an infinite distance;* each of which lines also meets the curve in one finite point whose distance is given by the equation

$$2\,(g \cos\theta + f \sin\theta)\,\rho + c = 0.$$

If we multiply by ρ^2 the equation

$$a \cos^2\theta + 2h \cos\theta \sin\theta + b \sin^2\theta = 0,$$

and substitute for $\rho \cos\theta$, $\rho \sin\theta$ their values x and y, we obtain for the equation of the two lines

$$ax^2 + 2hxy + by^2 = 0.$$

There are two directions in which lines can be drawn through *any* point to meet the curve at infinity, for by transformation of coordinates we can make that point the origin, and the preceding proof applies. Now it was proved (Art. 134) that a, h, b are unchanged by such a transformation; the directions are, therefore, always determined by the *same* quadratic

$$a \cos^2\theta + 2h \cos\theta \sin\theta + b \sin^2\theta = 0.$$

Hence, *if through any point two real lines can be drawn to meet the curve at infinity, parallel lines through any other point will meet the curve at infinity.*†

---

* The following processes apply equally if the original equation had been in oblique coordinates. We then substitute $m\rho$ for x, and $n\rho$ for y, where m is $\frac{\sin\theta}{\sin\omega}$ and n is $\frac{\sin(\omega - \theta)}{\sin\omega}$ (Art. 12); and proceed as in the text.

† This indeed is evident geometrically, since parallel lines may be considered as passing through the same point at infinity.

137. One of the most important questions we can ask, concerning the *form* of the curve represented by any equation, is, whether it be limited in every direction, or whether it extend in any direction to infinity. We have seen, in the case of the circle, that an equation of the second degree may represent a limited curve, while the case where it represents right lines shows us that it may also represent loci extending to infinity. It is necessary, therefore, to find a test whereby we may distinguish which class of locus is represented by any particular equation of the second degree.

With such a test we are furnished by the last article. For if the curve be limited in every direction, *no* radius vector drawn from the origin to the curve can have an infinite value; but we found in the last article that when the radius vector becomes infinite, we have $a + 2h \tan\theta + b \tan^2\theta = 0$.

(1) If now we suppose $h^2 - ab$ to be negative, the roots of this equation will be imaginary, and *no* real value of θ can be found which will render

$$a \cos^2\theta + 2h \cos\theta \sin\theta + b \sin^2\theta = 0.$$

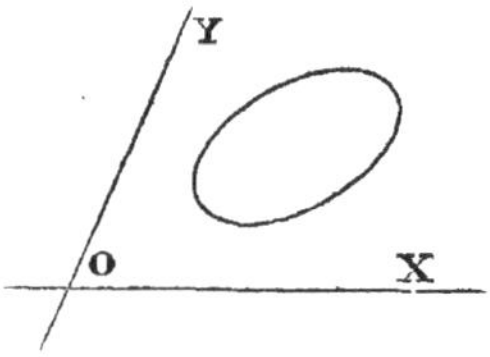

In this case, therefore, no real line can be drawn to meet the curve at infinity, and *the curve will be limited in every direction.* We shall show, in the next chapter, that its form is that represented in the figure. A curve of this class is called an *Ellipse.*

(2) If $h^2 - ab$ be *positive*, the roots of the equation

$$a + 2h \tan\theta + b \tan^2\theta = 0$$

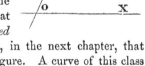

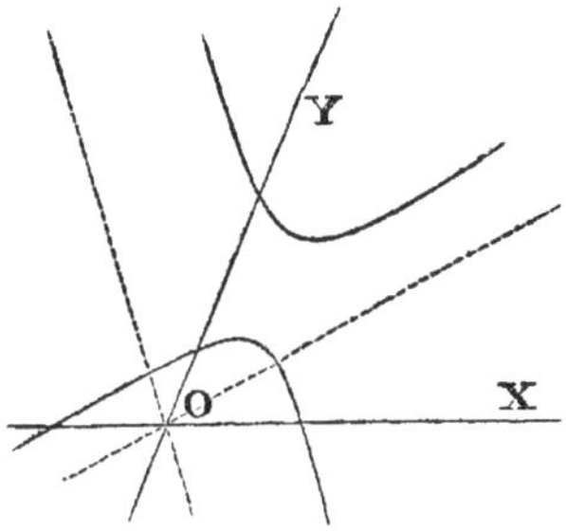

will be real; consequently there are two real values of θ which will render infinite the radius vector to the curve. Hence, two real lines ($ax^2 + 2hxy + by^2 = 0$) can, in this case, be drawn through the origin to meet the curve at infinity. A curve of this class is called a *Hyperbola*, and we shall show in the next chapter that its form is that represented in the figure.

(3) If $h^2 - ab = 0$, the roots of the equation

$$a + 2h \tan\theta + b \tan^2\theta = 0$$

will then be equal, and, therefore, the two directions in which a right line can be drawn to meet the curve at infinity will in this case coincide. A curve of this class is called a *Parabola*, and we shall (Chap. XII.) show that its form is that here represented. The condition here found may be otherwise expressed, by saying that the curve is a parabola when the first three terms of the equation form a perfect square.

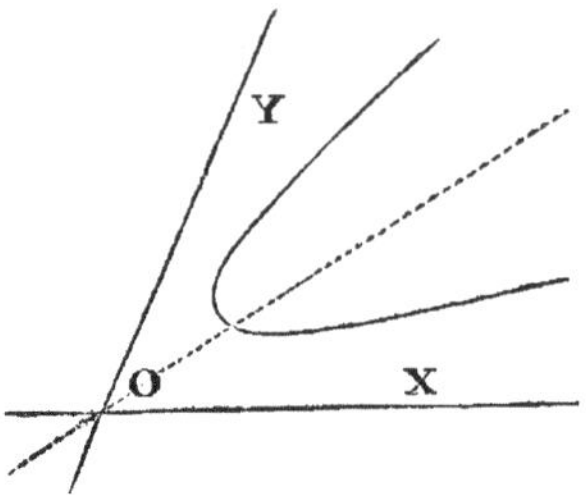

138. We find it convenient to postpone the deducing the figure of the curve from the equation until we have first, by transformation of coordinates, reduced the equation to its simplest form. The general truth, however, of the statements in the preceding article may be seen if we attempt to construct the figure represented by the equation in the manner explained (Art. 16). Solving for y in terms of x, we find (Art. 76)

$$by = -(hx + f) \pm \sqrt{\{(h^2 - ab)x^2 + 2(hf - bg)x + (f^2 - bc)\}}.$$

Now, since by the theory of quadratic equations, any quantity of the form $x^2 + px + q$ is equivalent to the product of two real or imaginary factors $(x - \alpha)(x - \beta)$, the quantity under the radical may be written $(h^2 - ab)(x - \alpha)(x - \beta)$. If then $h^2 - ab$ be negative, the quantity under the radical is negative (and therefore y imaginary), when the factors $x - \alpha$, $x - \beta$ are either both positive or both negative. Real values for y are only found when x is intermediate between α and β, and therefore the curve only exists in the space included between the lines $x = \alpha$, $x = \beta$ (see Ex. 3, p. 13). The case is the reverse when $h^2 - ab$ is positive. Then we get real values of y for any values of x, which make the factors $x - \alpha$, $x - \beta$ either both positive or both negative; but not so if one is positive and the other negative. The curve then consists of two branches stretching to infinity both in the positive and in the negative direction, but separated by an interval included by the lines $x = \alpha$, $x = \beta$, in which no part of the curve is found. If $h^2 - ab$ vanishes, the quantity under the radical is of the form either $x-\alpha$ or $\alpha-x$. In the one case we have real values of y, provided only that x is greater than α; in the other, provided only that it is less. The curve, therefore, consists of a single branch stretching to infinity either on the right or the left-hand side of the line $x=\alpha$.

If the roots α and β be imaginary, the quantity under the radical may be thrown into the form $(h^2-ab)\{(x-\gamma)^2+\delta^2\}$. If then h^2-ab is positive, the quantity under the radical is always positive, and lines parallel to the axis of y always meet the curve. Thus in the figure of the hyperbola, Art. 137, lines parallel to the axis of y always meet the curve, although lines parallel to the axis of x may not. On the other hand, if h^2-ab is negative, the quantity under the radical is always negative, and no real figure is represented by the equation.

Ex. 1. Construct, as in Art. 16, the figures of the following curves, and determine their species:

$3x^2+4xy+y^2-3x-2y+21=0.$ *Ans.* Hyperbola.

$5x^2+4xy+y^2-5x-2y-19=0.$ *Ans.* Ellipse.

$4x^2+4xy+y^2-5x-2y-10=0.$ *Ans.* Parabola.

Ex. 2. The circle is a particular case of the ellipse. For in the most general form of the equation of the circle, $a=b$, $h=a\cos\omega$ (Art. 81); and therefore h^2-ab is negative, being $=-a^2\sin^2\omega$.

Ex. 3. What is the species of the curve when $h=0$? *Ans.* An ellipse when a and b have the same sign, and a hyperbola when they have opposite signs.

Ex. 4. If either a or $b=0$, what is the species? *Ans.* A parabola if also $h=0$; otherwise a hyperbola. When $a=0$ the axis of x meets the curve at infinity; and when $b=0$, the axis of y.

Ex. 5. What is represented by

$$\frac{x^2}{a^2}-\frac{2xy}{ab}+\frac{y^2}{b^2}-\frac{2x}{a}-\frac{2y}{b}+1=0?$$

Ans. A parabola touching the axes at the points $x=a$, $y=b$.

139. If in a quadratic $Ax^2+2Bx+C=0$, the coefficient B vanishes, the roots are equal with opposite signs. This then will be the case with the equation

$$(a\cos^2\theta+2h\cos\theta\sin\theta+b\sin^2\theta)\,\rho^2+2\,(g\cos\theta+f\sin\theta)\,\rho+c=0,$$

if the radius vector be drawn in the direction determined by the equation $g\cos\theta+f\sin\theta=0$.

The points answering to the equal and opposite values of ρ are equidistant from the origin, and on opposite sides of it;

therefore the chord represented by the equation $gx+fy=0$ is bisected at the origin.

Hence, *through any given point can in general be drawn one chord which will be bisected at that point.*

140. There is one case, however, where more chords than one can be drawn, so as to be bisected, through a given point.

If, in the general equation, we had $g=0$, $f=0$, then the quantity $g\cos\theta+f\sin\theta$ would be $=0$, whatever were the value of θ; and we see, as in the last article, that in this case *every* chord drawn through the origin would be bisected. The origin would then be called the *centre* of the curve. Now, we can in general, by transforming the equation to a new origin, cause the coefficients g and f to vanish. Thus equating to nothing the values given (Art. 134) for the new g and f, we find that the coördinates of the new origin must fulfil the conditions

$$ax'+hy'+g=0,\ hx'+by'+f=0.$$

These two equations are sufficient to determine x' and y', and being *linear*, can be satisfied by only *one* value of x and y; hence, *conic sections have in general* one *and* only *one centre.* Its coordinates are found, by solving the above equations, to be

$$x'=\frac{hf-bg}{ab-h^2},\ y'=\frac{hg-af}{ab-h^2}.$$

In the ellipse and hyperbola $ab-h^2$ is always finite (Art. 137); but in the parabola $ab-h^2=0$, and the coordinates of the centre become infinite. The ellipse and hyperbola are hence often classed together as *central* curves, while the parabola is called a *non-central* curve. Strictly speaking, however, *every* curve of the second degree has a centre, although in the case of the parabola this centre is situated at an infinite distance.

141. *To find the locus of the middle points of chords, parallel to a given line, of a curve of the second degree.*

We saw (Art. 139) that a chord through the origin is bisected if $g\cos\theta+f\sin\theta=0$. Now, transforming the origin to any point, it appears, in like manner, that a parallel chord will be bisected at the new origin if the new g multiplied by $\cos\theta +$ the new f multiplied by $\sin\theta = 0$, or (Art. 134)

$$\cos\theta\,(ax' + hy' + g) + \sin\theta\,(hx' + by' + f) = 0.$$

This, therefore, is a relation which must be satisfied by the coordinates of the new origin, if it be the middle point of a chord making with the axis of x the angle θ. Hence the middle point of any parallel chord must lie on the right line

$$\cos\theta\,(ax + hy + g) + \sin\theta\,(hx + by + f) = 0,$$

which is, therefore, the required locus.

Every right line bisecting a system of parallel chords is called a *diameter*, and the lines which it bisects are called its *ordinates*.

The form of the equation shows (Art. 40) that every diameter must pass through the intersection of the two lines

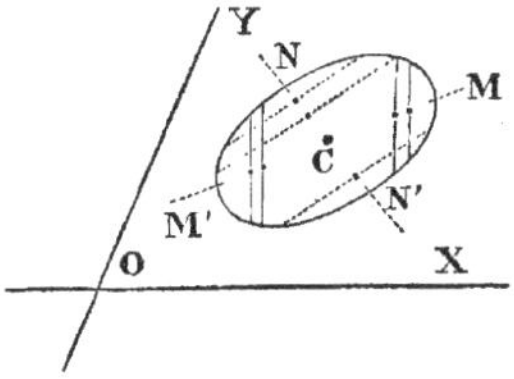

$ax + hy + g = 0$, and $hx + by + f = 0$;

but, these being the equations by which we determined the coordinates of the centre (Art. 140), we infer that *every diameter passes through the centre of the curve.*

It appears by making θ alternately $= 0$, and $= 90°$ in the above equation, that

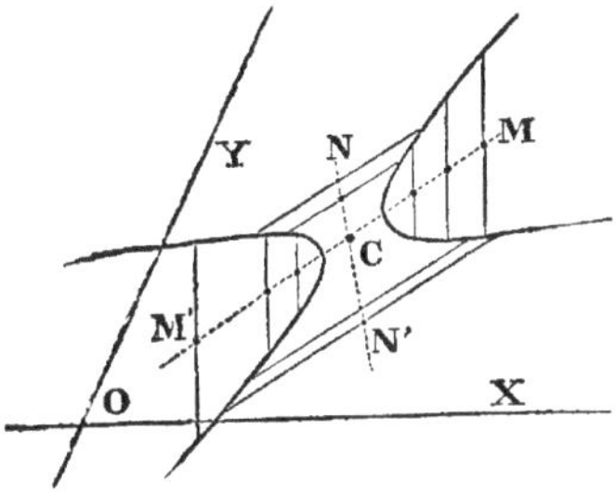

$$ax + hy + g = 0$$

is the equation of the diameter bisecting chords parallel to the axis of x, and that

$$hx + by + f = 0$$

is the equation of the diameter bisecting chords parallel to the axis of y.*

In the parabola $h^2 = ab$, or $\dfrac{a}{h} = \dfrac{h}{b}$, and hence the line

* The equation (Art. 138) which is of the form $by = -(hx + f) \pm R$ is most easily constructed by first laying down the line $hx + by + f$, and then taking on each ordinate MP of that line portions PQ, PQ', above and below P and equal to R. Thus also it appears that each ordinate is bisected by $hx + by + f$.

$ax + hy + g$ is parallel to the line $hx + by + f$; consequently, *all diameters of a parabola are parallel to each other.* This, indeed, is evident, since we have proved that all diameters of any conic section must pass through the centre, which, in the case of the parabola, is at an infinite distance, and since parallel right lines may be considered as meeting in a point at infinity.*

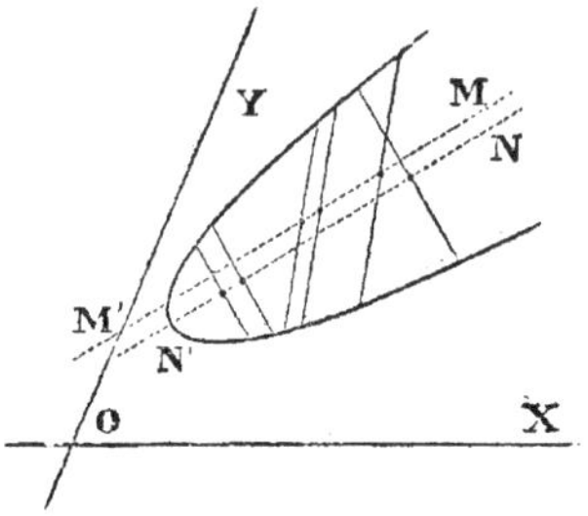

The familiar example of the circle will sufficiently illustrate to the beginner the nature of the diameters of curves of the second degree. He must observe, however, that diameters do not in general, as in the case of the circle, cut their ordinates at right angles. In the parabola, for instance, the direction of the diameter being invariable, while that of the ordinates may be any whatever, the angle between them *may take any possible value.*

142. *The direction of the diameters of a parabola is the same as that of the line through the origin which meets the curve at an infinite distance.*

For the lines through the origin which meet the curve at infinity are (Art. 136)

$$ax^2 + 2hxy + by^2 = 0,$$

or, writing for h its value $\sqrt{(ab)}$,

$$\{\sqrt{(a)}\,x + \sqrt{(b)}\,y\}^2 = 0.$$

But the diameters are parallel to $ax + hy = 0$ (by the last article), which, if we write for h the same value $\sqrt{(ab)}$, will also reduce to

$$\sqrt{(a)}\,x + \sqrt{(b)}\,y = 0.$$

Hence, every diameter of the parabola meets the curve once at infinity, and, therefore, can only meet it in one finite point.

* Hence, a portion of any conic section being drawn on paper, we can find its centre and determine its species. For if we draw any two parallel chords, and join their middle points, we have one diameter. In like manner we can find another diameter. Then, if these two diameters be parallel, the curve is a parabola; but if not, the point of intersection is the centre. It will be on the concave side when the curve is an ellipse, and on the convex when it is a hyperbola.

143. *If two diameters of a conic section be such that one of them bisects all chords parallel to the other, then, conversely, the second will bisect all chords parallel to the first.*

The equation of the diameter which bisects chords making an angle θ with the axis of x is (Art. 141)

$$(ax + hy + g) + (hx + by + f) \tan\theta = 0.$$

But (Art. 21) the angle which this line makes with the axis is θ' where

$$\tan\theta' = -\frac{a + h\tan\theta}{h + b\tan\theta},$$

whence $$b\tan\theta\tan\theta' + h(\tan\theta + \tan\theta') + a = 0.$$

And the symmetry of the equation shows that the chords making an angle θ' are also bisected by a diameter making an angle θ.

Diameters so related, that each bisects every chord parallel to the other, are called *conjugate diameters*.*

If in the general equation $h = 0$, the axes will be parallel to a pair of conjugate diameters. For the diameter bisecting chords parallel to the axis of x will, in this case, become $ax + g = 0$, and will, therefore, be parallel to the axis of y. In like manner, the diameter bisecting chords parallel to the axis of y will, in this case, be $by + f = 0$, and will, therefore, be parallel to the axis of x.

144. If in the general equation $c = 0$, the origin is on the curve (Art. 81); and accordingly one of the roots of the quadratic

$$(a\cos^2\theta + 2h\cos\theta\sin\theta + b\sin^2\theta)\rho^2 + 2(g\cos\theta + f\sin\theta)\rho = 0$$

is always $\rho = 0$. The second root will be also $\rho = 0$, or the radius vector will meet the curve at the origin in two coincident points, if $g\cos\theta + f\sin\theta = 0$. Multiplying this equation by ρ, we have the equation of the tangent at the origin, viz. $gx + fy = 0$.† The equation of the tangent at any other point on the curve may be found by first transforming the equation to that point as origin, and when the equation of the tangent has been then found, transforming it back to the original axes.

* It is evident that none but central curves can have conjugate diameters, since in the parabola the direction of all diameters is the same.

† The same argument proves that in an equation of any degree when the absolute term vanishes the origin is on the curve, and that then the terms of the first degree represent the tangent at the origin.

Ex. The point (1, 1) is on the curve

$$3x^2 - 4xy + 2y^2 + 7x - 5y - 3 = 0;$$

transform the equation to parallel axes through that point and find the tangent at it.

Ans. $9x - 5y = 0$ referred to the new axes, or $9(x-1) = 5(y-1)$ referred to the old.

If this method is applied to the general equation, we get for the tangent at any point $x'y'$ the same equation as that found by a different method (Art. 86), viz.

$$ax'x + h(x'y + y'x) + by'y + g(x + x') + f(y + y') + c = 0.$$

145. It was proved (Art. 89) that if it be required to draw a tangent to the curve from any point $x'y'$, not supposed to be on the curve, the points of contact are the intersections with the curve of a right line whose equation is identical in form with that last written, and which is called the polar of $x'y'$. Consequently, since every right line meets the curve in two points, *through any point $x'y'$ there can be drawn two real, coincident, or imaginary tangents to the curve.**

It was also proved (Art. 89) that the polar of the origin is $gx + fy + c = 0$. Now this line is evidently parallel to the chord $gx + fy$, which (Art. 139) is drawn through the origin so as to be bisected. But this last is plainly an ordinate of the diameter passing through the origin. Hence, *the polar of any point is parallel to the ordinates of the diameter passing through that point.* This includes as a particular case: *The tangent at the extremity of any diameter is parallel to the ordinates of that diameter.* Or again, in the case of central curves, since the ordinates of any diameter are parallel to the conjugate diameter, we infer that *the polar of any point on a diameter of a central curve is parallel to the conjugate diameter.*

146. The principal properties of poles and polars have been proved by anticipation in former chapters. Thus it was proved (Art. 98) that if a point A lie on the polar of B, then B lies on the polar of A. This may be otherwise stated: *If a point move along a fixed line* [the polar of B] *its polar passes through a fixed point* [B]; or, conversely, *If a line* [the polar of A] *pass*

* A curve is said to be of the nth *class* when through any point n tangents can be drawn to the curve. A conic is, therefore, a curve of the second degree and of the second class; but in higher curves the degree and class of a curve are commonly not the same.

through a fixed point, then the locus of its pole [A] *is a fixed right line.* Or, again, *The intersection of any two lines is the pole of the line joining their poles;* and, conversely, *The line joining any two points is the polar of the intersections of the polars of these points.* For if we take any two points on the polar of A, the polars of these points intersect in A.

It was proved (Art. 100) that *if two lines be drawn through any point, and the points joined where they meet the curve, the joining lines will intersect on the polar of that point.* Let the two lines coincide, and we derive, as a particular case of this, *If through a point O any line OR be drawn, the tangents at R' and R'' meet on the polar of O;* a property which might also be inferred from the last paragraph. For since $R'R''$, the polar of P, passes through O, P must lie on the polar of O.

And it was also proved (Ex. 3, p. 96), that if on any radius vector through the origin, OR be taken a harmonic mean between OR' and OR'', the locus of R is the polar of the origin; and therefore that, *any line drawn through a point is cut harmonically by the point, the curve, and the polar of the point;* as was also proved otherwise (Art. 91).

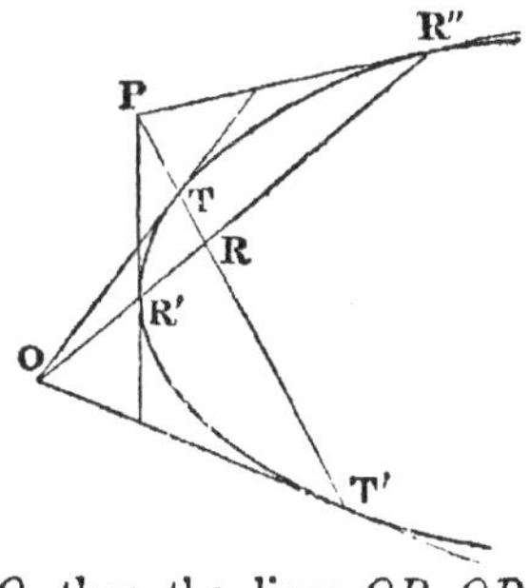

Lastly, we infer that if any line OR be drawn through a point O, and P the pole of that line be joined to O, then the lines OP, OR will form a harmonic pencil with the tangents from O. For since OR is the polar of P, $PTRT'$ is cut harmonically, and therefore OP, OT, OR, OT' form a harmonic pencil.

Ex. 1. If a quadrilateral $ABCD$ be inscribed in a conic section, any of the points E, F, O is the pole of the line joining the other two.

Since EC, ED are two lines drawn through the point E, and CD, AB, one pair of lines joining the points where they meet the conic, these lines must intersect on the polar of E; so must also AD and CB; therefore the line OF is the polar of E. In like manner it can be proved that EF is the polar of O and EO the polar of F.

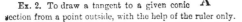

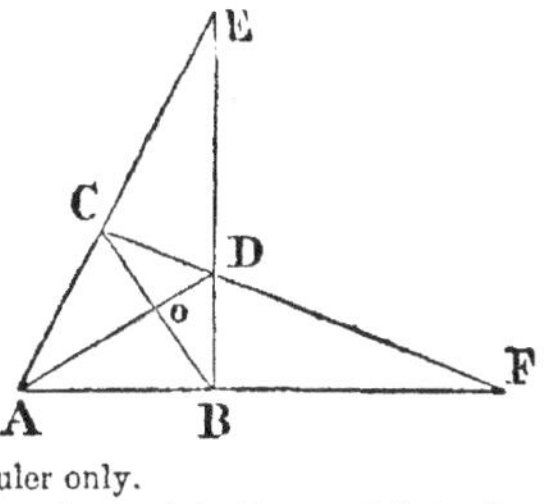

Ex. 2. To draw a tangent to a given conic section from a point outside, with the help of the ruler only.

Draw any two lines through the given point E, and complete the quadrilateral as in the figure, then the line OF will meet the conic in two points, which, being joined to E, will give the two tangents required.

Ex. 3. If a quadrilateral be circumscribed about a conic section, any diagonal is the polar of the intersection of the other two.

We shall prove this Example, as we might have proved Ex. 1, by means of the harmonic properties of a quadrilateral. It was proved (Ex. 1, p. 57) that EA, EO, EB, EF are a harmonic pencil. Hence, since EA, EB are, by hypothesis, two tangents to a conic section, and EF a line through their point of intersection, by Art. 146, EO must pass through the pole of EF; for the same reason, FO must pass through the pole of EF; this pole must, therefore, be O.

147. We have proved (Art. 92) that the equation of the pair of tangents to the curve from any point $x'y'$ is

$$(ax'^2+2hx'y'+by'^2+2gx'+2fy'+c)(ax^2+2hxy+by^2+2gx+2fy+c)$$
$$=\{ax'x+h(x'y+y'x)+by'y+g(x'+x)+f(y'+y)+c\}^2.$$

The equation of the pair of tangents through the origin may be derived from this by making $x'=y'=0$; or it may be got directly by the same process as that used Ex. 4, p. 78. If a radius vector through the origin touch the curve, the two values of ρ must be equal, which are given by the equation

$$(a\cos^2\theta+2h\cos\theta\sin\theta+b\sin^2\theta)\rho^2+2(g\cos\theta+f\sin\theta)\rho+c=0.$$

Now this equation will have equal roots if θ satisfy the equation

$$(a\cos^2\theta+2h\cos\theta\sin\theta+b\sin^2\theta)c=(g\cos\theta+f\sin\theta)^2.$$

Multiplying by ρ^2 we get the equation of the two tangents, viz.

$$(ac-g^2)x^2+2(ch-gf)xy+(bc-f^2)y^2=0.$$

This equation again will have equal roots; that is to say, the two tangents will coincide if

$$(ac-g^2)(bc-f^2)=(ch-fg)^2,$$

or
$$c(abc+2fgh-af^2-bg^2-ch^2)=0.$$

This will be satisfied if $c=0$, that is if the origin be on the curve. Hence, *any point on the curve may be considered as the intersection of two coincident tangents*, just as any tangent may be considered as the line joining two consecutive points.

The equation will have also equal roots if

$$abc+2fgh-af^2-bg^2-ch^2=0.$$

Now we obtained this equation (p. 72) as the condition that the equation of the second degree should represent two right lines. To explain why we should here meet with this equation again, it must be remarked that by a tangent we mean in general a line which meets the curve in two coincident points; if then the curve reduce to two right lines, the only line which can meet the locus in two coincident points is the line drawn to the point of intersection of these right lines, and since *two* tangents can always be drawn to a curve of the second degree, both tangents must in this case coincide with the line to the point of intersection.

148. *If through any point O two chords be drawn, meeting the curve in the points R', R'', S', S'', then the ratio of the rectangles* $\dfrac{OR'.OR''}{OS'.OS''}$ *will be constant, whatever be the position of the point O, provided that the* directions *of the lines OR, OS be constant.*

For, from the equation given to determine ρ in Art. 136, it appears that

$$OR'.OR'' = \frac{c}{a\cos^2\theta + 2h\cos\theta\sin\theta + b\sin^2\theta}.$$

In like manner

$$OS'.OS'' = \frac{c}{a\cos^2\theta' + 2h\cos\theta'\sin\theta' + b\sin^2\theta'};$$

hence $$\frac{OR'.OR''}{OS'.OS''} = \frac{a\cos^2\theta' + 2h\cos\theta'\sin\theta' + b\sin^2\theta'}{a\cos^2\theta + 2h\cos\theta\sin\theta + b\sin^2\theta}.$$

But this is a constant ratio; for a, h, b remain unaltered when the equation is transformed to parallel axes through any new origin (Art. 134), and θ, θ' are evidently constant while the direction of the radii vectores is constant.

The theorem of this Article may be otherwise stated thus: *If through two fixed points O and O' any two parallel lines OR and $O'\rho$ be drawn, then the ratio of the rectangles* $\dfrac{OR'.OR''}{O'\rho'.O'\rho''}$ *will be constant, whatever be the direction of these lines.*

For these rectangles are

$$\frac{c}{a\cos^2\theta + 2h\cos\theta\sin\theta + b\sin^2\theta},\ \frac{c'}{a\cos^2\theta + 2h\cos\theta\sin\theta + b\sin^2\theta}$$

(c' being the new absolute term when the equation is transferred to O' as origin); the ratio of these rectangles $= \dfrac{c}{c'}$, and is, therefore, independent of θ.

This theorem is the generalization of Euclid III. 35, 36.

149. The theorem of the last Article includes under it several particular cases, which it is useful to notice separately.

I. Let O' be the centre of the curve, then $O'\rho' = O'\rho''$ and the quantity $O'\rho'.O'\rho''$ becomes the square of the semi-diameter parallel to OR'. Hence, *The rectangles under the segments of two chords which intersect are to each other as the squares of the diameters parallel to those chords.*

II. Let the line OR be a tangent, then $OR' = OR''$, and the quantity $OR'.OR''$ becomes the square of the tangent; and, since two tangents can be drawn through the point O, we may extract the square root of the ratio found in the last paragraph, and infer that *Two tangents drawn through any point are to each other as the diameters to which they are parallel.*

III. Let the line OO' be a diameter, and OR, $O'\rho$ parallel to its ordinates, then $OR' = OR''$ and $O'\rho' = O'\rho''$. Let the diameter meet the curve in the points A, B, then $\dfrac{OR^2}{AO.OB} = \dfrac{O'\rho^2}{AO'.O'B}$. Hence, *The squares of the ordinates of any diameter are proportional to the rectangles under the segments which they make on the diameter.*

150. There is one case in which the theorem of Article 148 becomes no longer applicable, namely, when the line OS is parallel to one of the lines which meet the curve at infinity; the segment OS'' is then infinite, and OS only meets the curve in one finite point. We propose, in the present Article, to inquire whether, in this case, the ratio $\dfrac{OS'}{OR'.OR''}$ will be constant.

Let us, for simplicity, take the line OS for our axis of x, and OR for the axis of y. Since the axis of x is parallel to one of the lines which meet the curve at infinity, the coefficient a will $= 0$ (Art. 138, Ex. 4), and the equation of the curve will be of the form

$$2hxy + by^2 + 2gx + 2fy + c = 0.$$

Making $y = 0$, the intercept on the axis of x is found to be $OS' = -\dfrac{c}{2g}$; and, making $x = 0$, the rectangle under the intercepts on the axis of y is $= \dfrac{c}{b}$.

Hence $$\frac{OS'}{OR'.OR''} = -\frac{b}{2g}.$$

Now, if we transform the axes to parallel axes through any point $x'y'$ (Art. 134), b will remain unaltered, and the new $g = hy' + g$.

Hence the new ratio will be

$$-\frac{b}{2(hy'+g)}.$$

Now, if the curve be a parabola, $h = 0$, and this ratio is constant; hence, *If a line parallel to a given one meet any diameter* (Art. 142) *of a parabola, the rectangle under its segments is in a constant ratio to the intercept on the diameter.*

If the curve be a hyperbola, the ratio will only be constant while y' is constant; hence, *The intercepts made by two parallel chords of a hyperbola, on a given line meeting the curve at infinity, are proportional to the rectangles under the segments of the chords.*

*151. *To find the condition that the line $\lambda x + \mu y + \nu$ may touch the conic represented by the general equation.* Solving for y from $\lambda x + \mu y + \nu = 0$, and substituting in the equation of the conic, the abscissæ of the intersections of the line and curve are determined by the equation

$$(a\mu^2 - 2h\lambda\mu + b\lambda^2)\,x^2 + 2\,(g\mu^2 - h\mu\nu - f\mu\lambda + b\lambda\nu)\,x + (c\mu^2 - 2f\mu\nu + b\nu^2) = 0.$$

The line will touch when the quadratic has equal roots, or when

$$(a\mu^2 - 2h\lambda\mu + b\lambda^2)\,(c\mu^2 - 2f\mu\nu + b\nu^2) = (g\mu^2 - h\mu\nu - f\mu\lambda + b\lambda\nu)^2.$$

Multiplying out, the equation proves to be divisible by μ^2, and becomes

$$(bc - f^2)\,\lambda^2 + (ca - g^2)\,\mu^2 + (ab - h^2)\,\nu^2 + 2\,(gh - af)\,\mu\nu + 2\,(hf - bg)\,\nu\lambda + 2\,(fg - ch)\,\lambda\mu = 0.$$

We shall afterwards give other methods of obtaining this equation, which may be called the tangential equation of the curve. We shall often use abbreviations for the coefficients, and write the equation in the form

$$A\lambda^2 + B\mu^2 + C\nu^2 + 2F\mu\nu + 2G\nu\lambda + 2H\lambda\mu = 0.$$

The values of the coefficients will be more easily remembered by the help of the following rule. Let Δ denote the *discriminant* of the equation; that is to say, the function

$$abc + 2fgh + af^2 - bg^2 - ch^2,$$

whose vanishing is the condition that the equation may represent right lines. Then A is the derived function formed from Δ, regarding a as the variable; and B, C, $2F$, $2G$, $2H$ are the derived functions taken respectively with regard to b, c, f, g, h.

The coordinates of the centre (given Art. 140) may be written $\frac{G}{C}$, $\frac{F}{C}$.

Miscellaneous Examples.

Ex. 1. Form the equation of the conic making intercepts λ, λ', μ, μ' on the axes. Since if we make $y = 0$ or $x = 0$ in the equation, it must reduce to

$$x^2 - (\lambda + \lambda')\, x + \lambda\lambda' = 0, \quad y^2 - (\mu + \mu')\, y + \mu\mu' = 0;$$

the equation is

$$\mu\mu' x^2 + 2hxy + \lambda\lambda' y^2 - \mu\mu' (\lambda + \lambda')\, x - \lambda\lambda' (\mu + \mu')\, y + \lambda\lambda'\mu\mu' = 0,$$

and h is undetermined, unless another condition be given. Thus two parabolas can be drawn through the four given points; for in this case

$$h = \pm \sqrt{(\lambda\lambda'\mu\mu')}.$$

Ex. 2. Given four points on a conic, the polar of any fixed point passes through a fixed point. We may choose the axes so that the given points may lie two on each axis, and the equation of the curve is that found in Ex. 1. But the equation of the polar of any point $x'y'$ (Art. 145) involves the indeterminate h in the first degree, and, therefore, passes through a fixed point.

Ex. 3. Find the locus of the centre of a conic passing through four fixed points. The centre of the conic in Ex. 1 is given by the equations

$$2\mu\mu' x + 2hy - \mu\mu' (\lambda + \lambda') = 0, \quad 2\lambda\lambda' y + 2hx - \lambda\lambda' (\mu + \mu') = 0;$$

whence, eliminating the indeterminate h, the locus is

$$2\mu\mu' x^2 - 2\lambda\lambda' y^2 - \mu\mu' (\lambda + \lambda')\, x + \lambda\lambda' (\mu + \mu')\, y = 0,$$

a conic passing through the intersections of each of the three pairs of lines which can be drawn through the four points, and through the middle points of these lines. The locus will be a hyperbola when λ, λ' and μ, μ' have either both like or both unlike signs; and an ellipse in the contrary case. Thus it will be an ellipse when the two points on one axis lie on the same side of the origin, and on the other axis on opposite sides; in other words, when the quadrilateral formed by the four given points has a re-entrant angle. This is also geometrically evident; for a quadrilateral with a re-entrant angle evidently cannot be inscribed in a figure of the shape of the ellipse or parabola. The circumscribing conic must, therefore, always be a hyperbola, so that some vertices may lie in opposite branches. And since the centre of a hyperbola is never at infinity, the locus of centres is in this case an ellipse. In the other case, two positions of the centre will be at infinity, corresponding to the two parabolas which can be described through the given points.

CHAPTER XI.

EQUATIONS OF THE SECOND DEGREE REFERRED TO THE CENTRE AS ORIGIN.

152. In investigating the properties of the ellipse and hyperbola, we shall find our equations much simplified by choosing the centre for the origin of coordinates. If we transform the general equation of the second degree to the centre as origin, we saw (Art. 140) that the coefficients of x and y will $=0$ in the transformed equation, which will be of the form

$$ax^2 + 2hxy + by^2 + c' = 0.$$

It is sometimes useful to know the value of c' in terms of the coefficients of the first given equation. We saw (Art. 134) that

$$c' = ax'^2 + 2hx'y' + by'^2 + 2gx' + 2fy' + c,$$

where x', y' are the coordinates of the centre. The calculation of this may be facilitated by putting c' into the form

$$c' = (ax' + hy' + g)\,x' + (hx' + by' + f)\,y' + gx' + fy' + c.$$

The first two sets of terms are rendered $=0$ by the coordinates of the centre, and the last (Art. 140)

$$= g\frac{hf - bg}{ab - h^2} + f\frac{hg - af}{ab - h^2} + c = \frac{abc + 2fgh - af^2 - bg^2 - ch^2}{ab - h^2}.*$$

153. If the numerator of this fraction were $=0$, the transformed equation would be reduced to the form

$$ax^2 + 2hxy + by^2 = 0,$$

and would, therefore (Art. 73), represent two real or imaginary

---

* Observing that when f and g vanish the discriminant reduces to $c\,(ab - h^2)$, we can see that what has been here proved shows that transformation to parallel axes does not alter the value of the discriminant, a particular case of a theorem to be proved afterwards (Art. 371).

It is evident in like manner that the result of substituting $x'y'$, the coordinates of the centre, in the equation of the polar of any point $x''y''$, viz.

$$(ax' + hy' + g)\,x'' + (hx' + by' + f)\,y'' + gx' + fy' + c,$$

is the same as the result of substituting $x'y'$ in the equation of the curve. For the first two sets of terms vanish in both cases.

right lines, according as $ab - h^2$ is negative or positive. Hence, as we have already seen, p. 72, the condition that the general equation of the second degree should represent two right lines, is

$$abc + 2fgh - af^2 - bg^2 - ch^2 = 0.$$

For it must plainly be fulfilled, in order that when we transfer the origin to the point of intersection of the right lines, the absolute term may vanish.

Ex. 1. Transform $3x^2 + 4xy + y^2 - 5x - 6y - 3 = 0$ to the centre $(\frac{7}{2}, -4)$.

Ans. $12x^2 + 16xy + 4y^2 + 1 = 0$.

Ex. 2. Transform $x^2 + 2xy - y^2 + 8x + 4y - 8 = 0$ to the centre $(-3, -1)$.

Ans. $x^2 + 2xy - y^2 = 22$.

154. We have seen (Art. 136) that when θ satisfies the condition

$$a \cos^2\theta + 2h \cos\theta \sin\theta + b \sin^2\theta = 0,$$

the radius vector meets the curve at infinity, and also meets the curve in one other point, whose distance from the origin is

$$\rho = -\frac{c}{g \cos\theta + f \sin\theta}.$$

But if the origin be *the centre*, we have $g = 0$, $f = 0$, and this distance will *also* become infinite. Hence two lines can be drawn through the centre, which will meet the curve in *two coincident points* at infinity, and which therefore may be considered as tangents to the curve whose points of contact are at infinity. These lines are called the *asymptotes* of the curve; they are imaginary in the case of the ellipse, but real in that of the hyperbola. We shall show hereafter, that though the asymptotes do not meet the curve at any finite distance, yet the further they are produced the more nearly they approach the curve.

Since the points of contact of the two real or imaginary tangents drawn through the centre are at an infinite distance, the line joining these points of contact is altogether at an infinite distance. Hence, from our definition of poles and polars (Art. 89), *the centre may be considered as the pole of a line situated altogether at an infinite distance.* This inference may be confirmed from the equation of the polar of the origin, $gx + fy + c = 0$, which, if the centre be the origin, reduces to $c = 0$, an equation which (Art. 67) represents a line at infinity.

155. We have seen that by taking the centre for origin, the coefficients g and f in the general equation can be made to vanish; but the equation can be further simplified by taking a pair of conjugate diameters for axes, since then (Art. 143) h will vanish, and the equation be reduced to the form

$$ax^2 + by^2 + c = 0.$$

It is evident, now, that any line parallel to either axis is bisected by the other; for if we give to x any value, we obtain equal and opposite values for y. Now the angle between conjugate diameters is not in general right; but we shall show that there is always *one pair* of conjugate diameters which cut each other at right angles. These diameters are called the *axes* of the curves and the points where they meet it are called its *vertices*.

We have seen (Art. 143) that the angles made with the axis by two conjugate diameters are connected by the relation

$$b \tan\theta \tan\theta' + h(\tan\theta + \tan\theta') + a = 0.$$

But if the diameters are at right angles, $\tan\theta' = -\dfrac{1}{\tan\theta}$ (Art. 25). Hence

$$h \tan^2\theta + (a - b)\tan\theta - h = 0.$$

We have thus a quadratic equation to determine θ. Multiplying by ρ^2, and writing x, y, for $\rho\cos\theta$, $\rho\sin\theta$, we get

$$hx^2 - (a - b)xy - hy^2 = 0.$$

This is the equation of two real lines at right angles to each other (Art. 74); we perceive, therefore, that central curves have two, and only two, conjugate diameters at right angles to each other.

On referring to Art. 75 it will be found that the equation which we have just obtained for the *axes* of the curve is the same a that of the lines bisecting the internal and external angles between the real or imaginary lines represented by the equation

$$ax^2 + 2hxy + by^2 = 0.$$

The axes of the curve, therefore, are the diameters which bisect the angles between the asymptotes; and (note, p. 71) they will be real whether the asymptotes be real or imaginary; that is to say, whether the curve be an ellipse or a hyperbola.

156. We might have obtained the results of the last Article by the method of transformation of coordinates, since we can

thus prove directly that it is always possible to transform the equation to a pair of rectangular axes, such that the coefficient of xy in the transformed equation may vanish. Let the original axes be rectangular; then, if we turn them round through any angle θ, we have (Art. 9) to substitute for x, $x\cos\theta - y\sin\theta$, and for y, $x\sin\theta + y\cos\theta$; the equation will therefore become

$$a(x\cos\theta - y\sin\theta)^2 + 2h(x\cos\theta - y\sin\theta)(x\sin\theta + y\cos\theta)$$
$$+ b(x\sin\theta + y\cos\theta)^2 + c = 0$$

or, arranging the terms, we shall have

the new $a = a\cos^2\theta + 2h\cos\theta\sin\theta + b\sin^2\theta$;

the new $h = b\sin\theta\cos\theta + h(\cos^2\theta - \sin^2\theta) - a\sin\theta\cos\theta$;

the new $b = a\sin^2\theta - 2h\cos\theta\sin\theta + b\cos^2\theta$.

Now, if we put the new $h = 0$, we get the very same equation as in Art. 155, to determine $\tan\theta$. This equation gives us a simple expression for the angle made with the given axes by either axis of the curve, namely,

$$\tan 2\theta = \frac{2h}{a-b}.$$

157. When it is required to transform a given equation to the form $ax^2 + by^2 + c = 0$, and to calculate numerically the value of the new coefficients, our work will be much facilitated by the following theorem: *If we transform an equation of the second degree from one set of rectangular axes to another, the quantities $a + b$ and $ab - h^2$ will remain unaltered.*

The first part is proved immediately by adding the values of the new a and b (Art. 156), when we have

$$a' + b' = a + b.$$

To prove the second part, write the values in the last article

$$2a' = a + b + 2h\sin 2\theta + (a-b)\cos 2\theta,$$
$$2b' = a + b - 2h\sin 2\theta - (a-b)\cos 2\theta.$$

Hence $\quad 4a'b' = (a+b)^2 - \{2h\sin 2\theta + (a-b)\cos 2\theta\}^2.$

But $\quad 4h'^2 = \{2h\cos 2\theta - (a-b)\sin 2\theta\}^2$;

therefore $4(a'b' - h'^2) = (a+b)^2 - 4h^2 - (a-b)^2 = 4(ab - h^2)$.

When, therefore, we want to form the equation transformed to the *axes*, we have the new $h = 0$,

$$a' + b' = a + b, \quad a'b' = ab - h^2.$$

Having, therefore, the sum and the product of a' and b', we can form the quadratic which determines these quantities.

Ex. 1. Find the axes of the ellipse $14x^2 - 4xy + 11y^2 = 60$, and transform the equation to them.

The axes are (Art. 155) $4x^2 + 6xy - 4y^2 = 0$, or $(2x - y)(x + 2y) = 0$.

We have $a' + b' = 25$; $a'b' = 150$; $a' = 10$; $b' = 15$; and the transformed equation is $2x^2 + 3y^2 = 12$.

Ex. 2. Transform the hyperbola $11x^2 + 84xy - 24y^2 = 156$ to the axes.

$$a' + b' = -13, \quad a'b' = -2028; \quad a' = 39; \quad b' = -52.$$

Transformed equation is $3x^2 - 4y^2 = 12$.

Ex. 3. Transform $ax^2 + 2hxy + by^2 = c$ to the axes.

Ans. $(a + b - R)x^2 + (a + b + R)y^2 = 2c$, where $R^2 = 4h^2 + (a - b)^2$.

*158. Having proved that the quantities $a + b$ and $ab - h^2$ remain unaltered when we transform from one rectangular system to another, let us now inquire what these quantities become if we transform to an oblique system. We may retain the old axis of x, and if we take an axis of y inclined to it at an angle ω, then (Art. 9) we are to substitute $x + y \cos\omega$ for x, and $y \sin\omega$ for y. We shall then have

$$a' = a, \quad h' = a\cos\omega + h\sin\omega,$$
$$b' = a\cos^2\omega + 2h\cos\omega\sin\omega + b\sin^2\omega.$$

Hence, it easily follows

$$\frac{a' + b' - 2h'\cos\omega}{\sin^2\omega} = a + b, \quad \frac{a'b' - h'^2}{\sin^2\omega} = ab - h^2.$$

If, then, we transform the equation from one pair of axes to any other, the quantities $\dfrac{a + b - 2h\cos\omega}{\sin^2\omega}$ *and* $\dfrac{ab - h^2}{\sin^2\omega}$ *remain unaltered.*

We may, by the help of this theorem, transform to the *axes* an equation given in oblique coordinates, for we can still express the sum and product of the new a and b in terms of the old coefficients.

Ex. 1. If $\cos\omega = \frac{3}{5}$, transform to the axes $10x^2 + 6xy + 5y^2 = 10$.

$$a + b = \tfrac{285}{16}, \quad ab = \tfrac{1025}{16}, \quad a = 5, \quad b = \tfrac{205}{16}.$$

Ans. $16x^2 + 41y^2 = 32$.

Ex. 2. Transform to the axes $x^2 - 3xy + y^2 + 1 = 0$, where $\omega = 60°$.

Ans. $x^2 - 15y^2 = 3$.

Ex. 3. Transform $ax^2 + 2hxy + by^2 = c$ to the axes.

Ans. $(a + b - 2h\cos\omega - R)x^2 + (a + b - 2h\cos\omega + R)y^2 = 2c\sin^2\omega$, where $R^2 = \{2h - (a + b)\cos\omega\}^2 + (a - b)^2\sin^2\omega$.

*159. We add the demonstration of the theorems of the last two articles given by Professor Boole (*Cambridge Math. Jour.*, III. 1, 106, and New Series, VI. 87).

Let us suppose that we are transforming an equation from axes inclined at an angle ω, to any other axes inclined at an angle Ω; and that, on making the substitutions of Art. 9, the quantity $ax^2 + 2hxy + by^2$ becomes $a'X^2 + 2h'XY + b'Y^2$. Now we know that the effect of the same substitution will be to make the quantity $x^2 + 2xy\cos\omega + y^2$ become $X^2 + 2XY\cos\Omega + Y^2$, since either is the expression for the square of the distance of any point from the origin. It follows, then, that

$$ax^2 + 2hxy + by^2 + \lambda(x^2 + 2xy\cos\omega + y^2)$$
$$= a'X^2 + 2h'XY + b'Y^2 + \lambda(X^2 + 2XY\cos\Omega + Y^2).$$

And if we determine λ so that the first side of the equation may be a perfect square, the second must be a perfect square also. But the condition that the first side may be a perfect square is

$$(a + \lambda)(b + \lambda) = (h + \lambda\cos\omega)^2,$$

or λ must be one of the roots of the equation

$$\lambda^2 \sin^2\omega + (a + b - 2h\cos\omega)\lambda + ab - h^2 = 0.$$

We get a quadratic of like form to determine the value of λ, which will make the second side of the equation a perfect square; but since both sides become perfect squares for the *same* values of λ, these two quadratics must be identical. Equating, then, the coefficients of the corresponding terms, we have, as before,

$$\frac{a + b - 2h\cos\omega}{\sin^2\omega} = \frac{a' + b' - 2h'\cos\Omega}{\sin^2\Omega}; \quad \frac{ab - h^2}{\sin^2\Omega} = \frac{a'b' - h'^2}{\sin^2\Omega}.$$

Ex. 1. The sum of the squares of the reciprocals of two semi-diameters at right angles to each other is constant.

Let their lengths be α and β; then making alternately $x = 0$, $y = 0$, in the equation of the curve, we have $a\alpha^2 = c$, $b\beta^2 = c$, and the theorem just stated is only the geometrical interpretation of the fact that $a + b$ is constant.

Ex. 2. The area of the triangle formed by joining the extremities of two conjugate semi-diameters is constant.

The equation referred to two conjugate diameters is $\frac{x^2}{\alpha'^2} + \frac{y^2}{\beta'^2} = 1$, and since $\frac{ab - h^2}{\sin^2\omega}$ is constant, we have $\alpha'\beta'\sin\omega$ constant.

Ex. 3. The sum of the squares of two conjugate semi-diameters is constant.

Since $\frac{a + b - 2h\cos\omega}{\sin^2\omega}$ is constant, $\frac{1}{\sin^2\omega}\left(\frac{1}{\alpha'^2} + \frac{1}{\beta'^2}\right) = \frac{\alpha'^2 + \beta'^2}{\alpha'^2\beta'^2\sin^2\omega}$ is constant; and since $\alpha'\beta'\sin\omega$ is constant, so must $\alpha'^2 + \beta'^2$.

THE EQUATION REFERRED TO THE AXES.

160. We saw that the equation referred to the axes was of the form

$$Ax^2 + By^2 = C,$$

B being positive in the case of the ellipse, and negative in that of the hyperbola (Art. 138, Ex. 3). We have replaced the small letters by capitals, because we are about to use the letters a and b with a different meaning.

The equation of the ellipse may be written in the following more convenient form:

Let the intercepts made by the ellipse on the axes be $x = a$, $y = b$, then making $y = 0$ and $x = a$ in the equation of the curve, we have $Aa^2 = C$, and $A = \frac{C}{a^2}$. In like manner $B = \frac{C}{b^2}$. Substituting these values, the equation of the ellipse may be written

$$\frac{x^2}{a^2} + \frac{y^2}{b^2} = 1.$$

Since we may choose whichever axis we please for the axis of x, we shall suppose that we have chosen the axes so that a may be greater than b.

The equation of the hyperbola, which we saw only differs from that of the ellipse in the sign of the coefficient of y^2, may be written in the corresponding form:

$$\frac{x^2}{a^2} - \frac{y^2}{b^2} = 1.$$

The intercept on the axis of x is evidently $= \pm a$, but that on the axis of y, being found from the equation $y^2 = -b^2$, is imaginary; the axis of y, therefore, does not meet the curve in real points.

Since we have chosen for our axis of x the axis which meets the curve in real points, we are not in this case entitled to assume that a is greater than b.

161. *To find the polar equation of the ellipse, the centre being the pole.*

Write $\rho \cos\theta$ for x, and $\rho \sin\theta$ for y in the preceding equation, and we get

$$\frac{1}{\rho^2} = \frac{\cos^2\theta}{a^2} + \frac{\sin^2\theta}{b^2},$$

an equation which we may write in any of the equivalent forms,

$$\rho^2 = \frac{a^2b^2}{a^2\sin^2\theta + b^2\cos^2\theta} = \frac{a^2b^2}{b^2 + (a^2 - b^2)\sin^2\theta} = \frac{a^2b^2}{a^2 - (a^2 - b^2)\cos^2\theta}.$$

It is customary to use the following abbreviations:

$$a^2 - b^2 = c^2;\quad \frac{a^2 - b^2}{a^2} = e^2;$$

and the quantity e is called the *eccentricity* of the curve.

Dividing by a^2 the numerator and denominator of the fraction last found, we obtain the form most commonly used, viz.

$$\rho^2 = \frac{b^2}{1 - e^2\cos^2\theta}.$$

162. *To investigate the figure of the ellipse.*

The *least* value that $b^2 + (a^2 - b^2)\sin^2\theta$, the denominator in the value of ρ^2, can have, is when $\theta = 0$; therefore the *greatest* value of ρ is the intercept on the axis of x, and is $= a$.

Again, the *greatest* value of $b^2 + (a^2 - b^2)\sin^2\theta$ is when $\sin\theta = 1$, or $\theta = 90°$; hence, the *least* value of ρ is the intercept on the axis of y, and is $= b$. The greatest line, therefore, that can be drawn through the centre is the axis of x, and the least line the axis of y. From this property these lines are called the axis *major* and the axis *minor* of the curve.

It is plain that the smaller θ is, the greater ρ will be; hence, *the nearer any diameter is to the axis major, the greater it will be.* The form of the curve will, therefore, be that here represented.

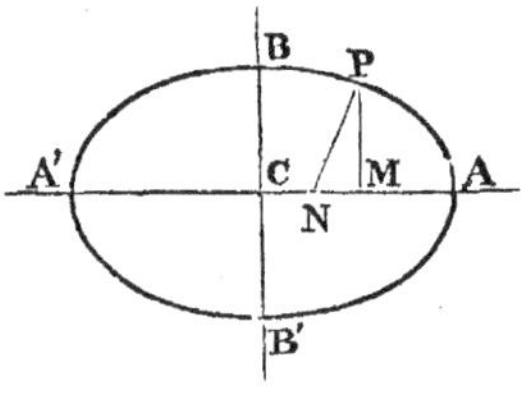

We obtain the same value of ρ whether we suppose $\theta = \alpha$, or $\theta = -\alpha$. Hence, *Two diameters which make equal angles with the axis will be equal.* And it is easy to show that the converse of this theorem is also true.

This property enables us, being given the centre of a conic, to determine its axes geometrically. For, describe any concentric circle intersecting the conic, then the semi-diameters drawn to the points of intersection will be equal; and by the theorem just proved, the axes of the conic will be the lines internally and externally bisecting the angle between them.

163. The equation of the ellipse can be put into another form, which will make the figure of the curve still more apparent. If we solve for y we get

$$y = \frac{b}{a}\sqrt{(a^2 - x^2)}.$$

Now, if we describe a concentric circle with the radius a its equation will be

$$y = \sqrt{(a^2 - x^2)}.$$

Hence we derive the following construction:

"*Describe a circle on the axis major, and take on each ordinate LQ a point P, such that LP may be to LQ in the constant ratio $b : a$, then the locus of P will be the required ellipse.*"

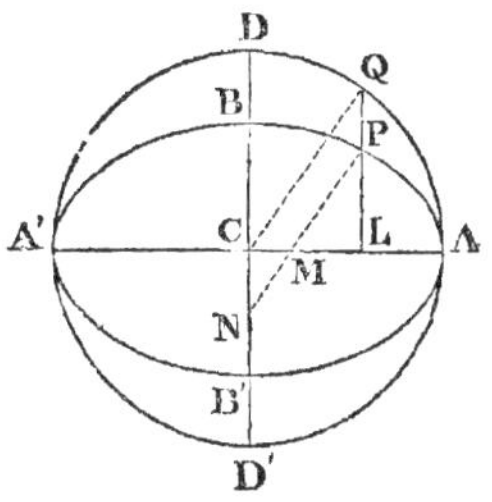

Hence the circle described on the axis major lies wholly *without* the curve. We might, in like manner, construct the ellipse by describing a circle on the axis minor and *increasing* each ordinate in the constant ratio $a : b$.

Hence the circle described on the axis minor lies wholly *within* the curve.

The equation of the circle is the particular form which the equation of the ellipse assumes when we suppose $b = a$.

164. *To find the polar equation of the hyperbola.*

Transforming to polar coordinates, as in Art. 161, we get

$$\rho^2 = \frac{a^2b^2}{b^2\cos^2\theta - a^2\sin^2\theta} = \frac{a^2b^2}{b^2 - (a^2 + b^2)\sin^2\theta} = \frac{a^2b^2}{(a^2 + b^2)\cos^2\theta - a^2}.$$

Since formulæ concerning the ellipse are altered to the corresponding formulæ for the hyperbola by changing the sign of b^2, we must in this case use the abbreviation c^2 for $a^2 + b^2$ and e^2 for $\frac{a^2 + b^2}{a^2}$, the quantity e being called the *eccentricity* of the hyperbola. Dividing then by a^2 the numerator and denominator of the last found fraction, we obtain the polar equation of the hyperbola, which only differs from that of the ellipse in the sign of b^2, viz.

$$\rho^2 = \frac{b^2}{e^2\cos^2\theta - 1}.$$

165. *To investigate the figure of the hyperbola.*

The terms axis major and axis minor not being applicable to the hyperbola (Art. 160), we shall call the axis of x the *transverse* axis, and the axis of y the *conjugate* axis.

Now $b^2 - (a^2 + b^2)\sin^2\theta$, the denominator in the value of ρ^2, will plainly be greatest when $\theta = 0$, therefore, in the same case, ρ will be least; or *the transverse axis is the shortest line which can be drawn from the centre to the curve.*

As θ increases, ρ continually increases, until

$$\sin\theta = \frac{b}{\sqrt{(a^2+b^2)}}, \quad \left(\text{or } \tan\theta = \frac{b}{a}\right),$$

when the denominator of the value of ρ becomes $=0$, and ρ becomes infinite. After this value of θ, ρ^2 becomes negative, and the diameters cease to meet the curve in real points, until again

$$\sin\theta = \frac{b}{\sqrt{(a^2+b^2)}}, \quad \left(\text{or } \tan\theta = -\frac{b}{a}\right),$$

when ρ again becomes infinite. It then decreases regularly as θ increases, until θ becomes $=180^\circ$, when it again receives its minimum value $=a$.

The form of the hyperbola, therefore, is that represented by the dark curve on the figure, next article.

166. We found that the axis of y does not meet the hyperbola in real points, since we obtained the equation $y^2 = -b^2$ to determine its point of intersection with the curve. We shall, however, still mark off on the axis of y portions CB, $CB' = \pm b$, and we shall find that the length CB has an important connexion with the curve, and may be conveniently called an axis of the curve.

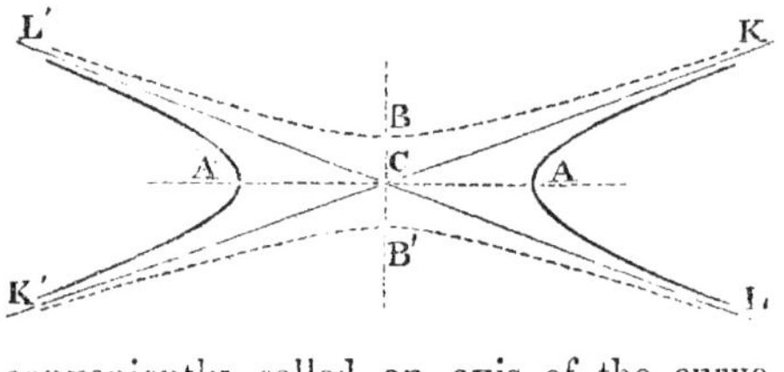

In like manner, if we obtained an equation to determine the length of any other diameter, of the form $\rho^2 = -R^2$, although this diameter cannot meet the curve, yet if we measure on it from the centre lengths $= \pm R$, these lines may be conveniently spoken of as diameters of the hyperbola.

The *locus* of the extremities of these diameters which do not meet the curve is, by changing the sign of ρ^2 in the equation of the curve, at once found to be

$$\frac{1}{\rho^2} = \frac{\sin^2\theta}{b^2} - \frac{\cos^2\theta}{a^2},$$

or

$$\frac{y^2}{b^2} - \frac{x^2}{a^2} = 1.$$

This is the equation of a hyperbola having the axis of y for the axis meeting it in real points, and the axis of x for the axis meeting it in imaginary points. It is represented by the dotted curve on the figure, and is called the hyperbola *conjugate* to the given hyperbola.

167. We proved (Art. 165) that the diameters answering to $\tan\theta = \pm\frac{b}{a}$ meet the curve at infinity; they are, therefore, the same as the lines called, in Art. 154, the *asymptotes* of the curve. They are the lines CK, CL on the figure, and evidently separate those diameters which meet the curve in real points from those which meet it in imaginary points. It is evident also that two conjugate hyperbolæ have the same asymptotes.

The expression $\tan\theta = \pm\frac{b}{a}$ enables us, being given the axes in magnitude and position, to find the asymptotes, for if we form a rectangle by drawing parallels to the axes through B and A, then the asymptote CK must be the diagonal of this rectangle.

Again $$\cos\theta = \frac{a}{\sqrt{(a^2+b^2)}} = \frac{1}{e}.$$

But, since the asymptotes make equal angles with the axis of x, the angle which they make with each other must be $= 2\theta$. Hence, *being given the eccentricity of a hyperbola, we are given the angle between the asymptotes*, which is double the angle whose secant is the eccentricity.

Ex. To find the eccentricity of a conic given by the general equation.

We can (Art. 74) write down the tangent of the angle between the lines denoted by $ax^2 + 2hxy + by^2 = 0$, and thence form the expression for the secant of its half; or we may proceed by the help of Art. 157, Ex. 3.

We have $$\frac{1}{\alpha^2} = \frac{a+b-R}{2c},\quad \frac{1}{\beta^2} = \frac{a+b+R}{2c},$$

where $$R^2 = 4h^2 + (a-b)^2 = 4h^2 - 4ab + (a+b)^2.$$

Hence $$\frac{1}{\beta^2} - \frac{1}{\alpha^2} = \frac{R}{c},\quad \frac{\alpha^2 - \beta^2}{\alpha^2} = \frac{2R}{a+b+R}.$$

CONJUGATE DIAMETERS.

168. We now proceed to investigate some of the properties of the ellipse and hyperbola. We shall find it convenient to consider both curves together, for, since their equations only differ in the sign of b^2, they have many properties in common which can be proved at the same time, by considering the sign of b^2 as indeterminate. We shall, in the following Articles, use the signs which apply to the ellipse. The reader may then obtain the corresponding formulæ for the hyperbola by changing the sign of b^2.

We shall first apply to the particular form $\frac{x^2}{a^2} + \frac{y^2}{b^2} = 1$, some of the results already obtained for the general equation. Thus (Art. 86) the equation of the tangent at any point $x'y'$ being got by writing $x'x$ and $y'y$ for x^2 and y^2 is

$$\frac{x'x}{a^2} + \frac{y'y}{b^2} = 1.$$

The proof given in general may be repeated for this particular case. The equation of the chord joining any two points on the curve is

$$\frac{(x-x')(x-x'')}{a^2} + \frac{(y-y')(y-y'')}{b^2} = \frac{x^2}{a^2} + \frac{y^2}{b^2} - 1,$$

or $$\frac{(x'+x'')x}{a^2} + \frac{(y'+y'')y}{b^2} = \frac{x'x''}{a^2} + \frac{y'y''}{b^2} + 1;$$

which, when $x', y' = x'', y''$, becomes the equation of the tangent already written.

The argument here used applies whether the axes be rectangular or oblique. Now if the axes be a pair of conjugate diameters, the coefficient of xy vanishes (Art. 143); the coefficients of x and y vanish, since the origin is the centre; and if a' and b' be the lengths of the intercepts on the axes, it is proved exactly, as in Art. 160, that the equation of the curve may be written

$$\frac{x^2}{a'^2} + \frac{y^2}{b'^2} = 1.$$

And it follows from this article that in the same case the equation of the tangent is

$$\frac{xx'}{a'^2} + \frac{yy'}{b'^2} = 1.$$

169. The equation of the polar, or line joining the points of contact of tangents from any point $x'y'$, is similar in form to the equation of the tangent (Arts. 88, 89), and is therefore

$$\frac{xx'}{a^2} + \frac{yy'}{b^2} = 1, \text{ or } \frac{xx'}{a'^2} + \frac{yy'}{b'^2} = 1;$$

the axes of coordinates in the latter case being any pair of conjugate diameters, in the former case the axes of the curve. In particular, the polar of any point on the axis of x is $\frac{xx'}{a'^2} = 1$. Hence the polar of any point P is found by drawing a diameter through the point, taking $CP.CP' =$ to the square of the semi-diameter, and then drawing through P' a parallel to the conjugate diameter. This includes, as a particular case, the theorem proved already (Art. 145), viz., *The tangent at the extremity of any diameter is parallel to the conjugate diameter.*

Ex. 1. To find the condition that $\lambda x + \mu y = 1$ may touch $\frac{x^2}{a^2} + \frac{y^2}{b^2} = 1$.

Comparing $\frac{xx'}{a^2} + \frac{yy'}{b^2} = 1$, $\lambda x + \mu y = 1$, we find $\frac{x'}{a} = \lambda a$, $\frac{y'}{b} = \mu b$, and $a^2\lambda^2 + b^2\mu^2 = 1$.

Ex. 2. To find the equation of the pair of tangents from $x'y'$ to the curve (see Art. 92).

$$\textit{Ans.} \left(\frac{x'^2}{a^2} + \frac{y'^2}{b^2} - 1\right)\left(\frac{x^2}{a^2} + \frac{y^2}{b^2} - 1\right) = \left(\frac{xx'}{a^2} + \frac{yy'}{b^2} - 1\right)^2.$$

Ex. 3. To find the angle ϕ between the pair of tangents from $x'y'$ to the curve.

When an equation of the second degree represents two right lines, the three highest terms being put $= 0$, denote two lines through the origin parallel to the two former; hence, the angle included by the first pair of right lines depends solely on the three highest terms of the general equation. Arranging, then, the equation found in the last Example, we find, by Art. 74,

$$\tan\phi = \frac{2ab\sqrt{\left(\frac{x'^2}{a^2} + \frac{y'^2}{b^2} - 1\right)}}{x'^2 + y'^2 - a^2 - b^2}.$$

Ex. 4. Find the locus of a point, the tangents through which intersect at right angles.

Equating to 0 the denominator in the value of $\tan\phi$, we find $x^2 + y^2 = a^2 + b^2$, the equation of a circle concentric with the ellipse. The locus of the intersection of tangents which cut at a given angle is, in general, a curve of the fourth degree.

170. *To find the equation, referred to the axes, of the diameter conjugate to that passing through any point $x'y'$ on the curve.*

The line required passes through the origin, and (Art. 169) is parallel to the tangent at $x'y'$; its equation is therefore

$$\frac{xx'}{a^2} + \frac{yy'}{b^2} = 0.$$

Let θ, θ' be the angles made with the axis of x by the original diameter and its conjugate; then plainly $\tan\theta = \frac{y'}{x'}$; and from the equation of the conjugate we have (Art. 21) $\tan\theta = -\frac{b^2x'}{a^2y'}$. Hence $\tan\theta\tan\theta' = -\frac{b^2}{a^2}$, as might also be inferred from Art. 143.

The corresponding relation for the hyperbola (see Art. 168) is

$$\tan\theta\tan\theta' = \frac{b^2}{a^2}.$$

171. Since in the ellipse $\tan\theta\tan\theta'$ is negative, if one of the angles θ, θ' be acute (and, therefore, its tangent positive), the other must be obtuse (and, therefore, its tangent negative). Hence, *conjugate diameters in the ellipse lie on different sides of the axis minor* (which answers to $\theta = 90°$).

In the hyperbola, on the contrary, $\tan\theta\tan\theta'$ is positive; therefore θ and θ' must be either both acute or both obtuse. Hence, *in the hyperbola, conjugate diameters lie on the same side of the conjugate axis.*

In the hyperbola, if $\tan\theta$ be less, $\tan\theta'$ must be greater than $\frac{b}{a}$, but (Art. 167) the diameter answering to the angle whose tangent is $\frac{b}{a}$, is the asymptote, which (by the same Article) separates those diameters which meet the curve from those which do not intersect it. Hence, *if one of two conjugate diameters meet a hyperbola in real points, the other will not.* Hence also it may be seen that each asymptote is its own conjugate.

172. *To find the coordinates $x''y''$ of the extremity of the diameter conjugate to that passing through $x'y'$.*

These coordinates are obviously found by solving for x and y between the equation of the conjugate diameter and that of the curve, viz.

$$\frac{xx'}{a^2} + \frac{yy'}{b^2} = 0, \quad \frac{x^2}{a^2} + \frac{y^2}{b^2} = 1,$$

Substituting in the second the values of x and y found from the first equation, and remembering that x', y' satisfy the equation of the curve, we find without difficulty

$$\frac{x''}{a} = \pm\frac{y'}{b}, \quad \frac{y''}{b} = \mp\frac{x'}{a}.$$

173. *To express the lengths of a diameter (a'), and its conjugate (b'), in terms of the abscissa of the extremity of the diameter.*

(1) We have $$a'^2 = x'^2 + y'^2.$$

But $$y'^2 = \frac{b^2}{a^2}(a^2 - x'^2).$$

Hence $$a'^2 = b^2 + \frac{a^2 - b^2}{a^2}x'^2 = b^2 + e^2x'^2.$$

(2) Again, we have

$$b'^2 = x''^2 + y''^2 = \frac{a^2}{b^2}y'^2 + \frac{b^2}{a^2}x'^2,$$

or $$= (a^2 - x'^2) + \frac{b^2}{a^2}x'^2;$$

hence $$b'^2 = a^2 - e^2x'^2.$$

From these values we have

$$a'^2 + b'^2 = a^2 + b^2;$$

or, *The sum of the squares of any pair of conjugate diameters of an ellipse is constant* (see Ex. 3, Art. 159).

174. In the hyperbola we must change the signs of b^2 and b'^2, and we get

$$a'^2 - b'^2 = a^2 - b^2,$$

or, *The difference of the squares of any pair of conjugate diameters of a hyperbola is constant.*

If in the hyperbola we have $a = b$, its equation becomes

$$x^2 - y^2 = a^2,$$

and it is called an *equilateral* hyperbola.

The theorem just proved shows that *every diameter of an equilateral hyperbola is equal to its conjugate.*

The asymptotes of the equilateral hyperbola being given by the equation

$$x^2 - y^2 = 0,$$

are at right angles to each other. Hence this hyperbola is often called a *rectangular* hyperbola.

The condition that the general equation of the second degree should represent an equilateral hyperbola is $a = -b$; for (Art. 74) this is the condition that the asymptotes $(ax^2 + 2hxy + by^2)$ should be at right angles to each other; but if the hyperbola be *re tangular* it must be *equilateral*, since (Art. 167) the tangent of half the angle between the asymptotes $= \frac{b}{a}$; therefore, if this angle $= 45°$, we have

$$b = a.$$

175. *To find the length of the perpendicular from the centre on the tangent.*

The length of the perpendicular from the origin on the line

$$\frac{xx'}{a^2} + \frac{yy'}{b^2} = 1,$$

is (Art. 23)
$$\frac{1}{\sqrt{\left(\frac{x'^2}{a^4} + \frac{y'^2}{b^4}\right)}} = \frac{ab}{\sqrt{\left(\frac{b^2x'^2}{a^2} + \frac{a^2y'^2}{b^2}\right)}};$$

but we proved (Art. 173) that

$$b'^2 = \frac{b^2x'^2}{a^2} + \frac{a^2y'^2}{b^2};$$

hence
$$p = \frac{ab}{b'}.$$

176. *To find the angle between any pair of conjugate diameters.*

The angle between the diameters is equal to the angle between either, and the tangent parallel to the other. Now

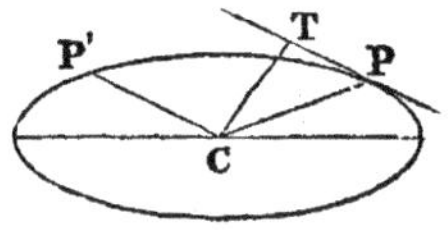

$$\sin CPT = \frac{CT}{CP} = \frac{p}{a'}.$$

Hence
$$\sin\phi \text{ (or } PCP') = \frac{ab}{a'b'}.$$

The equation $a'b' \sin\phi = ab$ proves that *the triangle formed by joining the extremities of conjugate diameters of an ellipse or hyperbola has a constant area* (see Art. 159, Ex. 2).

177. The sum of the squares of any two conjugate diameters of an ellipse being constant, their rectangle is a maximum when they are equal; and, therefore, in this case, $\sin\phi$ is a minimum; hence the acute angle between the two *equal* conjugate diameters is less (and, consequently, the obtuse angle greater) than the angle between any other pair of conjugate diameters.

The length of the equal conjugate diameters is found by making $a' = b'$ in the equation $a'^2 + b'^2 = a^2 + b^2$, whence a'^2 is half the sum of a^2 and b^2, and in this case

$$\sin\phi = \frac{2ab}{a^2 + b^2}.$$

The angle which either of the equi-conjugate diameters makes with the axis of x is found from the equation

$$\tan\theta \tan\theta' = -\frac{b^2}{a^2},$$

by making $\tan\theta = -\tan\theta'$; for any two equal diameters make equal angles with the axis of x on opposite sides of it (Art. 162).

Hence $$\tan\theta = \frac{b}{a}.$$

It follows, therefore, from Art. 167, that if an ellipse and hyperbola have the same axes in magnitude and position, then the asymptotes of the hyperbola will coincide with the equi-conjugate diameters of the ellipse.

The general equation of an ellipse, referred to two conjugate diameters (Art. 168), becomes $x^2 + y^2 = a'^2$, when $a' = b'$. We see, therefore, that, by taking the equi-conjugate diameters for axes, the equation of *any* ellipse may be put into the same form as the equation of the circle, $x^2 + y^2 = r^2$, but that in the case of the ellipse the angle between these axes will be *oblique*.

178. *To express the perpendicular from the centre on the tangent in terms of the angles which it makes with the axes.*

If we proceed to throw the equation of the tangent $\left(\frac{xx'}{a^2} + \frac{yy'}{b^2} = 1\right)$ into the form $x\cos\alpha + y\sin\alpha = p$ (Art. 23), we find immediately, by comparing these equations,

$$\frac{x'}{a^2} = \frac{\cos\alpha}{p}, \quad \frac{y'}{b^2} = \frac{\sin\alpha}{p}.$$

Substituting in the equation of the curve the values of x', y', hence obtained, we find

$$p^2 = a^2 \cos^2\alpha + b^2 \sin^2\alpha.*$$

The equation of the tangent may, therefore, be written

$$x \cos\alpha + y \sin\alpha - \sqrt{(a^2 \cos^2\alpha + b^2 \sin^2\alpha)} = 0.$$

Hence, by Art. 34, the perpendicular from any point $(x'y')$ on the tangent is

$$\sqrt{(a^2 \cos^2\alpha + b^2 \sin^2\alpha)} - x' \cos\alpha - y' \sin\alpha,$$

where we have written the formula so that the perpendiculars shall be positive when $x'y'$ is on the same side of the tangent as the centre.

Ex. To find the locus of the intersection of tangents which cut at right angles. Let p, p' be the perpendiculars on those tangents, then

$$p^2 = a^2 \cos^2\alpha + b^2 \sin^2\alpha, \quad p'^2 = a^2 \sin^2\alpha + b^2 \cos^2\alpha, \quad p^2 + p'^2 = a^2 + b^2.$$

But the square of the distance from the centre, of the intersection of two lines which cut at right angles, is equal to the sum of the squares of its distances from the lines themselves. The distance, therefore, is constant, and the required locus is a circle (see p. 166, Ex. 4).

179. The chords which join the extremities of any diameter to any point on the curve are called *supplemental chords.*

Diameters parallel to any pair of supplemental chords are conjugate.

For if we consider the triangle formed by joining the extremities of any diameter AB to any point on the curve D; since, by elementary geometry, the line joining the middle points of two sides must be parallel to the third, the diameter bisecting AD will be parallel to BD, and the diameter bisecting BD will be parallel to AD. The same thing may be proved analytically, by forming the equations of AD and BD, and showing that the product of the tangents of the angles made by these lines with the axis is $= -\dfrac{b^2}{a^2}$.

This property enables us to draw geometrically a pair of conjugate diameters making any angle with each other. For if we describe on any diameter a segment of a circle, containing the

* In like manner, $p^2 = a'^2 \cos^2\alpha + b'^2 \cos^2\beta$, α and β being the angles the perpendicular makes with any pair of conjugate diameters.

given angle, and join the points where it meets the curve to the extremities of the assumed diameter, we obtain a pair of supplemental chords inclined at the given angle, the diameters parallel to which will be conjugate to each other.

Ex. 1. Tangents at the extremities of any diameter are parallel.

Their equations are $$\frac{xx'}{a^2} + \frac{yy'}{b^2} = \pm 1.$$

This also follows from the first theorem of Art. 146, and from considering that the centre is the pole of the line at infinity (Art. 154).

Ex. 2. If any variable tangent to a central conic section meet two fixed parallel tangents, it will intercept portions on them, whose rectangle is constant, and equal to the square of the semi-diameter parallel to them.

Let us take for axes the diameter parallel to the tangents and its conjugate, then the equations of the curve and of the variable tangent will be

$$\frac{x^2}{a'^2} + \frac{y^2}{b'^2} = 1, \quad \frac{xx'}{a'^2} + \frac{yy'}{b'^2} = 1.$$

The intercepts on the fixed tangents are found by making x alternately $= \pm a'$ in the latter equation, and we get

$$y = \frac{b'^2}{y'}\left(1 \mp \frac{x'}{a'}\right);$$

and, therefore, their product is $$\frac{b'^4}{y'^2}\left(1 - \frac{x'^2}{a'^2}\right);$$

which, substituting for y'^2 from the equation of the curve, reduces to b'^2.

Ex. 3. The same construction remaining, the rectangle under the segments of the variable tangent is equal to the square of the semi-diameter parallel to it.

For, the intercept on either of the parallel tangents is to the adjacent segment of the variable tangent as the parallel semi-diameters (Art. 149); therefore, the rectangle under the intercepts of the fixed tangent is to the rectangle under the segments of the variable tangent as the *squares* of these semi-diameters; and, since the first rectangle is equal to the square of the semi-diameter parallel to it, the second rectangle must be equal to the square of the semi-diameter parallel to *it*.

Ex. 4. If any tangent meet any two conjugate diameters, the rectangle under its segments is equal to the square of the parallel semi-diameter.

Take for axes the semi-diameter parallel to the tangent and its conjugate; then the equations of any two conjugate diameters being (Art. 170)

$$y = \frac{y'}{x'}x, \quad \frac{xx'}{a'^2} + \frac{yy'}{b'^2} = 0,$$

the intercepts made by them on the tangent are found, by making $x = a'$, to be

$$y = \frac{y'}{x'}a', \text{ and } y = -\frac{b'^2}{a'}\frac{x'}{y'},$$

whose rectangle is evidently $= b'^2$.

We might, in like manner, have given a purely algebraical proof of Ex. 3.

Hence, also, if the centre be joined to the points where two parallel tangents meet any tangent, the joining lines will be conjugate diameters.

Ex. 5. Given, in magnitude and position, two conjugate semi-diameters, Oa, Ob, of a central conic, to determine the position of the axes.

The following construction is founded on the theorem proved in the last Example:—Through a the extremity of either diameter, draw a parallel to the other; it must of course be a tangent to the curve. Now, on Oa take a point P, such that the rectangle $Oa.aP = Ob^2$ (on the side remote from O for the ellipse, on the same side for the hyperbola), and describe a circle through O, P, having its centre on aC, then the lines OA, OB are the axes of the curve; for, since the rectangle $Aa.aB = Oa.aP = Ob^2$, the lines OA, OB are conjugate diameters, and since AB is a diameter of the circle, the angle AOB is right.

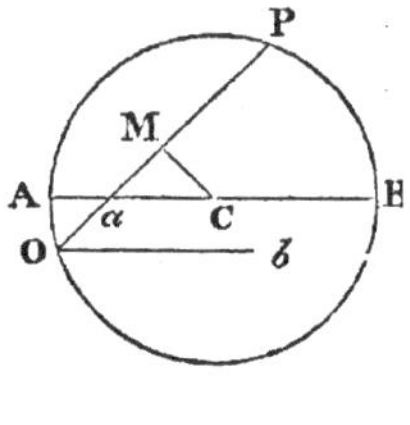

Ex. 6. Given any two semi-diameters, if from the extremity of each an ordinate be drawn to the other, the triangles so formed will be equal in area.

Ex. 7. Or if tangents be drawn at the extremity of each, the triangles so formed will be equal in area.

THE NORMAL.

180. A line drawn through any point of a curve perpendicular to the tangent at that point is called the *Normal*.

Forming, by Art. 32, the equation of a line drawn through $(x'y')$ perpendicular to $\left(\frac{xx'}{a^2} + \frac{yy'}{b^2} = 1\right)$, we find for the equation of the normal to a conic

$$\frac{x'}{a^2}(y - y') = \frac{y'}{b^2}(x - x'),$$

or

$$\frac{a^2x}{x'} - \frac{b^2y}{y'} = c^2,$$

c^2 being used, as in Art. 161, to denote $a^2 - b^2$.

Hence we can find the portion CN intercepted by the normal on either axis; for, making $y = 0$ in the equation just given, we find

$$x = \frac{c^2}{a^2}x', \text{ or } x = e^2x'.$$

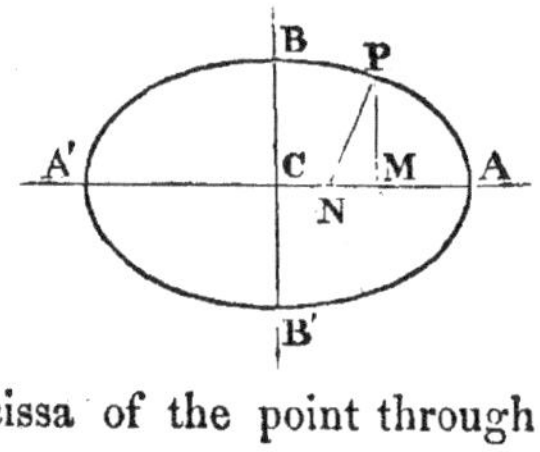

We can thus draw a normal to an ellipse from any point on the axis, for given CN we can find x', the abscissa of the point through which the normal is drawn.

The circle may be considered as an ellipse whose eccentricity $= 0$, since $c^2 = a^2 - b^2 = 0$. The intercept CN, therefore, is constantly $= 0$ in the case of the circle, or *every normal to a circle passes through its centre.*

181. The portion MN intercepted on the axis between the normal and ordinate is called the *Subnormal.* Its length is, by the last Article,

$$x' - \frac{c^2}{a^2}x' = \frac{b^2}{a^2}x'.$$

The normal, therefore, cuts the abscissa into parts which are in a constant ratio.

If a tangent drawn at the point P cut the axis in T, the intercept MT is, in like manner, called the *Subtangent.*

Since the whole length $CT = \frac{a^2}{x'}$ (Art. 169), the subtangent

$$= \frac{a^2}{x'} - x' = \frac{a^2 - x'^2}{x'}.$$

The length of the *normal* can also be easily found. For

$$PN^2 = PM^2 + NM^2 = y'^2 + \frac{b^4}{a^4}x'^2 = \frac{b^2}{a^2}\left(\frac{a^2}{b^2}y'^2 + \frac{b^2}{a^2}x'^2\right).$$

But if b' be the semi-diameter conjugate to CP, the quantity within the parentheses $= b'^2$ (Art. 173). Hence the length of the normal $PN = \frac{bb'}{a}$.

If the normal be produced to meet the axis minor, it can be proved, in like manner, that its length $= \frac{ab'}{b}$. Hence, *the rectangle under the segments of the normal is equal to the square of the conjugate semi-diameter.*

Again, we found (Art. 175) that the perpendicular from the centre on the tangent $= \frac{ab}{b'}$. Hence, *the rectangle under the normal and the perpendicular from the centre on the tangent is constant and equal to the square of the semi-axis minor.*

Thus, too, we can express the normal in terms of the angle it makes with the axis, for

$$PN = \frac{b^2}{p} = \frac{b^2}{\sqrt{(a^2\cos^2\alpha + b^2\sin^2\alpha)}} \text{ (Art. 178)}; = \frac{a(1-e^2)}{\sqrt{(1-e^2\sin^2\alpha)}}.$$

Ex. 1. To draw a normal to an ellipse or hyperbola passing through a given point.

The equation of the normal, $a^2x'y - b^2x'y = c^2x'y'$, expresses a relation between the coordinates $x'y'$ of any point on the curve, and xy the coordinates of any point on the normal at $x'y'$. We express that the point on the normal is known, and the point on the curve sought, by removing the accents from the coordinates of the latter point, and accentuating those of the former. Thus we find that the points on the curve, whose normals will pass through $(x'y')$ are the points of intersection of the given curve with the hyperbola

$$c^2xy = a^2x'y - b^2y'x.$$

Ex. 2. If through a given point on a conic any two lines at right angles to each other be drawn to meet the curve, the line joining their extremities will pass through a fixed point on the normal.

Let us take for axes the tangent and normal at the given point, then the equation of the curve must be of the form

$$ax^2 + 2hxy + by^2 + 2fy = 0,$$

(for $c = 0$, because the origin is on the curve, and $g = 0$ (Art. 144), because the tangent is supposed to be the axis of x, whose equation is $y = 0$).

Now, let the equation of *any* two lines through the origin be

$$x^2 + 2pxy + qy^2 = 0.$$

Multiply this equation by a, and subtract it from that of the curve, and we get

$$2(h - ap)xy + (b - aq)y^2 + 2fy = 0.$$

This (Art. 40) is the equation of a locus passing through the points of intersection of the lines and conic; but it may evidently be resolved into $y = 0$ (the equation of the tangent at the given point), and

$$2(h - ap)x + (b - aq)y + 2f = 0,$$

which must be the equation of the chord joining the extremities of the given lines.

The point where this chord meets the normal (the axis of y) is $y = \frac{2f}{aq - b}$; but if the lines are at right angles $q = -1$ (Art. 74), and the intercept on the normal has the constant length

$$= -\frac{2f}{a + b}.*$$

If the curve be an equilateral hyperbola, $a + b = 0$, and the line in question is constantly parallel to the normal. Thus then, if through any point on an equilateral hyperbola be drawn two chords at right angles, the perpendicular let fall on the line joining their extremities is the tangent to the curve.

Ex. 3. To find the coordinates of the intersection of the tangents at the points $x'y'$, $x''y''$.

The coordinates of the intersection of the lines

$$\frac{x'x}{a^2} + \frac{y'y}{b^2} = 1, \quad \frac{x''x}{a^2} + \frac{y''y}{b^2} = 1,$$

are
$$x = \frac{a^2(y' - y'')}{y'x'' - y''x'}, \quad y = \frac{b^2(x' - x'')}{x'y'' - y'x''}.$$

Ex. 4. To find the coordinates of the intersection of the normals at the points $x'y'$, $x''y''$.

Ans. $$x = \frac{(a^2 - b^2)x'x''X}{a^4}, \quad y = \frac{(b^2 - a^2)y'y''Y}{b^4},$$

---

* This theorem will be equally true if the lines be drawn so as to make with the normal angles the product of whose tangents is constant, for, in this case, q is constant, and, therefore, the intercept $\frac{2f}{aq - b}$ is constant.

where X, Y are the coordinates of the intersection of tangents, found in the last Example.

The values of X and Y may be written in other forms. Since by combining the equations

$$\frac{x'^2}{a^2}+\frac{y'^2}{b^2}=1,\quad \frac{x''^2}{a^2}+\frac{y''^2}{b^2}=1,$$

we get the results, $x'^2y''^2-y'^2x''^2=b^2(x'^2-x''^2)=-a^2(y'^2-y''^2)$;

hence $$X=\frac{x'y''+y'x''}{y'+y''},\quad Y=\frac{x'y''+y'x''}{x'+x''}.$$

We can also prove $$X=\frac{(x'+x'')}{1+\frac{x'x''}{a^2}+\frac{y'y''}{b^2}},\quad Y=\frac{(y'+y'')}{1+\frac{x'x''}{a^2}+\frac{y'y''}{b^2}}.$$

181 (a). Let CP, CQ be a pair of conjugate semi-diameters of an ellipse; let the normal PN meet CQ in R; take PD, PD' each equal to CQ; then the lengths of the lines CD, CD' are $a-b$, $a+b$ respectively.

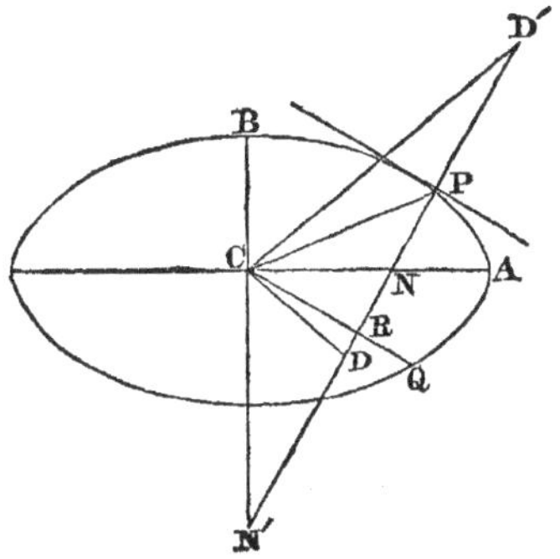

For

$$CD'^2=CP^2+PD'^2+2PD'.PR,$$

but

$$CP^2+PD'^2=a^2+b^2 \text{ (Art. 173)},$$

and $$2PD'.PR=2ab \text{ (Art. 175)}.$$

Hence $CD'^2=(a+b)^2$. Similarly for CD.

The axis-major bisects the angle DCD'. For the line

$$D'N=D'P+PN=b'+\frac{bb'}{a}=\frac{b'}{a}(a+b).$$

Similarly $DN=\dfrac{b'}{a}(a-b)$. At the point N, therefore, the base of the triangle DCD' is divided in the ratio of the sides, and, therefore, CN is the internal bisector of the vertical angle. In like manner, it is proved that CN' is the external bisector.

Hence then, being given two conjugate semi-diameters CP, CQ in magnitude and position, we are given the axes in magnitude and position. For we have only from P to let fall on CQ the perpendicular PR; to take PD, PD' each equal CQ; then the axes are in direction the bisectors of the angle DCD'; while their lengths are the sum and difference of CD, CD'.

THE FOCI.

182. If on the axis major of an ellipse we take two points equidistant from the centre whose common distance

$$= \pm \sqrt{(a^2 - b^2)}, \text{ or } = \pm c,$$

these points are called the *foci* of the curve.

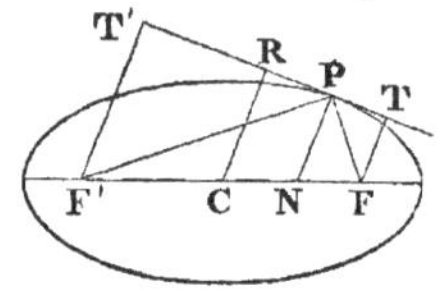

The foci of a hyperbola are two points on the transverse axis, at a distance from the centre still $= \pm c$, c being in the hyperbola

$$= \sqrt{(a^2 + b^2)}.$$

To express the distance of any point on an ellipse from the focus.

Since the coordinates of one focus are $(x = +c, y = 0)$, the square of the distance of any point from it

$$= (x' - c)^2 + y'^2 = x'^2 + y'^2 - 2cx' + c^2.$$

But (Art. 173)

$$x'^2 + y'^2 = b^2 + e^2x'^2, \text{ and } b^2 + c^2 = a^2.$$

Hence $$FP^2 = a^2 - 2cx' + e^2x'^2;$$

and recollecting that $c = ae$, we have

$$FP = a - ex'.$$

[We reject the value $(ex' - a)$ obtained by giving the other sign to the square root. For, since x' is less than a, and e less than 1, the quantity $ex' - a$ is constantly negative, and therefore does not concern us, as we are now considering, not the direction, but the absolute magnitude of the radius vector FP.]

We have, similarly, the distance from the other focus

$$F'P = a + ex',$$

since we have only to write $-c$ for $+c$ in the preceding formulæ.

Hence $$FP + F'P = 2a,$$

or, *The sum of the distances of any point on an ellipse from the foci is constant, and equal to the axis major.*

183. In applying the preceding proposition to the hyperbola, we obtain the same value for FP^2; but in extracting the square

root we must change the sign in the value of FP, for in the hyperbola x' is greater than a and e is greater than 1. Hence, $a - ex'$ is constantly negative; the absolute magnitude therefore of the radius vector is

$$FP = ex' - a.$$

In like manner $$F'P = ex' + a.$$

Hence $$F'P - FP = 2a.$$

Therefore, *in the hyperbola, the* difference *of the focal radii is constant, and equal to the transverse axis.*

The *rectangle* under the focal radii $= \pm(a^2 - e^2x^2)$, that is, (Art. 173) $= b'^2$.

184. The reader may prove the converse of the above results by seeking the locus of the vertex of a triangle, if the base and either sum or difference of sides be given.

Taking the middle point of the base ($= 2c$) for origin, the equation is

$$\sqrt{\{y^2 + (c + x)^2\}} \pm \sqrt{\{y^2 + (c - x)^2\}} = 2a,$$

which, when cleared of radicals, becomes

$$\frac{x^2}{a^2} + \frac{y^2}{a^2 - c^2} = 1.$$

Now, if the *sum* of the sides be given, since the sum must always be greater than the base, a is greater than c, therefore the coefficient of y^2 is positive, and the locus an *ellipse.*

If the *difference* be given, a is less than c, the coefficient of y^2 is negative, and the locus a *hyperbola.*

185. By the help of the preceding theorems we can describe an ellipse or hyperbola mechanically.

If the extremities of a thread be fastened at two fixed points F and F', it is plain that a pencil moved about so as to keep the thread always stretched will describe an ellipse whose foci are F and F', and whose axis major is equal to the length of the thread.

In order to describe a hyperbola, let a ruler be fastened at one extremity (F), and capable of moving round it, then if a thread, fastened to a fixed point F', and also to a fixed point on the ruler (R), be kept stretched by a ring at P, as the ruler is moved round, the point

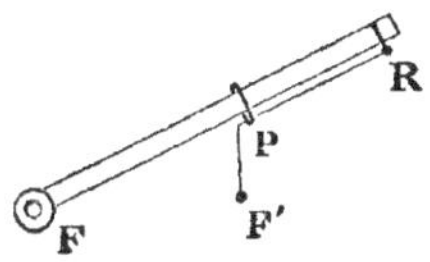

P will describe a hyperbola; for, since the sum of $F'P$ and PR is constant, the difference of FP and $F'P$ will be constant.

186. The polar of either focus is called the *directrix* of the conic section. The directrix must, therefore (Art. 169), be a line perpendicular to the axis major at a distance from the centre $= \pm \frac{a^2}{c}$.

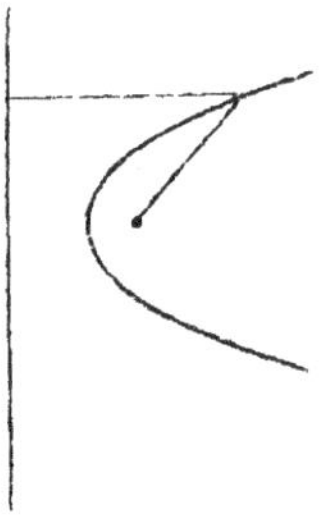

Knowing the distance of the directrix from the centre, we can find its distance from any point on the curve. It must be equal to

$$\frac{a^2}{c} - x', \text{ or } = \frac{a}{c}(a - ex') = \frac{1}{e}(a - ex').$$

But the distance of any point on the curve from the focus $= a - ex'$. Hence we obtain the important property, that *the distance of any point on the curve from the focus is in a constant ratio to its distance from the directrix*, viz. as e to 1.

Conversely, a conic section may be defined as the locus of a point whose distance from a fixed point (the focus) is in a constant ratio to its distance from a fixed line (the directrix). On this definition several writers have based the theory of conic sections. Taking the fixed line for the axis of x, the equation of the locus is at once written down

$$(x - x')^2 + (y - y')^2 = e^2y^2,$$

which it is easy to see will represent an ellipse, hyperbola, or parabola, according as e is less, greater than, or equal to 1.

Ex. If a curve be such that the distance of any point of it from a fixed point can be expressed as a rational function of the first degree of its coordinates, then the curve must be a conic section, and the fixed point its focus (see O'Brien's *Coordinate Geometry*, p. 85).

For, if the distance can be expressed

$$\rho = Ax + By + C,$$

since $Ax + By + C$ is proportional to the perpendicular let fall on the right line whose equation is ($Ax + By + C = 0$) the equation signifies that the distance of any point of the curve from the fixed point is in a constant ratio to its distance from this line.

187. *To find the length of the perpendicular from the focus on the tangent.*

The length of the perpendicular from the focus $(+c, 0)$ on

the line $\left(\frac{xx'}{a^2}+\frac{yy'}{b^2}=1\right)$ is, by Art. 34,

$$=\frac{1-\frac{cx'}{a^2}}{\sqrt{\left(\frac{x'^2}{a^4}+\frac{y'^2}{b^4}\right)}};$$

but, Art. 175, $$\sqrt{\left(\frac{x'^2}{a^4}+\frac{y'^2}{b^4}\right)}=\frac{b'}{ab}.$$

Hence (see fig. p. 177)

$$FT=\frac{b}{b'}(a-ex')=\frac{b}{b'}FP.$$

Likewise $$F'T'=\frac{b}{b'}(a+ex')=\frac{b}{b'}F'P.$$

Hence $$FT.F'T'=b^2 \text{ (since } a^2-e^2x'^2=b'^2\text{)},$$

or, *The rectangle under the focal perpendiculars on the tangent is constant, and equal to the square of the semi-axis minor.*

This property applies equally to the ellipse and the hyperbola.

188. *The focal radii make equal angles with the tangent.*

For we had $$FT=\frac{b}{b'}FP, \text{ or } \frac{FT}{FP}=\frac{b}{b'};$$

but $$\frac{FT}{FP}=\sin FPT.$$

Hence the sine of the angle which the focal radius vector FP makes with the tangent $=\frac{b}{b'}$. But we find, in like manner, the same value for $\sin F'PT'$, the sine of the angle which the other focal radius vector $F'P$ makes with the tangent.

The theorem of this article is true both for the ellipse and hyperbola, and, on looking at the figures, it is evident that the tangent to the ellipse is the *external* bisector of the angle between the focal radii, and the tangent to the hyperbola the *internal* bisector.

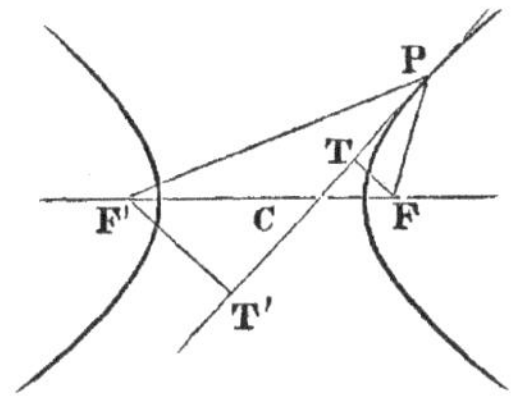

Hence, *if an ellipse and hyperbola, having the same foci, pass through the same point, they will cut each other at right angles,* that is to say, the tangent to the ellipse at that point will be at right angles to the tangent to the hyperbola.

Ex. 1. Prove analytically that confocal conics cut at right angles.

The coordinates of the intersection of the conics

$$\frac{x^2}{a^2}+\frac{y^2}{b^2}=1,\ \frac{x^2}{a'^2}+\frac{y^2}{b'^2}=1,$$

satisfy the relation obtained by subtracting the equations one from the other, viz.

$$\frac{(a^2-a'^2)\,x'^2}{a^2a'^2}+\frac{(b^2-b'^2)\,y'^2}{b^2b'^2}=0.$$

But if the conics be confocal, $a^2-a'^2=b^2-b'^2$, and this relation becomes

$$\frac{x'^2}{a^2a'^2}+\frac{y'^2}{b^2b'^2}=0.$$

But this is the condition (Art. 32) that the two tangents

$$\frac{xx'}{a^2}+\frac{yy'}{b^2}=1,\ \frac{xx'}{a'^2}+\frac{yy'}{b'^2}=1,$$

should be perpendicular to each other.

Ex. 2. Find the length of a line drawn through the centre parallel to either focal radius vector, and terminated by the tangent.

This length is found by dividing the perpendicular from the centre on the tangent $\left(\frac{ab}{b'}\right)$, by $\left(\frac{b}{b'}\right)$ the sine of the angle between the radius vector and tangent, and is therefore $=a$.

Ex. 3. Verify that the normal, which is a bisector of the angle between the focal radii, divides the distance between the foci into parts which are proportional to the focal radii (Euc. VI. 3). The distance of the foot of the normal from the centre is (Art. 180) $=e^2x'$. Hence its distances from the foci are $c+e^2x'$ and $c-e^2x'$, quantities which are evidently e times $a+ex'$ and $a-ex'$.

Ex. 4. To draw a normal to the ellipse from any point on the axis minor.

Ans. The circle through the given point and the two foci, will meet the curve at the point whence the normal is to be drawn.

189. Another important consequence may be deduced from the theorem of Art. 187, that the rectangle under the focal perpendiculars on the tangent is constant.

For, if we take any two tangents, we have (see figure, next page)

$$FT.F'T'=Ft.F't',\ \text{or}\ \frac{FT}{Ft}=\frac{F't'}{F'T'};$$

but $\frac{FT}{Ft}$ is the ratio of the sines of the parts into which the line FP divides the angle at P, and $\frac{F't'}{F'T'}$ is the ratio of the sines of the parts into which $F'P$ divides the same angle; we have, therefore, the angle $TPF = t'PF'$.

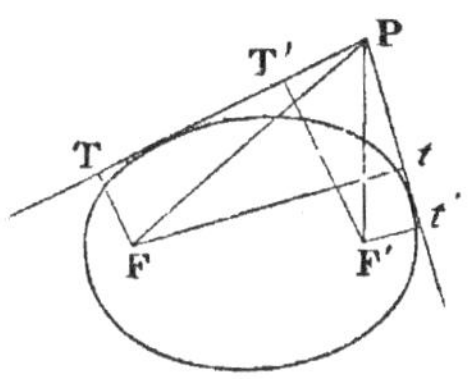

If we conceive a conic section to pass through P, having F and F' for foci, it was proved in Art. 188, that the tangent to it must be equally inclined to the lines FP, $F'P$; it follows, therefore, from the present Article, that it must be also equally inclined to PT, Pt; hence we learn that *if through any point* (P) *of a conic section we draw tangents* (PT, Pt) *to a* confocal *conic section, these tangents will be equally inclined to the tangent at* P.

190. *To find the locus of the foot of the perpendicular let fall from either focus on the tangent.*

The perpendicular from the focus is expressed in terms of the angles it makes with the axis by putting $x' = c$, $y' = 0$ in the formula of Art. 178, viz.,

$$p = \sqrt{(a^2 \cos^2 \alpha + b^2 \sin^2 \alpha)} - x' \cos \alpha - y' \sin \alpha.$$

Hence the polar equation of the locus is

$$\rho = \sqrt{(a^2 \cos^2 \alpha + b^2 \sin^2 \alpha)} - c \cos \alpha,$$

or $$\rho^2 + 2c\rho \cos \alpha + c^2 \cos^2 \alpha = a^2 \cos^2 \alpha + b^2 \sin^2 \alpha,$$

or $$\rho^2 + 2c\rho \cos \alpha = b^2.$$

This (Art. 95) is the polar equation of a circle whose centre is on the axis of x, at a distance from the focus $= -c$; the circle is, therefore, concentric with the curve. The radius of the circle is, by the same Article, $= a$.

Hence, *If we describe a circle having for diameter the transverse axis of an ellipse or hyperbola, the perpendicular from the focus will meet the tangent on the circumference of this circle.*

Or, conversely, *if from any point* F (see figure, p. 177) *we draw a radius vector* FT *to a given circle, and draw* TP *perpendicular to* FT, *the line* TP *will always touch a conic section, having* F *for its focus, which will be an ellipse or hyperbola, according as* F *is within or without the circle.*

It may be inferred from Art. 188, Ex. 2, that the line CT, whose length $= a$, is parallel to the focal radius vector $F'P$.

191. *To find the angle subtended at the focus by the tangent drawn to a central conic from any point* (xy).

Let the point of contact be $(x'y')$, the centre being the origin, then, if the radii from the focus F to the points (xy), $(x'y')$, be ρ, ρ', and make angles θ, θ', with the axis, it is evident that

$$\cos\theta = \frac{x+c}{\rho}, \quad \sin\theta = \frac{y}{\rho}; \quad \cos\theta' = \frac{x'+c}{\rho'}, \quad \sin\theta' = \frac{y'}{\rho'}.$$

Hence $$\cos(\theta - \theta') = \frac{(x+c)(x'+c) + yy'}{\rho\rho'};$$

but from the equation of the tangent we must have

$$\frac{xx'}{a^2} + \frac{yy'}{b^2} = 1.$$

Substituting this value of yy', we get

$$\rho\rho' \cos(\theta - \theta') = xx' + cx + cx' + c^2 - \frac{b^2}{a^2}xx' + b^2,$$

or $$= e^2xx' + cx + cx' + a^2 = (a + ex)(a + ex');$$

or, since $\rho' = a + ex'$, we have, (see O'Brien's *Coordinate Geometry*, p. 156),

$$\cos(\theta - \theta') = \frac{a + ex}{\rho}.$$

Since this value depends solely on the coordinates xy, and does not involve the coordinates of the point of contact, either tangent drawn from xy subtends the same angle at the focus. Hence, *The angle subtended at the focus by any chord is bisected by the line joining the focus to its pole.*

192. *The line joining the focus to the pole of any chord passing through it is perpendicular to that chord.*

This may be deduced as a particular case of the last Article, the angle subtended at the focus being in this case 180°; or directly as follows:—The equation of the perpendicular through any point $x'y'$ to the polar of that point $\left(\frac{xx'}{a^2} + \frac{yy'}{b^2} = 1\right)$ is, as in Art. 180,

$$\frac{a^2x}{x'} - \frac{b^2y}{y'} = c^2.$$

But if $x'y'$ be anywhere on the directrix, we have $x' = \frac{a^2}{c}$, and it will then be found that both the equation of the polar and that of the perpendicular are satisfied by the coordinates of the focus $(x = c, y = 0)$.

When in any curve we use polar coordinates, the portion intercepted by the tangent on a perpendicular to the radius vector drawn through the pole is called the *polar subtangent*. Hence the theorem of this Article may be stated thus: *The focus being the pole, the locus of the extremity of the polar subtangent is the directrix.*

It will be proved (Chap. XII.) that the theorems of this and the last Article are true also for the parabola.

Ex. 1. The angle is constant which is subtended at the focus, by the portion intercepted on a variable tangent between two fixed tangents.

By Art. 191, it is half the angle subtended by the chord of contact of the fixed tangents.

Ex. 2. If any chord PP' cut the directrix in D, then FD is the external bisector of the angle PFP'. For FT is the internal bisector (Art. 191); but D is the pole of FT (since it is the intersection of PP', the polar of T, with the directrix, the polar of F); therefore, DF is perpendicular to FT, and is therefore the external bisector.

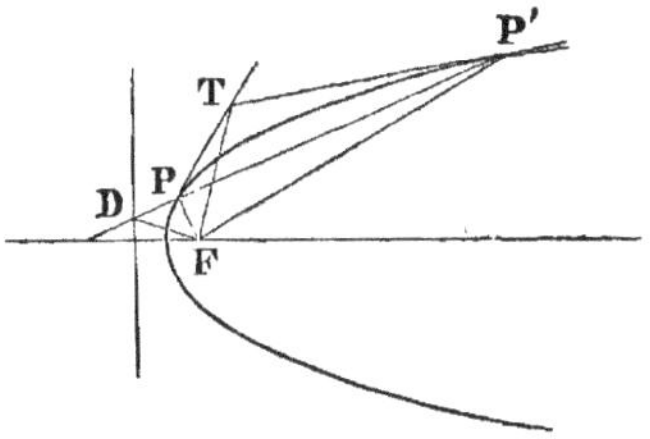

[The following theorems (communicated to me by the Rev. W. D. Sadleir) are founded on the analogy between the equations of the polar and the tangent.]

Ex. 3. If a point be taken anywhere on a fixed perpendicular to the axis, the perpendicular from it on its polar will pass through a fixed point on the axis. For the intercept made by the perpendicular will (as in Art. 180) be e^2x', and will therefore be constant when x' is constant.

Ex. 4. Find the lengths of the perpendicular from the centre and from the foci on the polar of $x'y'$.

Ex. 5. Prove $CM.PN' = b^2$. This is analogous to the theorem that the rectangle under the normal and the central perpendicular on tangent is constant.

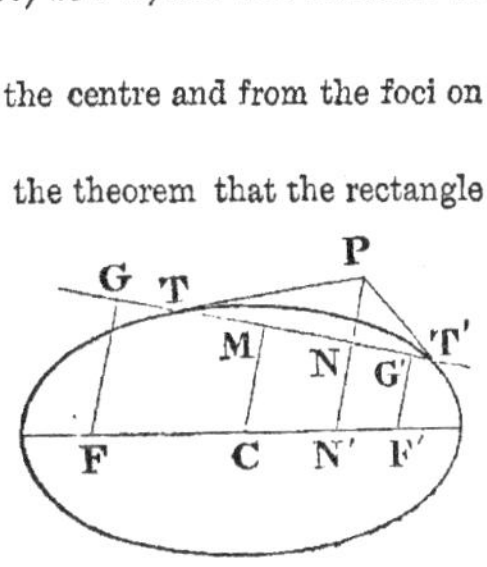

Ex. 6. Prove $PN'.NN' = \frac{b^2}{a^2}(a^2 - e^2x'^2)$. When P is on the curve this equation gives us the known expression for the normal $= \frac{bb'}{a}$ (Art. 181).

Ex. 7. Prove $FG.F'G' = CM.NN'$. When P is on the curve this theorem becomes $FG.F'G' = b^2$.

193. *To find the polar equation of the ellipse or hyperbola, the focus F' being the pole.*

The length of the focal radius vector (Art. 182) $= a - ex'$; but x' (being measured from the centre) $= \rho\cos\theta + c$.

Hence $$\rho = a - e\rho\cos\theta - ec,$$

or $$\rho = \frac{a(1-e^2)}{1+e\cos\theta} = \frac{b^2}{a}\cdot\frac{1}{1+e\cos\theta}.$$

The double ordinate at the focus is called the *parameter*; its half is found, by making $\theta = 90^\circ$ in the equation just given, to be $= \frac{b^2}{a} = a(1-e^2)$. The parameter is commonly denoted by the letter p. Hence the equation is often written

$$\rho = \frac{p}{2} \cdot \frac{1}{1 + e\cos\theta}.$$

The parameter is also called the *Latus Rectum*.

Ex. 1. The harmonic mean between the segments of a focal chord is constant, and equal to the semi-parameter.

For, if the radius vector FP, when produced backwards through the focus, meet the curve again in P', then FP being $\frac{p}{2} \cdot \frac{1}{1 + e\cos\theta}$, FP', which answers to $(\theta + 180^\circ)$, will $= \frac{p}{2} \cdot \frac{1}{1 - e\cos\theta}$.

Hence
$$\frac{1}{FP} + \frac{1}{FP'} = \frac{4}{p}.$$

Ex. 2. The rectangle under the segments of a focal chord is proportional to the whole chord.

This is merely another way of stating the result of the last Example; but it may be proved directly by calculating the quantities $FP \,.\, FP'$, and $FP + FP'$, which are easily seen to be respectively

$$\frac{b^4}{a^2} \frac{1}{1 - e^2\cos^2\theta}, \text{ and } \frac{2b^2}{a} \frac{1}{1 - e^2\cos^2\theta}.$$

Ex. 3. Any focal chord is a third proportional to the transverse axis and the parallel diameter.

For it will be remembered that the length of a semi-diameter making an angle θ with the transverse axis is (Art. 161)

$$R^2 = \frac{b^2}{1 - e^2\cos^2\theta}.$$

Hence the length of the chord $FP + FP'$ found in the last Example $= \frac{2R^2}{a}$.

Ex. 4. The sum of two focal chords drawn parallel to two conjugate diameters is constant.

For the sum of the squares of two conjugate diameters is constant (Art. 173).

Ex 5. The sum of the reciprocals of two focal chords at right angles to each other is constant.

194. The equation of the ellipse, referred to the vertex, is

$$\frac{(x-a)^2}{a^2} + \frac{y^2}{b^2} = 1,$$

or
$$y^2 = \frac{2b^2}{a} x - \frac{b^2}{a^2} x^2 = px - \frac{b^2}{a^2} x^2.$$

Hence, in the ellipse, the square of the ordinate *is less than* the rectangle under the parameter and abscissa.

The equation of the hyperbola is found in like manner,

$$y^2 = px + \frac{b^2}{a^2}x^2.$$

Hence, in the hyperbola, the square of the ordinate *exceeds* the rectangle under the parameter and abscissa.

We shall show, in the next chapter, that in the parabola these quantities are *equal*.

It was from this property that the names *parabola*, *hyperbola*, and *ellipse*, were first given (see Pappus, *Math. Coll.*, Book VII.).

CONFOCAL CONICS.*

194 (*a*). Since the distance between the foci is $2c$, where $c^2 = a^2 - b^2$, two concentric and coaxal conics will have the same foci when the difference of the squares of the axes is the same for both; and if we take the ellipse whose semi-axes are a and b, any conic will be confocal with it, whose equation is of the form

$$\frac{x^2}{a^2 \pm \lambda^2} + \frac{y^2}{b^2 \pm \lambda^2} = 1.$$

If we give the positive sign to λ^2, the confocal conic will be an ellipse; it will also be an ellipse when λ^2 is negative as long as it is less than b^2. When λ^2 is between b^2 and a^2, the curve will be a hyperbola, and when λ^2 is greater than a^2, the curve is imaginary. If $\lambda^2 = b^2$, the equation reducing itself to $y^2 = 0$, the axis of x is itself the limit which separates confocal ellipses from hyperbolas. But the two foci belong to this limit in a special sense. In fact, through a given point can in general be drawn two conics confocal to a given one, since we have a quadratic to determine λ^2, viz.

$$\frac{x'^2}{a^2 - \lambda^2} + \frac{y'^2}{b^2 - \lambda^2} = 1,$$

or $$\lambda^4 - \lambda^2 (a^2 + b^2 - x'^2 - y'^2) + a^2b^2 - b^2x'^2 - a^2y'^2 = 0.$$

When $y' = 0$, this quadratic becomes $(\lambda^2 - b^2)(\lambda^2 - a^2 + x'^2) = 0$, and *one* of its roots is $\lambda^2 = b^2$, but if we have also $x'^2 = a^2 - b^2$,

* This section may be omitted on a first reading.

the second root is also $\lambda^2 = b^2$, and therefore the two foci are in a special sense points corresponding to that value of λ^2.

If in the quadratic for λ^2 we substitute $\lambda^2 = a^2$, we get the positive result $(a^2 - b^2)x'^2$; if we substitute $\lambda^2 = b^2$ we get the negative result $(b^2 - a^2)y'^2$; if we substitute negative infinity we get a positive result; hence, one of the roots lies between a^2 and b^2, and the other is less than b^2; that is to say, one of the conics is a hyperbola and the other an ellipse, as is evident geometrically. In fact, through a given point P can clearly be described two conics having two given points F, F' for foci; viz. the ellipse, whose major axis is the sum of FP, $F'P$, and the hyperbola whose transverse axis is the difference of the same lines. Conversely, if a', a'' be the semi-axes major of the ellipse and hyperbola, FP and $F'P$ are $a' + a''$ and $a' - a''$.

194 (*b*). This theory can be made to furnish a kind of coordinate system which is sometimes employed; viz. any point P is known when we know the axes of the two conics, confocal to a given one, which can be drawn through it; and in terms of these axes can be expressed the ordinary coordinates of P, and the lengths of all other lines geometrically connected with it. Perhaps the easiest way of getting such expressions is to investigate anew the problem of drawing through P a conic with given foci, taking for unknown quantity the transverse axis of the conic. Then since c^2 is known, we write $a^2 - c^2$ for b^2, and have

$$\frac{x'^2}{a^2} + \frac{y'^2}{a^2 - c^2} = 1,$$

or

$$a^4 - a^2(x'^2 + y'^2 + c^2) + c^2x'^2 = 0.$$

In like manner, if b^2 had been taken as the unknown quantity we should have had

$$b^4 - b^2(x'^2 + y'^2 - c^2) - c^2y'^2 = 0.$$

The products of the roots of these equations are respectively $c^2x'^2$ and $-c^2y'^2$. Hence, we have at once expressions for the coordinates of the intersections of two confocal conics, viz. $c^2x'^2 = a'^2a''^2$, $c^2y'^2 = -b'^2b''^2$. The last value being negative, it follows that one of the values of b^2 is positive and the other negative; that is to say, that one of the conics is an ellipse and the other a hyperbola. Considering then b''^2 as containing implicitly a negative sign, the values we have obtained for the coordinates may be written symmetrically

$$x'^2 = \frac{a'^2a''^2}{a^2 - b^2}, \quad y'^2 = \frac{b'^2b''^2}{b^2 - a^2}.$$

194 (*c*). From the second term in either of the equations we get an expression for the square of the radius vector to the point *P*, viz.

$$x'^2 + y'^2 = a'^2 + a''^2 - c^2 = a'^2 + b''^2 = b'^2 + a''^2.$$

This also may be got by adding the expressions for x'^2 and y'^2 just found, since

$$a'^2a''^2 - b'^2b''^2 = a'^2(a''^2 - b''^2) + b''^2(a'^2 - b'^2),$$

and $$a'^2 - b'^2 = a''^2 - b''^2 = c^2.$$

The square of the semi-diameter of the ellipse conjugate to *CP* is given by the equation $\beta^2 = a'^2 + b'^2 - (a'^2 + b''^2)$, and is therefore $b'^2 - b''^2$ or $a'^2 - a''^2$.

If p' be the perpendicular on the tangent to the ellipse at *P*, we have $\beta p' = a'b'$, and therefore

$$p'^2 = \frac{a'^2b'^2}{a'^2 - a''^2}.$$

In like manner if p'' be the perpendicular on the tangent to the hyperbola we have

$$p''^2 = \frac{a''^2b''^2}{a''^2 - a'^2}.$$

The reader will observe the symmetry which exists between these values for p'^2, p''^2, and the values already found for x'^2, y'^2. If the two tangents at *P* be taken as axes of coordinates, p', p'' are the coordinates of the centre *C*. The analogy then between the values for p', p'' and those for x', y' may be stated as follows: With the point *P* as centre, two confocal conics may be described having the tangents at *P* as axes, and intersecting in *C*. The axes of the new system are a', a''; b', b''; and the tangents at *C* to the new system are the axes of the old system.

194 (d). Returning to the quadratic of 194 (a), if λ'^2, λ''^2 be the roots, we have $\lambda'^2\lambda''^2 = a^2b^2 - b^2x'^2 - a^2y'^2$. Now if $x'y'$ be a point external to $\frac{x^2}{a^2} + \frac{y^2}{b^2} = 1$, we have $\lambda'^2 = a'^2 - a^2$, $\lambda''^2 = a''^2 - a^2$; and it will be observed that λ''^2 is essentially negative, since the axis of any hyperbola of the system is less than that of any ellipse. Thus we have

$$\frac{x'^2}{a^2} + \frac{y'^2}{b^2} - 1 = \frac{(a'^2 - a^2)(a^2 - a''^2)}{a^2b^2}.$$

The expression given (Ex. 3, Art. 169) for the angle between the tangents to an ellipse from an external point may be thrown into the form

$$\tan\phi = \frac{2\sqrt{(a'^2 - a^2)(a^2 - a''^2)}}{(a'^2 - a^2) + (a''^2 - a^2)}.$$

Now, when we have a formula $\tan\phi = \frac{2\lambda\mu}{\lambda^2 - \mu^2}$, we have at once $\tan\frac{1}{2}\phi = \frac{\mu}{\lambda}$, or in the present case $= \sqrt{\left(\frac{a^2 - a''^2}{a'^2 - a^2}\right)}$.

We have seen (Art. 189) that the tangents PT, Pt are equally inclined to the tangent to the confocal ellipse at P, or, in other words, that that tangent is the external bisector of the angle TPt. If then that tangent make an angle ψ with PT, ψ will be the complement of $\frac{1}{2}\phi$, and we have

$$\sin\psi = \sqrt{\left(\frac{a'^2 - a^2}{a'^2 - a''^2}\right)}, \quad \cos\psi = \sqrt{\left(\frac{a^2 - a''^2}{a'^2 - a''^2}\right)}.$$

Cor. 1. We have always

$$a'^2\cos^2\psi + a''^2\sin^2\psi = a^2.$$

Cor. 2. If on the tangents PT, Pt be taken from P portions, equal respectively to the focal distances PF, PF', the length of the line joining their extremities will be $2a$. For if we consider the triangle whose sides are $a' + a''$, $a' - a''$ (see Art. 194a) and $2a$, and apply the ordinary trigonometric formula $\tan^2\frac{1}{2}C = \frac{(s-a)(s-b)}{s(s-c)}$, we find for the angle between the first two lines the same value as that just found for ϕ.

Cor. 3. If from a point P tangents be drawn to two fixed confocal ellipses, the ratio $(\sin\psi : \sin\psi')$ of the sines of the angles which these tangents make with the tangent to the confocal ellipse passing through P will be constant while P moves on that ellipse. For if a and A be the semi-axes of the interior ellipses, we have, from what has been just proved,

$$\frac{\sin\psi}{\sin\psi'} = \sqrt{\left(\frac{a'^2 - a^2}{a'^2 - A^2}\right)},$$

an expression not involving a''^2, and therefore the same for every point on the ellipse a'.

THE ASYMPTOTES.

195. We have hitherto discussed properties common to the ellipse and the hyperbola. There is, however, one class of properties of the hyperbola which have none corresponding to them in the ellipse, those, namely, depending on the asymptotes, which in the ellipse are imaginary.

We saw that the equation of the asymptotes was always obtained by putting the terms containing the highest powers of the variables $= 0$, the centre being the origin. Thus the equation of the curve, referred to any pair of conjugate diameters, being

$$\frac{x^2}{a'^2} - \frac{y^2}{b'^2} = 1,$$

that of the asymptotes is

$$\frac{x^2}{a'^2} - \frac{y^2}{b'^2} = 0, \text{ or } \frac{x}{a'} - \frac{y}{b'} = 0, \text{ and } \frac{x}{a'} + \frac{y}{b'} = 0.$$

Hence the asymptotes are parallel to the diagonals of the parallelogram, whose adjacent sides are any pair of conjugate semi-diameters. For, the equation of CT is $\frac{y}{x} = \frac{b'}{a'}$, and must, therefore, coincide with one asymptote, while the equation of AB $\left(\frac{x}{a'} + \frac{y}{b'} = 1\right)$ is parallel to the other (see Art. 167).

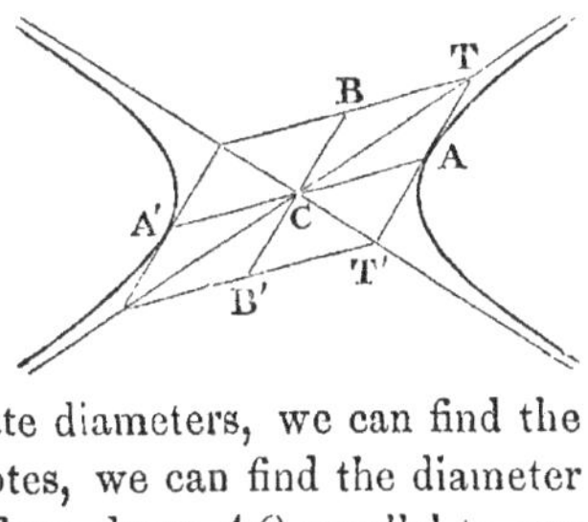

Hence, given any two conjugate diameters, we can find the asymptotes; or, given the asymptotes, we can find the diameter conjugate to any given one; for if we draw AO parallel to one asymptote, to meet the other, and produce it till $OB = AO$, we find B, the extremity of the conjugate diameter.

196. *The portion of any tangent intercepted by the asymptotes is bisected at the curve, and is equal to the conjugate diameter.*

This appears at once from the last Article, where we have proved $AT = b' = AT'$; or directly, taking for axes the diameter through the point and its conjugate, the equation of the asymptotes is

$$\frac{x^2}{a'^2} - \frac{y^2}{b'^2} = 0.$$

Hence, if we take $x = a'$, we have $y = \pm b'$; but the tangent at A being parallel to the conjugate diameter, this value of the ordinate is the intercept on the tangent.

197. *If any line cut a hyperbola, the portions DE, FG, intercepted between the curve and its asymptotes, are equal.*

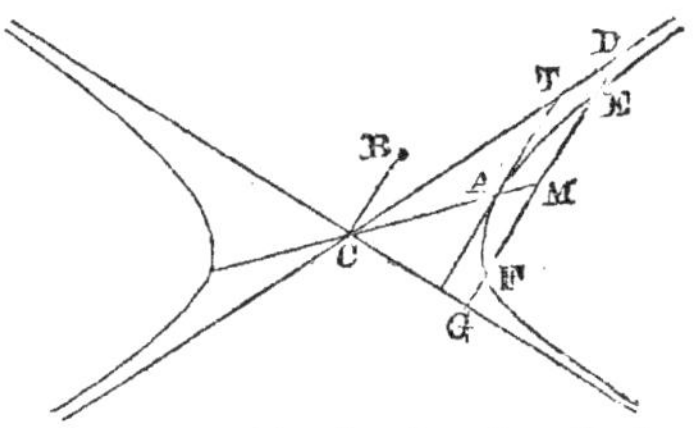

For, if we take for axes a diameter parallel to DG and its conjugate, it appears from the last Article that the portion DG is bisected by the diameter; so is also the portion EF; hence $DE = FG$.

The lengths of these lines can immediately be found, for, from the equation of the asymptotes $\left(\frac{x^2}{a'^2} - \frac{y^2}{b'^2} = 0\right)$, we have $y\,(= DM = MG) = \pm \frac{b'}{a'}x$.

Again, from the equation of the curve

we have $$y\,(= EM = FM) = \pm b' \sqrt{\left(\frac{x^2}{a'^2} - 1\right)}.$$

Hence $$DE\,(= FG) = b' \left\{\frac{x}{a'} - \sqrt{\left(\frac{x^2}{a'^2} - 1\right)}\right\},$$

and $$DF\,(= EG) = b' \left\{\frac{x}{a'} + \sqrt{\left(\frac{x^2}{a'^2} - 1\right)}\right\}.$$

198. From these equations it at once follows that the *rectangle $DE.DF$ is constant, and $= b'^2$.* Hence, the greater DF is, the smaller will DE be. Now, the further from the centre we draw DF the greater will it be, and it is evident from the value

given in the last article, that by taking x sufficiently large, we can make DF greater than any assigned quantity. Hence, *the further from the centre we draw any line, the less will be the intercept between the curve and its asymptote, and by increasing the distance from the centre, we can make this intercept less than any assigned quantity.*

199. If the asymptotes be taken for axes, the coefficients g and f of the general equation vanish, since the origin is the centre; and the coefficients a and b vanish, since the axes meet the curve at infinity (Art. 138, Ex. 4); hence the equation reduces to the form

$$xy = k^2.$$

The geometrical meaning of this equation evidently is, that *the area of the parallelogram formed by the coordinates is constant.*

The equation being given in the form $xy = k^2$, the equation of any chord is (Art. 86),

$$(x - x')(y - y'') = xy - k^2,$$

or

$$x'y + y''x = k^2 + x'y''.$$

Making $x' = x''$ and $y' = y''$, we find the equation of the tangent

$$x'y + y'x = 2k^2,$$

or (writing $x'y'$ for k^2)

$$\frac{x}{x'} + \frac{y}{y'} = 2.$$

From this form it appears that the intercepts made on the asymptotes by any tangent $= 2x'$ and $2y'$; their rectangle is, therefore, $4k^2$. Hence, *the triangle which any tangent forms with the asymptotes has a constant area, and is equal to double the area of the parallelogram formed by the coordinates.*

Ex. 1. If two fixed points $(x'y', x''y'')$ on a hyperbola be joined to any variable point on the curve $(x'''y''')$, the portion which the joining lines intercept on either asymptote is constant.

The equation of one of the joining lines being

$$x'''y + y'x = y'x''' + k^2,$$

the intercept made by it from the origin on the axis of x is found, by making $y = 0$, to be $x''' + x'$. Similarly the intercept from the origin made by the other joining line is $x''' + x''$, and the difference between these two $(x' - x'')$ is independent of the position of the point $x'''y'''$.

Ex. 2. Find the coordinates of the intersection of the tangents at $x'y'$, $x''y''$.
Solve for x and y from

$$x'y + y'x = 2k^2, \quad x''y + y''x = 2k^2,$$

and we find
$$x = \frac{2k^2 (x' - x'')}{x'y'' - y'x''},$$

which if we substitute for y', y'', $\frac{k^2}{x'}, \frac{k^2}{x''}$ becomes $\frac{2x'x''}{x' + x''}$.

Similarly
$$y = \frac{2y'y''}{y' + y''}.$$

200. *To express the quantity k^2 in terms of the lengths of the axes of the curve.*

Since the axis bisects the angle between the asymptotes, the coordinates of its vertex are found, by putting $x = y$ in the equation $xy = k^2$, to be $x = y = k$.

Hence, if θ be the angle between the axis and the asymptote

$$a = 2k \cos\theta,$$

(since a is the base of an isosceles triangle whose sides $= k$ and base angle $= \theta$), but (Art. 165)

$$\cos\theta = \frac{a}{\sqrt{(a^2 + b^2)}};$$

hence
$$k = \frac{\sqrt{(a^2 + b^2)}}{2}.$$

And the equation of the curve, referred to its asymptotes, is

$$xy = \frac{a^2 + b^2}{4}.$$

201. *The perpendicular from the focus on the asymptote is equal to the conjugate semi-axis b.*

For it is $CF \sin\theta$, but $CF = \sqrt{(a^2 + b^2)}$, and $\sin\theta = \frac{b}{\sqrt{(a^2 + b^2)}}$.

This might also have been deduced as a particular case of the property, that the product of the perpendiculars from the foci on any tangent is constant, and $= -b^2$. For the asymptote may be considered as a tangent, whose point of contact is at an infinite distance (Art. 154), and the perpendiculars from the foci on it are evidently equal to each other, and on opposite sides of it.

202. *The distance of the focus from any point on the curve is equal to the length of a line drawn through the point parallel to an asymptote to meet the directrix.*

For the distance from the focus is e times the distance from the directrix (Art. 186), and the distance from the directrix is to the length of the parallel line as $\cos\theta\left(=\frac{1}{e}\text{, Art. 167}\right)$ is to 1.

Hence has been derived a method of describing the hyperbola by continued motion. A ruler ABR, bent at B, slides along the fixed line DD'; a thread of a length $=RB$ is fastened at the two points R and F, while a ring at P keeps the thread always stretched; then, as the ruler is moved along, the point P will describe an hyperbola, of which F is a focus, DD' a directrix, and BR parallel to an asymptote, since PF must always $=PB$.

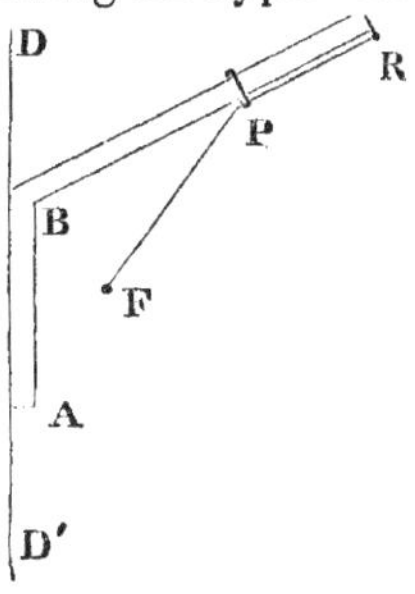

CHAPTER XII.

THE PARABOLA.

REDUCTION OF THE EQUATION.

203. THE equation of the second degree (Art. 137) will represent a parabola, when the first three terms form a perfect square, or when the equation is of the form

$$(\alpha x + \beta y)^2 + 2gx + 2fy + c = 0.$$

We saw (Art. 140) that we could not transform this equation so as to make the coefficients of x and y both to vanish. The form of the equation, however, points at once to another method of simplifying it. We know (Art. 34) that the quantities $\alpha x + \beta y$, $2gx + 2fy + c$, are respectively proportional to the lengths of perpendiculars let fall from the point (xy) on the right lines, whose equations are

$$\alpha x + \beta y = 0, \quad 2gx + 2fy + c = 0.$$

Hence, the equation of the parabola asserts that the square of the perpendicular from any point of the curve on the first of these lines is proportional to the perpendicular from the same point on the second line. Now if we transform our equation, making these two lines the new axes of coordinates, then since the new x and y are proportional to the perpendiculars from any point on the new axes, the transformed equation must be of the form $y^2 = px$.

The new origin is evidently a point on the curve; and since for every value of x we have two equal and opposite values of y, our new axis of x will be a diameter whose ordinates are parallel to the new axis of y. But the ordinate drawn at the extremity of any diameter touches the curve (Art. 145); therefore the new axis of y is a tangent at the origin. Hence the line $\alpha x + \beta y$ is the diameter passing through the origin, and $2gx + 2fy + c$ is the tangent at the point where this diameter meets the curve. And the equation of the curve referred to a diameter and tangent at its extremity, as axes, is of the form $y^2 = px$.

204. The new axes to which we were led in the last article are in general not rectangular. We shall now show that it is possible to transform the equation to the form $y^2 = px$, the new axes being rectangular. If we introduce the arbitrary constant k, it is easy to verify that the equation of the parabola may be written in the form

$$(\alpha x + \beta y + k)^2 + 2(g - \alpha k)x + 2(f - \beta k)y + c - k^2 = 0.$$

Hence, as in the last article, $\alpha x + \beta y + k$ is a diameter, $2(g - \alpha k)x + 2(f - \beta k)y + c - k^2$ is the tangent at its extremity, and if we take these lines as axes, the transformed equation is of the form $y^2 = px$. Now the condition that these new axes should be perpendicular is (Art. 25)

$$\alpha(g - \alpha k) + \beta(f - \beta k) = 0,$$

whence

$$k = \frac{\alpha g + \beta f}{\alpha^2 + \beta^2}.$$

Since we get a simple equation for k, we see that there is *one* diameter whose ordinates cut it perpendicularly, and this diameter is called the *axis* of the curve.

205. We might also have reduced the equation to the form $y^2 = px$ by direct transformation of coordinates. In Chap. XI. we reduced the general equation by first transforming to parallel axes through a new origin, and then turning round the axes so as to make the coefficient of xy vanish. We might equally well have performed this transformation in the opposite order; and in the case of the parabola this is more convenient, since we cannot, by transformation to a new origin, make the coefficients of x and y both vanish.

We take for our new axes the line $\alpha x + \beta y$, and the line perpendicular to it $\beta x - \alpha y$. Then since the new X and Y are to denote the lengths of perpendiculars from any point on the new axes, we have (Art. 34)

$$Y = \frac{\alpha x + \beta y}{\sqrt{(\alpha^2 + \beta^2)}}; \quad X = \frac{\beta x - \alpha y}{\sqrt{(\alpha^2 + \beta^2)}}.$$

If for shortness we write $\alpha^2 + \beta^2 = \gamma^2$, the formulæ of transformation become

$$\gamma Y = \alpha x + \beta y, \quad \gamma X = \beta x - \alpha y,$$

whence

$$\gamma x = \alpha Y + \beta X, \quad \gamma y = \beta Y - \alpha X.$$

Making these substitutions in the equation of the curve it becomes

$$\gamma^3 Y^2 + 2(g\beta - f\alpha) X + 2(g\alpha + f\beta) Y + \gamma c = 0.$$

Thus, by turning round the axes, we have reduced the equation to the form $\qquad b'y^2 + 2g'x + 2f'y + c' = 0.$

If we change now to parallel axes through any new origin $x'y'$, substituting $x + x'$, $y + y'$ for x and y, the equation becomes

$$b'y^2 + 2g'x + 2(b'y' + f')y + b'y'^2 + 2g'x' + 2f'y' + c' = 0.$$

The coefficient of x is thus unaltered by transformation, and therefore cannot in this way be made to vanish. But we can evidently determine x' and y', so that the coefficients of y and the absolute term may vanish, and the equation thus be reduced to $y^2 = px$. The actual values of the coordinates of the new origin are $y' = -\frac{f'}{b'}$, $x' = \frac{f'^2 - b'c'}{2g'b'}$; and p is evidently $-\frac{2g'}{b'}$, or in terms of the original coefficients

$$p = \frac{2(f\alpha - g\beta)}{(\alpha^2 + \beta^2)^{\frac{3}{2}}}.$$

When the equation of a parabola is reduced to the form $y^2 = px$, the quantity p is called the *parameter* of the diameter, which is the axis of x; and if the axes be rectangular, p is called the *principal parameter* (see Art. 194).

Ex. 1. Find the principal parameter of the parabola

$$9x^2 + 24xy + 16y^2 + 22x + 46y + 9 = 0.$$

First, if we proceed as in Art. 204, we determine $k = 5$. The equation may then be written

$$(3x + 4y + 5)^2 = 2(4x - 3y + 8).$$

Now if the distances of any point from $3x + 4y + 5$ and $4x - 3y + 8$ be Y and X, we have

$$5Y = 3x + 4y + 5, \quad 5X = 4x - 3y + 8,$$

and the equation may be written $\qquad Y^2 = \frac{2}{5}X.$

The process of Art. 205 is first to transform to the lines $3x + 4y$, $4x - 3y$ as axes, when the equation becomes

$$25Y^2 + 50Y - 10X + 9 = 0,$$

or

$$25(Y + 1)^2 = 10X + 16,$$

which becomes $Y^2 = \frac{2}{5}X$ when transformed to parallel axes through $(-\frac{8}{5}, -1)$.

Ex. 2. Find the parameter of the parabola

$$\frac{x^2}{a^2} - \frac{2xy}{ab} + \frac{y^2}{b^2} - \frac{2x}{a} - \frac{2y}{b} + 1 = 0. \qquad \textit{Ans. } \frac{4a^2b^2}{(a^2 + b^2)^{\frac{3}{2}}}.$$

This value may also be deduced directly by the help of the following theorem, which will be proved afterwards :—"The focus of a parabola is the foot of a perpendicular let fall from the intersection of two tangents which cut at right angles on their chord of contact;" and "The parameter of a conic is found by dividing four times the rectangle under the segments of a focal chord by the length of that chord" (Art. 193, Ex. 1).

Ex. 3. If a and b be the lengths of two tangents to a parabola which intersect at right angles, and m one quarter of the parameter, prove

$$\frac{a^{\frac{2}{3}}}{b^{\frac{4}{3}}} + \frac{b^{\frac{2}{3}}}{a^{\frac{4}{3}}} = \frac{1}{m^{\frac{2}{3}}}.$$

206. If in the original equation $g\beta = f\alpha$, the coefficient of x vanishes in the equation transformed as in the last article; and that equation $b'y^2 + 2f'y + c' = 0$, being equivalent to one of the form

$$b'(y-\lambda)(y-\mu) = 0,$$

represents two real, coincident, or imaginary lines parallel to the new axis of x.

We can verify that in this case the general condition that the equation should represent right lines is fulfilled. For this condition may be written

$$c(ab - h^2) = af^2 - 2hfg + bg^2.$$

But if we substitute for a, h, b, respectively, α^2, $\alpha\beta$, β^2, the left-hand side of the equation vanishes, and the right-hand side becomes $(f\alpha - g\beta)^2$. Writing the condition $f\alpha = g\beta$ in either of the forms $f\alpha^2 = g\alpha\beta$, $f\alpha\beta = g\beta^2$, we see that the general equation of the second degree represents two parallel right lines when $h^2 = ab$, and also either $af = hg$, or $fh = bg$.

*207. If the original axes were oblique, the equation is still reduced, as in Art. 205, by taking for our new axes the line $\alpha x + \beta y$, and the line perpendicular to it, whose equation is (Art. 26)

$$(\beta - \alpha\cos\omega)x - (\alpha - \beta\cos\omega)y = 0.$$

And if we write $\gamma^2 = \alpha^2 + \beta^2 - 2\alpha\beta\cos\omega$, the formulæ of transformation become, by Art. 34,

$$\gamma Y = (\alpha x + \beta y)\sin\omega, \quad \gamma X = (\beta - \alpha\cos\omega)x - (\alpha - \beta\cos\omega)y;$$

whence

$$\gamma x\sin\omega = (\alpha - \beta\cos\omega)Y + \beta X\sin\omega;$$

$$\gamma y\sin\omega = (\beta - \alpha\cos\omega)Y - \alpha X\sin\omega.$$

Making these substitutions, the equation becomes

$$\gamma^3 Y^2 + 2\sin^2\omega\,(g\beta - f\alpha)X$$
$$+ 2\sin\omega\,\{g(\alpha - \beta\cos\omega) + f(\beta - \alpha\cos\omega)\}Y + \gamma c\sin^2\omega = 0.$$

And the transformation to parallel axes proceeds as in Art. 205. The principal parameter is

$$p = -\frac{2g'}{b'} = \frac{2(f\alpha - g\beta)\sin^2\omega}{(\alpha^2 + \beta^2 - 2\alpha\beta\cos\omega)^{\frac{3}{2}}}.$$

Ex. Find the principal parameter of

$$\frac{x^2}{a^2} - \frac{2xy}{ab} + \frac{y^2}{b^2} - \frac{2x}{a} - \frac{2y}{b} + 1 = 0. \qquad \textit{Ans.}\ \frac{4a^2b^2\sin^2\omega}{(a^2 + b^2 + 2ab\cos\omega)^{\frac{3}{2}}}.$$

FIGURE OF THE CURVE.

208. From the equation $y^2 = px$ we can at once perceive the figure of the curve. It must be symmetrical on both sides of the axis of x, since every value for x gives two equal and opposite for y. None of it can lie on the negative side of the origin, since if we make x negative, y will be imaginary, and as we give increasing positive values to x, we obtain increasing values for y. Hence the figure of the curve is that here represented.

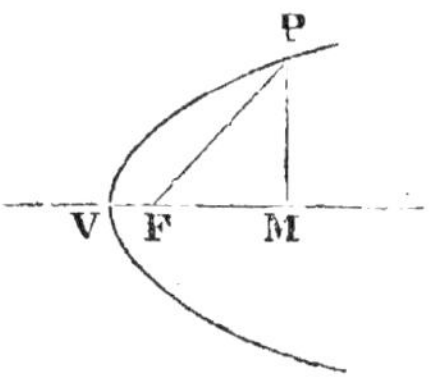

Although the parabola resembles the hyperbola in having infinite branches, yet there is an important difference between the nature of the infinite branches of the two curves. Those of the hyperbola, we saw, tend ultimately to coincide with two diverging right lines; but this is not true for the parabola, since, if we seek the points where any right line $(x = ky + l)$ meets the parabola $(y^2 = px)$, we obtain the quadratic

$$y^2 - pky - pl = 0,$$

whose roots can never be infinite as long as k and l are finite.

There is no finite right line which meets the parabola in two coincident points at infinity; for any diameter $(y = m)$, which meets the curve once at infinity (Art. 142), meets it once also in the point $x = \frac{m^2}{p}$; and although this value increases as m increases, yet it will never become infinite as long as m is finite.

209. The figure of the parabola may be more clearly conceived from the following theorem: If we suppose one vertex

and focus of an ellipse given, while its axis major increases without limit, the curve will ultimately become a parabola.

The equation of the ellipse referred to its vertex is (Art. 194)

$$y^2 = \frac{2b^2}{a} x - \frac{b^2}{a^2} x^2.$$

We wish to express b in terms of the distance $VF (= m)$, which we suppose fixed. We have $m = a - \sqrt{(a^2 - b^2)}$ (Art. 182), whence $b^2 = 2am - m^2$, and the equation becomes

$$y^2 = \left(4m - \frac{2m^2}{a}\right) x - \left(\frac{2m}{a} - \frac{m^2}{a^2}\right) x^2.$$

Now, if we suppose a to become infinite, all but the first term of the right-hand side of the equation will vanish, and the equation becomes

$$y^2 = 4mx,$$

the equation of a parabola.

A parabola may also be considered as an ellipse whose eccentricity is equal to 1. For $e^2 = 1 - \frac{b^2}{a^2}$. Now we saw that $\frac{b^2}{a^2}$, which is the coefficient of x^2 in the preceding equation, vanished as we supposed a increased, according to the prescribed conditions; hence e^2 becomes finally $= 1$.

THE TANGENT.

210. The equation of the chord joining two points on the curve is (Art. 86) $(y - y')(y - y'') = y^2 - px$,

or $$(y' + y'')\, y = px + y'y''.$$

And if we make $y'' = y'$, and for y'^2 write its equal px', we have the equation of the tangent

$$2y'y = p\,(x + x').$$

If in this equation we put $y = 0$, we get $x = -x'$, hence TM (see fig. next page), which is called the Subtangent, is bisected at the vertex.

These results hold equally if the axes of coordinates are oblique; that is to say, if the axes are any diameter and the tangent at its vertex, in which case we saw (Art. 203) that the equation of the parabola is still of the form $y^2 = p'x$.

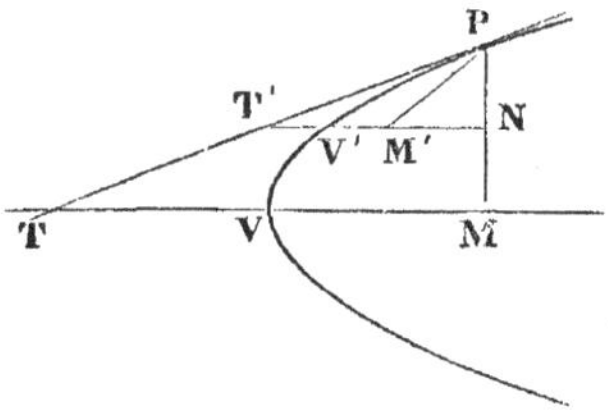

This Article enables us, therefore, to draw a tangent at any point on the parabola, since we have only to take $TV = VM$ and join PT; or again, having found this tangent, to draw an ordinate from P to any other diameter, since we have only to take $V'M' = T'V'$, and join PM'.

211. The equation of the polar of any point $x'y'$ is similar in form to that of the tangent (Art. 89), and is, therefore,

$$2y'y = p(x + x').$$

Putting $y = 0$, we find that the intercept made by this polar on the axis of x is $-x'$. Hence *the intercept which the polars of any two points cut off on the axis is equal to the intercept between perpendiculars from those points on that axis;* each of these quantities being equal to $(x' - x'')$.

DIAMETERS.

212. We have said that if we take for axes any diameter and the tangent at its extremity, the equation will be of the form $y^2 = p'x$.

We shall prove this again by actual transformation of the equation referred to rectangular axes ($y^2 = px$), because it is desirable to express the new p' in terms of the old p.

If we transform the equation $y^2 = px$ to parallel axes through any point $(x'y')$ on the curve, writing $x + x'$ and $y + y'$ for x and y, the equation becomes

$$y^2 + 2yy' = px.$$

Now if, preserving our axis of x, we take a new axis of y, inclined to that of x at an angle θ, we must substitute (Art. 9), $y \sin\theta$ for y, and $x + y\cos\theta$ for x, and our equation becomes

$$y^2 \sin^2\theta + 2y'y \sin\theta = px + py\cos\theta.$$

In order that this should reduce to the form $y^2 = px$, we must have

$$2y' \sin\theta = p\cos\theta, \text{ or } \tan\theta = \frac{p}{2y'}.$$

Now this is the very angle which the tangent makes with the axis of x, as we see from the equation

$$2y'y = p(x + x').$$

The equation, therefore, referred to a diameter and tangent, will take the form

$$y^2 = \frac{p}{\sin^2\theta} x, \text{ or } y^2 = p'x.$$

The quantity p' is called the parameter corresponding to the diameter $V'M'$, and we see that *the parameter of any diameter is inversely proportional to the square of the sine of the angle which its ordinates make with the axis*, since $p' = \frac{p}{\sin^2\theta}$.

We can express the parameter of any diameter in terms of the coordinates of its vertex, from the equation $\tan\theta = \frac{p}{2y'}$; hence

$$\sin\theta = \frac{p}{\sqrt{(p^2 + 4y'^2)}} = \sqrt{\left(\frac{p}{p + 4x'}\right)};$$

hence $$p' = p + 4x'.$$

THE NORMAL.

213. The equation of a line through $(x'y')$ perpendicular to the tangent $2yy' = p(x + x')$ is

$$p(y - y') + 2y'(x - x') = 0.$$

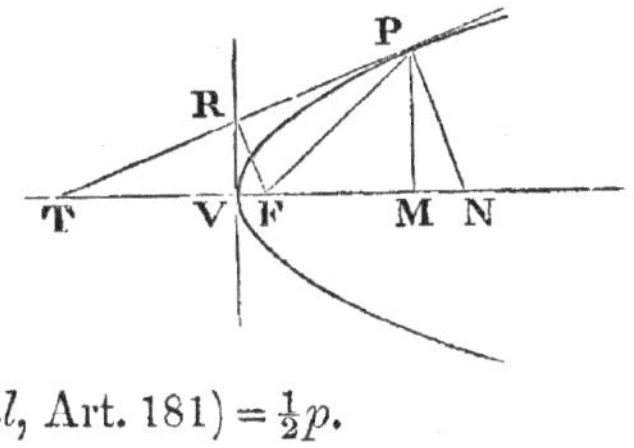

If we seek the intercept on the axis of x we have

$$x\ (= VN) = x' + \tfrac{1}{2}p;$$

and, since $VM = x'$, we must have

$$MN \text{ (the } \textit{subnormal}\text{, Art. 181)} = \tfrac{1}{2}p.$$

Hence *in the parabola the subnormal is constant, and equal to the semi-parameter.* The normal itself

$$= \sqrt{(PM^2 + MN^2)} = \sqrt{(y'^2 + \tfrac{1}{4}p^2)} = \sqrt{\{p(x' + \tfrac{1}{4}p)\}} = \tfrac{1}{2}\sqrt{(pp')}.$$

THE FOCUS.

214. A point situated on the axis of a parabola, at a distance from the vertex equal to one-fourth of the principal parameter, is called the *focus* of the curve. This is the point which, Art. 209, has led us to expect to find analogous to the focus of an ellipse; and we shall show, in the present section, that a parabola may in every respect be considered as an ellipse, having one of its foci at this distance and the other at infinity.

To avoid fractions we shall often, in the following Articles, use the abbreviation $m = \frac{1}{4}p$.

To find the distance of any point on the curve from the focus.

The coordinates of the focus being $(m, 0)$, the square of its distance from any point is

$$(x' - m)^2 + y'^2 = x'^2 - 2mx' + m^2 + 4mx' = (x' + m)^2.$$

Hence the distance of any point from the focus $= x' + m$.

This enables us to express more simply the result of Art. 212, and to say that *the parameter of any diameter is four times the distance of its extremity from the focus.*

215. The polar of the focus of a parabola is called the *directrix*, as in the ellipse and hyperbola.

Since the distance of the focus from the vertex $= m$, its polar is (Art. 211) a line perpendicular to the axis at the same distance on the other side of the vertex. The distance of any point from the directrix must, therefore, $= x' + m$.

Hence, by the last Article, *the distance of any point on the curve from the directrix is equal to its distance from the focus.*

We saw (Art. 186) that in the ellipse and hyperbola the distance from the focus is to the distance from the directrix in the constant ratio e to 1. We see, now, that this is true for the parabola also, since in the parabola $e = 1$ (Art. 209).

The method given for mechanically describing an hyperbola, Art. 202, can be adapted to the mechanical description of the parabola, by simply making the angle ABR a right angle.

216. *The point where any tangent cuts the axis, and its point of contact, are equally distant from the focus.*

For, the distance from the vertex of the point where the tangent cuts the axis $= x'$ (Art. 210), its distance from the focus is therefore $x' + m$.

217. *Any tangent makes equal angles with the axis and with the focal radius vector.*

This is evident from inspection of the isosceles triangle, which, in the last Article, we proved was formed by the axis, the focal radius vector, and the tangent.

This is only an extension of the property of the ellipse (Art. 188), that the angle $TPF = T'PF'$; for, if we suppose the focus F' to go off to infinity, the line PF' will become parallel to the axis, and $TPF' = PTF$. (See figure, p. 200)

Hence the tangent at the extremity of the focal ordinate cuts the axis at an angle of 45°.

218. *To find the length of the perpendicular from the focus on the tangent.*

The perpendicular from the point $(m, 0)$ on the tangent $\{yy' = 2m(x + x')\}$ is

$$= \frac{2m(x'+m)}{\sqrt{(y'^2+4m^2)}} = \frac{2m(x'+m)}{\sqrt{(4mx'+4m^2)}} = \sqrt{\{m(x'+m)\}}.$$

Hence (see fig., p. 202) FR is a mean proportional between FV and FP.

It appears, also, from this expression and from Art. 213 that FR is half the normal, as we might have inferred geometrically from the fact that $TF = FN$.

219. *To express the perpendicular from the focus in terms of the angles which it makes with the axis.*

We have

$$\cos\alpha = \sin FTR = (\text{Art. } 212) \sqrt{\left(\frac{m}{x'+m}\right)}.$$

Therefore (Art. 218)

$$FR = \sqrt{\{m(x'+m)\}} = \frac{m}{\cos\alpha}.$$

The equation of the tangent, *the focus being the origin*, can therefore be expressed

$$x\cos\alpha + y\sin\alpha + \frac{m}{\cos\alpha} = 0,$$

and hence we can express the perpendicular from any other point in terms of the angle it makes with the axis.

220. *The locus of the extremity of the perpendicular from the focus on the tangent is a right line.*

For, taking the focus for pole, we have at once the polar equation

$$\rho = \frac{m}{\cos\alpha}, \quad \rho\cos\alpha = m,$$

which obviously represents the tangent at the vertex.

Conversely, if from any point F we draw FR a radius vector to a right line VR, and draw PR perpendicular to it, the line FR will always touch a parabola having F for its focus.

We shall show hereafter how to solve generally questions of this class, where one condition less than is sufficient to determine a line is given, and it is required to find its *envelope*, that is to say, the curve which it always touches.

We leave, as a useful exercise to the reader, the investigation of the locus of the foot of the perpendicular by ordinary rectangular coordinates.

221. *To find the locus of the intersection of tangents which cut at right angles to each other.*

The equation of any tangent being (Art. 219)

$$x \cos^2 \alpha + y \sin \alpha \cos \alpha + m = 0;$$

the equation of a tangent perpendicular to this (that is, whose perpendicular makes an angle $= 90° + \alpha$ with the axis) is found by substituting $\cos\alpha$ for $\sin\alpha$, and $-\sin\alpha$ for $\cos\alpha$, or

$$x \sin^2 \alpha - y \sin \alpha \cos \alpha + m = 0.$$

α is eliminated by simply adding the equations, and we get

$$x + 2m = 0,$$

the equation of *the directrix*, since the distance of focus from directrix $= 2m$.

222. *The angle between any two tangents is half the angle between the focal radii vectores to their points of contact.*

For, from the isosceles PFT, the angle PTF, which the tangent makes with the axis, is half the angle PFN, which the focal radius makes with it. Now, the angle between any two tangents is equal to the difference of the angles they make with the axis, and the angle between two focal radii is equal to the difference of the angles which *they* make with the axis.

The theorem of the last Article follows as a particular case of the present theorem: for if two tangents make with each other an angle of 90°, the focal radii must make with each other an angle of 180°, therefore the two tangents must be drawn at the extremities of a chord through the focus, and, therefore, from the definition of the directrix, must meet on the directrix.

223. *The line joining the focus to the intersection of two tangents bisects the angle which their points of contact subtend at the focus.*

Subtracting one from the other, the equations of two tangents, viz.

$$x \cos^2\alpha + y \sin\alpha \cos\alpha + m = 0, \quad x \cos^2\beta + y \sin\beta \cos\beta + m = 0,$$

we find for the line joining their intersection to the focus,

$$x \sin(\alpha + \beta) - y \cos(\alpha + \beta) = 0.$$

This is the equation of a line making the angle $\alpha + \beta$ with the axis of x. But since α and β are the angles made with the axis by the perpendiculars on the tangent, we have $VFP = 2\alpha$ and $VFP' = 2\beta$; therefore the line making an angle with the axis $= \alpha + \beta$ must bisect the angle PFP'. This theorem may also be proved by calculating, as in Art. 191, the angle $(\theta - \theta')$ subtended at the focus by the tangent to a parabola from the point xy, when it will be found that $\cos(\theta - \theta') = \dfrac{x + m}{\rho}$, a value which, being independent of the coordinates of the point of contact, will be the same for each of the two tangents which can be drawn through xy. (See O'Brien's *Coordinate Geometry*, p. 156.)

Cor. 1. If we take the case where the angle $PFP' = 180°$, then PP' passes through the focus; the tangents TP, TP' will intersect on the directrix, and the angle $TFP = 90°$ (See Art. 192). This may also be proved directly by forming the equations of the polar of any point $(-m, y')$ on the directrix, and also the equation of the line joining that point to the focus. These two equations are

$$y'y = 2m(x - m), \quad 2m(y - y') + y'(x + m) = 0,$$

which obviously represent two right lines at right angles to each other.

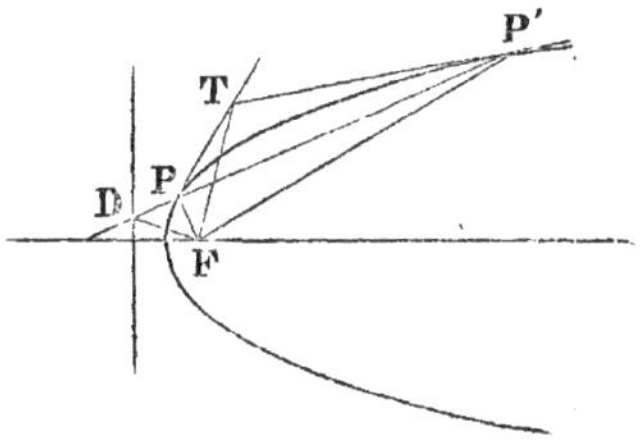

Cor. 2. If any chord PP' cut the directrix in D, then FD is the external bisector of the angle PFP'. This is proved as at p. 184.

Cor. 3. If any variable tangent to the parabola meet two fixed tangents, the angle subtended at the focus by the portion of the variable tangent intercepted between the fixed tangents is the supplement of the angle between the fixed tangents. For (see next figure) the angle QRT is half pFq (Art. 222), and, by the present Article, PFQ is obviously also half pFq, therefore PFQ is $= QRT$, or is the supplement of PRQ.

COR. 4. *The circle circumscribing the triangle formed by any three tangents to a parabola will pass through the focus.* For the circle described through PRQ must pass through F, since the angle contained in the segment PFQ will be the supplement of that contained in PRQ.

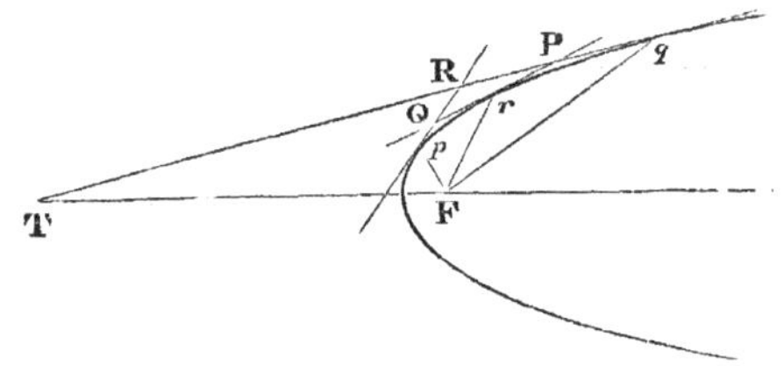

224. *To find the polar equation of the parabola, the focus being the pole.*

We proved (Art. 214) that the focal radius

$= x' + m = VM + m = FM + 2m = \rho \cos\theta + 2m.$

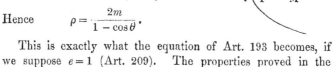

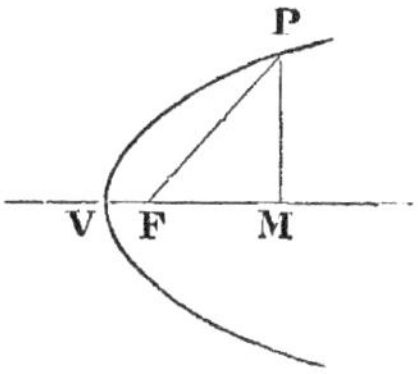

Hence $\rho = \dfrac{2m}{1 - \cos\theta}$.

This is exactly what the equation of Art. 193 becomes, if we suppose $e = 1$ (Art. 209). The properties proved in the Examples to Art. 193 are equally true of the parabola.

In this equation θ is supposed to be measured from the side FM; if we suppose it measured from the side FV, the equation becomes

$$\rho = \frac{2m}{1 + \cos\theta}.$$

This equation may be written

$$\rho \cos^2 \tfrac{1}{2}\theta = m,$$

or $$\rho^{\frac{1}{2}} \cos \tfrac{1}{2}\theta = (m)^{\frac{1}{2}},$$

and is, therefore, one of a class of equations

$$\rho^n \cos n\theta = a^n,$$

some of whose properties we shall mention hereafter.

CHAPTER XIII.

EXAMPLES AND MISCELLANEOUS PROPERTIES OF CONIC SECTIONS.

225. THE method of applying algebra to problems relating to conic sections is essentially the same as that employed in the case of the right line and circle, and will present no difficulty to any reader who has carefully worked out the Examples given in Chapters III. and VII. We, therefore, only think it necessary to select a few out of the great multitude of examples which lead to loci of the second order, and we shall then add some properties of conic sections, which it was not found convenient to insert in the preceding Chapters.

Ex. 1. Through a fixed point P is drawn a line LK (see fig., p. 40) terminated by two given lines. Find the locus of a point Q taken on the line, so that $PL = QK$.

Ex. 2. Two equal rulers AB, BC, are connected by a pivot at B; the extremity A is fixed, while the extremity C is made to traverse the right line AC; find the locus described by any fixed point P on BC.

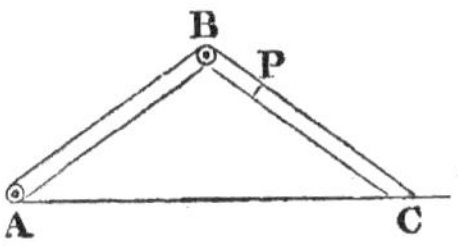

Ex. 3. Given base and the product of the tangents of the halves of the base angles of a triangle; find the locus of vertex.

Expressing the tangents of the half angles in terms of the sides, it will be found that the sum of sides is given; and, therefore, that the locus is an ellipse, of which the extremities of the base are the foci.

Ex. 4. Given base and sum of sides of a triangle; find the locus of the centre of the inscribed circle.

It may be immediately inferred, from the last example, and from Ex. 4, p. 47, that the locus is an ellipse, whose vertices are the extremities of the given base.

Ex. 5. Given base and sum of sides, find the locus of the intersection of bisectors of sides.

Ex. 6. Find the locus of the centre of a circle which makes given intercepts on two given lines.

Ex. 7. Find the locus of the centre of a circle which touches two given circles, or which touches a right line and a given circle.

Ex. 8. Find locus of centre of a circle which passes through a given point and makes a given intercept on a given line.

Ex. 9. Or which passes through a given point, and makes on a given line an intercept subtending a given angle at that point.

Ex. 10. Two vertices of a *given* triangle move along fixed right lines; find the locus of the third.

Ex. 11. A triangle ABC circumscribes a given circle; the angle at C is given, and B moves along a fixed line; find the locus of A.

Let us use polar coordinates, the centre O being the pole, and the angles being measured from the perpendicular on the fixed line; let the coordinates of A, B, be ρ, θ; ρ', θ'. Then we have $\rho' \cos\theta' = p$. But it is easy to see that the angle AOB is given ($= \alpha$). And since the perpendicular of the triangle AOB is given, we have

$$r = \frac{\rho\rho' \sin\alpha}{\sqrt{(\rho^2 + \rho'^2 - 2\rho\rho' \cos\alpha)}}.$$

But $\theta + \theta' = \alpha$; therefore the polar equation of the locus is

$$r^2 = \frac{p^2\rho^2 \sin^2\alpha}{\rho^2 \cos^2(\alpha - \theta) + p^2 - 2p\rho \cos\alpha \cos(\alpha - \theta)},$$

which represents a conic.

Ex. 12. Find the locus of the pole with respect to one conic A of any tangent to another conic B.

Let $\alpha\beta$ be any point of the locus, and $\lambda x + \mu y + \nu$ its polar with respect to the conic A, then (Art. 89) λ, μ, ν are functions of the first degree in α, β. But (Art. 151) the condition that $\lambda x + \mu y + \nu$ should touch B is of the second degree in λ, μ, ν. The locus is therefore a conic.

Ex. 13. Find the locus of the intersection of the perpendicular from a focus on any tangent to a central conic, with the radius vector from centre to the point of contact.

Ans. The corresponding directrix.

Ex. 14. Find the locus of the intersection of the perpendicular from the centre on any tangent, with the radius vector from a focus to the point of contact. *Ans.* A circle.

Ex. 15. Find the locus of the intersection of tangents at the extremities of conjugate diameters.

Ans. $\dfrac{x^2}{a^2} + \dfrac{y^2}{b^2} = 2.$

This is obtained at once by squaring and adding the equations of the two tangents, attending to the relations, Art. 172.

Ex. 16. Trisect a given arc of a circle. The points of trisection are found as the intersection of the circle with a hyperbola. See Ex. 7, p. 47.

Ex. 17. One of the two parallel sides of a trapezium is given in magnitude and position, and the other in magnitude. The sum of the remaining two sides is given; find the locus of the intersection of diagonals.

Ex. 18. One vertex of a parallelogram circumscribing an ellipse moves along one directrix; prove that the opposite vertex moves along the other, and that the two remaining vertices are on the circle described on the axis major as diameter.

226. We give in this Article some examples on the focal properties of conics.

Ex. 1. The distance of any point on a conic from the focus is equal to the whole length of the ordinate at that point, produced to meet the tangent at the extremity of the focal ordinate.

Ex. 2. If from the focus a line be drawn making a given angle with any tangent, find the locus of the point where it meets it.

Ex. 3. To find the locus of the pole of a fixed line with regard to a series of concentric and confocal conic sections.

We know that the pole of any line $\left(\dfrac{x}{m} + \dfrac{y}{n} = 1\right)$, with regard to the conic $\left(\dfrac{x^2}{a^2} + \dfrac{y^2}{b^2} = 1\right)$, is found from the equations $mx = a^2$ and $ny = b^2$ (Art. 169).

Now, if the foci of the conic are given, $a^2 - b^2 = c^2$ is given; hence, the locus of the pole of the fixed line is

$$mx - ny = c^2,$$

the equation of a right line perpendicular to the given line.

If the given line touch one of the conics, its pole will be the point of contact. Hence, given two confocal conics, if we draw any tangent to one and tangents to the second where this line meets it, these tangents will intersect on the normal to the first conic.

Ex. 4. Find the locus of the points of contact of tangents to a series of confocal ellipses from a fixed point on the axis major. *Ans.* A circle.

Ex. 5. The lines joining each focus to the foot of the perpendicular from the other focus on any tangent intersect on the corresponding normal and bisect it.

Ex. 6. The focus being the pole, prove that the polar equation of the chord through points whose angular coordinates are $\alpha + \beta$, $\alpha - \beta$, is

$$\frac{p}{2\rho} = e \cos\theta + \sec\beta \cos(\theta - \alpha).$$

This expression is due to Mr. Frost (*Cambridge and Dublin Math. Journal*, I., 68, cited by Walton, Examples, p. 375). It follows easily from Ex. 3, p. 37.

Ex. 7. The focus being the pole, prove that the polar equation of the tangent, at the point whose angular coordinate is α, is $\dfrac{p}{2\rho} = e\cos\theta + \cos(\theta - \alpha)$.

This expression is due to Mr. Davies (*Philosophical Magazine* for 1842, p. 192, cited by Walton, Examples, p. 368).

Ex. 8. If a chord PP' of a conic pass through a fixed point O, then

$$\tan \tfrac{1}{2}PFO \,.\, \tan \tfrac{1}{2}P'FO$$

is constant.

The reader will find an investigation of this theorem by the help of the equation of Ex. 6 (Walton's Examples, p. 377). I insert here the geometrical proof given by Mr. MacCullagh, to whom, I believe, the theorem is due. Imagine a point O taken anywhere on PP' (see figure p. 206), and let the distance FO be e' times the distance of O from the directrix: then, since the distances of P and O from the directrix are proportional to PD and OD, we have

$$\frac{FP}{FD} \div \frac{FO}{OD} = \frac{e}{e'}, \text{ or } \frac{\sin PDF}{\sin PFD} \div \frac{\sin ODF}{\sin OFD} = \frac{e}{e'}.$$

Hence (Art. 192)
$$\frac{\cos OFT}{\cos PFT} = \frac{e}{e'};$$

or, since (Art. 191) PFT is half the sum, and OFT half the difference, of PFO and $P'FO$,

$$\tan \tfrac{1}{2}PFO \,.\, \tan \tfrac{1}{2}P'FO = \frac{e - e'}{e + e'}.$$

It is obvious that the product of these tangents remains constant if O be not fixed, but be anywhere on a conic having the same focus and directrix as the given conic.

Ex. 9. To express the condition that the chord joining two points $x'y'$, $x''y''$ on the curve passes through the focus.

This condition may be expressed in several equivalent forms, two of the most useful of which are got by expressing that $\theta'' = \theta' + 180°$, where θ', θ'' are the angles made with the axis by the lines joining the focus to the points. The condition $\sin\theta'' = -\sin\theta'$ gives

$$\frac{y'}{a - ex'} + \frac{y''}{a - ex''} = 0; \quad a(y' + y'') = e(x'y'' + x''y').$$

The condition $\cos\theta'' = -\cos\theta'$ gives

$$\frac{x'-c}{a-ex'}+\frac{x''-c}{a-ex''}=0;\quad 2ex'x''-(a+ce)(x'+x'')+2ac=0.$$

Ex. 10. If normals be drawn at the extremities of any focal chord, a line drawn through their intersection parallel to the axis major will bisect the chord. [This solution is by Larrose, *Nouvelles Annales*, XIX. 85.]

Since each normal bisects the angle between the focal radii, the intersection of normals at the extremities of a focal chord is the centre of the circle inscribed in the triangle whose base is that chord, and sides the lines joining its extremities to the other focus. Now if a, b, c be the sides of a triangle whose vertices are $x'y'$, $x''y''$, $x'''y'''$, then, Ex. 6, p. 6, the coordinates of the centre of the inscribed circle are

$$x=\frac{ax'+bx''+cx'''}{a+b+c},\quad y=\frac{ay'+by''+cy'''}{a+b+c}.$$

In the present case the coordinates of the vertices are x', y'; x'', y''; $-c, 0$; and the lengths of opposite sides are $a+ex''$, $a+ex'$, $2a-ex'-ex''$. We have therefore

$$y=\frac{(a+ex')\,y''+(a+ex'')\,y'}{4a},$$

or, reducing by the first relation of the last Example, $y=\frac{1}{2}(y'+y'')$, which proves the theorem.

In like manner we have

$$x=\frac{(a+ex'')\,x'+(a+ex')\,x''-(2a-ex'-ex'')\,c}{4a},$$

which, reduced by the second relation, becomes

$$x=\frac{(a+ec)\,(x'+x'')-2ca}{2a}.$$

We could find, similarly, expressions for the coordinates of the intersection of tangents at the extremities of a focal chord, since this point is the centre of the circle exscribed to the base of the triangle just considered. The line joining the intersection of tangents to the corresponding intersection of normals evidently passes through a focus, being the bisector of the vertical angle of the same triangle.

Ex. 11. To find the locus of the intersection of normals at the extremities of a focal chord.

Let α, β be the coordinates of the middle point of the chord, and we have, by the last Example,

$$\alpha=\tfrac{1}{2}(x'+x'')=\frac{a^2(x+c)}{a^2+c^2};\quad \beta=\tfrac{1}{2}(y'+y'')=y.$$

If, then, we knew the equation of the locus described by $\alpha\beta$, we should, by making the above substitutions, have the equation of the locus described by xy. Now the polar equation of the locus of middle point, the focus being origin, is (Art. 193)

$$\rho=\tfrac{1}{2}(\rho'-\rho'')=\frac{-b^2}{a}\,\frac{e\cos\theta}{1-e^2\cos^2\theta},$$

which, transformed to rectangular axes, the centre being origin, becomes

$$b^2\alpha^2+a^2\beta^2=b^2c\alpha.$$

The equation of the locus sought is, therefore,

$$a^2b^2(x+c)^2+(a^2+c^2)^2y^2=b^2c\,(a^2+c^2)\,(x+c).$$

Ex. 12. If θ be the angle between the tangents to an ellipse from any point P, and if ρ, ρ' be the distances of that point from the foci, prove that $\cos\theta = \frac{\rho^2 + \rho'^2 - 4a^2}{2\rho\rho'}$ (see also Art. 194 *d*).

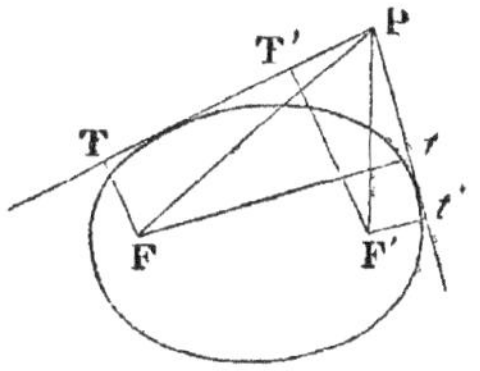

For (Art. 189)

$$\sin TPF . \sin tPF = \frac{FT . F'T'}{PF . PF'} = \frac{b^2}{\rho\rho'} .$$

But $\cos FPF' - \cos TPt = 2 \sin TPF . \sin tPF$;

and $$2\rho\rho' \cos FPF' = \rho^2 + \rho'^2 - 4c^2.$$

Ex. 13. If from any point O two lines be drawn to the foci (or touching any confocal conic) meeting the conic in R, R'; S, S'; then (see also Ex. 15, Art. 231)

$$\frac{1}{OR} - \frac{1}{OR'} = \frac{1}{OS} - \frac{1}{OS'} . \quad \text{[Mr. M. Roberts.]}$$

It appears from the quadratic, by which the radius vector is determined (Art. 136), that the difference of the reciprocals of the roots will be the same for two values of θ, which give the same value to

$$(ac - g^2) \cos^2\theta + 2 (ch - gf) \cos\theta \sin\theta + (bc - f^2) \sin^2\theta.$$

Now it is easy to see that $A \cos^2\theta + 2H \cos\theta \sin\theta + B \sin^2\theta$ has equal values for any two values of θ, which correspond to the directions of lines equally inclined to the two represented by $Ax^2 + 2Hxy + By^2 = 0$. But the function we are considering becomes $= 0$ for the direction of the two tangents through O (Art. 147); and tangents to any confocal are equally inclined to these tangents (Art. 189). It follows from this example that chords which touch a confocal conic are proportional to the squares of the parallel diameters (see Ex. 15, Art. 231).

227. We give in this Article some examples on the parabola. The reader will have no difficulty in distinguishing those of the examples of the last Article, the proofs of which apply equally to the parabola.

Ex. 1. Find the coordinates of the intersection of the two tangents at the points $x'y'$, $x''y''$, to the parabola $y^2 = px$.

Ans. $y = \frac{y' + y''}{2}$, $x = \frac{y'y''}{p}$.

Ex. 2. Find the locus of the intersection of the perpendicular from focus on tangent with the radius vector from vertex to the point of contact.

Ex. 3. The three perpendiculars of the triangle formed by three tangents intersect on the directrix (Steiner, Gergonne, *Annales*, XIX. 59; Walton, p. 119).

The equation of one of those perpendiculars is (Art. 32)

$$\frac{y'y''' - y'y''}{p}\left(x - \frac{y''y'''}{p}\right) + \frac{y''' - y''}{2}\left(y - \frac{y'' + y'''}{2}\right) = 0;$$

which, after dividing by $y''' - y''$, may be written

$$y'\left(x + \frac{p}{4}\right) - \frac{y'y''y'''}{p} + \frac{py}{2} - \frac{p\,(y' + y'' + y''')}{4} = 0.$$

The symmetry of the equation shows that the three perpendiculars intersect on the directrix at a height

$$y = \frac{2y'y''y'''}{p^2} + \frac{y' + y'' + y'''}{2} .$$

Ex. 4. The area of the triangle formed by three tangents is half that of the triangle formed by joining their points of contact (Gregory, *Cambridge Journal*, II. 16 Walton, p. 137. See also *Lessons on Higher Algebra*, Ex. 12, p. 15).

Substituting the coordinates of the vertices of the triangles in the expression of Art. 36, we find for the latter area $\frac{1}{2p}(y'-y'')(y''-y''')(y'''-y')$; and for the former area half this quantity.

Ex. 5. Find an expression for the radius of the circle circumscribing a triangle inscribed in a parabola.

The radius of the circle circumscribing a triangle, the lengths of whose sides are d, e, f, and whose area $= \Sigma$ is easily proved to be $\frac{def}{4\Sigma}$. But if d be the length of the chord joining the points $x''y''$, $x'''y'''$, and θ' the angle which this chord makes with the axis, it is obvious that $d \sin\theta' = y'' - y'''$. Using, then, the expression for the area found in the last Example, we have $R = \frac{p}{2\sin\theta'\sin\theta''\sin\theta'''}$. We might express the radius, also, in terms of the focal chords parallel to the sides of the triangle. For (Art. 193, Ex. 2) the length of a chord making an angle θ with the axis is $c = \frac{p}{\sin^2\theta}$. Hence $R^2 = \frac{c'c''c'''}{4p}$.

It follows from Art. 212 that c', c'', c''' are the parameters of the diameters which bisect the sides of the triangle.

Ex. 6. Express the radius of the circle circumscribing the triangle formed by three tangents to a parabola in terms of the angles which they make with the axis.

Ans. $R = \frac{p}{8\sin\theta'\sin\theta''\sin\theta'''}$; or $R^2 = \frac{p'p''p'''}{64p}$, where p', p'', p''' are the parameters of the diameters through the points of contact of the tangents (see Art. 212).

Ex. 7. Find the angle contained by the two tangents through the point $x'y'$ to the parabola $y^2 = 4mx$.

The equation of the pair of tangents is (as in Art. 92) found to be

$$(y'^2 - 4mx')(y^2 - 4mx) = \{yy' - 2m(x + x')\}^2.$$

A parallel pair of lines through the origin is

$$x'y^2 - y'xy + mx^2 = 0.$$

The angle contained by which is (Art. 74) $\tan\phi = \frac{\sqrt{(y'^2 - 4mx')}}{x' + m}$.

Ex. 8. Find the locus of intersection of tangents to a parabola which cut at a given angle.

Ans. The hyperbola, $y^2 - 4mx = (x + m)^2\tan^2\phi$, or $y^2 + (x - m)^2 = (x + m)^2\sec^2\phi$. From the latter form of the equation it is evident (see Art. 186) that the hyperbola has the same focus and directrix as the parabola, and that its eccentricity $= \sec\phi$.

Ex. 9. Find the locus of the foot of the perpendicular from the focus of a parabola on the normal.

The length of the perpendicular from $(m, 0)$ on $2m(y - y') + y'(x - x') = 0$ is

$$\frac{y'(x' + m)}{\sqrt{(y'^2 + 4m^2)}} = \sqrt{\{x'(x' + m)\}}.$$

But if θ be the angle made with the axis by the perpendicular (Art. 212)

$$\sin\theta = \sqrt{\left(\frac{m}{x' + m}\right)}, \quad \cos\theta = \sqrt{\left(\frac{x'}{x' + m}\right)}.$$

Hence the polar equation of the locus is

$$\rho = \frac{m\cos\theta}{\sin^2\theta} \quad \text{or } y^2 = mx.$$

Ex. 10. Find the coordinates of the intersection of the normals at the points $x'y'$, $x''y''$.

$$Ans.\ \ x = 2m + \frac{y'^2 + y'y'' + y''^2}{4m},\ \ y = -\frac{y'y''\,(y' + y'')}{8m^2}.$$

Or if α, β be the coordinates of the corresponding intersection of tangents, then (Ex. 1)

$$x = 2m + \frac{\beta^2}{m} - \alpha,\ \ y = -\frac{\alpha\beta}{m}.$$

Ex. 11. Find the coordinates of the points on the curve, the normals at which pass through a given point $x'y'$.

Solving between the equation of the normal and that of the curve, we find

$$2y^3 + (p^2 - 2px')\,y = p^2y',$$

and the three roots are connected by the relation $y_1 + y_2 + y_3 = 0$. The geometric meaning of this is, that the chord joining any two, and the line joining the third to the vertex, make equal angles with the axis.

Ex. 12. Find the locus of the intersection of normals at the extremities of chords which pass through a given point $x'y'$.

We have then the relation $\beta y' = 2m\,(x' + \alpha)$; and on substituting in the results of Ex. 10 the value of α derived from this relation we have

$$2mx + \beta y' = 4m^2 + 2\beta^2 + 2mx';\ \ 2m^2y = 2\beta mx' - \beta^2 y';$$

whence, eliminating β, we find

$$2\,\{2m\,(y - y') + y'\,(x - x')\}^2 = (4mx' - y'^2)\,(y'y + 2x'x - 4mx' - 2x'^2),$$

the equation of a parabola whose axis is perpendicular to the polar of the given point. If the chords be parallel to a fixed line, the locus reduces to a right line, as is also evident from Ex. 11.

Ex. 13. Find the locus of the intersection of normals at right angles to each other.

In this case $\alpha = -m,\ x = 3m + \frac{\beta^2}{m},\ y = \beta,\ y^2 = m\,(x - 3m)$.

Ex. 14. If the lengths of two tangents be a, b, and the angle between them ω, find the parameter.

Draw the diameter bisecting the chord of contact; then the parameter of that diameter is $p' = \frac{y^2}{x}$, and the principal parameter is $p = \frac{y^2 \sin^2\theta}{x} = \frac{\varpi^2 y^2}{4x^3}$ (where ϖ is the length of the perpendicular on the chord from the intersection of the tangents). But $2\varpi y = ab \sin\omega$, and $16x^2 = a^2 + b^2 + 2ab\cos\omega$; hence

$$p = \frac{4a^2b^2 \sin^2\omega}{(a^2 + b^2 + 2ab\cos\omega)^{\frac{3}{2}}}$$ (see p. 199).

Ex. 15. Show, from the equation of the circle circumscribing three tangents to a parabola, that it passes through the focus.

The equation of the circle circumscribing a triangle being (Art. 124)

$$\beta\gamma \sin A + \gamma\alpha \sin B + \alpha\beta \sin C = 0;$$

the absolute term in this equation is found (by writing at full length for α, $x\cos\alpha + y\sin\alpha - p$, &c.) to be $p'p''\sin(\beta - \gamma) + p''p\sin(\gamma - \alpha) + pp'\sin(\alpha - \beta)$. But if the line α be a tangent to a parabola, and the origin the focus, we have (Art. 219) $p = \frac{m}{\cos\alpha}$, and the absolute term

$$= \frac{m^2}{\cos\alpha\cos\beta\cos\gamma}\{\sin(\beta - \gamma)\cos\alpha + \sin(\gamma - \alpha)\cos\beta + \sin(\alpha - \beta)\cos\gamma\},$$

which vanishes identically.

Ex. 16. Find the locus of the intersection of tangents to a parabola, being given either (1) the product of sines, (2) the product of tangents, (3) the sum or (4) difference of cotangents of the angles they make with the axis.

Ans. (1) a circle, (2) a right line, (3) a right line, (4) a parabola.

228. We add a few miscellaneous examples.

Ex. 1. If an equilateral hyperbola circumscribe a triangle, it will also pass through the intersection of its perpendiculars (Brianchon and Poncelet; Gergonne, *Annales*, XI., 205; Walton, p. 283).

The equation of a conic meeting the axes in given points is (Ex. 1, p. 148)

$$\mu\mu' x^2 + 2hxy + \lambda\lambda' y^2 - \mu\mu' (\lambda + \lambda')\, x - \lambda\lambda' (\mu + \mu')\, y + \lambda\lambda'\mu\mu' = 0.$$

And if the axes be rectangular, this will represent an equilateral hyperbola (Art. 174) if $\lambda\lambda' = -\mu\mu'$. If, therefore, the axes be any side of the given triangle, and the perpendicular on it from the opposite vertex, the portions λ, λ', μ are given, therefore, μ' is also given; or the curve meets the perpendicular in the fixed point $y = -\dfrac{\lambda\lambda'}{\mu}$, which is (Ex. 7, p. 27) the intersection of the perpendiculars of the triangle.

Ex. 2. What is the locus of the centres of equilateral hyperbolas through three given points?

Ans. The circle through the middle points of sides (see Ex. 3, p. 153).

Ex. 3. A conic being given by the general equation, find the condition that the pole of the axis of x should lie on the axis of y, and *vice versa*. *Ans.* $hc = fg$.

Ex. 4. In the same case, what is the condition that an asymptote should pass through the origin? *Ans.* $af^2 - 2fgh + bg^2 = 0$.

Ex. 5. The circle circumscribing a triangle, self-conjugate with regard to an equilateral hyperbola (see Art. 99), passes through the centre of the curve. (Brianchon and Poncelet; Gergonne, XI. 210; Walton, p. 301). [This is a particular case of the theorem that the six vertices of two self-conjugate triangles lie on a conic (see Ex. 1, Art. 375).]

The condition of Ex. 3 being fulfilled, the equation of a circle passing through the origin and through the pole of each axis is

$$h\,(x^2 + 2xy\cos\omega + y^2) + fx + gy = 0,$$

or $$x\,(hx + by + f) + y\,(ax + hy + g) - (a + b - 2h\cos\omega)\, xy,$$

an equation which will evidently be satisfied by the coordinates of the centre, provided we have $a + b = 2h\cos\omega$, that is to say, provided the curve be an equilateral hyperbola (Arts. 74, 174).

Ex. 6. A circle described through the centre of an equilateral hyperbola, and through any two points, will also pass through the intersection of lines drawn through each of these points parallel to the polar of the other.

Ex. 7. Find the locus of the intersection of tangents which intercept a given length on a given fixed tangent.

The equation of the pair of tangents from a point $x'y'$ to a conic given by the general equation is given Art. 92. Make $y = 0$, and we have a quadratic whose roots are the intercepts on the axis of x.

Forming the difference of the roots of this equation, and putting it equal to a constant, we obtain the equation of the locus required, which will be in general of the fourth degree; but if $g^2 = ac$, the axis of x will touch the given conic, and the equation of the locus will become divisible by y^2, and will reduce to the second degree. We could, by the help of the same equation, find the locus of the intersection of tangents; if the sum, product, &c., of the intercepts on the axis be given.

Ex. 8. Given four tangents to a conic to find the locus of the centre. [The solution here given is by P. Serret, *Nouvelles Annales*, 2nd series, IV. 145.]

Take any axes, and let the equation of one of the tangents be $x\cos\alpha + y\sin\alpha - p = 0$, then α is the angle the perpendicular on the tangent makes with the axis of x; and if θ be the unknown angle made with the same axis by the axis major of the conic, then $\alpha - \theta$ is the angle made by the same perpendicular with the axis major. If then x and y be the coordinates of the centre, the formula of Art. 178 gives us

$$(x\cos\alpha + y\sin\alpha - p)^2 = a^2\cos^2(\alpha - \theta) + b^2\sin^2(\alpha - \theta).$$

We have four equations of this form from which we have to eliminate the three unknown quantities a^2, b^2, θ. Using for shortness the abbreviation α for $x\cos\alpha + y\sin\alpha - p$ (Art. 53), this equation expanded may be written

$$\alpha^2 = (a^2\cos^2\theta + b^2\sin^2\theta)\cos^2\alpha + 2(a^2 - b^2)\cos\theta\sin\theta\cos\alpha\sin\alpha + (a^2\sin^2\theta + b^2\cos^2\theta)\sin^2\alpha.$$

It appears then that the three quantities $a^2\cos^2\theta + b^2\sin^2\theta$, $(a^2 - b^2)\cos\theta\sin\theta$, $a^2\sin^2\theta + b^2\cos^2\theta$, may be eliminated linearly from the four equations; and the result comes out in the form of a determinant

$$\begin{vmatrix} \alpha^2, & \cos^2\alpha, & \cos\alpha\sin\alpha, & \sin^2\alpha \\ \beta^2, & \cos^2\beta, & \cos\beta\sin\beta, & \sin^2\beta \\ \gamma^2, & \cos^2\gamma, & \cos\gamma\sin\gamma, & \sin^2\gamma \\ \delta^2, & \cos^2\delta, & \cos\delta\sin\delta, & \sin^2\delta \end{vmatrix} = 0,$$

which expanded is of the form $A\alpha^2 + B\beta^2 + C\gamma^2 + D\delta^2 = 0$, where A, B, C, D are known constants. But this equation, though apparently of the second degree, is in reality only of the first; for if, before expanding the determinant, we write α^2, &c., at full length, the coefficients of x^2 are $\cos^2\alpha$, $\cos^2\beta$, $\cos^2\gamma$, $\cos^2\delta$; but these being the same as one column of the determinant, the part multiplied by x^2 vanishes on expansion. Similarly, the coefficients of the terms xy and y^2 vanish. The locus is therefore a right line. The geometrical determination of the line depends on principles to be proved afterwards; namely, that the polar of any point with regard to the conic is

$$A\alpha'\alpha + B\beta'\beta + C\gamma'\gamma + D\delta'\delta = 0;$$

and, therefore, that the polar of the point $\alpha\beta$ passes through $\gamma\delta$. But when a conic reduces to a line by the vanishing of the three highest terms in its equation, the polar of any point is a parallel line at double the distance from the point. Thus it is seen that the line represented by the equation bisects the lines joining the points $\alpha\beta$, $\gamma\delta$; $\alpha\gamma$, $\beta\delta$; $\alpha\delta$, $\beta\gamma$. Conversely, if we are given in any form the equations of four lines $\alpha = 0$, &c., the equation of the line joining the three middle points of diagonals of the quadrilateral may, in practice, be most easily formed by determining the constants so that $A\alpha^2 + B\beta^2 + C\gamma^2 + D\delta^2 = 0$ shall represent a right line.

Ex. 9. Given three tangents to a conic and the sum of the squares of the axes, find the locus of the centre. We have three equations as in the last example, and a fourth $a^2 + b^2 = k^2$, which may be written

$$k^2 = (a^2\cos^2\theta + b^2\sin^2\theta) + (a^2\sin^2\theta + b^2\cos^2\theta),$$

and, as before, the result appears in the form of a determinant

$$\begin{vmatrix} \alpha^2, & \cos^2\alpha, & \cos\alpha\sin\alpha, & \sin^2\alpha \\ \beta^2, & \cos^2\beta, & \cos\beta\sin\beta, & \sin^2\beta \\ \gamma^2, & \cos^2\gamma, & \cos\gamma\sin\gamma, & \sin^2\gamma \\ k^2, & 1, & 0, & 1 \end{vmatrix} = 0,$$

which expanded is of the form $A\alpha^2 + B\beta^2 + C\gamma^2 + D = 0$. It is seen, as in the last example, that the coefficient of xy vanishes in the expansion, and that the coefficients of x^2 and y^2 are the same. The locus is therefore a circle. Now if $A\alpha^2 + B\beta^2 + C\gamma^2 = 0$ represents a circle, it will afterwards appear that the centre is the intersection of perpendiculars of the triangle formed by the lines α, β, γ. The present equation therefore, which differs from this by a constant (Art. 81) represents a circle whose centre is the intersection of perpendiculars of the triangle formed by the three tangents.

If we consider the case of the equilateral hyperbola $a^2 + b^2 = 0$, we see that two equilateral hyperbolas can be described to touch four given lines, the centres being the intersections of the line joining the middle points of diagonals with any one of four circles whose centres are the intersections of perpendiculars of the four triangles formed by any three of the four given lines. From the fact that the four circles have two common points it follows that the four intersections of perpendiculars lie on a right line, perpendicular to the line joining middle points of diagonals (see Art. 268, Ex. 2).

Ex. 10. Given four points on a conic to find the locus of either focus. The distance of one of the given points from the focus (see Ex., Art. 186) satisfies the equation

$$\rho = Ax' + By' + C.$$

We have four such equations from which we can linearly eliminate A, B, C, and we get the determinant

$$\begin{vmatrix} \rho, & x', & y', & 1 \\ \rho', & x'', & y'', & 1 \\ \rho'', & x''', & y''', & 1 \\ \rho''', & x'''', & y'''', & 1 \end{vmatrix} = 0,$$

which expanded is of the form $l\rho + m\rho' + n\rho'' + p\rho''' = 0$. If we look to the actual values of the coefficients l, m, n, p, and their geometric meaning (Art. 36), this equation geometrically interpreted gives us a theorem of Möbius, viz.

$$OA \,.\, BCD + OC \,.\, ABD = OB \,.\, ACD + OD \,.\, ABC,$$

where O is the focus, and BCD the area of the triangle formed by three of the points (compare Art. 94). It is seen thus that $l + m + n + p = 0$. If we substitute for ρ its value $\sqrt{\{(x - x')^2 + (y - y')^2\}}$, &c., and clear of radicals, the equation of the locus, though apparently of the eighth, is found to be only of the sixth degree. In fact, we may clear of radicals by giving each radical its double sign, and multiplying together the eight factors $l\rho \pm m\rho' \pm n\rho'' \pm p\rho'''$; and then it is apparent that the highest powers in x and y will be $(x^2 + y^2)^4$ multiplied by the product of the factors $l \pm m \pm n \pm p$; and that these terms vanish in virtue of the relation $l + m + n + p = 0$.

If the four given points be on a circle, Mr. Sylvester has remarked that the locus breaks up into two of the third degree, as Mr. Burnside has thus shewn. We have by a theorem of Feuerbach's, given Art. 94,

$$l\rho^2 + m\rho'^2 + n\rho''^2 + p\rho'''^2 = 0.$$

We have then $\quad (l + m)\,(l\rho^2 + m\rho'^2) = (n + p)\,(n\rho''^2 + p\rho'''^2),$

$$(l\rho + m\rho')^2 = (n\rho'' + p\rho''')^2,$$

whence, subtracting $\quad lm\,(\rho - \rho')^2 = np\,(\rho'' - \rho''')^2,$

which obviously breaks up into factors.

THE ECCENTRIC ANGLE.*

229. It is always advantageous to express the position of a point on a curve, if possible, by *a single independent* variable,

* The use of this angle was recommended by Mr. O'Brien, *Cambridge Mathematical Journal*, vol. IV. p. 99.

rather than by the *two* coordinates $x'y'$. We shall, therefore, find it useful, in discussing properties of the ellipse, to make a substitution similar to that employed (Art. 102) in the case of the circle, and shall write

$$x' = a \cos\phi, \quad y' = b \sin\phi,$$

a substitution evidently consistent with the equation

$$\left(\frac{x'}{a}\right)^2 + \left(\frac{y'}{b}\right)^2 = 1.$$

The geometric meaning of the angle ϕ is easily explained. If we describe a circle on the axis major as diameter, and produce the ordinate at P to meet the circle at Q, then the angle $QCL = \phi$, for $CL = CQ \cos QCL$, or $x' = a \cos\phi$; and $PL = \frac{b}{a} QL$ (Art. 163); or, since $QL = a \sin\phi$, we have $y' = b \sin\phi$.

230. If we draw through P a parallel PN to the radius CQ, then $PM : CQ :: PL : QL :: b : a$,
but $CQ = a$, therefore $PM = b$.
PN parallel to CQ is, of course, $= a$.

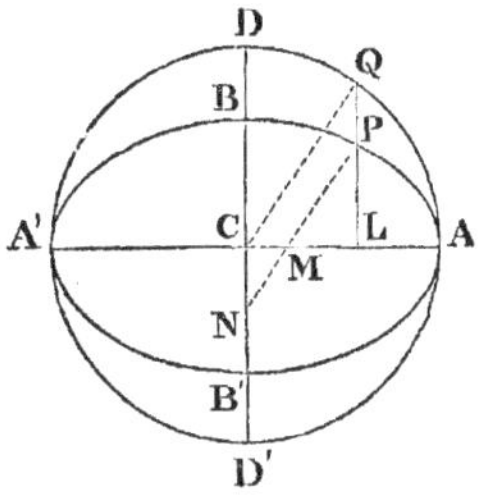

Hence, if from any point of an ellipse a line $= a$ be inflected to the minor axis, its intercept to the axis major $= b$. If the ordinate PQ were produced to meet the circle again in the point Q', it could be proved, in like manner, that a parallel through P to the radius CQ' is cut into parts of a constant length. Hence, conversely, if a line MN, of a constant length, move about in the legs of a right angle, and a point P be taken so that MP may be constant, the locus of P is an ellipse, whose axes are equal to MP and NP. (See Ex. 12, p. 47.)

On this principle has been constructed an instrument for describing an ellipse by continued motion, called the *Elliptic Compasses*. CA, CD' are two fixed rulers, MN a third ruler of a constant length, capable of sliding up and down between them, then a pencil fixed at any point of MN will describe an ellipse.

If the pencil be fixed at the middle point of MN, it will describe a circle. (O'Brien's *Coordinate Geometry*, p. 112.)

231. The consideration of the angle ϕ affords a simple method of constructing geometrically the diameter conjugate to a given one, for

$$\tan\theta = \frac{y'}{x'} = \frac{b}{a}\tan\phi.$$

Hence the relation

$$\tan\theta\tan\theta' = -\frac{b^2}{a^2} \text{ (Art. 170)}$$

becomes $$\tan\phi\tan\phi' = -1,$$

or $$\phi - \phi' = 90^\circ.$$

Hence we obtain the following construction. Let the ordinate at the given point P, when produced, meet the semicircle on the axis major at Q, join CQ, and erect CQ' perpendicular to it; then the perpendicular let fall on the axis from Q' will pass through P', a point on the conjugate diameter.

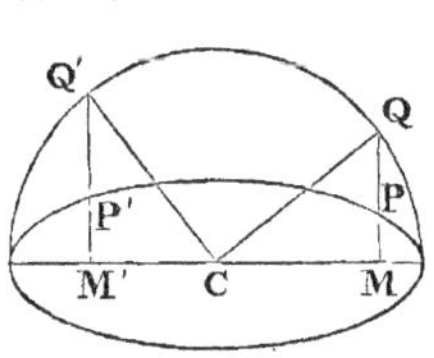

Hence, too, can easily be found the coordinates of P' given in Art. 172, for since

$$\cos\phi' = \sin\phi, \text{ we have } \frac{x''}{a} = \frac{y'}{b},$$

and since $$\sin\phi' = -\cos\phi, \text{ we have } \frac{y''}{b} = -\frac{x'}{a}.$$

From these values it appears that the areas of the triangles PCM, $P'CM'$ are equal.

Ex. 1. To express the lengths of two conjugate semi-diameters in terms of the angle ϕ. *Ans.* $a'^2 = a^2\cos^2\phi + b^2\sin^2\phi$; $b'^2 = a^2\sin^2\phi + b^2\cos^2\phi$.

Ex. 2. To express the equation of any chord of the ellipse in terms of ϕ and ϕ' (see p. 94).

Ans. $\frac{x}{a}\cos\frac{1}{2}(\phi+\phi') + \frac{y}{b}\sin\frac{1}{2}(\phi+\phi') = \cos\frac{1}{2}(\phi-\phi')$.

Ex. 3. To express similarly the equation of the tangent.

Ans. $\frac{x}{a}\cos\phi + \frac{y}{b}\sin\phi = 1$.

Ex. 4. To express the length of the chord joining two points α, β,

$$D^2 = a^2(\cos\alpha - \cos\beta)^2 + b^2(\sin\alpha - \sin\beta)^2,$$

$$D = 2\sin\tfrac{1}{2}(\alpha-\beta)\,\{a^2\sin^2\tfrac{1}{2}(\alpha+\beta) + b^2\cos^2\tfrac{1}{2}(\alpha+\beta)\}^{\frac{1}{2}}.$$

But (Ex. 1) the quantity between the parentheses is the semi-diameter conjugate to that to the point $\frac{1}{2}(\alpha+\beta)$; and (Ex. 2, 3) the tangent at the point $\frac{1}{2}(\alpha+\beta)$ is parallel to the chord joining the points α, β; hence, if b' denote the length of the semi-diameter parallel to the given chord, $D = 2b'\sin\frac{1}{2}(\alpha-\beta)$.

Ex. 5. To find the area of the triangle formed by three given points α, β, γ.

By Art. 36 we have

$$2\Sigma = ab\,\{\sin(\alpha-\beta)+\sin(\beta-\gamma)+\sin(\gamma-\alpha)\}$$
$$= ab\,\{2\sin\tfrac{1}{2}(\alpha-\beta)\cos\tfrac{1}{2}(\alpha-\beta) - 2\sin\tfrac{1}{2}(\alpha-\beta)\cos\tfrac{1}{2}(\alpha+\beta-2\gamma)\}$$
$$= 4ab\sin\tfrac{1}{2}(\alpha-\beta)\sin\tfrac{1}{2}(\beta-\gamma)\sin\tfrac{1}{2}(\gamma-\alpha)$$
$$\Sigma = 2ab\sin\tfrac{1}{2}(\alpha-\beta)\sin\tfrac{1}{2}(\beta-\gamma)\sin\tfrac{1}{2}(\gamma-\alpha).$$

Ex. 6. If the bisectors of sides of an inscribed triangle meet in the centre its area is constant.

Ex. 7. To find the radius of the circle circumscribing the triangle formed by three given points α, β, γ.

If d, e, f be the sides of the triangle formed by the three points,

$$R = \frac{def}{4\Sigma} = \frac{b'b''b'''}{ab},$$

where b', b'', b''' are the semi-diameters parallel to the sides of the triangle. If c', c'', c''' be the parallel focal chords, then (see Ex. 5, p. 213) $R^2 = \frac{c'c''c'''}{4p}$. (These expressions are due to Mr. MacCullagh, *Dublin Exam. Papers*, 1836, p. 22.)

Ex. 8. To find the equation of the circle circumscribing this triangle.

Ans. $$x^2+y^2 - \frac{2(a^2-b^2)x}{a}\cos\tfrac{1}{2}(\alpha+\beta)\cos\tfrac{1}{2}(\beta+\gamma)\cos\tfrac{1}{2}(\gamma+\alpha)$$
$$-\frac{2(b^2-a^2)y}{b}\sin\tfrac{1}{2}(\alpha+\beta)\sin\tfrac{1}{2}(\beta+\gamma)\sin\tfrac{1}{2}(\gamma+\alpha)$$
$$= \tfrac{1}{2}(a^2+b^2) - \tfrac{1}{2}(a^2-b^2)\{\cos(\alpha+\beta)+\cos(\beta+\gamma)+\cos(\gamma+\alpha)\}$$

From this equation the coordinates of the centre of this circle are at once obtained.

Ex. 9. The area of the triangle formed by three tangents is, by Art. 39,

$$ab\tan\tfrac{1}{2}(\alpha-\beta)\tan\tfrac{1}{2}(\beta-\gamma)\tan\tfrac{1}{2}(\gamma-\alpha).$$

Ex. 10. The area of the triangle formed by three normals is

$$\frac{c^4}{4ab}\tan\tfrac{1}{2}(\alpha-\beta)\tan\tfrac{1}{2}(\beta-\gamma)\tan\tfrac{1}{2}(\gamma-\alpha)\{\sin(\beta+\gamma)+\sin(\gamma+\alpha)+\sin(\alpha+\beta)\}^2,$$

consequently three normals meet in a point if

$$\sin(\beta+\gamma)+\sin(\gamma+\alpha)+\sin(\alpha+\beta)=0. \quad \text{[Mr. Burnside.]}$$

Ex. 11. To find the locus of the intersection of the focal radius vector FP with the radius of the circle CQ.

Let the central coordinates of P be $x'y'$, of O, xy, then we have, from the similar triangles, FON, FPM,

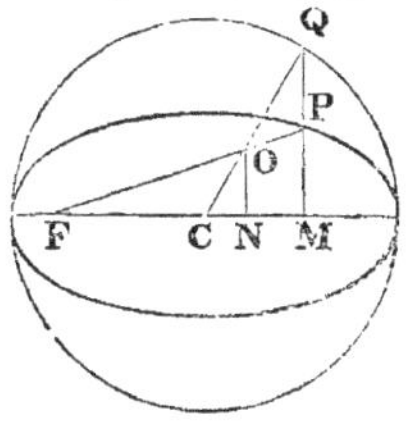

$$\frac{y}{x+c} = \frac{y'}{x'+c} = \frac{b\sin\phi}{a(e+\cos\phi)}.$$

Now, since ϕ is the angle made with the axis by the radius vector to the point O, we at once obtain the polar equation of the locus by writing $\rho\cos\phi$ for x, $\rho\sin\phi$ for y, and we find

$$\frac{\rho}{e+\rho\cos\phi} = \frac{b}{a(e+\cos\phi)},$$

or

$$\rho = \frac{bc}{c+(a-b)\cos\phi}.$$

Hence (Art. 193) the locus is an ellipse, of which C is one focus, and it can easily be proved that F is the other.

Ex. 12. The normal at P is produced to meet CQ; the locus of their intersection is a circle concentric with the ellipse.

The equation of the normal is

$$\frac{ax}{\cos\phi} - \frac{by}{\sin\phi} = c^2;$$

but we may, as in the last example, write $\rho\cos\phi$ and $\rho\sin\phi$ for x and y, and the equation becomes

$$(a-b)\,\rho = c^2, \text{ or } \rho = a + b.$$

Ex. 13. Prove that $\tan\frac{1}{2}PFC = \sqrt{\left(\frac{1-e}{1+e}\right)}\tan\frac{1}{2}\phi$.

Ex. 14. If from the vertex of an ellipse a radius vector be drawn to any point on the curve, find the locus of the point where a parallel radius through the centre meets the tangent at the point.

The tangent of the angle made with the axis by the radius vector to the vertex $= \frac{y'}{x'+a}$; therefore the equation of the parallel radius through the centre is

$$\frac{y}{x} = \frac{y'}{x'+a} = \frac{b\sin\phi}{a(1+\cos\phi)} = \frac{b}{a}\,\frac{1-\cos\phi}{\sin\phi};$$

or

$$\frac{y}{b}\sin\phi + \frac{x}{a}\cos\phi = \frac{x}{a},$$

and the locus of the intersection of this line with the tangent

$$\frac{y}{b}\sin\phi + \frac{x}{a}\cos\phi = 1,$$

is, obviously, $\frac{x}{a} = 1$, the tangent at the other extremity of the axis.

The same investigation will apply, if the first radius vector be drawn through any point of the curve, by substituting a' and b' for a and b; the locus will then be the tangent at the diametrically opposite point.

Ex. 15. The length of the chord of an ellipse which touches a confocal ellipse, the squares of whose semiaxes are $a^2 - h^2$, $b^2 - h^2$, is $\frac{2hb'^2}{ab}$ [Mr. Burnside].

The condition that the chord joining two points α, β should touch the confocal conic is

$$\frac{a^2-h^2}{a^2}\cos^2\tfrac{1}{2}(\alpha+\beta) + \frac{b^2-h^2}{b^2}\sin^2\tfrac{1}{2}(\alpha+\beta) = \cos^2\tfrac{1}{2}(\alpha-\beta),$$

or

$$\sin^2\tfrac{1}{2}(\alpha-\beta) = \frac{h^2}{a^2b^2}\{b^2\cos^2\tfrac{1}{2}(\alpha+\beta) + a^2\sin^2\tfrac{1}{2}(\alpha+\beta)\} = \frac{h^2}{a^2b^2}b'^2. \quad \text{(Ex. 4)}.$$

But the length of the chord is

$$2b'\sin\tfrac{1}{2}(\alpha-\beta) = \frac{2hb'^2}{ab}.$$

By the help of this Example several theorems concerning chords through a focus may be extended to chords touching confocal conics. Hence also is immediately derived a proof of Ex. 13, p. 212, for $OR.OR'$ is to $OS.OS'$ as the squares of the parallel diameters (Art. 149), and it is here proved that the chords $OR - OR'$, $OS - OS'$ are to each other in the same ratio.

232. The methods of the preceding Articles do not apply to the hyperbola. For the hyperbola, however, we may substitute

$$x' = a\sec\phi, \quad y' = b\tan\phi,$$

since

$$\left(\frac{x'}{a}\right)^2 - \left(\frac{y'}{b}\right)^2 = 1.$$

This angle may be represented geometrically by drawing a tangent MQ from the foot of the ordinate M to the circle described on the transverse axis, then the angle $QCM = \phi$, since

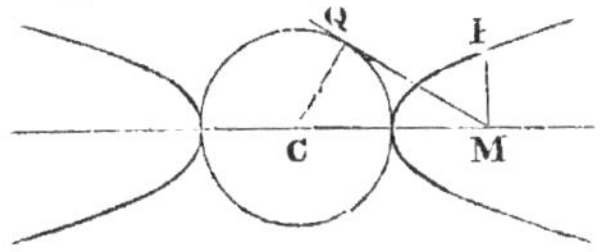

$$CM = CQ \sec QCM.$$

We have also $QM = a \tan \phi$, but $PM = b \tan \phi$. Hence, if from the foot of any ordinate of a hyperbola we draw a tangent to the circle described on the transverse axis, this tangent is in a constant ratio to the ordinate.

Ex. If any point on the conjugate hyperbola be expressed similarly $y'' = b \sec \phi'$, $x'' = a \tan \phi'$, prove that the relation connecting the extremities of conjugate diameters is $\phi = \phi'$. [Mr. Turner.]

SIMILAR CONIC SECTIONS.

233. Any two figures are said to be *similar and similarly placed* if radii vectores drawn to the first from a certain point O are in a constant ratio to parallel radii drawn to the second from another point o. If it be possible to find any two such points O and o, we can find an infinity of others; for, take any point C, draw oc parallel to OC, and in the constant ratio $\frac{op}{OP}$, then from the similar triangles OCP, ocp, cp is parallel to CP and in the given ratio. In like manner, any other radius vector through c can be proved to be proportional to the parallel radius through C.

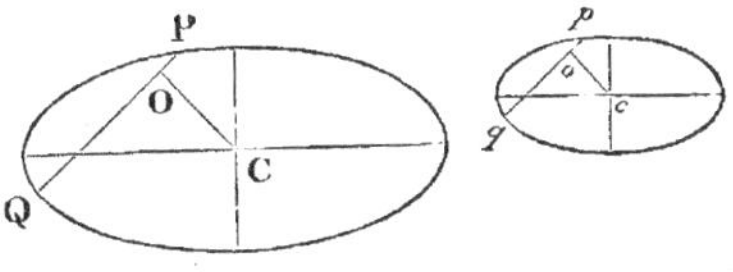

If two *central conic sections* be similar and similarly placed, all diameters of the one are proportional to the parallel diameters of the other, since the rectangles $OP.OQ$, $op.oq$ are proportional to the squares of the parallel diameters (Art. 149).

234. To find the condition that two conics, given by the general equations, should be similar and similarly placed.

Transforming to the centre of the first as origin, we find (Art. 152) that the square of any semi-diameter of the first is equal to a constant divided by $a \cos^2\theta + 2h \cos\theta \sin\theta + b \sin^2\theta$, and, in like manner, that the square of a parallel semi-diameter of the second is equal to another constant divided by

$$a' \cos^2\theta + 2h' \cos\theta \sin\theta + b' \sin^2\theta.$$

The ratio of the two cannot be independent of θ unless

$$\frac{a}{a'} = \frac{h}{h'} = \frac{b}{b'}.$$

Hence *two conic sections will be similar and similarly placed, if the coefficients of the highest powers of the variables are the same in both or only differ by a constant multiplier.*

235. It is evident that the directions of the axes of these conics must be the same, since the greatest and least diameters of one must be parallel to the greatest and least diameters of the other. If the diameter of one become infinite, so must also the parallel diameter of the other, that is to say, *the asymptotes of similar and similarly placed hyperbolas are parallel.* The same thing follows from the result of the last Article, since (Art. 154) the directions of the asymptotes are wholly determined by the highest terms of the equation.

Similar conics have the same eccentricity; for $\frac{a^2 - b^2}{a^2}$ must be $= \frac{m^2a^2 - m^2b^2}{m^2a^2}$. Similar and similarly placed conic sections have hence sometimes been defined as those whose axes are parallel, and which have the same eccentricity.

If two hyperbolas have parallel asymptotes they are similar, for their axes must be parallel, since they bisect the angles between the asymptotes (Art. 155), and the eccentricity wholly depends on the angle between the asymptotes (Art. 167).

236. Since the eccentricity of every parabola is $=1$, we should be led to infer that all parabolas are similar and similarly placed, the direction of whose axes is the same. In fact, the equation of one parabola, referred to its vertex, being $y^2 = px$, or

$$\rho = \frac{p \cos\theta}{\sin^2\theta},$$

it is plain that a parallel radius vector through the vertex of the other will be to this radius in the constant ratio $p' : p$.

Ex. 1. If on any radius vector to a conic section through a fixed point O, OQ be taken in a constant ratio to OP, find the locus of Q. We have only to substitute $m\rho$ for ρ in the polar equation, and the locus is found to be a conic similar to the given conic, and similarly placed.

The point O may be called *the centre of similitude* of the two conics; and it is obviously (see also Art. 115) the point where common tangents to the two conics intersect, since when the radii vectores OP, OP' to the first conic become equal, so must also OQ, OQ' the radii vectores to the other.

Ex. 2. If a pair of radii be drawn through a centre of similitude of two similar conics, the chords joining their extremities will be either parallel, or will meet on the chord of intersection of the conics.

This is proved precisely as in Art. 116.

Ex. 3. Given three conics, similar and similarly placed, their six centres of similitude will lie three by three on right lines (see figure, page 108).

Ex. 4. If any line cut two similar and concentric conics, its parts intercepted between the conics will be equal.

Any chord of the outer conic which touches the interior will be bisected at the point of contact.

These are proved in the same manner as the theorems at page 191, which are but particular cases of them; for the asymptotes of any hyperbola may be considered as a conic section similar to it, since the highest terms in the equation of the asymptotes are the same as in the equation of the curve.

Ex. 5. If a tangent drawn at any point P of the inner of two concentric and similar ellipses meet the outer in the points T and T', then any chord of the inner drawn through P is half the algebraic sum of the parallel chords of the outer through T and T'.

237. Two figures will be similar, although not similarly placed, if the proportional radii make a constant angle with each other, instead of being parallel; so that if we could imagine one of the figures turned round through the given angle, they would be then both similar and similarly placed.

To find the condition that two conic sections, given by the general equations, should be similar, even though not similarly placed.

We have only to transform the first equation to axes making any angle θ with the given axes, and examine whether any value can be assigned to θ which will make the new a, h, b proportional to a', h', b'. Suppose that they become ma', mh', mb'.

Now, the axes being supposed rectangular, we have seen (Art. 157) that the quantities $a+b$, $ab-h^2$, are unaltered by transformation of coordinates; hence we have

$$a+b=m(a'+b'),$$

$$ab-h^2=m^2(a'b'-h'^2),$$

and the required condition is evidently

$$\frac{ab - h^2}{(a+b)^2} = \frac{a'b' - h'^2}{(a' + b')^2}.$$

If the axes be oblique, it is seen in like manner (Art. 158) that the condition for similarity is

$$\frac{ab - h^2}{(a + b - 2h \cos \omega)^2} = \frac{a'b' - h'^2}{(a' + b' - 2h' \cos \omega)^2}.$$

It will be seen (Arts. 74, 151) that the condition found expresses that the angle between the (real or imaginary) asymptotes of the one curve is equal to that between those of the other.

THE CONTACT OF CONIC SECTIONS.

238. *Two curves of the mth and n^{th} degrees respectively intersect in mn points.*

For, if we eliminate either x or y between the equations, the resulting equation in the remaining variable will in general be of the mn^{th} degree (*Higher Algebra*, Art. 73). If it should happen that the resulting equation should appear to fall below the mn^{th} degree, in consequence of the coefficients of one or more of the highest powers vanishing, the curves would still be considered to intersect in mn points, one or more of these points being at infinity (see Art. 135). If account be thus taken of infinitely distant as well as of imaginary points, it may be asserted that the two curves *always* intersect in mn points. In particular *two conics always intersect in four points.* In the next Chapter some of the cases will be noticed where points of intersection of two conics are infinitely distant; at present we are about to consider the cases where two or more of them coincide.

Since four points may be connected by six lines, viz. 12, 34; 13, 24; 14, 23; *two conics have three pairs of chords of intersection.*

239. When two of the points of intersection coincide, the conics touch each other, and the line joining the coincident points is the common tangent. The conics will in this case meet in two real or imaginary points L, M distinct from the point of contact. This is called a *contact of the first order.* The contact is said to be *of the second order* when three of the points of intersection

coincide, as, for instance, if the point M move up until it coincide

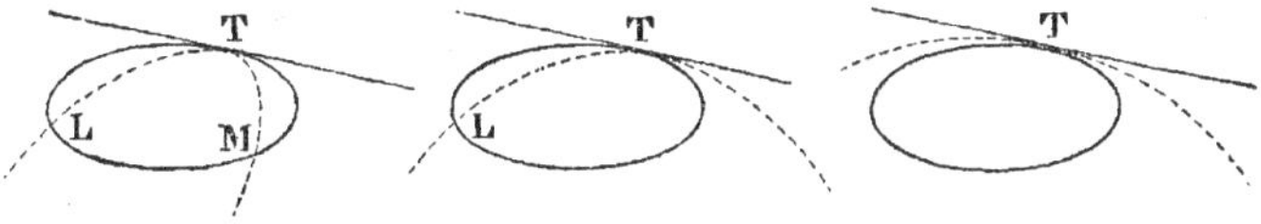

with T. Curves which have contact of an order higher than the first are also said to *osculate;* and it appears that conics which osculate must intersect in *one* other point. Contact *of the third order* is when two curves have four consecutive points common; and since two conics cannot have more than four points common, this is the highest order of contact they can have.

Thus, for example, the equations of two conics, both passing through the origin and having the line x for a common tangent are (Art. 144)

$$ax^2 + 2hxy + by^2 + 2gx = 0, \quad a'x^2 + 2h'xy + b'y^2 + 2g'x = 0.$$

And, as in Ex. 2, p. 175,

$$x\{(ab' - a'b)x + 2(hb' - h'b)y + 2(gb' - g'b)\} = 0,$$

represents a figure passing through their four points of intersection. The first factor represents the tangent which passes through the two coincident points of intersection, and the second factor denotes the line LM passing through the other two points. If now $gb' = g'b$, LM passes through the origin, and the conics have contact of the second order. If in addition $hb' = h'b$, the equation of LM reduces to $x = 0$; LM coincides with the tangent, and the conics have contact of the third order. In this last case, if we make by multiplication the coefficients of y^2 the same in both the equations, the coefficients of xy and x will also be the same, and the equations of the two conics may be reduced to the form

$$ax^2 + 2hxy + by^2 + 2gx = 0, \quad a'x^2 + 2hxy + by^2 + 2gx = 0.$$

240. Two conics may have *double contact* if the points of intersection 1, 2 coincide and also the points 3, 4. The condition that the pair of conics considered in the last Article should touch at a second point is found by expressing the condition that the line LM, whose equation is there given, should touch

either conic. Or, more simply, as follows: Multiply the equations by g' and g respectively, and subtract, and we get

$$(ag' - a'g)\, x^2 + 2\,(hg' - h'g)\, xy + (bg' - b'g)\, y^2 = 0,$$

which denotes the pair of lines joining the origin to the two points in which LM meets the conics. And these lines will coincide if

$$(ag' - a'g)\,(bg' - b'g) = (hg' - h'g)^2.$$

241. Since a conic can be found to satisfy any five conditions (Art. 133), a conic can be found to touch a given conic at a given point, and satisfy any three other conditions. If it have contact of the second order at the given point, it can be made to satisfy two other conditions; and if it have contact of the third order, it can be made to satisfy one other condition. Thus we can determine a *parabola* having contact of the third order at the origin with

$$ax^2 + 2hxy + by^2 + 2gx = 0.$$

Referring to the last two equations (Art. 239), we see that it is only necessary to write a' instead of a, where a' is determined by the equation $a'b = h^2$.

We cannot, in general, describe a *circle* to have contact of the third order with a given conic, because *two* conditions must be fulfilled in order that an equation should represent a circle; or, in other words, we cannot describe a circle through four consecutive points on a conic, since three points are sufficient to determine a circle. We can, however, easily find the equation of the circle passing through *three* consecutive points on the curve. This circle is called the *osculating circle*, or the *circle of curvature.*

The equation of the conic to oblique or rectangular axes being, as before,

$$ax^2 + 2hxy + by^2 + 2gx = 0,$$

that of any circle touching it at the origin is (Art. 84, Ex. 3)

$$x^2 + 2xy \cos\omega + y^2 - 2rx \sin\omega = 0.$$

Applying the condition $gb' = g'b$ (Art. 239), we see that the condition that the circle should osculate is

$$g = -rb \sin\omega, \text{ or } r = \frac{-g}{b \sin\omega}.*$$

The quantity r is called the *radius of curvature* of the conic at the point T.

242. *To find the radius of curvature at any point on a central conic.*

In order to apply the formula of the last Article the tangent at the point must be made the axis of y. Now the equation referred to a diameter through the point and its conjugate $\left(\frac{x^2}{a'^2} + \frac{y^2}{b'^2} = 1\right)$ is transferred to parallel axes through the given point, by substituting $x + a'$ for x, and becomes

$$\frac{x^2}{a'^2} + \frac{y^2}{b'^2} + \frac{2x}{a'} = 0.$$

Therefore, by the last Article, the radius of curvature is $\frac{b'^2}{a' \sin\omega}$. Now $a' \sin\omega$ is the perpendicular from the centre on the tangent, therefore the radius of curvature

$$= \frac{b'^2}{p}, \text{ or (Art. 175)} = \frac{b'^3}{ab}.$$

243. Let N denote the length of the normal PN, and let ψ denote the angle FPN between the normal and focal radius vector, then the radius of curvature is $\frac{N}{\cos^2\psi}$. For $N = \frac{bb'}{a}$ (Art. 181),

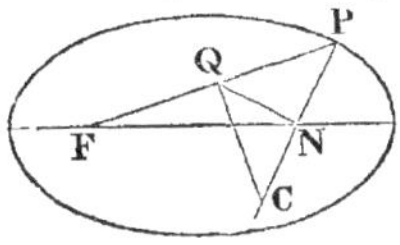

and $\cos\psi = \frac{b}{b'}$ (Art. 188), whence the truth of the formula is manifest.

---

* In the Examples which follow we find the absolute magnitude of the radius of curvature, without regard to sign. The sign, as usual, indicates the direction in which the radius is measured. For it indicates whether the given curve is osculated by a circle whose equation is of the form

$$x^2 + 2xy \cos\omega + y^2 \mp 2rx \sin\omega = 0,$$

the upper sign signifying one whose centre is in the positive direction of the axis of x; and the lower, one whose centre is in the negative direction. The formula in the text then gives a positive radius of curvature when the concavity of the curve is turned in the positive direction of the axis of x, and a negative radius when it is turned in the opposite direction.

Thus we have the following construction: Erect a perpendicular to the normal at the point where it meets the axis; and again at the point Q, where this perpendicular meets the focal radius, draw CQ perpendicular to it, then C will be the centre of curvature, and CP the radius of curvature.

244. Another useful construction is founded on the principle that *if a circle intersect a conic, its chords of intersection will make equal angles with the axis.* For the rectangles under the segments of the chords are equal (Euc. III. 35), and therefore the parallel diameters of the conic are equal (Art. 149), and therefore make equal angles with the axis (Art. 162).

Now, in the case of the circle of curvature, the tangent at T (see figure, p. 226) is one chord of intersection and the line TL the other; we have, therefore, only to draw TL, making the same angle with the axis as the tangent, and we have the point L; then the circle described through the points T, L, and, touching the conic at T, is the circle of curvature.

This construction shows that the osculating circle at either vertex has a contact of the third degree.

Ex. 1. Using the notation of the eccentric angle, find the condition that four points α, β, γ, δ should lie on the same circle (Joachimsthal, *Crelle*, XXXVI. 95).

The chord joining two of them must make the same angle with one side of the axis as the chord joining the other two does with the other; and the chords being

$$\frac{x}{a}\cos\tfrac{1}{2}(\alpha+\beta)+\frac{y}{b}\sin\tfrac{1}{2}(\alpha+\beta)=\cos\tfrac{1}{2}(\alpha-\beta);$$

$$\frac{x}{a}\cos\tfrac{1}{2}(\gamma+\delta)+\frac{y}{b}\sin\tfrac{1}{2}(\gamma+\delta)=\cos\tfrac{1}{2}(\gamma-\delta);$$

we have $\tan\tfrac{1}{2}(\alpha+\beta)+\tan\tfrac{1}{2}(\gamma+\delta)=0$; $\alpha+\beta+\gamma+\delta=0$; or $=2m\pi$.

Ex. 2. Find the coordinates of the point where the osculating circle meets the conic again.

We have $\alpha=\beta=\gamma$; hence $\delta=-3\alpha$; or $X=\dfrac{4x'^3}{a^2}-3x'$; $Y=\dfrac{4y'^3}{b^2}-3y'$.

Ex. 3. If the normals at three points α, β, γ meet in a point, the foot of the fourth normal from that point is given by the equation $\alpha+\beta+\gamma+\delta=(2m+1)\pi$.

Ex. 4. Find the equation of the chord of curvature TL.

Ans. $\dfrac{x}{a}\cos\alpha-\dfrac{y}{b}\sin\alpha=\cos 2\alpha$.

Ex. 5. There are three points on a conic whose osculating circles pass through a given point on the curve; these lie on a circle passing through the point, and form a triangle of which the centre of the curve is the intersection of bisectors of sides (Steiner, *Crelle*, XXXII. 300; Joachimsthal, *Crelle*, XXXVI. 95).

Here we are given δ, the point where the circle meets the curve again, and from the last Example the point of contact is $\alpha=-\tfrac{1}{3}\delta$. But since the sine and cosine of δ would not alter if δ were increased by 360°, we might also have $\alpha = -\frac{1}{3}\delta + 120°$, or $= -\frac{1}{3}\delta + 240°$, and, from Ex. 1, these three points lie on a circle passing through δ. If in the last Example we suppose X, Y given, since the cubics which determine x' and y' want the second terms, the sums of the three values of x' and of y' are respectively equal to nothing; and therefore (Ex. 4, p. 5) the origin is the intersection of the bisectors of sides of the triangle formed by the three points. It is easy to see that when the bisectors of sides of an inscribed triangle intersect in the centre, the normals at the vertices are the three perpendiculars of this triangle, and therefore meet in a point.

245. *To find the radius of curvature of a parabola.*

The equation referred to any diameter and tangent being $y^2 = p'x$, the radius of curvature (Art. 241) is $\frac{p'}{2\sin\theta}$, where θ is the angle between the axes. The expression $\frac{N}{\cos^2\psi}$, and the construction depending on it, hold for the parabola, since

$$N = \tfrac{1}{2}p'\sin\theta \text{ (Arts. 212, 213) and } \psi = 90° - \theta \text{ (Art. 217).}$$

Ex. 1. In all the conic sections the radius of curvature is equal to the cube of the normal divided by the square of the semi-parameter.

Ex. 2. Express the radius of curvature of an ellipse in terms of the angle which the normal makes with the axis.

Ex. 3. Find the lengths of the chords of the circle of curvature which pass through the centre or the focus of a central conic section. *Ans.* $\frac{2b'^2}{a'}$, and $\frac{2b'^2}{a}$.

Ex. 4. The focal chord of curvature of any conic is equal to the focal chord of the conic drawn parallel to the tangent at the point.

Ex. 5. In the parabola the focal chord of curvature is equal to the parameter of the diameter passing through the point.

246. *To find the coordinates of the centre of curvature of a central conic.*

These are evidently found by subtracting from the coordinates of the point on the conic the projections of the radius of curvature upon each axis. Now it is plain that this radius is to its projection on y as the normal to the ordinate y. We find the projection, therefore, of the radius of curvature on the axis of y $\left(\text{by multiplying the radius } \frac{b'^2}{p} \text{ by } \frac{y'}{N}\right) = \frac{b'^2y'}{b^2}$. The y of the centre of curvature then is $\frac{b^2 - b'^2}{b^2}y'$. But $b'^2 = b^2 + \frac{c^2}{b^2}y'^2$, therefore the y of the centre of curvature is $\frac{b^2 - a^2}{b^4}y'^3$. In like manner its x is $\frac{a^2 - b^2}{a^4}x'^3$.

We should have got the same values by making $\alpha = \beta = \gamma$ in Ex. 8, p. 220.

Or, again, the centre of the circle circumscribing a triangle is the intersection of perpendiculars to the sides at their middle points; and when the triangle is formed by three consecutive points on a curve, two sides are consecutive tangents to the curve, and the perpendiculars to them are the corresponding normals, and *the centre of curvature of any curve is the intersection of two consecutive normals.* Now if we make $x' = x'' = X$, $y' = y'' = Y$, in Ex. 4, p. 175, we obtain again the same values as those just determined.

247. *To find the coordinates of the centre of curvature of a parabola.*

The projection of the radius on the axis of y is found in like manner $\left(\text{by multiplying the radius of curvature } \frac{N}{\sin^2\theta} \text{ by } \frac{y'}{N}\right)$

$$= \frac{y'}{\sin^2\theta};$$

and subtracting this quantity from y' we have

$$Y = -\frac{y'}{\tan^2\theta} = -\frac{4y'^3}{p^2} \text{ (Art. 212).}$$

In like manner its X is $x' + \frac{p}{2\sin^2\theta} = x' + \frac{p + 4x'}{2} = 3x' + \frac{1}{2}p$.

The same values may be found from Ex. 10, p. 214.

248. The *evolute* of a curve is the locus of the centres of curvature of its different points. If it were required to find the evolute of a central conic, we should solve for $x'y'$ in terms of the x and y of the centre of curvature, and, substituting in the equation of the curve, should have $\left(\text{writing } \frac{c^2}{a} = A,\ \frac{c^2}{b} = B\right)$,

$$\frac{x^{\frac{2}{3}}}{A^{\frac{2}{3}}} + \frac{y^{\frac{2}{3}}}{B^{\frac{2}{3}}} = 1.$$

In like manner the equation of the evolute of a parabola is found to be

$$27py^2 = 16\,(x - \tfrac{1}{2}p)^3,$$

which represents a curve called the *semi-cubical parabola.*

CHAPTER XIV.

METHODS OF ABRIDGED NOTATION.

249. If $S=0$, $S'=0$ be the equations of two conics, then the equation of any conic passing through their four, real or imaginary, points of intersection can be expressed in the form $S=kS'$. For the form of this equation shows (Art. 40) that it denotes a conic passing through the four points common to S and S'; and we can evidently determine k so that $S=kS.$ shall be satisfied by the coordinates of any fifth point. It must then denote the conic determined by the five points.*

This will, of course, still be true if either or both the quantities S, S' be resolvable into factors. Thus $S=k\alpha\beta$, being evidently satisfied by the coordinates of the points where the right lines α, β meet S, represents a conic passing through the four points where S is met by this pair of lines; or, in other words, represents a conic having α and β for a pair of chords of intersection with S. If either α or β do not meet S in real points, it must still be considered as a chord of imaginary intersection, and will preserve many important properties in relation to the two curves, as we have already seen in the case of the circle (Art. 106). So, again, $\alpha\gamma=k\beta\delta$ denotes a conic circumscribing the quadrilateral $\alpha\beta\gamma\delta$, as we have already seen (Art. 122).† It is obvious that in what is here stated, α need not

---

* Since five conditions determine a conic, it is evident that the most general equation of a conic satisfying four conditions must contain one independent constant, whose value remains undetermined until a fifth condition is given. In like manner, the most general equation of a conic satisfying three conditions contains two independent constants, and so on. Compare the equations of a conic passing through three points or touching three lines (Arts. 124, 129).

If we are given any four conditions, in the expression of each of which the coefficients enter only in the first degree, the conic passes through four fixed points; for by eliminating all the coefficients but one, the equation of the conic is reduced to the form $S=kS'$.

† If $\alpha\beta$ be one pair of chords joining four points on a conic S, and $\gamma\delta$ another pair of chords, it is immaterial whether the general equation of a conic passing through the four points be expressed in any of the forms $S-k\alpha\beta$, $S-k\gamma\delta$, $\alpha\beta-k\gamma\delta$, where k is indeterminate; because, in virtue of the general principle, S is itself of the form $\alpha\beta-k\gamma\delta$.

be restricted, as at p. 53, to denote a line whose equation has been reduced to the form $x\cos\alpha + y\sin\alpha = p$; but that the argument holds if α denote a line expressed by the general equation.

250. There are three values of k, for which $S - kS'$ represents a pair of right lines. For the condition that this shall be the case, is found by substituting $a - ka'$, $b - kb'$, &c. for a, b, &c. in

$$abc + 2fgh - af^2 - bg^2 - ch^2 = 0,$$

and the result evidently is of the third degree in k, and is therefore satisfied by three values of k. If the roots of this cubic be k', k'', k''', then $S - k'S'$, $S - k''S'$, $S - k'''S'$, denote the three pairs of chords joining the four points of intersection of S and S' (Art. 238).

Ex. 1. What is the equation of a conic passing through the points where a given conic S meets the axes?

Here the axes $x = 0$, $y = 0$, are chords of intersection, and the equation must be of the form $S = kxy$, where k is indeterminate. See Ex. 1, Art. 151.

Ex. 2. Form the equation of the conic passing through five given points; for example (1, 2), (3, 5), (– 1, 4), (– 3, – 1) (– 4, 3). Forming the equations of the sides of the quadrilateral formed by the first four points, we see that the equation of the required conic must be of the form

$$(3x - 2y + 1)(5x - 2y + 13) = k(x - 4y + 17)(3x - 4y + 5).$$

Substituting in this, the coordinates of the fifth point (–4, 3), we obtain $k = -\frac{221}{15}$. Substituting this value and reducing the equation, it becomes

$$79x^2 - 320xy + 301y^2 + 1101x - 1665y + 1586 = 0.$$

251. The conics S, $S - k\alpha\beta$ will touch; or, in other words, two of their points of intersection will coincide; if either α or β touch S, or again, if α and β intersect in a point on S. Thus if $T = 0$ be the equation of the tangent to S at a given point on it $x'y'$, then $S = T(lx + my + n)$, is the most general equation of a conic touching S at the point $x'y'$; and if three additional conditions are given, we can complete the determination of the conic by finding l, m, n.

Three of the points of intersection will coincide if $lx + my + n$ pass through the point $x'y'$; and the most general equation of a conic osculating S at the point $x'y'$ is $S = T(lx + my - lx' - my')$. If it be required to find the equation of the osculating *circle*, we have only to express that the coefficient xy vanishes in this equation, and that the coefficient of $x^2 =$ that of y^2; when we have two equations which determine l and m.

The conics will have four consecutive points common if $lx + my + n$ coincide with T, so that the equation of the second conic is of the form $S = kT^2$. Compare Art. 239.

Ex. 1. If the axes of S be parallel to those of S', so will also the axes of $S - kS'$. For if the axes of coordinates be parallel to the axes of S, neither S nor S' will contain the term xy. If S' be a circle, the axes of $S - kS'$ are parallel to the axes of S. If $S - kS'$ represent a pair of right lines, its axes become the internal and external bisectors of the angles between them; and we have the theorem of Art. 244.

Ex. 2. If the axes of coordinates be parallel to the axes of S, and also to those of $S - k\alpha\beta$, then α and β are of the forms $lx + my + n$, $lx - my + n'$.

Ex. 3. To find the equation of the circle osculating a central conic. The equation must be of the form

$$\frac{x^2}{a^2} + \frac{y^2}{b^2} - 1 = \left(\frac{xx'}{a^2} + \frac{yy'}{b^2} - 1\right)(lx + my - lx' - my').$$

Expressing that the coefficient of xy vanishes, we reduce the equation to the form

$$\lambda\left(\frac{x^2}{a^2} + \frac{y^2}{b^2} - 1\right) = \left(\frac{xx'}{a^2} + \frac{yy'}{b^2} - 1\right)\left(\frac{xx'}{a^2} - \frac{yy'}{b^2} - \frac{x'^2}{a^2} + \frac{y'^2}{b^2}\right).$$

and expressing that the coefficient of $x^2 =$ that of y^2, we find $\lambda = \frac{b'^2}{b^2 - a^2}$, and the equation becomes

$$x^2 + y^2 - \frac{2(a^2 - b^2)\, x'^3 x}{a^4} - \frac{2(b^2 - a^2)\, y'^3 y}{b^4} + a'^2 - 2b'^2 = 0.$$

Ex. 4. To find the equation of the circle osculating a parabola.

Ans. $(p^2 + 4px')(y^2 - px) = \{2yy' - p(x + x')\}\{2yy' + px - 3px'\}$.

252. We have seen that $S = k\alpha\beta$ represents a conic passing through the four points $P, Q; p, q$, where α, β meet S; and it is evident that the closer to each other the lines α, β are, the nearer the point P is to p, and Q to q. Suppose that the lines α and β coincide, then the points P, p; Q, q coincide, and the second conic will touch the first at the points P, Q. Thus, then, the equation $S = k\alpha^2$ *represents a conic having double contact with S, α being the chord of contact.* Even if α do not meet S, it is to be regarded as the imaginary chord of contact of the conics S and $S - k\alpha^2$. In like manner $\alpha\gamma = k\beta^2$ represents a conic to which α and γ are tangents and β the chord of contact, as we have already seen (Art. 123). The equation of a conic having double contact with S at two given points $x'y'$, $x''y''$ may be also written in the

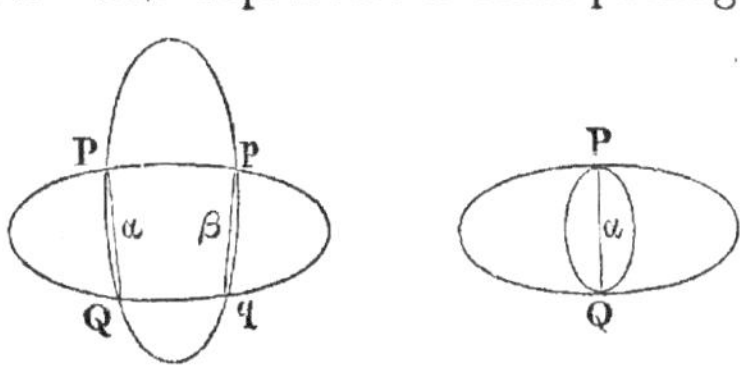

form $S = kTT'$, where T and T' represent the tangents at these points.

253. If the line α be parallel to an asymptote of the conic S, it will also be parallel to an asymptote of any conic represented by $S = k\alpha\beta$, which then denotes a system passing through three finite and one infinitely distant point. In like manner, if in addition β were parallel to the other asymptote, the system would pass through two finite and two infinitely distant points. Other forms which denote conics having points of intersection at infinity will be recognized by bearing in mind the principle (Art. 67) that the equation of an infinitely distant line is $0.x + 0.y + C = 0$; and hence (Art. 69) that an equation, apparently not homogeneous, may be made homogeneous in form, if in any of the terms which seem to be below the proper degree of the equation we replace one or more of the constant multipliers by $0.x + 0.y + C$. Thus, the equation of a conic referred to its asymptotes $xy = k^2$ (Art. 199) is a particular case of the form $\alpha\gamma = \beta^2$ referred to two tangents and the chord of contact (Arts. 123, 252). Writing the equation $xy = (0.x + 0.y + k)^2$, it is evident that the lines x and y are tangents, whose points of contact are at infinity (Art. 154).

254. Again, the equation of a parabola $y^2 = px$ is also a particular case of $\alpha\gamma = \beta^2$. Writing the equation $x(0.x + 0.y + p) = y^2$, the form of the equation shows, not only that the line x touches the curve, its point of contact being the point where x meets y, but also that the line at infinity touches the curve, its point of contact also being on the line y. The same inference may be drawn from the general equation of the parabola

$$(\alpha x + \beta y)^2 + (2gx + 2fy + c)(0.x + 0.y + 1) = 0,$$

which shews that both $2gx + 2fy + c$, and the line at infinity are tangents, and that the diameter $\alpha x + \beta y$ joins the points of contact. Thus, then, *every parabola has one tangent altogether at an infinite distance.* In fact, the equation which determines the direction of the points at infinity on a parabola is a perfect square (Art. 137); the two points of the curve at infinity therefore coincide; and therefore the line at infinity is to be regarded as a tangent (Art. 83).

Ex. The general equation

$$ax^2 + 2hxy + by^2 + 2gx + 2fy + c = 0$$

may be regarded as a particular case of the form (Art. 122) $\alpha\gamma = k\beta\delta$. For the first three terms denote two lines α, γ passing through the origin, and the last three terms denote the line at infinity β, together with the line δ, $2gx + 2fy + c$. The form of the equation then shows that the lines α, γ meet the curve at infinity, and also that δ represents the line joining the finite points in which $\alpha\gamma$ meet the curve.

255. In accordance with Art. 253, the equation $S = k\beta$ is to be regarded as a particular case of $S = \alpha\beta$, and denotes a system of conics passing through the two finite points where β meets S, and also through the two infinitely distant points where S is met by $0.x + 0.y + k$. Now it is plain that the coefficients of x^2, of xy, and of y^2, are the same in S and in $S - k\beta$, and therefore (Art. 234) that these equations denote conics similar and similarly placed. We learn, therefore, that *two conics similar and similarly placed meet each other in two infinitely distant points, and consequently only in two finite points.*

This is also geometrically evident when the curves are hyperbolas; for the asymptotes of similar conics are parallel (Art. 235), that is, they intersect at infinity; but each asymptote intersects its own curve at infinity; consequently the infinitely distant point of intersection of the two parallel asymptotes is also a point common to the two curves. Thus, on the figure, the infinitely distant point of meeting of the lines OX, Ox, and of the lines OY, Oy, are common to the curves. One of their finite points of intersection is shown on the figure, the other is on the opposite branches of the hyperbolas.

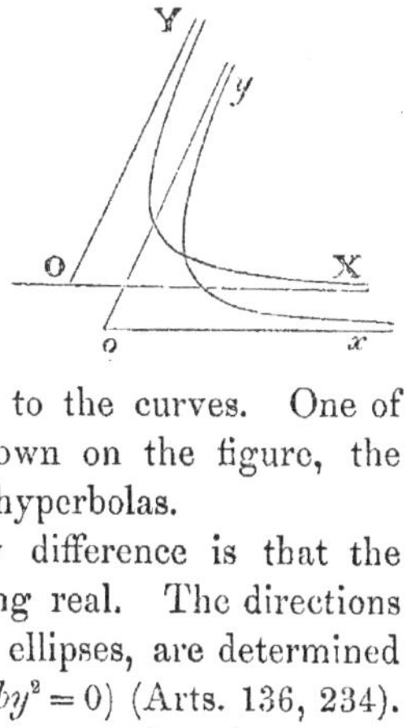

If the curves be ellipses, the only difference is that the asymptotes are imaginary instead of being real. The directions of the points at infinity, on two similar ellipses, are determined from the same equation $(ax^2 + 2hxy + by^2 = 0)$ (Arts. 136, 234). Now, although the roots of this equation are imaginary, yet they are, in both cases, *the same* imaginary roots, and therefore the curves are to be considered as having two imaginary points at infinity common. In fact, it was observed before, that even when the line α does not meet S in real points, it is to be regarded as a chord of imaginary intersection of S and $S-k\alpha\beta$, and this remains true when the line α is infinitely distant.

If the curves be parabolas, they are both touched by the line at infinity (Art. 254); but the direction of the point of contact, depending only on the first three terms of the equation, is the same for both. Hence, *two similar and similarly placed parabolas touch each other at infinity.* In short, the two infinitely distant points common to two similar conics are real, imaginary, or coincident, according as the curves are hyperbolas, ellipses, or parabolas.

256. The equation $S=k$, or $S=k(0.x+0.y+1)^2$ is manifestly a particular case of $S=k\alpha^2$, and therefore (Art. 252) denotes a conic having double contact with S, the chord of contact being at infinity. Now $S-k$ differs from S only in the constant term. Not only then are the conics similar and similarly placed, the first three terms being the same, but they are also concentric. For the coordinates of the centre (Art. 140) do not involve c, and therefore two conics whose equations differ only in the absolute term are concentric (see also Art. 81). Hence, *two similar and concentric conics are to be regarded as touching each other at two infinitely distant points.* In fact, the asymptotes of two such conics are not only parallel but coincident; they have therefore not only two points at infinity common, but also the tangents at those points; that is to say, the curves touch.

If the curves be parabolas, then, since the line at infinity touches both curves, S and $S-k^2$ have with each other, by Art. 251, a contact at infinity of the third order. Two parabolas whose equations differ only in the constant term will be equal to each other; for the curves $y^2=px$, $y^2=p(x+n)$ are obviously equal, and the equations transformed to any new axes will continue to differ only in the constant term. We have seen, too (Art. 205), that the expression for the parameter of a parabola does not involve the absolute term. The parabolas then, S and $S-k^2$ are equal, and we learn that *two equal and similarly placed parabolas whose axes are coincident may be considered as having with each other a contact of the third order at infinity.*

257. All circles are similar curves, the terms of the second degree being the same in all. It follows then, from the last

Articles, that *all circles pass through the same two imaginary points at infinity*, and on that account can never intersect in more than two finite points, and that *concentric circles touch each other in two imaginary points at infinity;* and on that account can never intersect in any finite point. It will appear hereafter that a multitude of theorems concerning circles are but particular cases of theorems concerning conics which pass through two fixed points.

258. It is important to notice the form $l^2\alpha^2 + m^2\beta^2 = n^2\gamma^2$, which denotes a conic with respect to which α, β, γ are the sides of a self-conjugate triangle (Art. 99). For the equation may be written in any of the forms

$$n^2\gamma^2 - m^2\beta^2 = l^2\alpha^2;\ n^2\gamma^2 - l^2\alpha^2 = m^2\beta^2;\ l^2\alpha^2 + m^2\beta^2 = n^2\gamma^2.$$

The first form shows that $n\gamma + m\beta$, $n\gamma - m\beta$ (which intersect in $\beta\gamma$) are tangents, and α their chord of contact. Consequently the point $\beta\gamma$ is the pole of α. Similarly from the second form $\gamma\alpha$ is the pole of β. It follows, then, that $\alpha\beta$ is the pole of γ; and this also appears from the third form, which shows that the two imaginary lines $l\alpha \pm m\beta\sqrt{(-1)}$ are tangents whose chord of contact is γ. Now these imaginary lines intersect in the real point $\alpha\beta$, which is therefore the pole of γ; although being within the conic, the tangents through it are imaginary.

It appears, in like manner, that

$$a\alpha^2 + 2h\alpha\beta + b\beta^2 = c\gamma^2$$

denotes a conic, such that $\alpha\beta$ is the pole of γ; for the left-hand side can be resolved into the product of factors representing lines which intersect in $\alpha\beta$.

Cor. If $l^2\alpha^2 + m^2\beta^2 = n^2\gamma^2$ denote a circle, its centre must be the intersection of perpendiculars of the triangle $\alpha\beta\gamma$. For the perpendicular let fall from any point on its polar must pass through the centre.

258*(*a*). If $x = 0$, $y = 0$ be any lines at right angles to each other through a focus, and γ the corresponding directrix, the equation of the curve is

$$x^2 + y^2 = e^2\gamma^2,$$

a particular form of the equation of Art. 258. Its form shows that the focus (xy) is the pole of the directrix γ, and that the

* This Article was numbered 279 in the previous editions.

polar of any point on the directrix is perpendicular to the line joining it to the focus (Art. 192); for y, the polar of $(x\gamma)$ is perpendicular to x, but x may be any line drawn through the focus.

The form of the equation shows that the two imaginary lines $x^2 + y^2$ are tangents drawn through the focus. Now, since these lines are the same whatever γ be, it appears that *all conics which have the same focus have two imaginary common tangents passing through this focus.* All conics, therefore, which have *both* foci common, have *four* imaginary common tangents, and may be considered as conics inscribed in the same quadrilateral. The imaginary tangents through the focus $(x^2 + y^2 = 0)$ are the same as the lines drawn to the two imaginary points at infinity on any circle (see Art. 257). Hence, we obtain the following general conception of foci: "Through each of the two imaginary points at infinity on any circle draw two tangents to the conic; these tangents will form a quadrilateral, two of whose vertices will be real and the foci of the curve, the other two may be considered as imaginary foci of the curve."

Ex. To find the foci of the conic given by the general equation. We have only to express the condition that $x - x' + (y - y')\sqrt{(-1)}$ should touch the curve. Substituting then in the formula of Art. 151, for λ, μ, ν respectively, 1, $\sqrt{(-1)}$, $-\{x' + y'\sqrt{(-1)}\}$; and equating separately the real and imaginary parts to cypher, we find that the foci are determined as the intersection of the two loci

$$C(x^2 - y^2) + 2Fy - 2Gx + A - B = 0, \quad Cxy - Fx - Gy + H = 0,$$

which denote two equilateral hyperbolas concentric with the given conic. Writing the equations

$$(Cx - G)^2 - (Cy - F)^2 = G^2 - AC - (F^2 - BC) = \Delta(a - b),$$

$$(Cx - G)(Cy - F) = FG - CH = \Delta h;$$

the coordinates of the foci are immediately given by the equations

$$(Cx - G)^2 = \tfrac{1}{2}\Delta(R + a - b); \quad (Cy - F)^2 = \tfrac{1}{2}\Delta(R + b - a),$$

where Δ has the same meaning as at p. 153, and R as at p. 158. If the curve is a parabola, $C = 0$, and we have to solve two linear equations which give

$$(F^2 + G^2)x = FH + \tfrac{1}{2}(A - B)G; \quad (F^2 + G^2)y = GH + \tfrac{1}{2}(B - A)F.$$

259. We proceed to notice some inferences which follow on interpreting, by the help of Art. 34, the equations we have already used. Thus (see Arts. 122, 123) the equation $\alpha\gamma = k\beta^2$ implies that *the product of the perpendiculars from any point of a conic on two fixed tangents is in a constant ratio to the square of the perpendicular on their chord of contact.*

The equation $\alpha\gamma = k\beta\delta$, similarly interpreted, leads to the important theorem: *The product of the perpendiculars let fall from any point of a conic on two opposite sides of an inscribed quadrilateral is in a constant ratio to the product of the perpendiculars let fall on the other two sides.*

From this property we at once infer, that *the anharmonic ratio of a pencil, whose sides pass through four fixed points of a conic, and whose vertex is any variable point of it, is constant.*

For the perpendicular

$$\alpha = \frac{OA.OB.\sin AOB}{AB}, \quad \gamma = \frac{OC.OD.\sin COD}{CD}, \text{ \&c.}$$

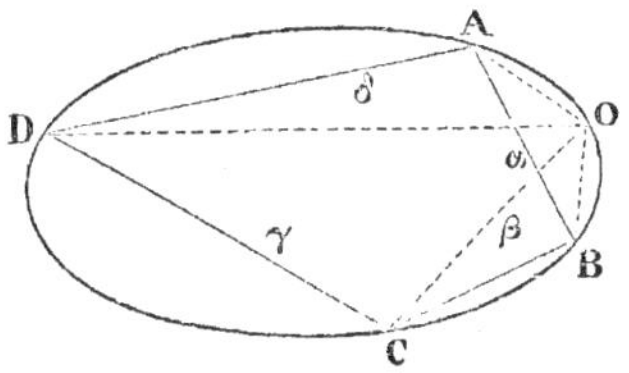

Now if we substitute these values in the equation $\alpha\gamma + k\beta\delta$, the continued product $OA.OB.OC.OD$ will appear on both sides of the equation, and may therefore be suppressed, and there will remain

$$\frac{\sin AOB.\sin COD}{\sin BOC.\sin AOD} = k.\frac{AB.CD}{BC.AD},$$

but the right-hand member of this equation is constant, while the left-hand member is the anharmonic ratio of the pencil OA, OB, OC, OD.

The consequences of this theorem are so numerous and important that we shall devote a section of another chapter to develope them more fully.

260. If $S = 0$ be the equation to a circle, then (Art. 90) S is the square of the tangent from any point xy to the circle; hence $S - k\alpha\beta = 0$ (the equation of a conic whose chords of intersection with the circle are α and β) expresses that *the locus of a point, such that the square of the tangent from it to a fixed circle is in a constant ratio to the product of its distances from two fixed lines, is a conic passing through the four points in which the fixed lines intersect the circle.*

This theorem is equally true whatever be the magnitude of the circle, and whether the right lines meet the circle in real or imaginary points; thus, for example, if the circle be infinitely small, *the locus of a point, the square of whose distance from a fixed point is in a constant ratio to the product of its distances from two fixed lines, is a conic section;* and the fixed lines may be considered as chords of imaginary intersection of the conic with an infinitely small circle whose centre is the fixed point.

261. Similar inferences can be drawn from the equation $S - k\alpha^2 = 0$, where S is a circle. We learn that *the locus of a point, such that the tangent from it to a fixed circle is in a constant ratio to its distance from a fixed line, is a conic touching the circle at the two points where the fixed line meets it;* or, conversely, that *if a circle have double contact with a conic, the tangent drawn to the circle from any point on the conic is in a constant ratio to the perpendicular from the point on the chord of contact.*

In the particular case where the circle is infinitely small, we obtain the fundamental property of the focus and directrix, and we infer that *the focus of any conic may be considered as an infinitely small circle, touching the conic in two imaginary points situated on the directrix.*

262. In general, *if in the equation of any conic the coordinates of any point be substituted, the result will be proportional to the rectangle under the segments of a chord drawn through the point parallel to a given line.**

For (Art. 148) this rectangle

$$= \frac{c'}{a\cos^2\theta + 2h\cos\theta\sin\theta + b\sin^2\theta},$$

where, by Art. 134, c' is the result of substituting in the equation the coordinates of the point; if, therefore, the angle θ be constant, this rectangle will be proportional to c'.

Ex. 1. If two conics have double contact, the square of the perpendicular from any point of one upon the chord of contact is in a constant ratio to the rectangle under the segments of that perpendicular made by the other.

Ex. 2. If a line parallel to a given one meets two conics in the points P, Q, p, q, and we take on it a point O, such that the rectangle $OP \cdot OQ$ may be to $Op \cdot Oq$ in a constant ratio, the locus of O is a conic through the points of intersection of the given conics.

Ex. 3. The diameter of the circle circumscribing the triangle formed by two tangents to a central conic and their chord of contact is $\frac{b'b''}{p}$; where b', b'' are the semi-diameters parallel to the tangents, and p is the perpendicular from the centre on the chord of contact. [Mr. Burnside].

* This is equally true for curves of any degree.

It will be convenient to suppose the equation divided by such a constant that the result of substituting the coordinates of the centre shall be unity. Let t', t'' be the lengths of the tangents, and let S' be the result of substituting the coordinates of their intersection; then

$$t'^2 : b'^2 :: S' : 1, \quad t''^2 : b''^2 :: S' : 1.$$

But also if ϖ be the perpendicular on the chord of contact from the vertex of the triangle, it is easy to see, attending to the remark, Note, p. 154,

$$\varpi : p :: S' : 1.$$

Hence $$\frac{t't''}{\varpi} = \frac{b'b''}{p}.$$

But the left-hand side of this equation, by Elementary Geometry, represents the diameter of the circle circumscribing the triangle.

Ex. 4. The expression (Art. 242) for the radius of curvature may be deduced if in the last example we suppose the two tangents to coincide, in which case the diameter of the circle becomes the radius of curvature (see Art. 398); or also from the following theorem due to Mr. Roberts: If n, n' be the lengths of two intersecting normals; p, p' the corresponding central perpendiculars on tangents; b' the semi-diameter parallel to the chord joining the two points on the curve, then $np + n'p' = 2b'^2$. For if S' be the result of substituting in the equation the coordinates of the middle point of the chord, ϖ, ϖ' the perpendiculars from that point on the tangents, and 2β the length of the chord, then it can be proved, as in the last example, that $\beta^2 = b'^2 S'$, $\varpi = pS'$, $\varpi' = p'S'$, and it is very easy to see that $n\varpi + n'\varpi' = 2\beta^2$.

263. *If two conics have each double contact with a third, their chords of contact with the third conic, and a pair of their chords of intersection with each other, will all pass through the same point, and will form a harmonic pencil.*

Let the equation of the third conic be $S = 0$, and those of the first two conics,

$$S + L^2 = 0, \quad S + M^2 = 0.$$

Now, on subtracting these equations, we find $L^2 - M^2 = 0$, which represents a pair of chords of intersection ($L \pm M = 0$) passing through the intersection of the chords of contact (L and M), and forming a harmonic pencil with them (Art. 57).

Ex. 1. The chords of contact of two conics with their common tangents pass through the intersection of a pair of their common chords. This is a particular case of the preceding, S being supposed to reduce to two right lines.

Ex. 2. The diagonals of any inscribed, and of the corresponding circumscribed quadrilateral, pass through the same point, and form a harmonic pencil. This is also a particular case of the preceding, S being any conic, and $S + L^2$, $S + M^2$ being supposed to reduce to right lines. The proof may also be stated thus: Let t_1, t_2, c_1; t_3, t_4, c_2 be two pairs of tangents and the corresponding chords of contact. In other words, c_1, c_2 are diagonals of the corresponding inscribed quadrilateral. Then the equation of S may be written in either of the forms

$$t_1t_2 - c_1^2 = 0, \quad t_3t_4 - c_2^2 = 0.$$

The second equation must therefore be identical with the first, or can only differ from it by a constant multiplier. Hence $t_1t_2 - \lambda t_3t_4$ must be identical with $c_1^2 - \lambda c_2^2$. Now $c_1^2 - \lambda c_2^2 = 0$ represents a pair of right lines passing through the intersection of c_1, c_2, and harmonically conjugate with them; and the equivalent form shows that these right lines join the points t_1t_3, t_2t_4 and t_1t_4, t_2t_3. For $t_1t_2 - \lambda t_3t_4 = 0$ must denote a locus passing through these points.

Ex. 3. If 2α, 2β, 2γ, 2δ be the eccentric angles of four points on a central conic, form the equation of the diagonals of the quadrilateral formed by their tangents. Here we have

$$t_1 = \frac{x}{a}\cos 2\alpha + \frac{y}{b}\sin 2\alpha - 1, \quad t_2 = \frac{x}{a}\cos 2\beta + \frac{y}{b}\sin 2\beta - 1,$$

$$c_1 = \frac{x}{a}\cos(\alpha+\beta) + \frac{y}{b}\sin(\alpha+\beta) - \cos(\alpha-\beta),$$

and we easily verify

$$t_1t_2 - c_1^2 = -\sin^2(\alpha-\beta)\left\{\frac{x^2}{a^2} + \frac{y^2}{b^2} - 1\right\}.$$

Hence reasoning, as in the last example, we find for the equations of the diagonals

$$\frac{c_1}{\sin(\alpha-\beta)} = \pm \frac{c_2}{\sin(\gamma-\delta)}.$$

264. *If three conics have each double contact with a fourth, six of their chords of intersection will pass three by three through the same points*, thus forming the sides and diagonals of a quadrilateral.

Let the conics be

$$S + L^2 = 0, \quad S + M^2 = 0, \quad S + N^2 = 0.$$

By the last Article the chords will be

$$L - M = 0, \quad M - N = 0, \quad N - L = 0;$$
$$L + M = 0, \quad M + N = 0, \quad N - L = 0;$$
$$L + M = 0, \quad M - N = 0, \quad N + L = 0;$$
$$L - M = 0, \quad M + N = 0, \quad N + L = 0.$$

As in the last Article, we may deduce hence many particular theorems, by supposing one or more of the conics to break up into right lines. Thus, for example, if S break up into right lines, it represents two common tangents to $S + M^2$, $S + N^2$; and if L denote any right line through the intersection of those common tangents, then $S + L^2$ also breaks up into right lines, and represents any two right lines passing through the intersection of the common tangents. Hence, *if through the intersection of the common tangents of two conics we draw any pair of right lines, the chords of each conic joining the extremities of those lines will meet on one of the common chords of the conics.* This is the extension of Art 116. Or, again, *tangents at the extremities of either of these right lines will meet on one of the common chords.*

265. If $S+L^2$, $S+M^2$, $S+N^2$, all break up into pairs of right lines, they will form a hexagon circumscribing S, the chords of intersection will be diagonals of that hexagon, and we get Brianchon's theorem: "*The three opposite diagonals of every hexagon circumscribing a conic intersect in a point.*" By the *opposite* diagonals we mean (if the sides of the hexagon be numbered 1, 2, 3, 4, 5, 6) the lines joining (1, 2) to (4, 5), (2, 3) to (5, 6), and (3, 4) to (6, 1); and by changing the order in which we take the sides we may consider the same lines as forming a number (sixty) of different hexagons, for each of which the present theorem is true. The proof may also be stated as in Ex. 2, Art. 263. If

$$t_1t_4 - c_4^2 = 0, \quad t_2t_5 - c_2^2 = 0, \quad t_3t_6 - c_3^2 = 0,$$

be equivalent forms of the equation of S, then $c_1 = c_2 = c_3$ represents three intersecting diagonals.*

266. *If three conic sections have one chord common to all, their three other chords will pass through the same point.*

Let the equation of one be $S=0$, and of the common chord $L=0$, then the equations of the other two are of the form

$$S+LM=0, \quad S+LN=0,$$

which must have, for their intersection with each other,

$$L(M-N)=0;$$

but $M-N$ is a line passing through the point (MN).

According to the remark in Art. 257, this is only an extension of the theorem (Art. 108), that the radical axes of three circles meet in a point. For three circles have one chord (the line at infinity) common to all, and the radical axes are their other common chords.

---

* Mr. Todhunter has with justice objected to this proof, that since no rule is given which of the diagonals of $t_1t_4t_2t_5$ is $c_1 = + c_2$, all that is in strictness proved is that the lines joining (1, 2) to (4, 5) and (2, 3) to (5, 6) intersect *either* on the line joining (3, 4) to (6, 1), or on that joining (1, 3) to (4, 6). But if the latter were the case the triangles 123, 456 would be homologous (see Ex. 3, p. 59), and therefore the intersections 14, 25, 36 on a right line; and if we suppose five of these tangents fixed, the sixth instead of touching a conic would pass through a fixed point.

The theorem of Art. 264 may be considered as a still further extension of the same theorem, and three conics which have each double contact with a fourth may be considered as having four radical centres, through each of which pass three of their common chords.

The theorem of this Article may, as in Art. 108, be otherwise enunciated: *Given four points on a conic section, its chord of intersection with a fixed conic passing through two of these points will pass through a fixed point.*

Ex. 1. If through one of the points of intersection of two conics we draw any line meeting the conics in the points P, p, and through any other point of intersection B a line meeting the conics in the points Q, q, then the lines PQ, pq will meet on CD, the other chord of intersection. This is got by supposing one of the conics to reduce to the pair of lines OA, OB.

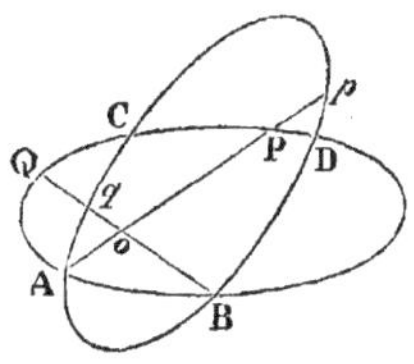

Ex. 2. If two right lines, drawn through the point of contact of two conics, meet the curves in points P, p, Q, q, then the chords PQ, pq will meet on the chord of intersection of the conics.

This is also a particular case of a theorem given in Art. 264, since one intersection of common tangents to two conics which touch reduces to the point of contact (Cor., Art. 117).

267. The equation of a conic circumscribing a quadrilateral ($\alpha\gamma = k\beta\delta$) furnishes us with a proof of "Pascal's theorem," that *the three intersections of the opposite sides of any hexagon inscribed in a conic section are in one right line.*

Let the vertices be *abcdef*, and let $ab = 0$ denote the equation of the line joining the points a, b; then, since the conic circumscribes the quadrilateral *abcd*, its equation must be capable of being put into the form

$$ab \, . \, cd - bc \, . \, ad = 0.$$

But since it also circumscribes the quadrilateral *defa*, the same equation must be capable of being expressed in the form

$$de \, . \, fa - ef \, . \, ad = 0.$$

From the identity of these expressions, we have

$$ab \, . \, cd - de \, . \, fa = (bc - ef) \, ad.$$

Hence, we learn that the left-hand side of this equation (which from its form represents a figure circumscribing the quadrilateral formed by the lines ab, de, cd, af) is resolvable into two factors, which must therefore represent the diagonals of that quadrilateral. But ad is evidently the diagonal which joins the vertices

a and d, therefore $bc - ef$ must be the other, and must join the points (ab, de), (cd, af); and since from its form it denotes a line through the point (bc, ef), it follows that these three points are in one right line.

268. We may, as in the case of Brianchon's theorem, obtain a number of different theorems concerning the same six points, according to the different orders in which we take them. Thus, since the conic circumscribes the quadrilateral $bcef$, its equation can be expressed in the form

$$be.cf - bc.ef = 0.$$

Now, from identifying this with the first form given in the last Article, we have

$$ab.cd - be.cf = (ad - ef)\, bc;$$

whence, as before, we learn that the three points (ab, cf), (cd, be), (ad, ef) lie in one right line, viz. $ad - ef = 0$.

In like manner, from identifying the second and third forms of the equation of the conic, we learn that the three points (de, cf), (fa, be), (ad, bc) lie in one right line, viz. $bc - ad = 0$. But the three right lines

$$bc - ef = 0,\quad ef - ad = 0,\quad ad - bc = 0,$$

meet in a point (Art. 41). Hence we have Steiner's theorem, that "the three Pascal's lines which are obtained by taking the vertices in the orders respectively, *abcdef*, *adcfeb*, *afcbed*, meet in a point." For some further developments on this subject we refer the reader to the note at the end of the volume.

Ex. 1. If a, b, c be three points on a right line; a', b', c' three points on another line, then the intersections $(bc', b'c)$, $(ca', c'a)$, $(ab', a'b)$ lie in a right line. This is a particular case of Pascal's theorem. It remains true if the second line be at infinity and the lines ba', ca' be parallel to a given line, and similarly for cb', ab'; ac', bc'.

Ex. 2. From four lines can be made four triangles, by leaving out in turn one line: the four intersections of perpendiculars of these triangles lie in a right line. Let a, b, c, d be the right lines; a', b', c', d' lines perpendicular to them; then the theorem follows by applying the last example to the three points of intersection of a, b, c with d, and the three points at infinity on a', b', c'.*

* This proof was given me independently by Prof. De Morgan and by Mr. Burnside. The theorem itself, of which another proof has been given p. 217, may also be deduced from Steiner's theorem, Ex. 3, p. 212. For the four intersections of perpendiculars must lie on the directrix of the parabola, which has the four lines for tangents. The line joining the middle points of diagonals is parallel to the axis (see Ex. 1, p. 212). It follows in the same way from Cor. 4, p. 207, that the circles circumscribing the four triangles pass through the same point, viz. the focus of the same parabola. If we are

Ex. 3. Steiner's theorem, that the perpendiculars of the triangle formed by three tangents to a parabola intersect on the directrix is a particular case of Brianchon's theorem. For let the three tangents be a, b, c; let three tangents perpendicular to them be a', b', c', and let the line at infinity, which is also a tangent (Art. 254) be ∞. Then consider the six tangents a, b, c, c', ∞, a'; and the lines joining ab, $c'\infty$; bc, $a'\infty$; cc', aa' meet in a point. The first two are perpendiculars of the triangle, and the last is the directrix on which intersect every pair of rectangular tangents (Art. 221). This proof is by Mr. John C. Moore.

Ex. 4. Given five tangents to a conic, to find the point of contact of any. Let $ABCDE$ be the pentagon formed by the tangents; then, if AC and BE intersect in O, DO passes through the point of contact of AB. This is derived from Brianchon's theorem by supposing two sides of the hexagon to be indefinitely near, since any tangent is intersected by a consecutive tangent at its point of contact (Art. 147).

269. Pascal's theorem enables us, given five points A, B, C, D, E, to construct a conic; for if we draw any line AP through

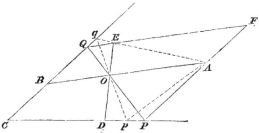

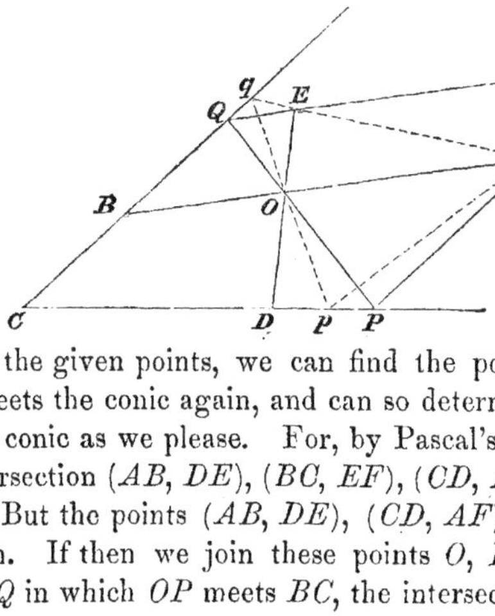

one of the given points, we can find the point F in which that line meets the conic again, and can so determine as many points on the conic as we please. For, by Pascal's theorem, the points of intersection (AB, DE), (BC, EF), (CD, AF) are in one right line. But the points (AB, DE), (CD, AF) are by hypothesis known. If then we join these points O, P, and join to E the point Q in which OP meets BC, the intersection of QE with AP determines F. In other words, *F is the vertex of a triangle FPQ whose sides pass through the fixed points A, E, O, and whose base angles P, Q move along the fixed lines CD, CB* (see Ex. 3, p. 42). The theorem was stated in this form by MacLaurin.

Ex. 1. Given five points on a conic, to find its centre. Draw AP parallel to BC and determine the point F. Then AF and BC are two parallel chords and the line joining their middle points is a diameter. In like manner, by drawing QE parallel to CD we can find another diameter, and thus the centre.

---

given five lines, M. Auguste Miquel has proved (see Catalan's *Théorèmes et Problèmes de Géométrie Elémentaire*, p. 93) that the foci of the five parabolas which have four of the given lines for tangents lie on a circle (see *Higher Plane Curves*, Art. 146).

Ex. 2. Given five points on a conic, to draw the tangent at any one of them. The point F must then coincide with A, and the line QF drawn through E must therefore take the position qA. The tangent therefore must be pA.

Ex. 3. Investigate by trilinear coordinates (Art. 62) Mac Laurin's method of generating conics. In other words, find the locus of the vertex of a triangle whose sides pass through fixed points and base angles move on fixed lines. Let α, β, γ be the sides of the triangle formed by the fixed points, and let the fixed lines be $l\alpha + m\beta + n\gamma = 0$, $l'\alpha + m'\beta + n'\gamma = 0$. Let the base be $\alpha = \mu\beta$. Then the line joining to $\beta\gamma$, the intersection of the base with the first fixed line, is

$$(l\mu + m)\beta + n\gamma = 0.$$

And the line joining to $\alpha\gamma$, the intersection of the base with the second line, is

$$(l'\mu + m')\alpha + n'\mu\gamma = 0.$$

Eliminating μ from the last two equations, the equation of the locus is found to be

$$lm'\alpha\beta = (m\beta + n\gamma)(l'\alpha + n'\gamma),$$

a conic passing through the points

$$\beta\gamma,\ \gamma\alpha,\ (\alpha,\ l\alpha + m\beta + n\gamma),\ (\beta,\ l'\alpha + m'\beta + n'\gamma).$$

EQUATION REFERRED TO TWO TANGENTS AND THEIR CHORD.

270. It much facilitates computation (Art. 229) when the position of a point on a curve can be expressed by a single variable; and this we are able to do in the case of two of the principal forms of equations of conics already given. First, let L, M be any two tangents and R their chord of contact. Then the equation of the conic (Art. 252) is $LM = R^2$; and if $\mu L = R$ be the equation of the line joining LR to any point on the curve (which we shall call the point μ), then substituting in the equation of the curve, we get $M = \mu R$ and $\mu^2 L = M$ for the equations of the lines joining the same point to MR and to LM. Any two of these three equations therefore will determine a point on the conic.

The equation of the chord joining two points on the curve μ, μ', is

$$\mu\mu' L - (\mu + \mu') R + M = 0.$$

For it is satisfied by either of the suppositions

$$(\mu L = R,\ \mu R = M),\quad (\mu' L = R,\ \mu' R = M).$$

If μ and μ' coincide we get the equation of the tangent, viz.

$$\mu^2 L - 2\mu R + M = 0.$$

Conversely, if the equation of a right line $(\mu^2 L - 2\mu R + M = 0)$ involve an indeterminate μ *in the second degree*, the line will always touch the conic $LM = R^2$.

271. *To find the equation of the polar of any point.*

The coordinates L', M', R' of the point substituted in the equation of either tangent through it give the result

$$\mu^2L' - 2\mu R' + M' = 0.$$

Now at the point of contact $\mu^2 = \frac{M}{L}$, and $\mu = \frac{R}{L}$ (Art. 270). Therefore the coordinates of the point of contact satisfy the equation

$$ML' - 2RR' + LM' = 0,$$

which is that of the polar required.

If the point had been given as the intersection of the lines $aL = R$, $bR = M$, it is found by the same method that the equation of the polar is

$$abL - 2aR + M = 0.$$

272. In applying these equations to examples it is useful to take notice that, if we eliminate R between the equations of two tangents

$$\mu^2L - 2\mu R + M = 0, \quad \mu'^2L - 2\mu' R + M = 0,$$

we get $\mu\mu' L = M$ for the equation of the line joining LM to the intersection of these tangents. Hence, if we are given the product of two μ's, $\mu\mu' = a$, the intersection of the corresponding tangents lies on the fixed line $aL = M$. In the same case, substituting a for $\mu\mu'$ in the equation of the chord joining the points, we see that that chord passes through the fixed point $(aL + M, R)$.

Again, since the equation of the line joining any point μ to LM is $\mu^2L = M$, the points $+\mu$, $-\mu$ lie on a right line passing through LM.

Lastly, if $LM = R^2$, $LM = R'^2$ be the equations of two conics having L, M for common tangents, then since the equation $\mu^2L = M$ does not involve R or R', the line joining the point $+\mu$ on one conic to either of the points $\pm\mu$ on the other, passes through LM the intersection of common tangents. We shall say that the point $+\mu$ on the one conic *corresponds directly* to the point $+\mu$ and *inversely* to the point $-\mu$ on the other. And we shall say that the chord joining any two points on one conic corresponds to the chord joining the corresponding points on the other.

Ex. 1. Corresponding chords of two conics intersect on one of the chords of intersection of the conics.

The conics $LM - R^2$, $LM - R'^2$ have $R^2 - R'^2$ for a pair of common chords. But the chords

$$\mu\mu' L - (\mu + \mu') R + M = 0, \quad \mu\mu' L - (\mu + \mu') R' + M = 0,$$

evidently intersect on $R - R'$. And if we change the signs of μ, μ' in the second equation, they intersect on $R + R'$.

Ex. 2. A triangle is circumscribed to a given conic; two of its vertices move on fixed right lines; to find the locus of the third.

Let us take for lines of reference the two tangents through the intersection of the fixed lines, and their chord of contact. Let the equations of the fixed lines be

$$aL - M = 0, \quad bL - M = 0,$$

while that of the conic is $$LM - R^2 = 0.$$

Now we proved (Art. 272) that two tangents which meet on $aL - M$ must have the product of their μ's $= a$; hence, if one side of the triangle touch at the point μ, the others will touch at the points $\frac{a}{\mu}$, $\frac{b}{\mu}$, and their equations will be

$$\frac{a^2}{\mu^2} L - 2\frac{a}{\mu} R + M = 0, \quad \frac{b^2}{\mu^2} L - 2\frac{b}{\mu} R + M = 0.$$

μ can easily be eliminated from the last two equations, and the locus of the vertex is found to be

$$LM = \frac{4ab}{(a+b)^2} R^2,$$

the equation of a conic having double contact with the given one along the line R*.

Ex. 3. To find the envelope of the base of a triangle, inscribed in a conic, and whose two sides pass through fixed points.

Take the line joining the fixed points for R, let the equation of the conic be $LM = R^2$, and those of the lines joining the fixed points to LM be

$$aL - M = 0, \quad bL - M = 0.$$

Now, it was proved (Art. 272) that the extremities of any chord passing through $(aL - M, R)$ must have the product of their μ's $= a$. Hence, if the vertex be μ, the base angles must be $\frac{a}{\mu}$ and $\frac{b}{\mu}$, and the equation of the base must be

$$abL - (a + b)\mu R + \mu^2 M = 0.$$

The base must, therefore (Art. 270), always touch the conic

$$LM = \frac{(a+b)^2}{4ab} R^2,$$

a conic having double contact with the given one along the line joining the given points.

Ex. 4. To inscribe in a conic section a triangle whose sides pass through three given points.

Two of the points being assumed as in the last Example, we saw that the equation of the base must be

$$abL - (a + b)\mu R + \mu^2 M = 0.$$

---

* This reasoning holds even when the point LM is within the conic, and therefore the tangents L, M imaginary. But it may also be proved by the methods of the next section, that when the equation of the conic is $L^2 + M^2 = R^2$, that of the locus is of the form $L^2 + M^2 = k^2R^2$.

Now, if this line pass through the point $cL - R = 0$, $dR - M = 0$, we must have

$$ab - (a+b)\mu c + \mu^2 cd = 0,$$

an equation sufficient to determine μ.

Now, at the point μ we have $\mu L = R$, $\mu^2 L = M$; hence the coordinates of this point must satisfy the equation

$$abL - (a+b)cR + cdM = 0.$$

The question, therefore, admits of two solutions, for either of the points in which this line meets the curve may be taken for the vertex of the required triangle. The geometric construction of this line is given Art. 297, Ex. 7.

Ex. 5. The base of a triangle touches a given conic, its extremities move on two fixed tangents to the conic, and the other two sides of the triangle pass through fixed points; find the locus of the vertex.

Let the fixed tangents be L, M, and the equation of the conic $LM = R^2$. Then the point of intersection of the line L with any tangent $(\mu^2 L - 2\mu R + M)$ will have its coordinates L, R, M respectively proportional to 0, 1, 2μ. And (by Art. 65) the equation of the line joining this point to any fixed point $L'R'M'$ will be

$$LM' - L'M = 2\mu(LR' - L'R).$$

Similarly, the equation of the line joining the fixed point $L''R''M''$ to the point $(2, \mu, 0)$, which is the intersection of the line M with the same tangent, is

$$2(RM'' - R''M) = \mu(LM'' - L''M).$$

Eliminating μ, the locus of the vertex is found to be

$$(LM' - L'M)(LM'' - L''M) = 4(LR' - L'R)(RM'' - R''M),$$

the equation of a conic through the two given points.

273. The chord joining the points $\mu\tan\phi$, $\mu\cot\phi$ (where ϕ is any constant angle) will always touch a conic having double contact with the given one. For (Art. 270) the equation of the chord is

$$\mu^2 L - \mu R(\tan\phi + \cot\phi) + M = 0,$$

which, since $\tan\phi + \cot\phi = 2\operatorname{cosec} 2\phi$, is the equation of a tangent to $LM\sin^2 2\phi = R^2$ at the point μ on that conic. It can be proved, in like manner, that the locus of the intersection of tangents at the points $\mu\tan\phi$, $\mu\cot\phi$ is the conic $LM = R^2\sin^2 2\phi$.

Ex. If in Ex. 5, Art. 272, the extremities of the base lie on any conic having double contact with the given conic, and passing through the given points, find the locus of the vertex.

Let the conics be

$$LM - R^2 = 0, \quad LM\sin^2 2\phi - R^2 = 0,$$

then, if any line touch the latter at the point μ, it will meet the former in the points $\mu\tan\phi$ and $\mu\cot\phi$; and if the fixed points are μ', μ'', the equations of the sides are

$$\mu\mu'\tan\phi L - (\mu' + \mu\tan\phi)R + M = 0,$$

$$\mu\mu''\cot\phi L - (\mu'' + \mu\cot\phi)R + M = 0.$$

Eliminating μ, the locus is found to be

$$(M - \mu'R)(\mu''L - R) = \tan^2\phi\,(M - \mu''R)(\mu'L - R).$$

274. *Given four points of a conic, the anharmonic ratio of the pencil joining them to any fifth point is constant* (Art. 259).

The lines joining four points μ', μ'', μ''', μ'''' to any fifth point μ, are

$$\mu'\ (\mu L - R) + (M - \mu R) = 0,\quad \mu''\ (\mu L - R) + (M - \mu R) = 0,$$
$$\mu'''\ (\mu L - R) + (M - \mu R) = 0,\quad \mu''''\ (\mu L - R) + (M - \mu R) = 0,$$

and their anharmonic ratio is (Art. 58)

$$\frac{(\mu' - \mu'')\ (\mu''' - \mu'''')}{(\mu' - \mu''')(\mu'' - \mu'''')},$$

and is, therefore, independent of the position of the point μ.

We shall, for brevity, use the expression, "the anharmonic ratio of four points of a conic," when we mean the anharmonic ratio of a pencil joining those points to any fifth point on the curve.

275. *Four fixed tangents cut any fifth in points whose anharmonic ratio is constant.*

Let the fixed tangents be those at the points μ', μ'', μ''', μ'''', and the variable tangent that at the point μ; then the anharmonic ratio in question is the same as that of the pencil joining the four points of intersection to the point LM. But (Art. 272) the equations of the joining lines are

$$\mu'\mu L - M = 0,\quad \mu''\mu L - M = 0,\quad \mu'''\mu L - M = 0,\quad \mu''''\mu L - M = 0,$$

a system (Art. 59) *homographic* with that found in the last Article, and whose anharmonic ratio is therefore the same. Thus, then, the anharmonie ratio of four tangents is the same as that of their points of contact.

276. The expression given (Art. 274) for the anharmonic ratio of four points on a conic μ', μ'', μ''', μ'''' remains unchanged if we alter the sign of each of these quantities; hence (Art. 272) *if we draw four lines through any point LM, the anharmonic ratio of four of the points* (μ', μ'', μ''', μ'''') *where these lines meet the conic, is equal to the anharmonic ratio of the other four points* ($-\mu'$, $-\mu''$, $-\mu'''$, $-\mu''''$) *where these lines meet the conic.*

For the same reason, *the anharmonic ratio of four points on one conic is equal to that of the four corresponding points on another;* since corresponding points have the same μ (Art. 272). Again, the expression (Art. 274) remains unaltered, if we multiply each

μ either by $\tan\phi$ or $\cot\phi$; hence, we obtain a theorem of Mr. Townsend's, "*If two conics have double contact, the anharmonic ratio of four of the points in which any four tangents to the one meet the other is the same as that of the other four points in which the four tangents meet the curve, and also the same as that of the four points of contact.*

277. Conversely, given three fixed chords of a conic aa', bb', cc'; a fourth chord dd', such that the anharmonic ratio of $abcd$ is equal to that of $a'b'c'd'$, will always touch a certain conic having double contact with the given one. For let a, b, c, a', b', c' denote the values of μ for the six given fixed points, and μ, μ' those for the extremity of the variable chord, then the equation

$$\frac{(a-b)(c-\mu)}{(a-c)(b-\mu)} = \frac{(a'-b')(c'-\mu')}{(a'-c')(b'-\mu')},$$

when cleared of fractions, may, for brevity, be written

$$A\mu\mu' + B\mu + C\mu' + D = 0,$$

where A, B, C, D are known constants. Solving for μ' from this equation, and substituting in the equation of the chord

$$\mu\mu'L - (\mu + \mu')R + M = 0,$$

it becomes

$$\mu(B\mu + D)L + R\{\mu(A\mu + C) - (B\mu + D)\} - M(A\mu + C) = 0,$$

or $\mu^2(BL + AR) + \mu\{DL + (C - B)R - AM\} - (DR + CM) = 0$,

which (Art. 270) always touches

$$\{DL + (C - B)R - AM\}^2 + 4(BL + AR)(CM + DR) = 0,$$

an equation which may be written in the form

$$4(BC - AD)(LM - R^2) + \{DL + (B + C)R + AM\}^2 = 0,$$

showing that it has double contact with the given conic.

In the particular case when $B = C$, the relation connecting μ, μ' becomes

$$A\mu\mu' + B(\mu + \mu') + D = 0,$$

which (Art. 51) expresses that the chord $\mu\mu'L - (\mu + \mu')R + M$ passes through a fixed point.

EQUATION REFERRED TO THE SIDES OF A SELF-CONJUGATE TRIANGLE.

278. The equation referred to the sides of a self-conjugate triangle $l^2\alpha^2 + m^2\beta^2 = n^2\gamma^2$ (Art. 258) also allows the position of

any point to be expressed by a single indeterminate. For if we write $l\alpha = n\gamma \cos\phi$, $m\beta = n\gamma \sin\phi$, then, as at pp. 94, 219, the chord joining any two points is

$$l\alpha \cos\tfrac{1}{2}(\phi + \phi') + m\beta \sin\tfrac{1}{2}(\phi + \phi') = n\gamma \cos\tfrac{1}{2}(\phi - \phi'),$$

and the tangent at any point is

$$l\alpha \cos\phi + m\beta \sin\phi = n\gamma.$$

If for symmetry we write the equation of the conic

$$a\alpha^2 + b\beta^2 + c\gamma^2 = 0,$$

then it may be derived from the last equation, that the equation of the tangent at any point $\alpha'\beta'\gamma'$ is

$$a\alpha\alpha' + b\beta\beta' + c\gamma\gamma' = 0,$$

and the equation of the polar of any point $\alpha'\beta'\gamma'$ is necessarily of the same form (Art. 89). Comparing the equation last written with $\lambda\alpha + \mu\beta + \nu\gamma = 0$, we see that the coordinates of the pole of the last line are $\frac{\lambda}{a}, \frac{\mu}{b}, \frac{\nu}{c}$; and, since the pole of any tangent is on the curve, the condition that $\lambda\alpha + \mu\beta + \nu\gamma$ may touch the conic is $\frac{\lambda^2}{a} + \frac{\mu^2}{b} + \frac{\nu^2}{c} = 0$. When this condition is fulfilled the conic is evidently touched by all the four lines $\lambda\alpha \pm \mu\beta \pm \nu\gamma$, and the lines of reference are the diagonals of the quadrilateral formed by these lines (see Ex. 3, Art. 146). In like manner, if the condition be fulfilled $a\alpha'^2 + b\beta'^2 + c\gamma'^2 = 0$, the conic passes through the four points $\alpha', \pm\beta', \pm\gamma'$.

Ex. 1. Find the locus of the pole of a given line $\lambda\alpha + \mu\beta + \nu\gamma$ with regard to a conic which passes through four fixed points $\alpha', \pm\beta', \pm\gamma'$.

Ans. $\frac{\lambda\alpha'^2}{\alpha} + \frac{\mu\beta'^2}{\beta} + \frac{\nu\gamma'^2}{\gamma} = 0.$

Ex. 2. Find the locus of the pole of a given line $\lambda\alpha + \mu\beta + \nu\gamma$, with regard to a conic which touches four fixed lines $l\alpha \pm m\beta \pm n\gamma$. *Ans.* $\frac{l^2\alpha}{\lambda} + \frac{m^2\beta}{\mu} + \frac{n^2\gamma}{\nu} = 0.$

These examples also give the locus of centre; since the centre is the pole of the line at infinity $\alpha \sin A + \beta \sin B + \gamma \sin C$.

Ex. 3. What is the equation of the circle having the triangle of reference for a self-conjugate triangle? *Ans.* (See Ex. 2, Art. 128) $\alpha^2 \sin 2A + \beta^2 \sin 2B + \gamma^2 \sin 2C = 0$.

It is easy to see (see Art. 258) that the centre of the circle is the intersection of perpendiculars of the triangle, the square of the radius being the rectangle under the segments of any of the perpendiculars (taken with a positive sign when the triangle is obtuse angled, and with a negative sign when it is acute angled). In the latter case, therefore, the circle is imaginary.

280*. The equation (Art. 258 (*a*)) $x^2+y^2=e^2\gamma^2$ (where the origin is a focus and γ the corresponding directrix) is a particular case of that just considered. The tangents through (γ, x) to the curve are evidently $e\gamma+x$ and $e\gamma-x$. If, therefore, the curve be a parabola, $e=1$; and the tangents are the internal and external bisectors of the angle (γx). Hence, "tangents to a parabola from any point on the directrix are at right angles to each other."

In general, since $x=e\gamma\cos\phi$, $y=e\gamma\sin\phi$, we have

$$\frac{y}{x}=\tan\phi;$$

or ϕ expresses the angle which any radius vector makes with x.

Hence we can find the envelope of a chord which subtends a constant angle at the focus, for the chord

$$x\cos\tfrac{1}{2}(\phi+\phi')+y\sin\tfrac{1}{2}(\phi+\phi')=e\gamma\cos\tfrac{1}{2}(\phi-\phi'),$$

if $\phi-\phi'$ be constant, must, by the present section, always touch

$$x^2+y^2=e^2\gamma^2\cos^2\tfrac{1}{2}(\phi-\phi'),$$

a conic having the same focus and directrix as the given one.

281. The line joining the focus to the intersection of two tangents is found by subtracting

$$x\cos\phi+y\sin\phi-e\gamma=0,$$

$$x\cos\phi'+y\sin\phi'-e\gamma=0,$$

to be $$x\sin\tfrac{1}{2}(\phi+\phi')-y\cos\tfrac{1}{2}(\phi+\phi')=0,$$

the equation of a line making an angle $\tfrac{1}{2}(\phi+\phi')$ with the axis of x, and therefore *bisecting the angle between the focal radii.*

The line joining to the focus the point where the chord of contact meets the directrix is

$$x\cos\tfrac{1}{2}(\phi+\phi')+y\sin\tfrac{1}{2}(\phi+\phi')=0,$$

a line evidently *at right angles to the last.*

To find the locus of the intersection of tangents at points which subtend a given angle 2δ *at the focus.*

By an elimination precisely the same as that in Ex. 2, Art. 102, the equation of the locus is found to be $(x^2+y^2)\cos^2\delta=e^2\gamma^2$,

* Art. 279 of the older editions is now numbered Art. 258 (*a*).

which represents a conic having the same focus and directrix as the given one, and whose eccentricity $= \dfrac{e}{\cos \delta}$.

If the curve be a parabola, the angle between the tangents is in this case given. For the tangent $(x \cos \phi + y \sin \phi - \gamma)$ bisects the angle between $x \cos \phi + y \sin \phi$ and γ. The angle between the tangents is, therefore, half the angle between $x \cos\phi + y \sin\phi$ and $x \cos \phi' + y \sin \phi'$, or $= \frac{1}{2}(\phi - \phi')$. Hence, *the angle between two tangents to a parabola is half the angle which the points of contact subtend at the focus;* and again, *the locus of the intersection of tangents to a parabola, which contain a given angle, is a hyperbola with the same focus and directrix, and whose eccentricity is the secant of the given angle*, or whose asymptotes contain double the given angle (Art. 167).

282. *Any two conics have a common self-conjugate triangle.* For (see Ex. 1, p. 148) if the conics intersect in the points A, B, C, D, the triangle formed by the points E, F, O, in which each pair of common chords intersect, is self-conjugate with regard to either conic. And if the sides of this triangle be α, β, γ, the equations of the conics can be expressed in the form

$$a\alpha^2 + b\beta^2 + c\gamma^2 = 0, \quad a'\alpha^2 + b'\beta^2 + c'\gamma^2 = 0.$$

We shall afterwards discuss the analytical problem of reducing the equations of the conics to this form. If the conics intersect in four imaginary points, the lines α, β, γ are still real. For it is obvious that any equation with real coefficients which is satisfied by the coordinates $x' + x'' \sqrt{(-1)}$, $y' + y'' \sqrt{(-1)}$, will also be satisfied by $x' - x'' \sqrt{(-1)}$, $y' - y'' \sqrt{(-1)}$, and that the line joining these points is real. Hence the four imaginary points common to two conics consist of two pairs $x' \pm x'' \sqrt{(-1)}$, $y' \pm y'' \sqrt{(-1)}$; $x''' \pm x'''' \sqrt{(-1)}$, $y''' \pm y'''' \sqrt{(-1)}$. Two of the common chords are real and four imaginary. But the equations of these imaginary chords are of the form $L \pm M \sqrt{(-1)}$, $L' \pm M' \sqrt{(-1)}$, intersecting in two real points LM, $L'M'$. Consequently the three points E, F, O are all real.

If the conics intersect in two real and two imaginary points, two of the common chords are real, viz. those joining the two real and two imaginary points; and the other four common chords are imaginary. And since each of the imaginary chords passes through one of the two real points, it can have no other real point on it. Therefore, in this case, one of the three points E, F, O is real and the other two imaginary; and one of the sides of the self-conjugate triangle is real and the other two imaginary.

Ex. 1. Find the locus of vertex of a triangle whose base angles move along one conic, and whose sides touch another. [The following solution is Mr. Burnside's.] Let the conic touched by the sides be $x^2 + y^2 - z^2$, and the other $ax^2 + by^2 - cz^2$. Then, as at Ex. 1, p. 94, the coordinates of the intersection of tangents at points α, γ are $\cos\frac{1}{2}(\alpha + \gamma)$, $\sin\frac{1}{2}(\alpha + \gamma)$, $\cos\frac{1}{2}(\alpha - \gamma)$; and the conditions of the problem give

$$a \cos^2 \tfrac{1}{2}(\alpha + \gamma) + b \sin^2 \tfrac{1}{2}(\alpha + \gamma) = c \cos^2 \tfrac{1}{2}(\alpha - \gamma);$$

or $$(a + b - c) + (a - b - c) \cos\alpha \cos\gamma + (b - c - a) \sin\alpha \sin\gamma = 0.$$

In l ke manner

$$(a + b - c) + (a - b - c) \cos\beta \cos\gamma + (b - c - a) \sin\beta \sin\gamma = 0,$$

whence $$(a + b - c) \cos\tfrac{1}{2}(\alpha + \beta) = (b + c - a) \cos\tfrac{1}{2}(\alpha - \beta) \cos\gamma,$$

$$(a + b - c) \sin\tfrac{1}{2}(\alpha + \beta) = (a + c - b) \cos\tfrac{1}{2}(\alpha - \beta) \sin\gamma;$$

and, since the coordinates of the point whose locus we seek are $\cos\frac{1}{2}(\alpha + \beta)$, $\sin\frac{1}{2}(\alpha + \beta)$, $\cos\frac{1}{2}(\alpha - \beta)$, the equation of the locus is

$$\frac{x^2}{(b + c - a)^2} + \frac{y^2}{(c + a - b)^2} = \frac{z^2}{(a + b - c)^2}.$$

Ex. 2. A triangle is inscribed in the conic $x^2 + y^2 = z^2$, and two sides touch the conic $ax^2 + by^2 = cz^2$; find the envelope of the third side.

Ans. $(ca + ab - bc)^2 x^2 + (ab + bc - ca)^2 y^2 = (bc + ca - ab)^2 z^2$.

ENVELOPES.

283. If the equation of a right line involve an indeterminate quantity in any degree, and if we give to that indeterminate a series of different values, the equation represents a series of different lines, all of which touch a certain curve, which is called the *envelope* of the system of lines. We shall illustrate the general method of finding the equation of an envelope by proving, independently of Art. 270, that the line $\mu^2 L - 2\mu R + M$, where μ is indeterminate, always touches the curve $LM - R^2$. The point of intersection of the lines answering to the values μ and $\mu + k$ is determined by the two equations

$$\mu^2 L - 2\mu R + M = 0, \quad 2(\mu L - R) + kL = 0;$$

the second equation being derived from the first by substituting $\mu + k$ for μ, erasing the terms which vanish in virtue of the first equation, and then dividing by k. The smaller k is, the more nearly does the second line approach to coincidence with the first; and if we make $k=0$, we find that the point of meeting of the first line with a consecutive line of the system is determined by the equations

$$\mu^2 L - 2\mu R + M = 0,\quad \mu L - R = 0;$$

or, what comes to the same thing, by the equations

$$\mu L - R = 0,\quad \mu R - M = 0.$$

Now since any point on a curve may be considered as the intersection of two of its consecutive tangents (Art. 147), the point where any line meets its envelope is the same as that where it meets a consecutive tangent to the envelope; and therefore the two equations last written determine the point on the envelope which has the line $\mu^2 L - 2\mu R + M$ for its tangent. And by eliminating μ between the equations we get the equation of the locus of all the points on the envelope, namely $LM = R^2$.

A similar argument will prove, even if L, M, R do not represent right lines, that the curve represented by $\mu^2 L - 2\mu R + M$ always touches the curve $LM = R^2$.

The envelope of $L\cos\phi + M\sin\phi - R$, where ϕ is indeterminate, may be either investigated directly in like manner, or may be reduced to the preceding by assuming $\tan\frac{1}{2}\phi = \mu$, when on substituting

$$\cos\phi = \frac{1-\mu^2}{1+\mu^2},\quad \sin\phi = \frac{2\mu}{1+\mu^2},$$

and clearing of fractions, we get an equation in which μ only enters in the second degree.

284. We might also proceed as follows: The line

$$\mu^2 L - 2\mu R + M$$

is obviously a tangent to a curve *of the second class* (see note, p. 147); for only two lines of the system can be drawn through a given point: namely, those answering to the values of μ determined by the equation

$$\mu^2 L' - 2\mu R' + M' = 0,$$

where L', R', M' are the results of substituting the coordinates of the given point in L, R, M. Now these values of μ will evidently coincide, or the point will be the intersection of two consecutive tangents, if its coordinates satisfy the equation $LM - R^2$. And, generally, if the indeterminate μ enter algebraically and in the nth degree, into the equation of a line, the line will touch a curve of the nth class, whose equation is found by expressing the condition that the equation in μ shall have equal roots.

Ex. 1. The vertices of a triangle move along the three fixed lines α, β, γ, and two of the sides pass through two fixed points $\alpha'\beta'\gamma'$, $\alpha''\beta''\gamma''$, find the envelope of the third side. Let $\alpha + \mu\beta$ be the line joining to $\alpha\beta$ the vertex which moves along γ, then the equations of the sides through the fixed points are

$$\gamma'(\alpha + \mu\beta) - (\alpha' + \mu\beta')\gamma = 0, \quad \gamma''(\alpha + \mu\beta) - (\alpha'' + \mu\beta'')\gamma = 0.$$

And the equation of the base is

$$(\alpha' + \mu\beta')\gamma''\alpha + (\alpha'' + \mu\beta'')\mu\gamma'\beta - (\alpha' + \mu\beta')(\alpha'' + \mu\beta'')\gamma = 0,$$

for it can be easily verified that this passes through the intersection of the first line with α, and of the second line with β. Arranging according to the powers of μ, we find for the envelope

$$(\alpha\beta'\gamma'' + \beta\gamma'\alpha'' - \gamma\alpha'\beta'' - \gamma\alpha''\beta')^2 = 4\alpha'\beta''(\alpha\gamma'' - \alpha''\gamma)(\beta\gamma' - \beta'\gamma).$$

This example may also be solved by arranging according to the powers of α, the equation in Ex. 3, p. 49.

Ex. 2. Find the envelope of a line such that the product of the perpendiculars on it from two fixed points may be constant.

Take for axes the line joining the fixed points and a perpendicular through its middle point, so that the coordinates of the fixed points may be $y = 0$, $x = \pm c$; then if the variable line be $y - mx + n = 0$, we have by the condition of the question

$$(n + mc)(n - mc) = b^2(1 + m^2),$$

or
$$n^2 = b^2 + b^2m^2 + c^2m^2,$$

but
$$n^2 = y^2 - 2mxy + m^2x^2,$$

therefore
$$m^2(x^2 - b^2 - c^2) - 2mxy + y^2 - b^2 = 0;$$

and the envelope is
$$x^2y^2 = (x^2 - b^2 - c^2)(y^2 - b^2),$$

or
$$\frac{x^2}{b^2 + c^2} + \frac{y^2}{b^2} = 1.$$

Ex. 3. Find the envelope of a line such that the sum of the squares of the perpendiculars on it from two fixed points may be constant. *Ans.* $\dfrac{2x^2}{b^2 - 2c^2} + \dfrac{2y^2}{b^2} = 1.$

Ex. 4. Find the envelope if the difference of squares of perpendiculars be given.
Ans. A parabola.

Ex. 5. Through a fixed point O any line OP is drawn to meet a fixed line; to find the envelope of PQ drawn so as to make the angle OPQ constant.

Let OP make the angle θ with the perpendicular on the fixed line, and its length is $p\sec\theta$; but the perpendicular from O on PQ makes a fixed angle β with OP, therefore its length is $= p\sec\theta\cos\beta$; and since this perpendicular makes an angle $= \theta + \beta$ with the perpendicular on the fixed line, if we assume the latter for the axis of x, the equation of PQ is

$$x\cos(\theta + \beta) + y\sin(\theta + \beta) = p\sec\theta\cos\beta,$$

or $$x\cos(2\theta+\beta)+y\sin(2\theta+\beta)=2p\cos\beta-x\cos\beta-y\sin\beta,$$
an equation of the form $$L\cos\phi+M\sin\phi=R,$$
whose envelope, therefore, is
$$x^2+y^2=(x\cos\beta+y\sin\beta-2p\cos\beta)^2,$$
the equation of a parabola having the point O for its focus.

Ex. 6. Find the envelope of the line $\frac{A}{\mu}+\frac{B}{\mu'}=1$, where the indeterminates are connected by the relation $\mu+\mu'=C$.

We may substitute for μ', $C-\mu$, and clear of fractions; the envelope is thus found to be $$A^2+B^2+C^2-2AB-2AC-2BC=0,$$
an equation to which the following form will be found to be equivalent,
$$\pm\sqrt{A}\pm\sqrt{B}\pm\sqrt{C}=0.$$
Thus, for example,—Given vertical angle and sum of sides of a triangle to find the envelope of base.

The equation of the base is $$\frac{x}{a}+\frac{y}{b}=1,$$
where $a+b=c$.

The envelope is, therefore,
$$x^2+y^2-2xy-2cx-2cy+c^2=0,$$
a parabola touching the sides x and y.

In like manner,—Given in position two conjugate diameters of an ellipse, and the sum of their squares, to find its envelope.

If in the equation $$\frac{x^2}{a'^2}+\frac{y^2}{b'^2}=1,$$
we have $a'^2+b'^2=c^2$, the envelope is
$$x\pm y\pm c=0.$$
The ellipse, therefore, must always touch four fixed right lines.

285. *If the coefficients in the equation of any right line* $\lambda\alpha+\mu\beta+\nu\gamma$ *be connected by any relation of the second order in* λ, μ, ν,
$$A\lambda^2+B\mu^2+C\nu^2+2F\mu\nu+2G\nu\lambda+2H\lambda\mu=0,$$
the envelope of the line is a conic section. Eliminating ν between the equation of the right line and the given relation, we have
$$(A\gamma^2-2G\gamma\alpha+C\alpha^2)\lambda^2+2(H\gamma^2-F\gamma\alpha-G\gamma\beta+C\alpha\beta)\lambda\mu$$
$$+(B\gamma^2-2F\gamma\beta+C\beta^2)\mu^2=0,$$
and the envelope is
$$(A\gamma^2-2G\gamma\alpha+C\alpha^2)(B\gamma^2-2F\gamma\beta+C\beta^2)=(H\gamma^2-F\gamma\alpha-G\gamma\beta+C\alpha\beta)^2.$$
Expanding this equation, and dividing by γ^2, we get
$$(BC-F^2)\alpha^2+(CA-G^2)\beta^2+(AB-H^2)\gamma^2$$
$$2+(GH-AF)\beta\gamma+2(HF-BG)\gamma\alpha+2(FG-CH)\alpha\beta=0.$$

The result of this article may be stated thus: *Any tangential equation of the second order in* λ, μ, ν *represents a conic, whose trilinear equation is found from the tangential by exactly the same process that the tangential is found from the trilinear.*

For it is proved (as in Art. 151) that the condition that $\lambda\alpha + \mu\beta + \nu\gamma$ shall touch

$$a\alpha^2 + b\beta^2 + c\gamma^2 + 2f\beta\gamma + 2g\gamma\alpha + 2h\alpha\beta = 0,$$

or, in other words, the tangential equation of that conic is

$$(bc - f^2)\lambda^2 + (ca - g^2)\mu^2 + (ab - h^2)\nu^2$$
$$+ 2(gh - af)\mu\nu + 2(hf - bg)\nu\lambda + 2(fg - ch)\lambda\mu = 0.$$

Conversely, the envelope of a line whose coefficients λ, μ, ν fulfil the condition last written, is the conic $a\alpha^2 + \&c. = 0$; and this may be verified by the equation of this article. For, if we write for A, B, &c., $bc - f^2$, $ca - g^2$, &c., the equation $(BC - F^2)\alpha^2 + \&c. = 0$ becomes

$$(abc + 2fgh - af^2 - bg^2 - ch^2)(a\alpha^2 + b\beta^2 + c\gamma^2 + 2f\beta\gamma + 2g\gamma\alpha + 2h\alpha\beta) = 0.$$

Ex. 1. We may deduce, as particular cases of the above, the results of Arts. 127, 130, namely, that the envelope of a line which fulfils the condition $\frac{F}{\lambda} + \frac{G}{\mu} + \frac{H}{\nu} = 0$ is $\sqrt{(F\alpha)} + \sqrt{(G\beta)} + \sqrt{(H\gamma)} = 0$; and of one which fulfils the condition

$$\sqrt{(F\lambda)} + \sqrt{(G\mu)} + \sqrt{(H\nu)} = 0 \text{ is } \frac{F}{\alpha} + \frac{G}{\beta} + \frac{H}{\gamma} = 0.$$

Ex. 2. What is the condition that $\lambda\alpha + \mu\beta + \nu\gamma$ should meet the conic given by the general equation in real points?

Ans. The line meets in real points when the quantity $(bc - f^2)\lambda^2 + \&c.$ is negative; in imaginary points when this quantity is positive; and touches when it vanishes.

Ex. 3. What is the condition that the tangents drawn through a point $\alpha'\beta'\gamma'$ should be real?

Ans. The tangents are real when the quantity $(BC - F^2)\alpha'^2 + \&c.$ is negative; or, in other words, when the quantities $abc + 2fgh + \&c.$ and $a\alpha'^2 + b\beta'^2 + \&c.$ have opposite signs. The point will be inside the conic and the tangents imaginary when these quantities have like signs.

286. It is proved, as at Art. 76, that if the condition be fulfilled, $$ABC + 2FGH - AF^2 - BG^2 - CH^2 = 0,$$

then the equation

$$A\lambda^2 + B\mu^2 + C\nu^2 + 2F\mu\nu + 2G\nu\lambda + 2H\lambda\mu = 0$$

may be resolved into two factors, and is equivalent to one of the form $$(\alpha'\lambda + \beta'\mu + \gamma'\nu)(\alpha''\lambda + \beta''\mu + \gamma''\nu) = 0.$$

And since the equation is satisfied if either factor vanish, it denotes (Art. 51) that the line $\lambda\alpha + \mu\beta + \nu\gamma$ passes through one or other of two fixed points.

If, as in the last article, we write for A, $bc - f^2$, &c., it will be found that the quantity $ABC + 2FGH +$ &c. is the square of $abc + 2fgh +$ &c.

Ex. If a conic pass through two given points and have double contact with a fixed conic, the chord of contact passes through one or other of two fixed points. For let S be the fixed conic, and let the equation of the other be $S = (\lambda\alpha + \mu\beta + \nu\gamma)^2$. Then substituting the coordinates of the two given points, we have

$$S' = (\lambda\alpha' + \mu\beta' + \nu\gamma')^2;\quad S'' = (\lambda\alpha'' + \mu\beta'' + \nu\gamma'')^2;$$

whence $$(\lambda\alpha' + \mu\beta' + \nu\gamma')\sqrt{(S'')} = \pm(\lambda\alpha'' + \mu\beta'' + \nu\gamma'')\sqrt{(S')},$$

showing that $\lambda\alpha + \mu\beta + \nu\gamma$ passes through one or other of two fixed points, since S', S'' are known constants.

287. To find the equation of a conic having double contact with two given conics, S and S'. Let E and F be a pair of their chords of intersection, so that $S - S' = EF$; then

$$\mu^2E^2 - 2\mu(S + S') + F^2 = 0$$

represents a conic having double contact with S and S'; for it may be written

$$(\mu E + F)^2 = 4\mu S,\ \text{ or }\ (\mu E - F)^2 = 4\mu S'.$$

Since μ is of the second degree, we see that through any point can be drawn *two* conics of this system; and there are *three* such systems, since there are three pairs of chords E, F. If S' break up into right lines, there are only two pairs of chords distinct from S', and but two systems of touching conics. And when both S and S' break up into right lines there is but one such system.

Ex. Find the equation of a conic touching four given lines.

Ans. $\mu^2E^2 - 2\mu(AC + BD) + F^2 = 0$,

where A, B, C, D are the sides; E, F the diagonals, and $AC - BD = EF$. Or more symmetrically if L, M, N be the diagonals, $L \pm M \pm N$ the sides,

$$\mu^2L^2 - \mu(L^2 + M^2 - N^2) + M^2 = 0.$$

For this always touches $4L^2M^2 - (L^2 + M^2 - N^2)^2$

$$= (L + M + N)(M + N - L)(L + N - M)(M + L - N).$$

Or, again, the equation may be written $N^2 = \dfrac{L^2}{\cos^2\phi} + \dfrac{M^2}{\sin^2\phi}$ (see Art. 278).

288. The equation of a conic having double contact with two *circles* assumes a simpler form, viz.

$$\mu^2 - 2\mu(C + C') + (C - C')^2 = 0.$$

The chords of contact of the conic with the circles are found to be $C - C' + \mu = 0$, and $C - C' - \mu = 0$, which are therefore parallel to each other, and equidistant from the radical axis of the circles. This equation may also be written in the form

$$\sqrt{C} \pm \sqrt{C'} = \sqrt{\mu}.$$

Hence, *the locus of a point, the sum or difference of whose tangents to two given circles is constant, is a conic having double contact with the two circles.* If we suppose both circles infinitely small, we obtain the fundamental property of the foci of the conic.

If μ be taken equal to the square of the intercept between the circles on one of their common tangents, the equation denotes a pair of common tangents to the circles.

Ex. 1. Solve by this method the Examples (Arts. 113, 114) of finding common tangents to circles.

Ans. Ex. 1. $\sqrt{C} + \sqrt{C'} = 4$ or $= 2$. *Ans.* Ex. 2. $\sqrt{C} + \sqrt{C'} = 1$ or $= \sqrt{-79}$.

Ex. 2. Given three circles; let L, L' be a pair of common tangents to C', C''; M, M' to C'', C; N, N' to C, C'; then if L, M, N meet in a point, so will L', M', N'.*

Let the equations of the pairs of common tangents be

$$\sqrt{C'} + \sqrt{C''} = t,\ \sqrt{C''} + \sqrt{C} = t',\ \sqrt{C} + \sqrt{C'} = t''.$$

Then the condition that L, M, N should meet in a point is $t' \pm t = t''$; and it is obvious that when this condition is fulfilled, L', M', N' also meet in a point.

Ex. 3. Three conics having double contact with a given one are met by three common chords, which do not pass all through the same point, in six points which lie on a conic. Consequently, if three of these points lie in a right line, so do the other three. Let the three conics be $S - L^2$, $S - M^2$, $S - N^2$; and the common chords $L + M$, $M + N$, $N + L$, then the truth of the theorem appears from inspection of the equation

$$S + MN + NL + LM = (S - L^2) + (L + M)(L + N).$$

---

* This principle is employed by Steiner in his solution of Malfatti's problem, viz. "To inscribe in a triangle three circles which touch each other and each of which touches two sides of the triangle." Steiner's construction is, "Inscribe circles in the triangles formed by each side of the given triangle and the two adjacent bisectors of angles; these circles having three common tangents meeting in a point will have three other common tangents meeting in a point, and these are common tangents to the circles required. For a geometrical proof of this by Dr. Hart, see *Quarterly Journal of Mathematics*, vol. I., p. 219. We may extend the problem by substituting for the word "circles," "conics having double contact with a given one." In this extension, the theorem of Ex. 3, or its reciprocal, takes the place of Ex. 2.

GENERAL EQUATION OF THE SECOND DEGREE.

289. There is no conic whose equation may not be written in the form

$$a\alpha^2 + b\beta^2 + c\gamma^2 + 2f\beta\gamma + 2g\gamma\alpha + 2h\alpha\beta = 0.$$

For this equation is obviously of the second degree; and since it contains five independent constants, we can determine these constants so that the curve which it represents may pass through five given points, and therefore coincide with any given conic. The trilinear equation just written includes the ordinary Cartesian equation, if we write x and y for α and β, and if we suppose the line γ at infinity, and therefore write $\gamma = 1$ (see Art. 69, and note p. 72).

In like manner the equation of every curve of any degree may be expressed as a homogeneous function of α, β, γ. For it can readily be proved that the number of terms in the *complete* equation of the nth order between two variables is the same as the number of terms in the *homogeneous* equation of the nth order between three variables. The two equations then, containing the same number of constants, are equally capable of representing any particular curve.

290. Since the coordinates of any point on the line joining two points $\alpha'\beta'\gamma'$, $\alpha''\beta''\gamma''$ are (Art. 66) of the form $l\alpha' + m\alpha''$, $l\beta' + m\beta''$, $l\gamma' + m\gamma''$, we can find the points where this joining line meets any curve by substituting these values for α, β, γ, and then determining the ratio $l : m$ by means of the resulting equation.* Thus (see Art. 92) the points where the line meets a conic are determined by the quadratic

$$l^2 (a\alpha'^2 + b\beta'^2 + c\gamma'^2 + 2f\beta'\gamma' + 2g\gamma'\alpha' + 2h\alpha'\beta')$$
$$+ 2lm \{a\alpha'\alpha'' + b\beta'\beta'' + c\gamma'\gamma''$$
$$+ f(\beta'\gamma'' + \beta''\gamma') + g(\gamma'\alpha'' + \gamma''\alpha') + h(\alpha'\beta'' + \alpha''\beta')\}$$
$$+ m^2 (a\alpha''^2 + b\beta''^2 + c\gamma''^2 + 2f\beta''\gamma'' + 2g\gamma''\alpha'' + 2h\alpha''\beta'') = 0;$$

or, as we may write it for brevity, $l^2S' + 2lmP + m^2S'' = 0$. When the point $\alpha'\beta'\gamma'$ is on the curve, S' vanishes, and the quadratic reduces to a simple equation. Solving it for $l : m$,

* This method was introduced by Joachimsthal.

we see that the coordinates of the point where the conic is met again by the line joining $\alpha''\beta''\gamma''$ to a point on the conic $\alpha'\beta'\gamma'$, are $S''\alpha' - 2P\alpha''$, $S''\beta' - 2P\beta''$, $S''\gamma' - 2P\gamma''$. These coordinates reduce to $\alpha'\beta'\gamma'$ if the condition $P = 0$ be fulfilled. Writing this at full length, we see that if $\alpha''\beta''\gamma''$ satisfy the equation

$$a\alpha\alpha' + b\beta\beta' + c\gamma\gamma' + f(\beta\gamma' + \beta'\gamma) + g(\gamma\alpha' + \gamma'\alpha) + h(\alpha\beta' + \alpha'\beta) = 0,$$

then the line joining $\alpha''\beta''\gamma''$ to $\alpha'\beta'\gamma'$ meets the curve in two points coincident with $\alpha'\beta'\gamma'$; in other words, $\alpha''\beta''\gamma''$ lies on the tangent at $\alpha'\beta'\gamma'$. The equation just written is therefore the equation of the tangent.

291. Arguing, as at Art. 89, from the symmetry between $\alpha\beta\gamma$, $\alpha'\beta'\gamma'$ of the equation just found, we infer that when $\alpha'\beta'\gamma'$ is not supposed to be on the curve, the equation represents the polar of that point. The same conclusion may be drawn from observing, as at Art. 91, that $P = 0$ expresses the condition that the line joining $\alpha'\beta'\gamma'$, $\alpha''\beta''\gamma''$ shall be cut harmonically by the curve. The equation of the polar may be written

$$\alpha'(a\alpha + h\beta + g\gamma) + \beta'(h\alpha + b\beta + f\gamma) + \gamma'(g\alpha + f\beta + c\gamma) = 0.$$

But the quantities which multiply α', β', γ' respectively, are half the differential coefficients of the equation of the conic with respect to α, β, γ. We shall for shortness write S_1, S_2, S_3, instead of $\frac{dS}{d\alpha}$, $\frac{dS}{d\beta}$, $\frac{dS}{d\gamma}$; and we see that the equation of the polar is

$$\alpha'S_1 + \beta'S_2 + \gamma'S_3 = 0.$$

In particular, if β', γ' both vanish, the polar of the point $\beta\gamma$ is S_1, or *the equation of the polar of the intersection of two of the lines of reference is the differential coefficient of the equation of the conic considered as a function of the third.* The equation of the polar being unaltered by interchanging $\alpha\beta\gamma$, $\alpha'\beta'\gamma'$, may also be written $\alpha S_1' + \beta S_2' + \gamma S_3' = 0$.

292. When a conic breaks up into two right lines, the polar of any point whatever passes through the intersection of the right lines. Geometrically, it is evident that the locus of harmonic means of radii drawn through the point is the fourth harmonic to the pair of lines and the line joining their intersection to the given point. And we might also infer, from the

formula of the last article, that the polar of any point with respect to the pair of lines $\alpha\beta$ is $\beta'\alpha+\alpha'\beta$, the harmonic conjugate with respect to α, β of $\beta'\alpha-\alpha'\beta$, the line joining $\alpha\beta$ to the given point. If then the general equation represent a pair of lines, the polars of the three points $\beta\gamma$, $\gamma\alpha$, $\alpha\beta$,

$$a\alpha+h\beta+g\gamma=0,\ h\alpha+b\beta+f\gamma=0,\ g\alpha+f\beta+c\gamma=0,$$

are three lines meeting in a point. Expressing, as in Art. 38, the condition that this should be the case, by eliminating α, β, γ between these equations, we get the condition, already found by other methods, that the equation should represent right lines, which we now see may be written in the form of a determinant,

$$\begin{vmatrix} a, & h, & g \\ h, & b, & f \\ g, & f, & c \end{vmatrix} = 0;$$

or, expanded, $\quad abc+2fgh-af^2-bg^2-ch^2=0.$

The left-hand side of this equation is called *the discriminant** of the equation of the conic. We shall denote it in what follows by the letter Δ.

293. To find the coordinates of the pole of any line $\lambda\alpha+\mu\beta+\nu\gamma$. Let $\alpha'\beta'\gamma'$ be the sought coordinates, then we must have

$$a\alpha'+h\beta'+g\gamma'=\lambda,\ h\alpha'+b\beta'+f\gamma'=\mu,\ g\alpha'+f\beta'+c\gamma'=\nu.$$

Solving these equations for α', β', γ', we get

$$\Delta\alpha'=\lambda\,(bc-f^2)+\mu\,(fg-ch)+\nu\,(hf-bg),$$
$$\Delta\beta'=\lambda\,(fg-ch)+\mu\,(ca-g^2)+\nu\,(gh-af),$$
$$\Delta\gamma'=\lambda\,(hf-bg)+\mu\,(gh-af)+\nu\,(ab-h^2);$$

or, if we use A, B, C,† &c. in the same sense as in Art. 151, we find the coordinates of the pole respectively proportional to

$$A\lambda+H\mu+G\nu,\ H\lambda+B\mu+F\nu,\ G\lambda+F\mu+C\nu.$$

Since the pole of any tangent to a conic is a point on that tangent, we can get the condition that $\lambda\alpha+\mu\beta+\nu\gamma$ may touch the conic, by expressing the condition that the coordinates just found satisfy $\lambda\alpha+\mu\beta+\nu\gamma=0$. We find thus, as in Art. 285,

$$A\lambda^2+B\mu^2+C\nu^2+2F\mu\nu+2G\nu\lambda+2H\lambda\mu=0.$$

---

* See *Lessons on Modern Higher Algebra*, Lesson XI.

† A, B, C, &c. are the *minors* of the determinant of the last article.

If we write this equation $\Sigma = 0$, it will be observed that the coordinates of the pole are Σ_1, Σ_2, Σ_3, that is to say, the differential coefficients of Σ with respect to λ, μ, ν. Just, then, as the equation of the polar of any point is $\alpha S_1' + \beta S_2' + \gamma S_3' = 0$, so the condition that $\lambda\alpha + \mu\beta + \nu\gamma$ may pass through the pole of $\lambda'\alpha + \mu'\beta + \nu'\gamma$ (or, in other words, the tangential equation of this pole) is $\lambda\Sigma_1' + \mu\Sigma_2' + \nu\Sigma_3' = 0$. And again, the condition that two lines $\lambda\alpha + \mu\beta + \nu\gamma$, $\lambda'\alpha + \mu'\beta + \nu'\gamma$ may be conjugate with respect to the conic, that is to say, may be such that the pole of either lies on the other, may obviously be written in either of the equivalent forms

$$\lambda'\Sigma_1 + \mu'\Sigma_2 + \nu'\Sigma_3 = 0,\ \lambda\Sigma_1' + \mu\Sigma_2' + \nu\Sigma_3' = 0.$$

From the manner in which Σ was here formed, it appears that Σ is the result of eliminating α', β', γ', ρ between the equations

$$a\alpha' + h\beta' + g\gamma' + \rho\lambda = 0,\ h\alpha' + b\beta' + f\gamma' + \rho\mu = 0,$$
$$g\alpha' + f\beta' + c\gamma' + \rho\nu = 0,\ \lambda\alpha' + \mu\beta' + \nu\gamma' = 0;$$

in other words, that Σ may be written as the determinant

$$\begin{vmatrix} \lambda, & \mu, & \nu, & 0 \\ a, & h, & g, & \lambda \\ h, & b, & f, & \mu \\ g, & f, & c, & \nu \end{vmatrix} = A\lambda^2 + B\mu^2 + C\nu^2 + 2F\mu\nu + 2G\nu\lambda + 2H\lambda\mu.$$

Ex. 1. To find the coordinates of the pole of $\lambda\alpha + \mu\beta + \nu\gamma$ with respect to $\sqrt{(l\alpha)} + \sqrt{(m\beta)} + \sqrt{(n\gamma)}$. The tangential equation in this case (Art. 130) being

$$l\mu\nu + m\nu\lambda + n\lambda\mu = 0,$$

the coordinates of the pole are

$$\alpha' = m\nu + n\mu,\quad \beta' = n\lambda + l\nu,\quad \gamma' = l\mu + m\lambda.$$

Ex. 2. To find the locus of the pole of $\lambda\alpha + \mu\beta + \nu\gamma$ with respect to a conic being given three tangents, and one other condition.*

Solving the preceding equations for l, m, n, we find l, m, n proportional to

$$\lambda\,(\mu\beta' + \nu\gamma' - \lambda\alpha'),\quad \mu\,(\nu\gamma' + \lambda\alpha' - \mu\beta'),\quad \nu\,(\lambda\alpha' + \mu\beta' - \nu\gamma').$$

Now $\sqrt{(l\alpha)} + \sqrt{(m\beta)} + \sqrt{(n\gamma)}$ denotes a conic touching the three lines α, β, γ; and any fourth condition establishes a relation between l, m, n, in which, if we substitute the values just found, we shall have the locus of the pole of $\lambda\alpha + \mu\beta + \nu\gamma$. If we write for λ, μ, ν the sides of the triangle of reference a, b, c, we shall have the locus of the pole of the line at infinity $a\alpha + b\beta + c\gamma$, that is, the locus of centre. Thus the condition that the conic should touch $A\alpha + B\beta + C\gamma$ being $\dfrac{l}{A} + \dfrac{m}{B} + \dfrac{n}{C} = 0$

* The method here used is taken from Hearn's *Researches on Conic Sections.*

(Art. 130), we infer that the locus of the pole of $\lambda\alpha + \mu\beta + \nu\gamma$ with respect to a conic touching the four lines α, β, γ, $A\alpha + B\beta + C\gamma$, is the right line

$$\frac{\lambda(\mu\beta + \nu\gamma - \lambda\alpha)}{A} + \frac{\mu(\nu\gamma + \lambda\alpha - \mu\beta)}{B} + \frac{\nu(\lambda\alpha + \mu\beta - \nu\gamma)}{C} = 0.$$

Or, again, since the condition that the conic should pass through $\alpha'\beta'\gamma'$ is $\sqrt{(l\alpha')} + \sqrt{(m\beta')} + \sqrt{(n\gamma')} = 0$, the locus of the pole of $\lambda\alpha + \mu\beta + \nu\gamma$ with respect to a conic which touches the three lines α, β, γ, and passes through a point $\alpha'\beta'\gamma'$, is

$$\sqrt{\{\lambda\alpha'(\mu\beta + \nu\gamma - \lambda\alpha)\}} + \sqrt{\{\mu\beta'(\nu\gamma + \lambda\alpha - \mu\beta)\}} + \sqrt{\{\nu\gamma'(\lambda\alpha + \mu\beta - \nu\gamma)\}} = 0,$$

which denotes a conic touching $\mu\beta + \nu\gamma - \lambda\alpha$, $\nu\gamma + \lambda\alpha - \mu\beta$, $\lambda\alpha + \mu\beta - \nu\gamma$. In the case where the locus of centre is sought, these three lines are the lines joining the middle points of the sides of the triangle formed by α, β, γ.

Ex. 3. To find the coordinates of the pole of $\lambda\alpha + \mu\beta + \nu\gamma$ with respect to $l\beta\gamma + m\gamma\alpha + n\alpha\beta$. The tangential equation in this case being, Art. 127,

$$l^2\lambda^2 + m^2\mu^2 + n^2\nu^2 - 2mn\mu\nu - 2nl\nu\lambda - 2lm\lambda\mu = 0,$$

the coordinates of the pole are

$$\alpha' = l(l\lambda - m\mu - n\nu),\ \beta' = m(m\mu - n\nu - l\lambda),\ \gamma' = n(n\nu - l\lambda - m\mu),$$

whence $m\gamma' + n\beta' = -2lmn\lambda$, $n\alpha' + l\gamma' = -2lmn\mu$, $l\beta' + m\alpha' = -2lmn\nu$;

and, as in the last example, we find l, m, n respectively proportional to

$$\alpha'(\mu\beta' + \nu\gamma' - \lambda\alpha'),\ \beta'(\nu\gamma' + \lambda\alpha' - \mu\beta'),\ \gamma'(\lambda\alpha' + \mu\beta' - \nu\gamma').$$

Thus, then, since the condition that a conic circumscribing $\alpha\beta\gamma$ should pass through a fourth point $\alpha'\beta'\gamma'$ is $\frac{l}{\alpha'} + \frac{m}{\beta'} + \frac{n}{\gamma'} = 0$, the locus of the pole of $\lambda\alpha + \mu\beta + \nu\gamma$, with regard to a conic passing through the four points, is

$$\frac{\alpha}{\alpha'}(\mu\beta + \nu\gamma - \lambda\alpha) + \frac{\beta}{\beta'}(\nu\gamma + \lambda\alpha - \mu\beta) + \frac{\gamma}{\gamma'}(\lambda\alpha + \mu\beta - \nu\gamma) = 0,$$

which, when the locus of centre is sought, denotes a conic passing through the middle points of the sides of the triangle. The condition that the conic should touch $A\alpha + B\beta + C\gamma$ being $\sqrt{(Al)} + \sqrt{(Bm)} + \sqrt{(Cn)} = 0$, the locus of the pole of $\lambda\alpha + \mu\beta + \nu\gamma$, with regard to a conic passing through three points and touching a fixed line, is

$$\sqrt{\{A\alpha(\mu\beta + \nu\gamma - \lambda\alpha)\}} + \sqrt{\{B\beta(\nu\gamma + \lambda\alpha - \mu\beta)\}} + \sqrt{C\gamma(\lambda\alpha + \mu\beta - \nu\gamma)} = 0,$$

which, in general, represents a curve of the fourth degree.

294. If $\alpha''\beta''\gamma''$ be any point on any of the tangents drawn to a curve from a fixed point $\alpha'\beta'\gamma'$, the line joining $\alpha'\beta'\gamma'$, $\alpha''\beta''\gamma''$ meets the curve in two coincident points, and the equation in $l:m$ (Art. 290), which determines the points where the joining line meets the curve, will have equal roots.

To find, then, the equation of all the tangents which can be drawn through $\alpha'\beta'\gamma'$, we must substitute $l\alpha + m\alpha'$, $l\beta + m\beta'$, $l\gamma + m\gamma'$ in the equation of the curve, and form the condition that the resulting equation in $l:m$ shall have equal roots.

Thus (see Art. 92) the equation of the pair of tangents to a conic is $SS' = P^2$, where

$$S = a\alpha^2 + \&c.,\quad S' = a\alpha'^2 + \&c.,\quad P = a\alpha\alpha' + \&c.$$

This equation may also be written in another form; for since any point on either tangent through $\alpha'\beta'\gamma'$ evidently possesses the property that the line joining it to $\alpha'\beta'\gamma'$ touches the curve, we have only to express the condition that the line joining two points (Art. 65)

$$\alpha(\beta'\gamma'' - \beta''\gamma') + \beta(\gamma'\alpha'' - \gamma''\alpha') + \gamma(\alpha'\beta'' - \alpha''\beta') = 0$$

should touch the curve, and then consider $\alpha''\beta''\gamma''$ variable, when we shall have the equation of the pair of tangents. In other words, we are to substitute $\beta\gamma' - \beta'\gamma$, $\gamma\alpha' - \gamma'\alpha$, $\alpha\beta' - \alpha'\beta$ for λ, μ, ν in the condition of Art. 285,

$$A\lambda^2 + B\mu^2 + C\nu^2 + 2F\mu\nu + 2G\nu\lambda + 2H\lambda\mu = 0.$$

Attending to the values given (Art. 285) for A, B, &c., it may easily be verified that

$$(a\alpha^2 + \&c.)(a\alpha'^2 + \&c.) - (a\alpha\alpha' + \&c.)^2 = A(\beta\gamma' - \beta'\gamma)^2 + \&c.$$

Ex. To find the locus of intersection of tangents which cut at right angles to a conic given by the general equation (see Ex. 4, p. 169).

We see now that the equation of the pair of tangents through any point (Art. 147) may also be written

$$A(y - y')^2 + B(x - x')^2 + C(xy' - yx')^2$$
$$- 2F(x - x')(xy' - yx') + 2G(y - y')(xy' - x'y) - 2H(x - x')(y - y') = 0.$$

This will represent two right lines at right angles when the sum of the coefficients of x^2 and y^2 vanishes, which gives for the equation of the locus

$$C(x^2 + y^2) - 2Gx - 2Fy + A + B = 0.$$

This circle has been called the director circle of the conic. When the curve is a parabola, $C = 0$, and we see that the equation of the directrix is $Gx + Fy = \frac{1}{2}(A + B)$.

295. It follows, as a particular case of the last, that the pairs of tangents from $\beta\gamma$, $\gamma\alpha$, $\alpha\beta$ are

$$B\gamma^2 + C\beta^2 - 2F\beta\gamma,\quad C\alpha^2 + A\gamma^2 - 2G\gamma\alpha,\quad A\beta^2 + B\alpha^2 - 2H\alpha\beta,$$

as indeed might be seen directly by throwing the equation of the curve into the form

$$(a\alpha + h\beta + g\gamma)^2 + (C\beta^2 + B\gamma^2 - 2F\beta\gamma) = 0.$$

Now if the pair of tangents through $\beta\gamma$ be $\beta - k\gamma$, $\beta - k'\gamma$, it appears from these expressions that $kk' = \dfrac{B}{C}$, and that the corresponding quantities for the other pairs of tangents are $\frac{C}{A}$, $\frac{A}{B}$, and these three multiplied together are $= 1$. Hence, recollecting the meaning of k (Art. 54), we learn that if A, F, B, D, C, E be the angles of a circumscribing hexagon,

$$\frac{\sin EAB.\sin FAB.\sin FBC.\sin DBC.\sin DCA.\sin ECA}{\sin EAC.\sin FAC\ \sin FBA.\sin DBA.\sin DCB.\sin ECB} = 1.$$

Hence also three pairs of lines will touch the same conic if their equations can be thrown into the form

$M^2 + N^2 + 2f'MN = 0$, $N^2 + L^2 + 2g'NL = 0$, $L^2 + M^2 + 2h'LM = 0$,

for the equations of the three pairs of tangents, already found can be thrown into this form by writing $L\sqrt{(A)}$ for α, &c.

296. If we wish to form the equations of the lines joining to $\alpha'\beta'\gamma'$ all the points of intersection of two curves, we have only to substitute $l\alpha + m\alpha'$, $l\beta + m\beta'$, $l\gamma + m\gamma'$ in both equations, and eliminate $l : m$ from the resulting equations. For any point on any of the lines in question evidently possesses the property that the line joining it to $\alpha'\beta'\gamma'$ meets both curves in the same point; therefore the equations in $l : m$, which determine the points where one of these lines meets both curves, must have a common root; and therefore the result of elimination between them is satisfied. Thus, the equation of the pair of lines joining to $\alpha'\beta'\gamma'$ the points where any right line L meets S, is $L'^2S - 2LL'P + L^2S' = 0$. If the point $\alpha'\beta'\gamma'$ be on the curve the equation reduces to $L'S - 2LP = 0$.

Ex. A chord which subtends a right angle at a given point on the curve passes through a fixed point (Ex. 2, Art. 181). We use the general equation, and by the formula last given, form the equation of the lines joining the given point to the intersection of the conic with $\lambda x + \mu y + \nu$. The coordinates being supposed rectangular, these lines will be at right angles if the sum of the coefficients of x^2 and y^2 vanish, which gives the condition

$$(\lambda x' + \mu y' + \nu)(a + b) = 2(a\lambda x' + b\mu y').$$

And since λ, μ, ν enter in the first degree, the chord passes through a fixed point, viz. $\frac{b-a}{b+a}x'$, $\frac{a-b}{a+b}y'$. If the point on the curve vary, this other point will describe a conic. If the angle subtended at the given point be not a right angle, or if the angle be a right angle, but the given point not on the curve, the condition found in like manner will contain λ, μ, ν in the second degree, and the chord will envelope a conic.

297. Since the equation of the polar of a point involves the coefficients of the equation in the first degree, if an indeterminate

enter in the first degree into the equation of a conic it will enter in the first degree into the equation of the polar. Thus, if P and P' be the polars of a point with regard to two conics S, S', then the polar of the same point with regard to $S+kS'$ will be $P+kP'$. For

$$(a+ka')\,\alpha\alpha' + \&c. = a\alpha\alpha' + \&c. + k\,\{a'\alpha\alpha' + \&c.\}.$$

Hence, *given four points on a conic, the polar of any given point passes through a fixed point* (Ex. 2, Art. 151).

If Q and Q' be the polars of another point with regard to S and S', then the polar of this second point with regard to $S+kS'$ is $Q+kQ'$. Thus, then (see Art. 59), the polars of two points with regard to a system of conics through four points form two homographic pencils of lines.

Given two homographic pencils of lines, the locus of the intersection of the corresponding lines of the pencils is a conic through the vertices of the pencils. For, if we eliminate k between $P+kP'$, $Q+kQ'$, we get $PQ'=P'Q$. In the particular case under consideration, the intersection of $P+kP'$, $Q+kQ'$ is the pole with respect to $S+kS'$ of the line joining the two given points. And we see that, *given four points on a conic, the locus of the pole of a given line is a conic* (Ex. 1, Art. 278).

If an indeterminate enter in the second degree into the equation of a conic, it must also enter in the second degree into the equation of the polar of a given point, which will then envelope a conic. Thus, if a conic have double contact with two fixed conics, the polar of a fixed point will envelope one of three fixed conics; for the equation of each system of conics in Art. 287 contains μ in the second degree.

We shall in another chapter enter into fuller details respecting the general equation, and here add a few examples illustrative of the principles already explained.

Ex. 1. A point moves along a fixed line; find the locus of the intersection of its polars with regard to two fixed conics. If the polars of any two points $\alpha'\beta'\gamma'$, $\alpha''\beta''\gamma''$ on the given line with respect to the two conics be P', P''; Q', Q''; then any other point on the line is $\lambda\alpha'+\mu\alpha''$, $\lambda\beta'+\mu\beta''$, $\lambda\gamma'+\mu\gamma''$; and its polars $\lambda P'+\mu P''$ $\lambda Q'+\mu Q''$, which intersect on the conic $P'Q''=P''Q'$.

Ex. 2. The anharmonic ratio of four points on a right line is the same as that of their four polars.

For the anharmonic ratio of the four points

$$l\alpha' + m\alpha'',\; l'\alpha' + m'\alpha'',\; l''\alpha' + m''\alpha'',\; l'''\alpha' + m'''\alpha'',$$

is evidently the same as that of the four lines

$$lP' + mP'', \quad l'P' + m'P'', \quad l''P' + m''P'', \quad l'''P' + m'''P''.$$

Ex. 3. To find the equation of the pair of tangents at the points where a conic S is met by the line γ.

The equation of the polar of any point on γ is (Art. 291) $\alpha' S_1 + \beta' S_2 = 0$. But the points where γ meets the curve are found by making $\gamma = 0$ in the general equation, whence

$$a\alpha'^2 + 2h\alpha'\beta' + b\beta'^2 = 0.$$

Eliminating α', β' between these equations, we get for the equation of the pair of tangents

$$aS_2^2 - 2hS_1S_2 + bS_1^2 = 0.$$

Thus the equation of the asymptotes of a conic (given by the Cartesian equation) is

$$a\left(\frac{dS}{dy}\right)^2 - 2h\left(\frac{dS}{dx}\right)\left(\frac{dS}{dy}\right) + b\left(\frac{dS}{dx}\right)^2 = 0,$$

for the asymptotes are the tangents at the points where the curve is met by the line at infinity z.

Ex. 4. Given three points on a conic: if one asymptote pass through a fixed point, the other will envelope a conic touching the sides of the given triangle. If t_1, t_2 be the asymptotes, and $a\alpha + b\beta + c\gamma$ the line at infinity, the equation of the conic is $t_1t_2 = (a\alpha + b\beta + c\gamma)^2$. But since it passes through $\beta\gamma$, $\gamma\alpha$, $\alpha\beta$, the equation must not contain the terms α^2, β^2, γ^2. If therefore t_1 be $\lambda\alpha + \mu\beta + \nu\gamma$, t_2 must be $\frac{a^2}{\lambda}\alpha + \frac{b^2}{\mu}\beta + \frac{c^2}{\nu}\gamma$; and if t_2 pass through $\alpha'\beta'\gamma'$, then (Ex. 1, Art. 285) t_1 touches $a\sqrt{(\alpha\alpha')} + b\sqrt{(\beta\beta')} + c\sqrt{(\gamma\gamma')} = 0$. The same argument proves that if a conic pass through three fixed points, and if one of its chords of intersection with a conic given by the general equation $a\alpha^2 + \&c. = 0$ be $\lambda\alpha + \mu\beta + \nu\gamma$, the other will be $\frac{a}{\lambda}\alpha + \frac{b}{\mu}\beta + \frac{c}{\nu}\gamma$.

Ex. 5. Given a self conjugate triangle with regard to a conic: if one chord of intersection with a fixed conic (given by the general equation) pass through a fixed point, the other will envelope a conic [Mr. Burnside]. The terms $\alpha\beta$, $\beta\gamma$, $\gamma\alpha$ are now to disappear from the equation, whence if one chord be $\lambda\alpha + \mu\beta + \nu\gamma$, the other is found to be

$$\lambda\alpha(\mu g + \nu h - \lambda f) + \mu\beta(\nu h + \lambda f - \mu g) + \nu\gamma(\lambda f + \mu g - \nu h).$$

Ex. 6. A and A' ($\alpha_1\beta_1\gamma_1$, $\alpha_2\beta_2\gamma_2$) are the points of contact of a common tangent to two conics U, V; P and P' are variable points, one on each conic; find the locus of C, the intersection of AP, $A'P'$, if PP' pass through a fixed point O on the common tangent [Mr. Williamson].

Let P and Q denote the polars of $\alpha_1\beta_1\gamma_1$, $\alpha_2\beta_2\gamma_2$, with respect to U and V respectively; then (Art. 290) if $\alpha\beta\gamma$ be the coordinates of C, those of the point P where AC meets the conic again, are $U\alpha_1 - 2P\alpha$, $U\beta_1 - 2P\beta$, $U\gamma_1 - 2P\gamma$; and those of the point P' are, in like manner, $V\alpha_2 - 2Q\alpha$, &c. If the line joining these points pass through O, which we choose as the intersection of α, β, we must have

$$\frac{U\alpha_1 - 2P\alpha}{U\beta_1 - 2P\beta} = \frac{V\alpha_2 - 2Q\alpha}{V\beta_2 - 2Q\beta};$$

and when A, A', O are unrestricted in position, the locus is a curve of the fourth order. If, however, these points be in a right line, we may choose this for the line α, and making α_1 and $\alpha_2 = 0$, the preceding equation becomes divisible by α, and reduces to the curve of the third order $PV\beta_2 = QU\beta_1$. Further, if the given points are points of contact of a common tangent, P and Q represent the same line; and another factor divides out of the equation which reduces to one of the form $U = kV$, representing a conic through the intersection of the given conics.

Ex. 7. To inscribe in a conic, given by the general equation, a triangle whose sides pass through the three points $\beta\gamma$, $\gamma\alpha$, $\alpha\beta$. We shall, as before, write S_1, S_2, S_3 for the three quantities, $a\alpha + h\beta + g\gamma$, $h\alpha + b\beta + f\gamma$, $g\alpha + f\beta + c\gamma$. Now we have seen, in general, that the line joining any point on the curve $\alpha\beta\gamma$ to another point $\alpha'\beta'\gamma'$ meets the curve again in a point, whose coordinates are $S'\alpha - 2P'\alpha'$, $S'\beta - 2P'\beta'$, $S'\gamma - 2P'\gamma$. Now if the point $\alpha'\beta'\gamma'$ be the intersection of lines β, γ, we may take $\alpha' = 1$, $\beta' = 0$, $\gamma' = 0$, which gives $S' = a$, $P' = S_1$, and the coordinates of the point where the line joining $\alpha\beta\gamma$ to $\beta\gamma$ meets the curve, are $a\alpha - 2S_1$, $a\beta$, $a\gamma$. In like manner, the line joining $\alpha\beta\gamma$ to $\gamma\alpha$, meets the curve again in $b\alpha$, $b\beta - 2S_2$, $b\gamma$. The line joining these two points will pass through $\alpha\beta$, if

$$\frac{a\alpha - 2S_1}{a\beta} = \frac{b\alpha}{b\beta - 2S_2};$$

or, reducing

$$2S_1S_2 = a\alpha S_2 + b\beta S_1,$$

which is the condition to be fulfilled by the coordinates of the vertex. Writing in this equation $a\alpha = S_1 - h\beta - g\gamma$, $b\beta = S_2 - h\alpha - f\gamma$, it becomes

$$h\,(\alpha S_1 + \beta S_2) + \gamma\,(fS_1 + gS_2) = 0.$$

But since $\alpha\beta\gamma$ is on the curve, $\alpha S_1 + \beta S_2 + \gamma S_3 = 0$, and the equation last written reduces to

$$\gamma\,(fS_1 + gS_2 - hS_3) = 0.$$

Now the factor γ may be set aside as irrelevant to the geometric solution of the problem; for although either of the points where γ meets the curve fulfils the condition which we have expressed analytically, namely, that if it be joined to $\beta\gamma$ and to $\gamma\alpha$, the joining lines meet the curve again in points which lie on a line with $\alpha\beta$; yet, since these joining lines coincide, they cannot be sides of a triangle. The vertex of the sought triangle is therefore either of the points where the curve is met by $fS_1 + gS_2 - hS_3$. It can be verified immediately that $fS_1 = gS_2 = hS_3$ denote the lines joining the corresponding vertices of the triangles $\alpha\beta\gamma$, $S_1S_2S_3$. Consequently (see Ex. 2, Art. 60), the line $fS_1 + gS_2 - hS_3$ is constructed as follows: "Form the triangle DEF whose sides are the polars of the given points A, B, C; let the lines joining the corresponding vertices of the two triangles meet the opposite sides of the polar triangle in L, M, M; then the lines LM, MN, NL pass through the vertices of the required triangles."

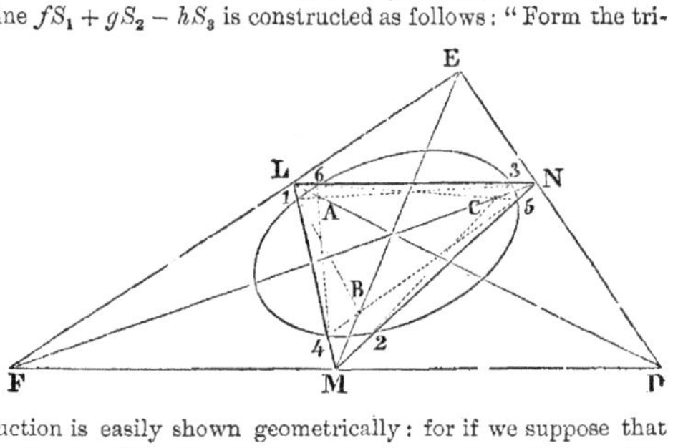

The truth of this construction is easily shown geometrically: for if we suppose that we have drawn the two triangles 123, 456 which can be drawn through the points A, B, C; then applying Pascal's theorem to the hexagon 123456, we see that the line BC passes through the intersection of 16, 34. But this latter point is the pole of AL (Ex. 1, Art. 146). Conversely, then, AL passes through the pole of BC, and L is on the polar of A (Ex. 1, Art. 146).

This construction becomes indeterminate if the triangle is selfconjugate in which case the problem admits of an infinity of solutions.

N N.

Ex. 8. If two conics have double contact, any tangent to the one is cut harmonically at its point of contact, the points where it meets the other, and where it meets the chord of contact.

If in the equation $S + R^2 = 0$, we substitute $l\alpha' + m\alpha''$, $l\beta' + m\beta''$, $l\gamma' + m\gamma''$, for α, β, γ, (where the points $\alpha'\beta'\gamma'$, $\alpha''\beta''\gamma''$ satisfy the equation $S = 0$), we get

$$(lR' + mR'')^2 + 2lmP = 0.$$

Now, if the line joining $\alpha'\beta'\gamma'$, $\alpha''\beta''\gamma''$, touch $S + R^2$, this equation must be a perfect square; and it is evident that the only way this can happen is if $P = -2R'R''$, when the equation becomes $(lR' - mR'')^2 = 0$; when the truth of the theorem is manifest.

Ex. 9. Find the equation of the conic touching five lines, viz. $\alpha, \beta, \gamma, A\alpha + B\beta + C\gamma$, $A'\alpha + B'\beta + C'\gamma$.

Ans. $(l\alpha)^{\frac{1}{2}} + (m\beta)^{\frac{1}{2}} + (n\gamma)^{\frac{1}{2}}$, where l, m, n are determined by the conditions

$$\frac{l}{A} + \frac{m}{B} + \frac{n}{C} = 0, \quad \frac{l}{A'} + \frac{m}{B'} + \frac{n}{C'} = 0.$$

Ex. 10. Find the equation of the conic touching the five lines, $\alpha, \beta, \gamma, \alpha + \beta + \gamma$, $2\alpha + \beta - \gamma$.

We have $l + m + n = 0$, $\frac{1}{2}l + m - n = 0$: hence the required equation is

$$2(-\alpha)^{\frac{1}{2}} + (3\beta)^{\frac{1}{2}} + (\gamma)^{\frac{1}{2}} = 0.$$

Ex. 11. Find the equation of the conic touching α, β, γ, at their middle points.

Ans. $(a\alpha)^{\frac{1}{2}} + (b\beta)^{\frac{1}{2}} + (c\gamma)^{\frac{1}{2}} = 0.$

Ex. 12. Find the condition that $(l\alpha)^{\frac{1}{2}} + (m\beta)^{\frac{1}{2}} + (n\gamma)^{\frac{1}{2}} = 0$ should represent a parabola.

Ans. The curve touches the line at infinity when $\frac{l}{a} + \frac{m}{b} + \frac{n}{c} = 0.$

Ex. 13. To find the locus of the focus of a parabola touching α, β, γ.

Generally, if the coordinates of one focus of a conic inscribed in the triangle $\alpha\beta\gamma$ be $\alpha'\beta'\gamma'$, the lines joining it to the vertices of the triangle will be

$$\frac{\alpha}{\alpha'} = \frac{\beta}{\beta'}, \quad \frac{\beta}{\beta'} = \frac{\gamma}{\gamma'}, \quad \frac{\gamma}{\gamma'} = \frac{\alpha}{\alpha'},$$

and since the lines to the other focus make equal angles with the sides of the triangle (Art. 189), these lines will be (Art. 55)

$$\alpha'\alpha = \beta'\beta, \quad \beta'\beta = \gamma'\gamma, \quad \gamma'\gamma = \alpha'\alpha;$$

and the coordinates of the other focus may be taken $\frac{1}{\alpha'}, \frac{1}{\beta'}, \frac{1}{\gamma'}$.

Hence, if we are given the equation of any locus described by one focus, we can at once write down the equation of the locus described by the other; and if the second focus be at infinity, that is, if $\alpha'' \sin A + \beta'' \sin B + \gamma'' \sin C = 0$, the first must lie on the circle $\frac{\sin A}{\alpha'} + \frac{\sin B}{\beta'} + \frac{\sin C}{\gamma'} = 0$. The coordinates of the focus of a parabola at infinity are $\frac{l}{\sin^2 A}, \frac{m}{\sin^2 B}, \frac{n}{\sin^2 C}$, since (remembering the relation in Ex. 12) these values satisfy both the equations,

$$\alpha \sin A + \beta \sin B + \gamma \sin C = 0, \quad \sqrt{l\alpha} + \sqrt{m\beta} + \sqrt{n\gamma} = 0.$$

The coordinates, then, of the finite focus are $-\frac{\sin^2 A}{l}, \frac{\sin^2 B}{m}, \frac{\sin^2 C}{n}$.

Ex. 14. To find the equation of the directrix of this parabola.

Forming, by Art. 291, the equation of the polar of the point whose coordinates have just been given, we find

$$l\alpha (\sin^2 B + \sin^2 C - \sin^2 A) + m\beta (\sin^2 C + \sin^2 A - \sin^2 B) + n\gamma (\sin^2 A + \sin^2 B - \sin^2 C) = 0,$$

or $$l\alpha \sin B \sin C \cos A + m\beta \sin C \sin A \cos B + n\gamma \sin A \sin B \cos C = 0.$$

Substituting for n from Ex. 12, the equation becomes

$$l \sin B \sin C (\alpha \cos A - \gamma \cos C) + m \sin C \sin A (\beta \cos B - \gamma \cos C) = 0;$$

hence the directrix always passes through the intersection of the perpendiculars of the triangle (see Ex. 3, Art. 54).

Ex. 15. Given four tangents to a conic find the locus of the foci. Let the four tangents be α, β, γ, δ; then, since any line can be expressed in terms of three others, these must be connected by an identical relation $a\alpha + b\beta + c\gamma + d\delta = 0$. This relation must be satisfied, not only by the coordinates of one focus $\alpha'\beta'\gamma'\delta'$, but also by those of the other $\frac{1}{\alpha'}$, $\frac{1}{\beta'}$, $\frac{1}{\gamma'}$, $\frac{1}{\delta'}$. The locus is therefore the curve of the third degree.

$$\frac{a}{\alpha} + \frac{b}{\beta} + \frac{c}{\gamma} + \frac{d}{\delta} = 0.$$

CHAPTER XV.

THE PRINCIPLE OF DUALITY; AND THE METHOD OF RECIPROCAL POLARS.

298. The methods of abridged notation, explained in the last chapter, apply equally to tangential equations. Thus, if the constants λ, μ, ν in the equation of a line be connected by the relation

$$(a\lambda + b\mu + c\nu)(a'\lambda + b'\mu + c'\nu) = (a''\lambda + b''\mu + c''\nu)(a'''\lambda + b'''\mu + c'''\nu),$$

the line (Art. 285) touches a conic. Now it is evident that one line which satisfies the given relation is that whose λ, μ, ν are determined by the equations

$$a\lambda + b\mu + c\nu = 0, \quad a''\lambda + b''\mu + c''\nu = 0.$$

That is to say, the line joining the points which these last equations represent (Art. 70), touches the conic in question. If then α, β, γ, δ represent equations of points, (that is to say, functions of the first degree in λ, μ, ν) $\alpha\gamma = k\beta\delta$ is the tangential equation of a conic touched by the four lines $\alpha\beta$, $\beta\gamma$, $\gamma\delta$, $\delta\alpha$. More generally, if S and S' in tangential coordinates represent any two curves, $S - kS'$ represents a curve touched by every tangent common to S and S'. For, whatever values of λ, μ, ν make both $S = 0$ and $S' = 0$, must also make $S - kS' = 0$. Thus, then, if S represent a conic, $S - k\alpha\beta$ represents a conic having common with S the pairs of tangents drawn from the points α, β. Again, the equation $\alpha\gamma = k\beta^2$ represents a conic such that the two tangents which can be drawn from the point α coincide with the line $\alpha\beta$; and those which can be drawn from γ coincide with the line $\gamma\beta$. The points α, γ are therefore on this conic, and β is the pole of the line joining them. In like manner, $S - \alpha^2$ represents a conic having double contact with S, and the tangents at the points of contact meet in α; or, in other words, α is the pole of the chord of contact.

So again, the equation $\alpha\gamma = k^2\beta^2$ may be treated in the same manner as at Art. 270, and any point on the curve may be represented by $\mu^2\alpha + 2\mu k\beta + \gamma$, while the tangent at that point joins the points $\mu\alpha + k\beta$, $\mu k\beta + \gamma$.*

Ex. 1. To find the locus of the centre of conics touching four given lines. Let $\Sigma = 0$, $\Sigma' = 0$ be the tangential equations of any two conics touching the four lines; then, by Art. 298, the tangential equation of any other is $\Sigma + k\Sigma' = 0$. And (see Art. 151) the coordinates of the centre are $\frac{G+kG'}{C+kC'}$, $\frac{F+kF'}{C+kC'}$, the form of which shows (Art. 7) that the centre of the variable conic is on the line joining the centres of the two assumed conics, whose coordinates are $\frac{G}{C}$, $\frac{F}{C}$; $\frac{G'}{C'}$, $\frac{F'}{C'}$; and that it divides the distance between them in the ratio $C : kC'$.

Ex. 2. To find the locus of the foci of conics touching four given lines. We have only in the equations (Ex. Art. 258*a*) which determine the foci to substitute $A + kA'$ for A, &c., and then eliminate k between them, when we get the result in the form

$$\{C(x^2 - y^2) + 2Fy - 2Gx + A - B\}\{C'xy - F'x - G'y + H'\}$$
$$= \{C'(x^2 - y^2) + 2F'y - 2G'x + A' - B'\}\{Cxy - Fx - Gy + H\}.$$

This represents a curve of the third degree (see Ex. 15, p. 275), the terms of higher order mutually destroying. If, however, Σ and Σ' be parabolas, $\Sigma + k\Sigma'$ denotes a system of parabolas having three tangents common. We have then C and C' both $= 0$, and the locus of foci reduces to a circle. Again, if the conics be concentric, taking the centre as origin, we have F, F', G, G' all $= 0$. In this case $\Sigma + k\Sigma'$ represents a system of conics touching the four sides of a parallelogram and the locus of foci is an equilateral hyperbola.†

Ex. 3. The director circles of conics touching four fixed lines have a common radical axis. This is apparent from what was proved, p. 270, that the equation of the director circle is a linear function of the coefficients A, B, &c., and that therefore when we substitute $A + kA'$ for A, &c. it will be of the form $S + kS' = 0$. This theorem includes as a particular case, "The circles having for diameters the three diagonals of a complete quadrilateral have a common radical axis."

299. Thus we see (as in Art. 70) that each of the equations used in the last chapter is capable of a double interpretation, according as it is considered as an equation in trilinear or in tangential coordinates. And the equations used in the last chapter, to establish any theorem, will, if interpreted as equations

---

* In other words, if in any system $x'y'z'$, $x''y''z''$, be the coordinates of any two points on a conic, and $x'''y'''z'''$ those of the pole of the line joining them, the coordinates of any point on the curve may be written

$$\mu^2x' + 2\mu kx''' + x'',\ \mu^2y' + 2\mu ky''' + y'',\ \mu^2z' + 2\mu kz''' + z'',$$

while the tangent at that point divides the two fixed tangents in the ratios $\mu : k$, $\mu k : 1$. When $k = 1$, the curve is a parabola. Want of space prevents us from giving illustrations of the great use of this principle in solving examples. The reader may try the question :—To find the locus of the point where a tangent meeting two fixed tangents is cut in a given ratio.

† It is proved in like manner that the locus of foci of conics passing through four fixed points, which is in general of the sixth degree, reduces to the fourth when the points form a parallelogram.

in tangential coordinates, yield another theorem, the *reciprocal* of the former. Thus (Art. 266) we proved that if three conics (S, $S+LM$, $S+LN$) have two points (S, L) common to all, the chords in each case joining the remaining common points (M, N, $M-N$), will meet in a point. Consider these as tangential equations, and the pair of tangents drawn from L is common to the three conics, while M, N, $M-N$ denote in each case the point of intersection of the other two common tangents. We thus get the theorem, "If three conics have two tangents common to all, the intersections in each case of the remaining pair of common tangents, lie in a right line." Every theorem *of position* (that is to say, one not involving the magnitudes of lines or angles) is thus twofold. From each theorem another can be derived by suitably interchanging the words "point" and "line"; and the same equations differently interpreted will establish either theorem. We shall in this chapter give an account of the geometrical method by which the attention of mathematicians was first called to this "principle of duality."*

300. Being given a fixed conic section (U) and any curve (S), we can generate another curve (s) as follows: draw any tangent to S, and take its pole with regard to U; the locus of this pole will be a curve s, which is called the *polar curve* of S with regard to U. The conic U, with regard to which the pole is taken, is called the *auxiliary* conic.

We have already met with a particular example of polar curves (Ex. 12, Art. 225), where we proved that the polar curve of one conic section with regard to another is always a curve of the second degree.

We shall for brevity say that a point *corresponds* to a line when we mean that the point is the pole of that line with regard to U. Thus, since it appears from our definition that every point of s is the pole with regard to U of some tangent to S, we shall

---

* The method of reciprocal polars was introduced by M. Poncelet, whose account of it will be found in Crelle's *Journal*, vol. IV. M. Plücker, in his "System der Analytischen Geometrie," 1835, presented the principle of duality in the purely analytical point of view, from which the subject is treated at the beginning of this chapter. But it was Möbius who, in his "Barycentrische Calcul," 1827, had made the important step of introducing a system of coordinates in which the position of a right line was indicated by coordinates and that of a point by an equation.

briefly express this relation by saying that every point of s *corresponds* to some tangent of S.

301. *The point of intersection of two tangents to S will correspond to the line joining the corresponding points of s.*

This follows from the property of the conic U, that the point of intersection of any two lines is the pole of the line joining the poles of these two lines (Art. 146).

Let us suppose that in this theorem the two tangents to S are indefinitely near, then the two corresponding points of s will also be indefinitely near, and the line joining them will be a tangent to s; and since any tangent to S intersects the consecutive tangent at its point of contact, the last theorem becomes for this case: *If any tangent to S correspond to a point on s, the point of contact of that tangent to S will correspond to the tangent through the point on s.*

Hence we see that the relation between the curves is *reciprocal*, that is to say, that the curve S might be generated from s in precisely the same manner that s was generated from S. Hence the name "reciprocal polars."

302. We are now able, being given any theorem of position concerning any curve S, to deduce another concerning the curve s. Thus, for example, if we know that a number of points connected with the figure S lie on one right line, we learn that the corresponding lines connected with the figure s meet in a point (Art. 146), and *vice versâ;* if a number of points connected with the figure S lie on a conic section, the corresponding lines connected with s will touch the polar of that conic with regard to U; or, in general, if the *locus* of any point connected with S be any curve S', the *envelope* of the corresponding line connected with s is s', the reciprocal polar of S'.

303. *The degree of the polar reciprocal of any curve is equal to the class of the curve* (see note, Art. 145), *that is, to the number of tangents which can be drawn from any point to that curve.*

For the degree of s is the same as the number of points in which any line cuts s; and to a number of points on s, lying on a right line, correspond *the same number* of tangents to S passing through the point corresponding to that line. Thus, if S be a conic section, two, and only two, tangents, real or imaginary, can be drawn to it from any point (Art. 145); therefore, any line meets s in two, and only two points, real or imaginary; we may thus infer, independently of Ex. 12, Art. 225, that the reciprocal of any conic section is a curve of the second degree.

304. We shall exemplify, in the case where S and s are conic sections, the mode of obtaining one theorem from another by this method. We know (Art. 267) that "if a hexagon be *inscribed* in S, whose *sides* are A, B, C, D, E, F, then the *points* of intersection, AD, BE, CF, are *in one right line*." Hence we infer, that "if a hexagon be *circum*scribed about s, whose *vertices* are a, b, c, d, e, f, then the lines, ad, be, cf, will *meet in a point*" (Art. 265). Thus we see that Pascal's theorem and Brianchon's are reciprocal to each other, and it was thus, in fact, that the latter was first obtained.

In order to give the student an opportunity of rendering himself expert in the application of this method, we shall write in parallel columns some theorems, together with their reciprocals. The beginner ought carefully to examine the force of the argument by which the one is inferred from the other, and he ought to attempt to form for himself the reciprocal of each theorem before looking at the reciprocal we have given. He will soon find that the operation of forming the reciprocal theorem will reduce itself to a mere mechanical process of interchanging the words "point" and "line," "inscribed" and "circumscribed," "locus" and "envelope," &c.

If two vertices of a triangle move along fixed right lines, while the sides pass each through a fixed point, the locus of the third vertex is a conic section. (Art. 269).	If two sides of a triangle pass through fixed points, while the vertices move on fixed right lines, the envelope of the third side is a conic section.
If, however, the points through which the sides pass lie in one right line, the locus will be a right line. (Ex. 2. p. 41).	If the lines on which the vertices move meet in a point, the third side will pass through a fixed point.
In what other case will the locus be a right line? (Ex. 3, p. 42).	In what other case will the third side pass through a fixed point? (p. 49).

If two conics touch, their reciprocals will also touch; for the first pair have a point common, and also the tangent at that point common, therefore the second pair will have a tangent common and its point of contact also common. So likewise if two conics have double contact their reciprocals will have double contact.

If a triangle be circumscribed to a conic section, two of whose vertices move on fixed lines, the locus of the third vertex is a conic section, having double contact with the given one. (Ex. 2, p. 250).

If a triangle be inscribed in a conic section, two of whose sides pass through fixed points, the envelope of the third side is a conic section, having double contact with the given one. (Ex. 3, p. 250).

305. We proved (Art. 301, see figure, p. 282) if to two points P, P', on S, correspond the tangents pt, $p't'$, on s, that the tangents at P and P' will correspond to the points of contact p, p', and therefore Q, the intersection of these tangents, will correspond to the chord of contact pp'. Hence we learn that *to any point Q, and its polar PP', with respect to S, correspond a line pp' and its pole q with respect to s.*

Given two points on a conic, and two of its tangents, the line joining the points of contact of those tangents passes through one or other of two fixed points. (Ex., Art. 286, p. 262).

Given two tangents and two points on a conic, the point of intersection of the tangents at those points will move along one or other of two fixed right lines.

Given four points on a conic, the polar of a fixed point passes through a fixed point. (Ex. 2, p. 153).

Given four tangents to a conic, the locus of the pole of a fixed right line is a right line. (Ex. 2, p. 254).

Given four points on a conic, the locus of the pole of a fixed right line is a conic section. (Ex. 1, p. 254).

Given four tangents to a conic, the envelope of the polar of a fixed point is a conic section.

The lines joining the vertices of a triangle to the opposite vertices of its polar triangle with regard to a conic meet in a point. (Art. 99).

The points of intersection of each side of any triangle, with the opposite side of the polar triangle, lie in one right line.

Inscribe in a conic a triangle whose sides pass through three given points. (Ex. 7, Art. 297, p. 273).

Circumscribe about a conic a triangle whose vertices rest on three given lines.

306. Given two conics, S and S', and their two reciprocals, s and s'; to the four points A, B, C, D common to S and S' correspond the four tangents a, b, c, d common to s and s', and to the six chords of intersection of S and S', AB, CD; AC, BD; AD, BC correspond the six intersections of common tangents to s and s'; ab, cd; ac, bd; ad, bc.*

If three conics have two common tangents, or if they have each double contact with a fourth, their six chords of intersection will pass three by three through the same points. (Art. 264).

If three conics have two points common, or if they have each double contact with a fourth, the six points of intersection of common tangents lie three by three on the same right lines.

Or, in other words, three conics, having each double contact with a fourth, may be

Or three conics, having each double contact with a fourth, may be considered

* A system of four points connected by six lines is accurately called a *quadrangle*, as a system of four lines intersecting in six points is called a quadrilateral.

considered as having four radical centres.

as having four axes of similitude. (See Art. 117, of which this theorem is an extension).

If through the point of contact of two conics which touch, any chord be drawn, tangents at its extremities will meet on the common chord of the two conics.

If from any point on the tangent at the point of contact of two conics which touch, a tangent be drawn to each, the line joining their points of contact will pass through the intersection of common tangents to the conics.

If through an intersection of common tangents of two conics any two chords be drawn, lines joining their extremities will intersect on one or other of the common chords of the two conics. (Ex. 1, p. 250).

If on a common chord of two conics, any two points be taken, and from these tangents be drawn to the conics, the diagonals of the quadrilateral so formed will pass through one or other of the intersections of common tangents to the conics.

If A and B be two conics having each double contact with S, the chords of contact of A and B with S, and their chords of intersection with each other, meet in a point, and form a harmonic pencil. (Art. 263).

If A and B be two conics having each double contact with S, the intersections of the tangents at their points of contact with S, and the intersections of tangents common to A and B, lie in one right line, which they divide harmonically.

If A, B, C be three conics, having each double contact with S, and if A and B both touch C, the tangents at the points of contact will intersect on a common chord of A and B.

If A, B, C be three conics, having each double contact with S, and if A and B both touch C, the line joining the points of contact will pass through an intersection of common tangents of A and B.

307. We have hitherto supposed the auxiliary conic U to be any conic whatever. It is most common, however, to suppose this conic a circle; and hereafter, when we speak of polar curves, we intend the reader to understand polars *with regard to a circle*, unless we expressly state otherwise.

We know (Art. 88) that the polar of any point with regard to a circle is perpendicular to the line joining this point to the centre, and that the distances of the point and its polar are, when multiplied together, equal to the square of the radius; hence the relation between polar curves with regard to a circle is often stated as follows: *Being given any point O, if from it we let fall a perpendicular OT on any tangent to a curve S, and produce it until the rectangle $OT.Op$ is equal to a constant k^2, then the locus of the point p is a curve s, which is called the polar reciprocal of S.* For this is evidently

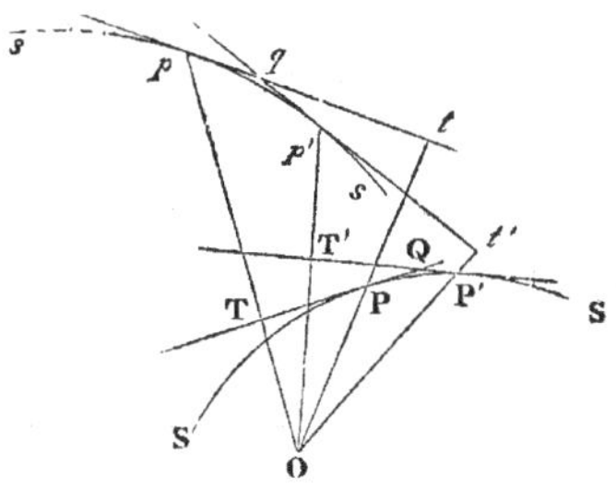

equivalent to saying that p is the pole of PT, with regard to a circle whose centre is O and radius k. We see, therefore (Art. 301), that the tangent pt will correspond to the point of contact P, that is to say, that OP will be perpendicular to pt, and that $OP.Ot = k^2$.

It is easy to show that a change in the magnitude of k will affect only the *size* and not the *shape* of s, which is all that in most cases concerns us. In this manner of considering polars, all mention of the circle may be suppressed, and s may be called the reciprocal of S *with regard to the point* O. We shall call this point the *origin*.

The advantage of using the circle for our auxiliary conic chiefly arises from the two following theorems, which are at once deduced from what has been said, and which enable us to transform, by this method, not only theorems of position, but also theorems involving the magnitude of lines and angles:

The distance of any point P from the origin is the reciprocal of the distance from the origin of the corresponding line pt.

The angle TQT' between any two lines TQ, $T'Q$, is equal to the angle pOp' subtended at the origin by the corresponding points p, p'; for Op is perpendicular to TQ, and Op' *to* $T'Q$.

We shall give some examples of the application of these principles when we have first investigated the following problem:

308. *To find the polar reciprocal of one circle with regard to another.* That is to say, to find the locus of the pole p with regard to the circle (O) of any tangent PT to the circle (C). Let MN be the polar of the point C with regard to O, then having the points C, p, and their polars MN, PT, we have, by Art. 101, the ratio $\dfrac{OC}{CP} = \dfrac{Op}{pN}$, but the first ratio is constant, since both OC and CP are constant; hence the

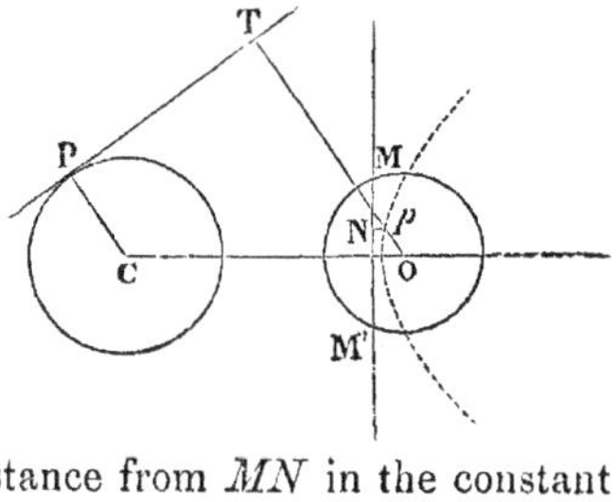

distance of p from O is to its distance from MN in the constant ratio $OC : CP$; its locus is therefore a conic, of which O is a focus, MN the corresponding directrix, and whose eccentricity is OC

divided by CP. Hence the eccentricity is greater, less than, or $= 1$, according as O is without, within, or on the circle C.

Hence the *polar reciprocal of a circle is a conic section, of which the origin is the focus, the line corresponding to the centre is the directrix, and which is an ellipse, hyperbola, or parabola, according as the origin is within, without, or on the circle.*

309. We shall now deduce some properties concerning angles, by the help of the last theorem given in Art. 307.

Any two tangents to a circle make equal angles with their chord of contact.	The line drawn from the focus to the intersection of two tangents bisects the angle subtended at the focus by their chord of contact. (Art. 191).

For the angle between one tangent PQ (see fig., p. 282) and the chord of contact PP' is equal to the angle subtended at the focus by the corresponding points p, q; and similarly, the angle $QP'P$ is equal to the angle subtended by p', q; therefore, since $QPP' = QP'P$, $pOq = p'Oq$.

Any tangent to a circle is perpendicular to the line joining its point of contact to the centre.	Any point on a conic, and the point where its tangent meets the directrix, subtend a right angle at the focus.

This follows as before, recollecting that the directrix of the conic answers to the centre of the circle.

Any line is perpendicular to the line joining its pole to the centre of the circle.	Any point and the intersection of its polar with the directrix subtend a right angle at the focus.
The line joining any point to the centre of a circle makes equal angles with the tangents through that point.	If the point where any line meets the directrix be joined to the focus, the joining line will bisect the angle between the focal radii to the points where the given line meets the curve.
The locus of the intersection of tangents to a circle, which cut at a given angle, is a concentric circle.	The envelope of a chord of a conic, which subtends a given angle at the focus, is a conic having the same focus and the same directrix.
The envelope of the chord of contact of tangents which cut at a given angle is a concentric circle.	The locus of the intersection of tangents, whose chord subtends a given angle at the focus, is a conic having the same focus and directrix.
If from a fixed point tangents be drawn to a series of concentric circles, the locus of the points of contact will be a circle passing through the fixed point, and through the common centre.	If a fixed line intersect a series of conics having the same focus and same directrix, the envelope of the tangents to the conics, at the points where this line meets them, will be a conic having the same focus, and touching both the fixed line and the common directrix.

In the latter theorem, if the fixed line be at infinity, we find the envelope of the asymptotes of a series of hyperbolas, having the same focus and same directrix, to be a parabola having the same focus and touching the common directrix.

If two chords at right angles to each other be drawn through any point on a circle, the line joining their extremities passes through the centre.

The locus of the intersection of tangents to a parabola which cut at right angles is the directrix.

We say a parabola, for, the point through which the chords of the circle are drawn being taken for origin, the polar of the circle is a parabola (Art. 308).

The envelope of a chord of a circle which subtends a given angle at a given point on the curve is a concentric circle.

The locus of the intersection of tangents to a parabola, which cut at a given angle, is a conic having the same focus and the same directrix.

Given base and vertical angle of a triangle, the locus of vertex is a circle passing through the extremities of the base.

Given in position two sides of a triangle, and the angle subtended by the base at a given point, the envelope of the base is a conic, of which that point is a focus, and to which the two given sides will be tangents.

The locus of the intersection of tangents to an ellipse or hyperbola which cut at right angles is a circle.

The envelope of any chord of a conic which subtends a right angle at any fixed point is a conic, of which that point is a focus.

"If from any point on the circumference of a circle perpendiculars be let fall on the sides of any inscribed triangle, their three feet will lie in one right line" (Art. 125).

If we take the fixed point for origin, to the triangle *inscribed* in a *circle* will correspond a triangle *circumscribed* about a *parabola;* again, to the foot of the perpendicular on any line corresponds a line through the corresponding point perpendicular to the radius vector from the origin. Hence, "If we join the focus to each vertex of a triangle circumscribed about a parabola, and erect perpendiculars at the vertices to the joining lines, those perpendiculars will pass through the same point." If, therefore, a circle be described, having for diameter the radius vector from the focus to this point, it will pass through the vertices of the circumscribed triangle. Hence, *Given three tangents to a parabola, the locus of the focus is the circumscribing circle* (p. 207).

The locus of the foot of the perpendicular (or of a line making a constant angle with the tangent) from the focus

If from any point a radius vector be drawn to a circle, the envelope of a perpendicular to it at its extremity (or of a of an ellipse or hyperbola on the tangent is a circle

line making a constant angle with it) is a conic having the fixed point for its focus.

310. Having sufficiently exemplified in the last Article the method of transforming theorems involving angles, we proceed to show that theorems involving the magnitude of lines *passing through the origin* are easily transformed by the help of the first theorem in Art. 307. For example, the sum (or, in some cases, the difference, if the origin be without the circle) of the perpendiculars let fall from the origin on any pair of parallel tangents to a circle is constant, and equal to the diameter of the circle.

Now, to two parallel lines correspond two points on a line passing through the origin. Hence, "the sum of the reciprocals of the segments of any focal chord of an ellipse is constant."

We know (p. 185) that this sum is four times the reciprocal of the parameter of the ellipse, and since we learn from the present example that it only depends on the diameter, and not on the position of the reciprocal circle, we infer that *the reciprocals of equal circles, with regard to any origin, have the same parameter.*

The rectangle under the segments of any chord of a circle through the origin is constant.

The rectangle under the perpendiculars let fall from the focus on two parallel tangents is constant.

Hence, given the tangent from the origin to a circle, we are given the conjugate axis of the reciprocal hyperbola.

Again, the theorem that the sum of the focal distances of any point on an ellipse is constant may be expressed thus:

The sum of the distances from the focus of the points of contact of parallel tangents is constant.

The sum of the reciprocals of perpendiculars let fall from any point within a circle on two tangents, whose chord of contact passes through the point, is constant.

311. If we are given any homogeneous equation connecting the perpendiculars PA, PB, &c. let fall from a variable point P on fixed lines, we can transform it so as to obtain a relation connecting the perpendiculars ap, bp' &c., let fall from the fixed points a, b, &c., which correspond to the fixed lines, on the variable line which corresponds to P. For we have only to divide the equation by a power of OP, the distance of P from the origin, and then, by Art. 101, substitute for each term

$\frac{PA}{OP}$, $\frac{ap}{Oa}$. For example, if PA, PB, PC, PD be the perpendiculars let fall from any point of a conic on the sides of an inscribed quadrilateral, $PA.PC = kPB.PD$ (Art. 259). Dividing each factor by OP, and substituting, as above, we have $\frac{ap}{Oa} \cdot \frac{cp''}{Oc} = k\frac{bp'}{Ob} \cdot \frac{dp'''}{Od}$; and Oa, Ob, Oc, Od being constant, we infer that *if a fixed quadrilateral be circumscribed to a conic, the product of the perpendiculars let fall from two opposite vertices on any variable tangent is in a constant ratio to the product of the perpendiculars let fall from the other two vertices.*

The product of the perpendiculars from any point of a conic on two fixed tangents is in a constant ratio to the square of the perpendicular on their chord of contact. (Art. 259).

The product of the perpendiculars from two fixed points of a conic on any tangent, is in a constant ratio to the square of the perpendicular on it, from the intersection of tangents at those points.

If, however, the origin be taken on the chord of contact, the reciprocal theorem is "the intercepts, made by any variable tangent on two parallel tangents, have a constant rectangle."

The product of the perpendiculars on any tangent of a conic from two fixed points (the foci) is constant.

The square of the radius vector from a fixed point to any point on a conic, is in a constant ratio to the product of the perpendiculars let fall from that point of the conic on two fixed right lines.

Generally, since every equation in trilinear coordinates is a homogeneous relation between the perpendiculars from a point on three fixed lines, we can transform it by the method of this article, so as to obtain a relation connecting λ, μ, ν, the perpendiculars let fall from three fixed points on any tangent to the reciprocal curve, which may be regarded as a kind of tangential equation* of that curve. Thus the general trilinear equation of a conic becomes, when transformed,

$$a\frac{\lambda^2}{\rho^2} + b\frac{\mu^2}{\rho'^2} + c\frac{\nu^2}{\rho''^2} + 2f\frac{\mu\nu}{\rho'\rho''} + 2g\frac{\nu\lambda}{\rho''\rho} + 2h\frac{\lambda\mu}{\rho\rho'} = 0,$$

where ρ, ρ', ρ'' are the distances of the origin from the vertices of the new triangle of reference. Or, conversely, if we are given any relation of the second degree $A\lambda^2 + \&c. = 0$, con-

* See Appendix on Tangential Equations.

necting the three perpendiculars λ, μ, ν, the trilinear equation of the reciprocal curve is

$$A\frac{\alpha^2}{\alpha'^2}+B\frac{\beta^2}{\beta'^2}+C\frac{\gamma^2}{\gamma'^2}+2F\frac{\beta\gamma}{\beta'\gamma'}+2G\frac{\gamma\alpha}{\gamma'\alpha'}+2H\frac{\alpha\beta}{\alpha'\beta'}=0,$$

where α', β', γ' are the trilinear coordinates of the origin.

Ex. 1. Given the focus and a triangle circumscribing a conic, the perpendiculars let fall from its vertices on any tangent to the conic are connected by the relation

$$\sin\theta\,\frac{\rho}{\lambda}+\sin\theta'\,\frac{\rho'}{\mu}+\sin\theta''\,\frac{\rho''}{\nu}=0,$$

where θ, θ', θ'' are the angles the sides of the triangle subtend at the focus. This is obtained by forming the reciprocal of the trilinear equation of the circle circumscribing a triangle. If the centre of the inscribed circle be taken as focus, we have $\theta=90^\circ+\frac{1}{2}A$, $\rho\sin\frac{1}{2}A=r$, whence the tangential equation, on this system, of the inscribed circle is

$$\mu\nu\cot\tfrac{1}{2}A+\nu\lambda\cot\tfrac{1}{2}B+\lambda\mu\cot\tfrac{1}{2}C=0.$$

In the case of any of the exscribed circles two of the cotangents are replaced by tangents.

Ex. 2. Given the focus and a triangle inscribed in a conic, the perpendiculars let fall from its vertices on any tangent are connected by the relation

$$\sin\tfrac{1}{2}\theta\sqrt{\left(\frac{\lambda}{\rho}\right)}+\sin\tfrac{1}{2}\theta'\sqrt{\left(\frac{\mu}{\rho'}\right)}+\sin\tfrac{1}{2}\theta''\sqrt{\left(\frac{\nu}{\rho''}\right)}=0.$$

The tangential equation of the circumscribing circle takes the form

$$\sin A\sqrt{(\lambda)}+\sin B\sqrt{(\mu)}+\sin C\sqrt{(\nu)}=0.$$

Ex. 3. Given focus and three tangents the trilinear equation of the conic is

$$\sin\theta\sqrt{\left(\frac{\alpha}{\alpha'}\right)}+\sin\theta'\sqrt{\left(\frac{\beta}{\beta'}\right)}+\sin\theta''\sqrt{\left(\frac{\gamma}{\gamma'}\right)}=0.$$

This is obtained by reciprocating the equation of the circumscribing circle last found.

Ex. 4. In like manner, from Ex. 1, we find that given focus and three points the trilinear equation is

$$\tan\tfrac{1}{2}\theta\,\frac{\alpha'}{\alpha}+\tan\tfrac{1}{2}\theta'\,\frac{\beta'}{\beta}+\tan\tfrac{1}{2}\theta''\,\frac{\gamma'}{\gamma}=0.$$

312. Very many theorems concerning magnitude may be reduced to theorems concerning lines cut harmonically or anharmonically, and are transformed by the following principle: *To any four points on a right line correspond four lines passing through a point, and the anharmonic ratio of this pencil is the same as that of the four points.*

This is evident, since each leg of the pencil drawn from the origin to the given points is perpendicular to one of the corresponding lines. We may thus derive the anharmonic properties of conics in general from those of the circle.

The anharmonic ratio of the pencil joining four points on a conic to a variable fifth is constant.

The anharmonic ratio of the point in which four fixed tangents to a conic cut any fifth variable tangent is constant.

The first of these theorems is true for the circle, since all the angles of the pencil are constant, therefore the second is true for all conics. The second theorem is true for the circle, since the angles which the four points subtend at the centre are constant, therefore the first theorem is true for all conics. By observing the angles which correspond in the reciprocal figure to the angles which are constant in the case of the circle, the student will perceive that the angles which the four points of the variable tangent subtend at either focus are constant, and that the angles are constant which are subtended at the focus by the four points in which any inscribed pencil meets the directrix.

313. The anharmonic ratio of a line is not the only relation concerning the magnitude of lines which can be expressed in terms of the angles subtended by the lines at a fixed point. For, if there be any relation which, by substituting (as in Art. 56) for each line AB involved in it, $\frac{OA.OB.\sin AOB}{OP}$, can be reduced to a relation between the sines of angles subtended at a given point O, this relation will be equally true for any transversal cutting the lines joining O to the points A, B, &c.; and by taking the given point for origin a reciprocal theorem can be easily obtained. For example, the following theorem, due to Carnot, is an immediate consequence of Art. 148: "If any conic meet the side AB of any triangle in the points c, c'; BC in a, a'; AC in b, b'; then the ratio

$$\frac{Ac.Ac'.Ba.Ba'.Cb.Cb'}{Ab.Ab'.Bc.Bc'.Ca.Ca'} = 1."$$

Now, it will be seen that this ratio is such that we may substitute for each line Ac the sine of the angle AOc, which it subtends at any fixed point; and if we take the reciprocal of this theorem, we obtain the theorem given already Art. 295.

314. Having shown how to form the reciprocals of particular theorems, we shall add some general considerations respecting reciprocal conics.

We proved (Art. 308) that the reciprocal of a circle is an ellipse, hyperbola, or parabola, according as the origin is within, without, or on the curve; we shall now extend this conclusion to all the conic sections. It is evident that, the nearer any line or point is to the origin, the farther the *corresponding* point or line will be; that if any line passes through the origin, the corresponding point must be at an infinite distance; and that the line corresponding to the origin itself must be altogether at an infinite distance. To two tangents, therefore, through the origin on one figure, will correspond two points at an infinite distance on the other; hence, if two *real* tangents can be drawn from the origin, the reciprocal curve will have two *real* points at infinity, that is, it will be a hyperbola; if the tangents drawn from the origin be imaginary, the reciprocal curve will be an ellipse; if the origin be on the curve, the tangents from it coincide, therefore the points at infinity on the reciprocal curve coincide, that is, the reciprocal curve will be a parabola. Since the line at infinity corresponds to the origin, we see that, if the origin be a point on one curve, the line at infinity will be a tangent to the reciprocal curve; and we are again led to the theorem (Art. 254) that *every parabola has one tangent situated at an infinite distance.*

315. To the points of contact of two tangents through the origin must correspond the tangents at the two points at infinity on the reciprocal curve, that is to say, the asymptotes of the reciprocal curve. The eccentricity of the reciprocal hyperbola depending solely on the angle between its asymptotes, depends therefore on the angle between the tangents drawn from the origin to the original curve.

Again, the intersection of the asymptotes of the reciprocal curve (*i.e.* its centre) corresponds to the chord of contact of tangents from the origin to the original curve. We met with a particular case of this theorem when we proved that to the centre of a circle corresponds the directrix of the reciprocal conic, for the directrix is the polar of the origin which is the focus of that conic.

Ex. 1. The reciprocal of a parabola with regard to a point on the directrix is an equilateral hyperbola. (See Art. 221).

Ex. 2. Prove that the following theorems are reciprocal:

The intersection of perpendiculars of a triangle circumscribing a parabola is a point on the directrix.

The intersection of perpendiculars of a triangle inscribed in an equilateral hyperbola lies on the curve.

Ex. 3. Derive the last from Pascal's theorem. (See Ex. 3, p. 247).

Ex. 4. The axes of the reciprocal curve are parallel to the tangent and normal of a conic drawn through the origin confocal with the given one. For the axes of the reciprocal curve must be parallel to the internal and external bisectors of the angle between the tangents drawn from the origin to the given curve. The theorem stated follows by Art. 189.

316. Given two circles, we can find an origin such that the reciprocals of both shall be *confocal* conics. For, since the reciprocals of all circles must have one focus (the origin) common; in order that the other focus should be common, it is only necessary that the two reciprocal curves should have the same centre, that is, that the polar of the origin with regard to both circles should be the same, or that the origin should be one of the two points determined in Art. 111. Hence, given a *system* of circles, as in Art. 109, their reciprocals with regard to one of these limiting points will be a *system* of confocal conics.

The reciprocals of any two conics will, in like manner, be concentric if taken with regard to any of the three points (Art. 282) whose polars with regard to the curves are the same.

Confocal conics cut at right angles (Art. 188).	The common tangent to two circles subtends a right angle at either limiting point.
The tangents from any point to two confocal conics are equally inclined to each other. (Art. 189).	If any line intersect two circles, its two intercepts between the circles subtend equal angles at either limiting point.
The locus of the pole of a fixed line with regard to a series of confocal conics is a line perpendicular to the fixed line. (Art. 226, Ex. 3).	The polar of a fixed point, with regard to a series of circles having the same radical axis, passes through a fixed point; and the two points subtend a right angle at either limiting point.

317. We may mention here that the method of reciprocal polars affords a simple solution of the problem, "to describe a circle touching three given circles." The locus of the centre of a circle touching *two* of the given circles (1), (2), is evidently a hyperbola, of which the centres of the given circles are the foci, since the problem is at once reduced to—"Given base and difference of sides of a triangle." Hence (Art. 308) the polar of the centre with regard to either of the given circles (1) will always touch a circle which can be easily constructed. In like manner, the polar of the centre of any circle touching (1) and (3) must also touch a given circle. Therefore, if we draw a common tangent to the two circles thus determined, and take the pole

of this line with respect to (1), we have the centre of the circle touching the three given circles.

318. *To find the equation of the reciprocal of a conic with regard to its centre.*

We found, in Art. 178, that the perpendicular on the tangent could be expressed in terms of the angles it makes with the axes,

$$p^2 = a^2 \cos^2\theta + b^2 \sin^2\theta.$$

Hence the polar equation of the reciprocal curve is

$$\frac{k^4}{\rho^2} = a^2 \cos^2\theta + b^2 \sin^2\theta,$$

or

$$\frac{a^2x^2}{k^4} + \frac{b^2y^2}{k^4} = 1,$$

a concentric conic, whose axes are the reciprocals of the axes of the given conic.

319. *To find the equation of the reciprocal of a conic with regard to any point $(x'y')$.*

The length of the perpendicular from any point is (Art. 178)

$$p = \frac{k^2}{\rho} = \sqrt{(a^2 \cos^2\theta + b^2 \sin^2\theta)} - x' \cos\theta - y' \sin\theta\,;$$

therefore the equation of the reciprocal curve is

$$(xx' + yy' + k^2)^2 = a^2x^2 + b^2y^2.$$

320. *Given the reciprocal of a curve with regard to the origin of coordinates, to find the equation of its reciprocal with regard to any point $(x'y')$.*

If the perpendicular from the origin on the tangent be P, the perpendicular from any other point is (Art. 34)

$$P - x' \cos\theta - y' \sin\theta,$$

and therefore the polar equation of the locus is

$$\frac{k^2}{\rho} = \frac{k^2}{R} - x' \cos\theta - y' \sin\theta\,;$$

hence $\dfrac{k^2}{R} = \dfrac{x'x + y'y + k^2}{\rho}$ and $\dfrac{R\cos\theta}{k^2} = \dfrac{\rho\cos\theta}{xx' + yy' + k^2}$;

we must therefore substitute, in the equation of the given reciprocal, $\dfrac{k^2x}{xx' + yy' + k^2}$ for x, and $\dfrac{k^2y}{xx' + yy' + k^2}$ for y.

The effect of this substitution may be very simply written as follows: Let the equation of the reciprocal with regard to the origin be

$$u_n + u_{n-1} + u_{n-2} + \&c. = 0,$$

where u_n denotes the terms of the nth degree, &c., then the reciprocal with regard to any point is

$$u_n + u_{n-1}\left(\frac{xx' + yy' + k^2}{k^2}\right) + u_{n-2}\left(\frac{xx' + yy' + k^2}{k^2}\right)^2 + \&c. = 0,$$

a curve of the same degree as the given reciprocal.

321. *To find the reciprocal with respect to $x^2 + y^2 - k^2$ of the conic given by the general equation.*

We find the locus of a point whose polar $xx' + yy' - k^2$ shall touch the given conic by writing x', y', $-k^2$ for λ, μ, ν in the tangential equation (Art. 151). The reciprocal is therefore

$$Ax^2 + 2Hxy + By^2 - 2Gk^2x - 2Fk^2y + Ck^4 = 0.$$

Thus, if the curve be a parabola, C or $ab - h^2 = 0$, and the reciprocal passes through the origin. We can, in like manner, verify by this equation other properties proved already geometrically. If we had, for symmetry, written $k^2 = -z^2$, and looked for the reciprocal with regard to the curve $x^2 + y^2 + z^2 = 0$, the polar would have been $xx' + yy' + zz'$, and the equation of the reciprocal would have been got by writing x, y, z for λ, μ, ν in the tangential equation. In like manner, the condition that $\lambda x + \mu y + \nu z$ may touch any curve, may be considered as the equation of its reciprocal with regard to $x^2 + y^2 + z^2$.

A tangential equation of the nth degree always represents a curve of the nth class; since if we suppose $\lambda x + \mu y + \nu z$ to pass through a fixed point, and therefore have $\lambda x' + \mu y' + \nu z' = 0$; eliminating ν between this equation and the given tangential equation, we have an equation of the nth degree to determine $\lambda : \mu$; and therefore n tangents can be drawn through the given point.

322. Before quitting the subject of reciprocal polars, we wish to mention a class of theorems, for the transformation of which M. Chasles has proposed to take as the auxiliary conic a *parabola* instead of a *circle*. We proved (Art. 211) that the intercept made on the axis of the parabola between any two

lines is equal to the intercept between perpendiculars let fall on the axis from the poles of these lines. This principle then enables us readily to transform theorems which relate to the magnitude of lines measured parallel to a fixed line. We shall give one or two specimens of the use of this method, premising that to two tangents parallel to the axis of the auxiliary parabola correspond the two points at infinity on the reciprocal curve, and that consequently the curve will be a hyperbola or ellipse, according as these tangents are real or imaginary. The reciprocal will be a parabola if the axis pass through a point at infinity on the original curve.

"Any variable tangent to a conic intercepts on two parallel tangents, portions whose rectangle is constant."

To the two points of contact of parallel tangents answer the asymptotes of the reciprocal hyperbola, and to the intersections of those parallel tangents with any other tangent answer parallels to the asymptotes through any point; and we obtain, in the first instance, that the asymptotes and parallels to them through any point on the curve intercept on any fixed line portions whose rectangle is constant. But this is plainly equivalent to the theorem: "The rectangle under parallels drawn to the asymptotes from any point on the curve is constant."

Chords drawn from two fixed points of a hyperbola to a variable third point intercept a constant length on the asymptote. (Art. 199, Ex. 1).	If any tangent to a parabola meet two fixed tangents, perpendiculars from its extremities on the tangent at the vertex will intercept a constant length on that line.

This method of parabolic polars is plainly very limited in its application.

CHAPTER XVI.

HARMONIC AND ANHARMONIC PROPERTIES OF CONICS.*

323. THE harmonic and anharmonic properties of conic sections admit of so many applications in the theory of these curves, that we think it not unprofitable to spend a little time in pointing out to the student the number of particular theorems either directly included in the general enunciations of these properties, or which may be inferred from them without much difficulty.

The cases which we shall most frequently consider are when one of the four points of the right line, whose anharmonic ratio we are examining, is at an infinite distance. The anharmonic ratio of four points, A, B, C, D, being in general (Art. 56) $= \frac{AB}{BC} \div \frac{AD}{DC}$ reduces to the simple ratio $-\frac{AB}{BC}$ when D is at an infinite distance, since then AD ultimately $= -DC$. If the line be cut harmonically, its anharmonic ratio $= -1$; and if D be at an infinite distance $AB = BC$, and AC is bisected. The reader is supposed to be acquainted with the geometric investigation of these and the other fundamental theorems connected with anharmonic section.

324. We commence with the theorem (Art. 146): "If any line through a point O meet a conic in the points R', R'', and the polar of O in R, the line $OR'RR''$ is cut harmonically."

First. Let R'' be at an infinite distance; then the line OR must be bisected at R'; that is, *if through a fixed point a line be drawn parallel to an asymptote of an hyperbola, or to a diameter of a parabola, the portion of this line between the fixed point and its polar will be bisected by the curve* (Art. 211).

* The fundamental property of anharmonic pencils was given by Pappus, *Math. Coll.* VII. 129. The name "anharmonic" was given by Chasles in his *History of Geometry*, from the notes to which the following pages have been developed. Further details will be found in his *Traité de Géométrie Supérieure;* and in his recently published *Treatise on Conics*. The anharmonic relation, however, had been studied by Möbius in his *Barycentric Calculus*, 1827, under the name of "Doppelschnitts-verhältniss." Later writers use the name "Doppelverhältniss."

Secondly. Let R be at an infinite distance, and $R'R''$ must be bisected at O; that is, *if through any point a chord be drawn parallel to the polar of that point, it will be bisected at the point.*

If the polar of O be at infinity, every chord through that point meets the polar at infinity, and is therefore bisected at O. Hence this point is the centre, or *the centre may be considered as a point whose polar is at infinity* (Art. 154).

Thirdly. Let the fixed point itself be at an infinite distance, then all the lines through it will be parallel, and will be bisected on the polar of the fixed point. Hence *every diameter of a conic may be considered as the polar of the point at infinity in which its ordinates are supposed to intersect.*

This also follows from the equation of the polar of a point (Art. 145)

$$(ax + hy + g) + (hx + by + f)\frac{y'}{x'} + \frac{gx + fy + c}{x'} = 0.$$

Now, if $x'y'$ be a point at infinity on the line $my = nx$, we must make $\frac{y'}{x'} = \frac{n}{m}$, and x' infinite, and the equation of the polar becomes
$$m(ax + hy + g) + n(hx + by + f) = 0,$$
a diameter conjugate to $my = nx$ (Art. 141).

325. Again, it was proved (Art. 146) that the two tangents through any point, any other line through the point, and the line to the pole of this last line, form a harmonic pencil.

If now one of the lines through the point be a diameter, the other will be parallel to its conjugate, and since the polar of any point on a diameter is parallel to its conjugate, we learn that the portion between the tangents of any line drawn parallel to the polar of the point is bisected by the diameter through it.

Again, let the point be the centre, the two tangents will be the asymptotes. Hence *the asymptotes, together with any pair of conjugate diameters, form a harmonic pencil,* and the portion of any tangent intercepted between the asymptotes is bisected by the curve (Art. 196).

326. The anharmonic property of the points of a conic (Art. 259) gives rise to a much greater variety of particular theorems. For, the four points on the curve may be any whatever, and

either one or two of them may be at an infinite distance; the fifth point O, to which the pencil is drawn, may be also either at an infinite distance, or may coincide with one of the four points, in which latter case one of the legs of the pencil will be the tangent at that point; then, again, we may measure the anharmonic ratio of the pencil by the segments on *any* line drawn across it, which we may, if we please, draw parallel to one of the legs of the pencil, so as to reduce the anharmonic ratio to a simple ratio.

The following examples being intended as a practical exercise to the student in developing the consequences of this theorem, we shall merely state the points whence the pencil is drawn, the line on which the ratio is measured, and the resulting theorem, recommending to the reader a closer examination of the manner in which each theorem is inferred from the general principle.

We use the abbreviation $\{O.ABCD\}$ to denote the anharmonic ratio of the pencil OA, OB, OC, OD.

Ex. 1. $\{A.ABCD\} = \{B.ABCD\}$.

Let these ratios be estimated by the segments on the line CD; let the tangents at A, B meet CD in the points T, T', and let the chord AB meet CD in K, then the ratios are

$$\frac{TK.DC}{TD.KC} = \frac{KT'.DC}{KD.T'C},$$

that is, if any chord CD meet two tangents in T, T', and their chord of contact in K,

$$KC.KT'.TD = KD.TK.T'C.$$

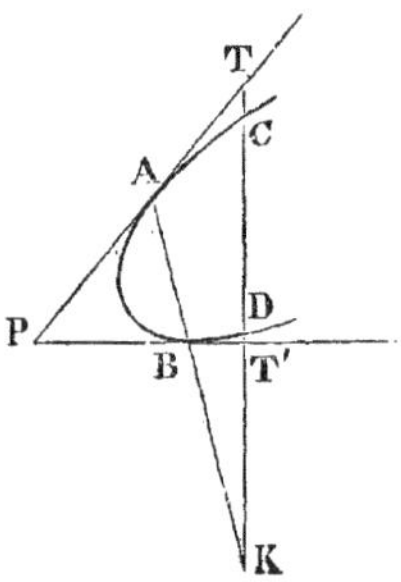

(The reader must be careful, in this and the following examples, to take the points of the pencil *in the same order* on both sides of the equation. Thus, on the left-hand side of this equation we took K second, because it answers to the leg OB of the pencil; on the right hand we take K first, because it answers to the leg OA).

Ex. 2. Let T and T' coincide, then

$$KC.TD = -KD.TC,$$

or, any chord through the intersection of two tangents is cut harmonically by the chord of contact.

Ex. 3. Let T' be at an infinite distance, or the secant CD drawn parallel to PT', and it will be found that the ratio will reduce to

$$TK^2 = TC.TD.$$

Ex. 4. Let one of the points be at an infinite distance, then $\{O.ABC\infty\}$ is constant. Let this ratio be estimated on the line $C\infty$. Let the lines AO, BO cut $C\infty$ in a, b; then the ratio of the pencil will reduce to $\frac{Ca}{Cb}$; and we learn, that if two fixed points, A, B, on a hyperbola or parabola, be joined to any variable point O, and the joining lines meet a fixed parallel to an asymptote (if the curve be a hyperbola), or a diameter (if the curve be a parabola), in a, b, then the ratio $Ca : Cb$ will be constant.

Ex. 5. If the same ratio be estimated on any other parallel line, lines inflected from any three fixed points to a variable point, on a hyperbola or parabola, cut a fixed parallel to an asymptote or diameter, so that $ab : ac$ is constant.

Ex. 6. It follows from Ex. 4, that if the lines joining A, B to any fourth point O' meet $C\infty$ in a', b', we must have

$$\frac{ab}{a'b'} = \frac{aC}{a'C}.$$

Now let us suppose the point C to be also at an infinite distance, the line $C\infty$ becomes an asymptote, the ratio $ab : a'b'$ becomes one of equality, and lines joining two fixed points to any variable point on the hyperbola intercept on either asymptote a constant portion (Art. 199, Ex. 1).

Ex. 7. $$\{A.ABC\infty\} = \{B.ABC\infty\}.$$

Let these ratios be estimated on $C\infty$; then if the tangents at A, B, cut $C\infty$ in a, b, and the chord of contact AB in K, we have

$$\frac{Ca}{CK} = \frac{CK}{Cb}$$

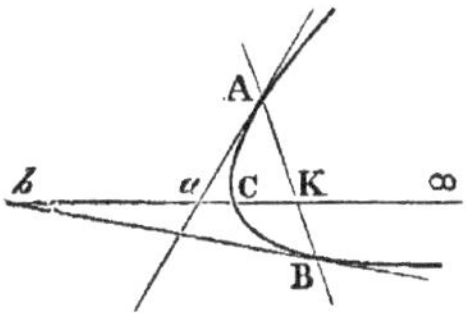

(observing the caution in Ex. 1). Or, if any paralle to an asymptote of a hyperbola, or a diameter of a parabola, cut two tangents and their chord of contact, the intercept from the curve to the chord is a geometric mean between the intercepts from the curve to the tangents. Or, conversely, if a line ab, parallel to a given one, meet the sides of a triangle in the points a, b, K, and there be taken on it a point C such that $CK^2 = Ca.Cb$, the locus of C will be a parabola, if Cb be parallel to the bisector of the base of the triangle (Art. 211), but otherwise a hyperbola, to an asymptote of which ab is parallel.

Ex. 8. Let two of the fixed points be at infinity,

$$\{\infty.AB\infty\infty'\} = \{\infty'.AB\infty\infty'\};$$

the lines $\infty\infty$, $\infty'\infty'$, are the two asymptotes, while $\infty\infty'$ is altogether at infinity. Let these ratios be estimated on the diameter OA; let this line meet the parallels to the asymptotes $B\infty$, $B\infty'$, in a and a'; then the ratios become $\frac{OA}{Oa} = \frac{Oa'}{OA}$. Or, parallels to the asymptotes through any point on a hyperbola cut any semi-diameter, so that it is a mean proportional between the segments on it from the centre.

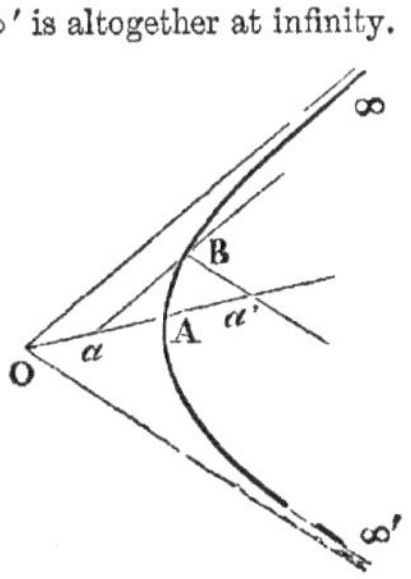

Hence, conversely, if through a fixed point O a line be drawn cutting two fixed lines, Ba, Ba', and a point A taken on it so that OA is a mean between Oa, Oa', the locus of A is a hyperbola, of which O is the centre, and Ba, Ba', parallel to the asymptotes.

Ex. 9. $$\{\infty.AB\infty\infty'\} = \{\infty'.AB\infty\infty'\}.$$

Let the segments be measured on the asymptotes, and we have $\frac{Oa}{Ob} = \frac{Ob'}{Oa'}$, ($O$ being the centre), or the rectangle under parallels to the asymptotes through any point on the curve is constant (we invert the second ratio for the reason given in Ex. 1).

327. We next examine some particular cases of the anharmonic property of the tangents to a conic (Art. 275).

Ex. 1. This property assumes a very simple form, if the curve be a parabola, for one tangent to a parabola is always at an infinite distance (Art. 254). Hence three fixed tangents to a parabola cut any fourth in the points A, B, C, so that $AB : AC$ is always constant. If the variable tangents coincide in turn with each of the given tangents, we obtain the theorem,

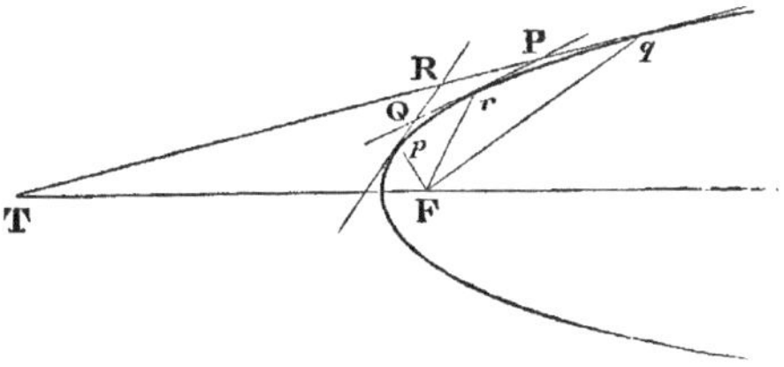

$$\frac{pQ}{QR} = \frac{RP}{Pq} = \frac{Qr}{rP}.$$

Ex. 2. Let two of the four tangents to an ellipse or hyperbola be parallel to each other, and let the variable tangent coincide alternately with each of the parallel tangents. In the first case the ratio is

$$\frac{Ab}{Ac}, \text{ and in the second } \frac{Dc'}{Db'}.$$

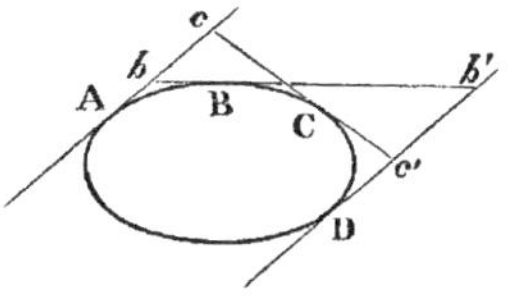

Hence the rectangle $Ab \,.\, Db'$ is constant.

It may be deduced from the anharmonic property of the points of a conic, that if the lines joining any point on the curve O to A, D, meet the parallel tangents in the points b, b', then the rectangle $Ab \,.\, Db'$ will be constant.

328. We now proceed to give some examples of problems easily solved by the help of the anharmonic properties of conics.

Ex. 1. To prove MacLaurin's method of generating conic sections (p. 248), viz.—To find the locus of the vertex V of a triangle whose sides pass through the points A, B, C, and whose base angles move on the fixed lines Oa, Ob.

Let us suppose four such triangles drawn, then since the pencil $\{C \,.\, aa'a''a'''\}$ is the same pencil as $\{C \,.\, bb'b''b'''\}$, we have

$$\{aa'a''a'''\} = \{bb'b''b'''\},$$

and, therefore,

$$\{A \,.\, aa'a''a'''\} = \{B \,.\, bb'b''b'''\};$$

or, from the nature of the question,

$$\{A \,.\, VV'V''V'''\} = \{B \,.\, VV'V''V'''\};$$

and therefore A, B, V, V', V'', V''' lie on the same conic section. Now if the first three triangles be fixed, it is evident that the locus of V''' is the conic section passing through $ABVV'V''$.

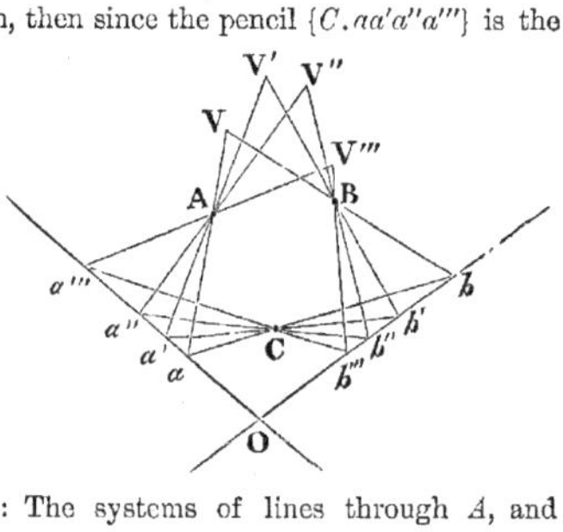

Or the reasoning may be stated thus: The systems of lines through A, and through B, being both homographic with the system through C, are homographic with each other; and therefore (Art. 297) the locus of the intersection of corresponding lines is a conic through A and B. The following examples are, in like manner, illustrations of the application of this principle of Art. 297.

Ex. 2. M. Chasles has showed that the same demonstration will hold if the side ab, instead of passing through the fixed point C, touch any conic which touches Oa, Ob; for then any four positions of the base cut Oa, Ob, so that

$$\{aa'a''a'''\} = \{bb'b''b'''\} \text{ (Art. 275)},$$

and the rest of the proof proceeds the same as before.

Ex. 3. Newton's method of generating conic sections:—Two angles of constant magnitude move about fixed points P, Q; the intersection of two of their sides traverses the right line AA'; then the locus of V, the intersection of their other two sides, will be a conic passing throug P, Q.

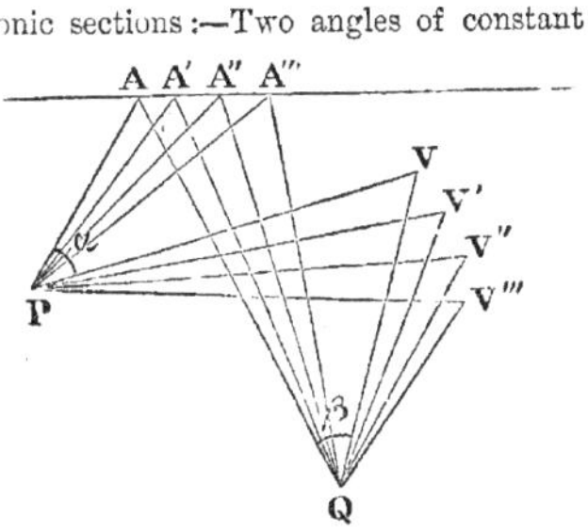

For, as before, take four positions of the angles, then

$$\{P.AA'A''A'''\} = \{Q.AA'A''A'''\};$$

but $\{P.AA'A''A'''\} = \{P.VV'V''V'''\}$,

$$\{Q.AA'A''A'''\} = \{Q.VV'V''V'''\},$$

since the angles of the pencils are the same; therefore

$$\{P.VV'V''V'''\} = \{Q.VV'V''V'''\};$$

and, therefore, as before, the locus of V''' is a conic through P, Q, V, V', V''.

Ex 4. M. Chasles has extended this method of generating conic sections, by supposing the point A, instead of moving on a right line, to move on any conic passing through the points P, Q; for we shall still have

$$\{P.AA'A''A'''\} = \{Q.AA'A''A'''\}.$$

Ex. 5. The demonstration would be the same if, in place of the angles APV, AQV being constant, APV and AQV cut off constant intercepts each on one of two fixed lines, for we should then prove the pencil

$$\{P.AA'A''A'''\} = \{P.VV'V''V'''\},$$

because both pencils cut off intercepts of the same length on a fixed line.

Thus, also, given base of a triangle and the intercept made by the sides on any fixed line, we can prove that the locus of vertex is a conic section.

Ex. 6. We may also extend Ex. 1, by supposing the extremities of the line ab to move on any conic section passing through the points AB, for, taking four positions of the triangle, we have, by Art. 276,

$$\{aa'a''a'''\} = \{bb'b''b'''\};$$

therefore, $$\{A.aa'a''a'''\} = \{B.bb'b''b'''\},$$

and the rest of the proof proceeds as before.

Ex. 7. The base of a triangle passes through C, the intersection of common tangents to two conic sections; the extremities of the base ab lie one on each of the conic sections, while the sides pass through fixed points A, B, one on each of the conics; the locus of the vertex is a conic through A, B.

The proof proceeds exactly as before, depending now on the second theorem proved, Art. 276. We may mention that this theorem of Art. 276 admits of a simple geometrical proof. Let the pencil $\{O.ABCD\}$ be drawn from points corresponding to $\{o.abcd\}$. Now, the lines OA, oa, intersect at r on one of the common chords of the conics; in like manner, BO, bo intersect in r' on the same chord, &c.; hence $\{rr'r''r'''\}$ measures the anharmonic ratio of both these pencils.

Ex. 8. In Ex. 6 the base instead of passing through a fixed point C, may be supposed to touch a conic having double contact with the given conic (see Art. 276).

Ex. 9. If a polygon be inscribed in a conic, all whose sides but one pass through fixed points, the envelope of that side will be a conic having double contact with the given one.

For, take any four positions of the polygon, then if a, b, c, &c. be the vertices of the polygon, we have

$$\{aa'a''a'''\} = \{bb'b''b'''\} = \{cc'c''c'''\}, \text{ \&c.}$$

The problem is, therefore, reduced to that of Art. 277,—"Given three pairs of points, $aa'a''$, $dd'd''$, to find the envelope of $a'''d'''$, such that

$$\{aa'a''a'''\} = \{dd'd''d'''\}."$$

Ex. 10. To inscribe in a conic section a polygon, all whose sides shall pass through fixed points.

If we assume any point (a) at random on the conic for the vertex of the polygon, and form a polygon whose sides pass through the given points, the point z, where the last side meets the conic, will not in general coincide with a. If we make four such attempts to inscribe the polygon, we must have, as in the last example,

$$\{aa'a''a'''\} = \{zz'z''z'''\}.$$

Now, if the last attempt were successful, the point a''' would coincide with z''', and the problem is reduced to—"Given three pairs of points, $aa'a''$, $zz'z''$, to find a point K such that

$$\{Kaa'a''\} = \{Kzz'z''\}."$$

Now if we make $az''a'za''z'$ the vertices of an inscribed hexagon (in the order here given, taking an a and z alternately, and so that az, $a'z'$, $a''z''$, may be opposite vertices), then either of the points in which the line joining the intersections of opposite sides meets the conic may be taken for the point K. For, in the figure, the points ACE are $aa'a''$, DFB are $zz'z''$; and if we take the sides in the order $ABCDEF$, L, M, N are the intersections of opposite sides. Now, since $\{KPNL\}$ measures both $\{D.KACE\}$ and $\{A.KDFB\}$, we have

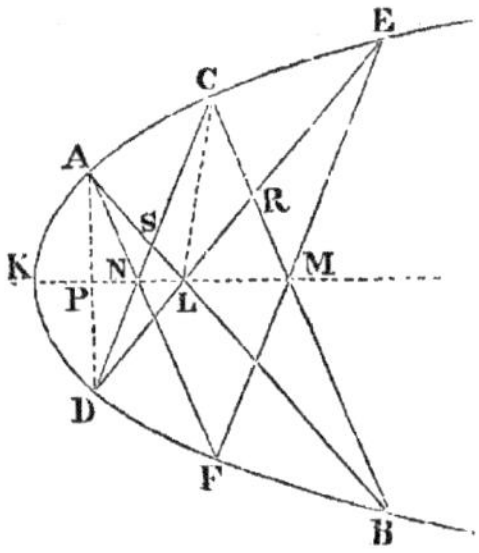

$$\{KACE\} = \{KDFB\}. \quad \text{Q.E.D.*}$$

It is easy to see, from the last example, that K is a point of contact of a conic having double contact with the given conic, to which az, $a'z'$, $a''z''$ are tangents, and that we have therefore just given the solution of the question. "To describe a conic touching three given lines, and having double contact with a given conic."

Ex. 11. The anharmonic property affords also a simple proof of Pascal's theorem, alluded to in the last example.

We have $\{E.CDFB\} = \{A.CDFB\}$. Now, if we examine the segments made by the first pencil on BC, and by the second on DC, we have

$$\{CRMB\} = \{CDNS\}.$$

* This construction for inscribing a polygon in a conic is due to M. Poncelet (*Traité des Propriétés Projectives*, p. 351). The demonstration here used is Mr. Townsend's. It shows that Poncelet's construction will equally solve the problem, "To inscribe a polygon in a conic, each of whose sides shall touch a conic having double contact with the given conic." The conics touched by the sides may be all different.

Now, if we draw lines from the point L to each of these points, we form two pencils which have the three legs, CL, DE, AB, common, therefore the fourth legs NL, LM, must form one right line. In like manner, Brianchon's theorem is derived from the anharmonic property of the tangents.

Ex. 12. Given four points on a conic, $ADFB$, and two fixed lines through any one of them, DC, DE, to find the envelope of the line CE joining the points where those fixed lines again meet the curve.

The vertices of the triangle CEM move on the fixed lines DC, DE, NL, and two of its sides pass through the fixed points, B, F; therefore, the third side envelopes a conic section touching DC, DE (by the reciprocal of MacLaurin's mode of generation).

Ex. 13. Given four points on a conic $ABDE$, and two fixed lines, AF, CD, passing each through a different one of the fixed points, the line CF joining the points where the fixed lines again meet the curve will pass through a fixed point.

For the triangle CFM has two sides passing through the fixed points B, E, and the vertices move on the fixed lines AF, CD, NL, which fixed lines meet in a point, therefore (p. 280) CF passes through a fixed point.

The reader will find in the Chapter on Projection how the last two theorems are suggested by other well-known theorems. (See Ex. 3 and 4, Art. 355).

Ex. 14. The anharmonic ratio of any four diameters of a conic is equal to that of their four conjugates. This is a particular case of Ex. 2, Art. 297, that the anharmonic ratio of four points on a line is the same as that of their four polars. We might also prove it directly, from the consideration that the anharmonic ratio of four chords proceeding from any point of the curve is equal to that of the supplemental chords (Art. 179).

Ex. 15. A conic circumscribes a given quadrangle, to find the locus of its centre. (Ex. 3, Art. 151).

Draw diameters of the conic bisecting the sides of the quadrangle, their anharmonic ratio is equal to that of their four conjugates, but this last ratio is given, since the conjugates are parallel to the four given lines; hence the locus is a conic passing through the middle points of the given sides. If we take the cases where the conic breaks up into two right lines, we see that the intersections of the diagonals, and also those of the opposite sides, are points in the locus, and therefore that these points lie on a conic passing through the middle points of the sides and of the diagonals.

329. We think it unnecessary to go through the theorems, which are only the polar reciprocals of those investigated in the last examples; but we recommend the student to form the polar reciprocal of each of these theorems, and then to prove it directly by the help of the anharmonic property of the *tangents* of a conic. Almost all are embraced in the following theorem:

If there be any number of points a, b, c, d, &c. on a right line, and a homographic system a', b', c', d', &c. on another line, the lines joining corresponding points will envelope a conic. For if we construct the conic touched by the two given lines and by three lines aa', bb', cc', then, by the anharmonic property of the tangents of a conic, any other of the lines dd' must touch the same conic.* The theorem here proved is the reciprocal of that proved Art. 297, and may also be established by interpreting tangentially the equations there used. Thus, if P, P'; Q, Q' represent tangentially two pairs of corresponding points, $P+\lambda P'$, $Q+\lambda Q'$ represent any other pair of corresponding points; and the line joining them touches the curve represented by the tangential equation of the second order, $PQ' = P'Q$.

Ex. Any transversal through a fixed point P meets two fixed lines OA, OA', in the points AA'; and portions of given length Aa, $A'a'$ are taken on each of the given lines; to find the envelope of aa'. Here, if we give the transversal four positions, it is evident that $\{ABCD\} = \{A'B'C'D'\}$, and that $\{ABCD\} = \{abcd\}$, and $\{A'B'C'D'\} = \{a'b'c'd'\}$.

330. Generally when the envelope of a moveable line is found by this method to be a conic section, it is useful to take notice whether in any particular position the moveable line can be altogether at an infinite distance, for if it can, the envelope is a parabola (Art. 254). Thus, in the last example the line aa' cannot be at an infinite distance, unless in some position AA' can be at an infinite distance, that is, unless P is at an infinite distance. Hence we see that in the last example, if the transversal, instead of passing through a fixed point, were parallel to a given line, the envelope would be a parabola. In like manner, the nature of the locus of a moveable point is often at once perceived by observing particular positions of the moveable point, as we have illustrated in the last example of Art. 328.

331. If we are given any system of points on a right line we can form a homographic system on another line, and such that three points taken arbitrarily a', b', c' shall correspond to three given points a, b, c of the first line. For let the distances of the given points on the first line measured from any fixed

* In the same case if P, P' be two fixed points, it follows from the last article that the locus of the intersection of Pd, $P'd'$ is a conic through P, P'. We saw (Art. 277) that if a, b, c, d, &c., a', b', c', d' be two homographic systems of points *on a conic*, that is to say, such that $\{abcd\}$ always $= \{a'b'c'd'\}$, the envelope of dd' is a conic having double contact with the given one. In the same case, if P, P' be fixed points *on the conic*, the locus of the intersection of Pd, $P'd'$ is a conic through P, P'. Again, two conics are cut by the tangents of any conic having double contact with both, in homographic systems of points, or such that $\{abcd\} = \{a'b'c'd'\}$ (Art. 276); but it is not true conversely, that if we have two homographic systems of points *on different conics*, the lines joining corresponding points necessarily envelope a conic.

origin on the line be a, b, c, and let the distance of any variable point on the line measured from the same origin be x. Similarly let the distances of the points on the second line from any origin on that line be a', b', c', x', then, as in Art. 277, we have the equation

$$\frac{(a-b)(c-x)}{(a-c)(b-x)} = \frac{(a'-b')(c'-x')}{(a'-c')(b'-x')},$$

which expanded is of the form

$$Axx' + Bx + Cx' + D = 0.*$$

This equation enables us to find a point x' in the second line corresponding to any assumed point x on the first line, and such that $\{abcx\} = \{a'b'c'x'\}$. If this relation be fulfilled, the line joining the points x, x' envelopes a conic touching the two given lines; and this conic will be a parabola if $A = 0$, since then x' is infinite when x is infinite.

The result at which we have arrived may be stated conversely thus: *Two systems of points connected by any relation will be homographic, if to one point of either system always corresponds one, and but one, point of the other.* For evidently an equation of the form

$$Axx' + Bx + Cx' + D = 0$$

is the most general relation between x and x' that we can write down, which gives a simple equation whether we seek to determine x in terms of x' or *vice versa*. And when this relation is fulfilled, the anharmonic ratio of four points of the first system is equal to that of the four corresponding points of the second. For the anharmonic ratio $\dfrac{(x-y)(z-w)}{(x-z)(y-w)}$ is unaltered

---

* M. Chasles states the matter thus: The points x, x' belong to homographic systems, if a, b, a', b' being fixed points, the ratios of the distances $ax : bx$, $a'x' : b'x'$, be connected by a linear relation, such as

$$\lambda \frac{ax}{bx} + \mu \frac{a'x'}{b'x'} + \nu = 0.$$

Denoting, as above, the distances of the points from fixed origins, by a, b, x; a', b', x', this relation is

$$\lambda \frac{a-x}{b-x} + \mu \frac{a'-x'}{b'-x'} + \nu = 0,$$

which, expanded, gives a relation between x and x' of the form

$$Axx' + Bx + Cx' + D = 0.$$

if instead of x we write $-\frac{Bx+D}{Ax+C}$, and make similar substitutions for y, z, w.

332. *The distances from the origin of a pair of points A, B on the axis of x being given by the equation, $ax^2 + 2hx + b = 0$, and those of another pair of points A', B' by $a'x^2 + 2h'x + b' = 0$, to find the condition that the two pairs should be harmonically conjugate.*

Let the distances from the origin of the first pair of points be α, β; and of the second α', β'; then the condition is

$$\frac{AA'}{A'B} = -\frac{AB'}{B'B}; \text{ or } \frac{\alpha-\alpha'}{\alpha'-\beta} = -\frac{\alpha-\beta'}{\beta'-\beta};$$

which expanded may be written

$$(\alpha + \beta)(\alpha' + \beta') = 2\alpha\beta + 2\alpha'\beta'.$$

But $\alpha + \beta = -\frac{2h}{a}$, $\alpha\beta = \frac{b}{a}$; $(\alpha' + \beta') = -\frac{2h'}{a'}$, $\alpha'\beta' = \frac{b'}{a'}$.

The required condition is therefore

$$ab' + a'b - 2hh' = 0.*$$

It is proved, similarly, that the same is the condition that the pairs of lines

$$a\alpha^2 + 2h\alpha\beta + b\beta^2, \; a'\alpha^2 + 2h'\alpha\beta + b'\beta^2,$$

should be harmonically conjugate.

333. If a pair of points $ax^2 + 2hx + b$, be harmonically conjugate with a pair $a'x^2 + 2h'x + b'$, and also with another pair $a''x^2 + 2h''x + b''$, it will be harmonically conjugate with every pair given by the equation

$$(a'x^2 + 2h'x + b') + \lambda(a''x^2 + 2h''x + b'') = 0.$$

For evidently the condition

$$a(b' + \lambda b'') + b(a' + \lambda a'') - 2h(h' + \lambda h'') = 0,$$

will be fulfilled if we have separately

$$ab' + ba' - 2hh' = 0, \quad ab'' + ba'' - 2hh'' = 0.$$

* It can be proved that the anharmonic ratio of the system of four points will be given, if $(ab' + a'b - 2hh')^2$ be in a given ratio to $(ab - h^2)(a'b' - h'^2)$.

334. *To find the locus of a point such that the tangents from it to two given conics may form a harmonic pencil.*

If four lines form a harmonic pencil they will cut any of the lines of reference harmonically. Now take the second form (Art. 294) of the equation of a pair of tangents from a point to a curve given by the general trilinear equation, and make $\gamma = 0$ when we get

$$(C\beta'^2 + B\gamma'^2 - 2F\beta'\gamma')\,\alpha^2 - 2\,(C\alpha'\beta' - F\alpha'\gamma' - G\beta'\gamma' + H\gamma'^2)\,\alpha\beta$$
$$+ (C\alpha'^2 + A\gamma'^2 - 2G\alpha'\gamma')\,\beta^2 = 0.$$

We have a corresponding equation to determine the pair or points where the line γ is met by the pair of tangents from $\alpha'\beta'\gamma'$ to a second conic. Applying then the condition of Art. 332 we find that the two pairs of points on γ will form a harmonic system, provided that $\alpha'\beta'\gamma'$ satisfies the equation

$$(C\beta^2 + B\gamma^2 - 2F\beta\gamma)\,(C'\alpha^2 + A'\gamma^2 - 2G'\alpha\gamma)$$
$$+ (C\alpha^2 + A\gamma^2 - 2G\alpha\gamma)\,(C'\beta^2 + B'\gamma^2 - 2F'\beta\gamma)$$
$$= 2\,(C\alpha\beta - F\alpha\gamma - G\beta\gamma + H\gamma^2)\,(C'\alpha\beta - F'\alpha\gamma - G'\beta\gamma + H'\gamma^2).$$

On expansion the equation is found to be divisible by γ^2, and the equation of the locus is found to be

$$(BC'+B'C-2FF')\alpha^2+(CA'+C'A-2GG')\beta^2+(AB'+A'B-2HH')\gamma^2$$
$$+2(GH'+G'H-AF'-A'F)\,\beta\gamma+2(HF'+H'F-BG'-B'G)\,\gamma\alpha$$
$$+ 2\,(FG'+F'G-CH'-C'H)\,\alpha\beta = 0;$$

a conic having important relations to the two conics, which will be treated of further on. If the anharmonic ratio of the four tangents be given, the locus is the curve of the fourth degree $F^2 = kSS'$, where S, S', F, denote the two given conics, and that now found.

335. *To find the condition that the line* $\lambda\alpha + \mu\beta + \nu\gamma$ *should be cut harmonically by the two conics.* Eliminating γ between this equation and that of the first conic, the points of intersection are found to satisfy the equation

$$(c\lambda^2 + a\nu^2 - 2g\lambda\nu)\,\alpha^2 + 2\,(c\lambda\mu - f\lambda\nu - g\mu\nu + h\nu^2)\,\alpha\beta$$
$$+ (c\mu^2 + b\nu^2 - 2f\mu\nu)\,\beta^2 = 0.$$

We have a similar equation satisfied for the points where the line meets the second conic; applying then the condition of

Art. 332, we find, precisely as in the last article, that the required condition is

$$(bc' + b'c - 2ff')\lambda^2 + (ca' + c'a - 2gg')\mu^2 + (ab' + a'b - 2hh')\nu^2$$
$$+ 2(gh' + g'h - af' - a'f)\mu\nu + 2(hf' + h'f - bg' - b'g)\nu\lambda$$
$$+ 2(fg' + f'g - ch' - c'h)\lambda\mu = 0.$$

The line consequently envelopes a conic.*

INVOLUTION.

336. Two systems of points a, b, c, &c., a', b', c', &c., situated *on the same right line*, will be homographic (Art. 331) if the distances measured from any origin, of two corresponding points, be connected by a relation of the form

$$Axx' + Bx + Cx' + D = 0.$$

Now this equation not being symmetrical between x and x', the point which corresponds to any point of the line considered as belonging to the first system, will in general not be the same as that which corresponds to it considered as belonging to the second system. Thus, to a point at a distance x considered as belonging to the first system, corresponds a point at the distance $-\dfrac{Bx + D}{Ax + C}$; but considered as belonging to the second system, corresponds $-\dfrac{Cx + D}{Ax + B}$.

Two homographic systems situated on the same line are said to form a system *in involution*, when to any point of the line the same point corresponds whether it be considered as belonging to the first or second system. That this should be the case it is evidently necessary and sufficient that we should have $B = C$ in the preceding equation, in order that the relation connecting x and x' may be symmetrical. We shall find it

* If substituting in the equations of two conics U, V, for α, $\lambda\alpha + \mu\alpha'$, &c. we obtain results

$$\lambda^2 U + 2\lambda\mu P + \mu^2 U', \quad \lambda^2 V + 2\lambda\mu Q + \mu^2 V',$$

then it is easy to see, as above, that $UV' + U'V - 2PQ$, represents the pair of lines which can be drawn through $\alpha'\beta'\gamma'$, so as to be cut harmonically by the conics. In the same case (Art. 296), the equation of the system of four lines joining $\alpha'\beta'\gamma'$ to the intersections of the conics, is

$$(UV' + U'V - 2PQ)^2 = 4(UU' - P^2)(VV' - Q^2).$$

$UU' - P^2$ and $VV' - Q^2$ denote the pairs of tangents from $\alpha'\beta'\gamma'$ to the conics.

convenient to write the relation connecting any two corresponding points $$Axx' + H(x + x') + B = 0;$$
and if the distances from the origin of a pair of corresponding points be given by the equation
$$ax^2 + 2hx + b = 0,$$
we must have $$Ab + Ba - 2Hh = 0.$$

337. It appears from what has been said that a system in involution consists of a number of pairs of points on a line a, a'; b, b'; &c., and such that the anharmonic ratio of any four is equal to that of their four conjugates. The expression of this equality gives a number of relations connecting the mutual distances of the points. Thus, from $\{abca'\} = \{a'b'c'a\}$, we have
$$\frac{ab.ca'}{aa'.bc} = \frac{a'b'.c'a}{a'a.b'c'},$$
or $$ab.ca'.b'c' = -a'b'.c'a.bc.$$
The development of such relations presents no difficulty.

338. The relation $Axx' + H(x + x') + B = 0$, connects the distances of two corresponding points *from any origin chosen arbitrarily;* but by a proper choice of origin this relation can be simplified. Thus, if the distances be measured from a point at the distance $x = \alpha$, the given relation becomes
$$A(x + \alpha)(x' + \alpha) + H(x + x' + 2\alpha) + B = 0;$$
or $$Axx' + (H + A\alpha)(x + x') + A\alpha^2 + 2H\alpha + B = 0.$$
And if we determine α, so that $H + A\alpha = 0$, the relation reduces to $xx' =$ constant. The point thus determined is called the centre of the system; and we learn that *the product of the distances from the centre of two corresponding points is constant.*

339. Since, in general, the point corresponding to any point x is $-\frac{Hx + B}{Ax + H}$; when $Ax + H = 0$, the corresponding point is infinitely distant: or *the centre is the point whose conjugate is infinitely distant.* The same thing appears from the relation $\{abcc'\} = \{a'b'c'c\}$, or
$$\frac{ac.bc'}{ac'.bc} = \frac{a'c'.b'c}{a'c.b'c'}.$$

Let c' be infinitely distant, bc' ultimately $= ac'$, and $a'c' = b'c'$, and this relation becomes $ac.a'c = bc.b'c$; or, in other words, the product of the distances from c of two conjugate points is constant. The relation connecting the distances from the centre may be either $ca.ca' = + k^2$ or $ca.ca' = - k^2$. In the one case two conjugate points lie on the same side of the centre; in the other case they lie on opposite sides.

340. A point which coincides with its conjugate is called a *focus* of the system. There are plainly two foci f, f' equidistant from the centre on either side of it, whose common distance from the centre c is given by the equation $cf^2 = \pm k^2$. Thus, when k^2 is taken with a positive sign, that is, when two conjugate points always lie on the same side of the centre, the foci are real. In the opposite case they are imaginary. By writing $x = x'$ in the general relation connecting corresponding points, we see that in general the distances of the foci from any origin are given by the equation

$$Ax^2 + 2Hx + B = 0.$$

341. We have seen (Art. 336) that if a pair of corresponding points be given by the equation $ax^2 + 2hx + b = 0$, we must have $Ab + Ba - 2Hh = 0$. Now this equation signifies (see Art. 332) that *any two corresponding points are harmonically conjugate with the two foci*. The same inference may be drawn from the relation $\{aff'a'\} = \{a'ff'a\}$, which gives

$$\frac{af.a'f'}{aa'.ff'} = \frac{a'f.af'}{a'a.ff'}; \text{ or } \frac{fa}{f'a} = -\frac{fa'}{f'a'};$$

or the distance between the foci ff' is divided internally and externally at a and a' into parts which are in the same ratio.

COR. When one focus is at infinity, the other bisects the distance between two conjugate points; and it follows hence that in this case the distance ab between any two points of the system is equal to $a'b'$, the distance between their conjugates.

342. *Two pairs of points determine a system in involution.* We may take arbitrarily two pairs of points

$$ax^2 + 2hx + b, \; a'x^2 + 2h'x + b',$$

and we can then determine A, H, B from the equations

$$Ab + Ba - 2Hh = 0, \quad Ab' + Ba' - 2Hh' = 0.$$

We see, as in Art. 333, that any other pair of points in involution with the two given pairs may be represented by an equation of the form

$$(ax^2 + 2hx + b) + \lambda (a'x^2 + 2h'x + b') = 0,$$

since, when A, H, B are determined so as to satisfy the two equations written above, they must also satisfy

$$A (b + \lambda b') + B (a + \lambda a') - 2H (h + \lambda h') = 0.*$$

The actual values of A, B, H, found by solving these equations, are $2(ah' - a'h)$, $2(hb' - h'b)$, $ab' - a'b$. Consequently the foci of the system determined by the given pairs of points, are given by the equation

$$(ah' - a'h) x^2 + (ab' - a'b) x + (hb' - h'b) = 0.$$

This may be otherwise written if we make the equations homogeneous by introducing a new variable y, and write

$$U = ax^2 + 2hxy + by^2, \quad V = a'x^2 + 2h'xy + b'y^2.$$

The equation which determines the foci is then

$$\frac{dU}{dx}\frac{dV}{dy} - \frac{dU}{dy}\frac{dV}{dx} = 0.$$

The foci of a system given by two pairs of points a, a'; b, b' may be also found as follows, from the consideration that $\{afba'\} = \{a'fb'a\}$, or

$$\frac{af.ba'}{a'f.ba} = \frac{a'f.b'a}{af.b'a'};$$

whence $\qquad af^2 : a'f^2 :: ab.ab' : a'b.a'b';$

or f is the point where aa' is cut either internally or externally in a certain given ratio.

343. The relation connecting six points in involution is of the class noticed in Art. 313, and is such that the same relations

---

* It easily follows from this, that the condition that three pairs of points $ax^2 + 2hx + b$, $a'x^2 + 2h'x + b'$, $a''x^2 + 2h''x + b''$ should belong to a system in involution, is the vanishing of the determinant

$$\begin{vmatrix} a, & h & b \\ a', & h', & b' \\ a'', & h'', & b'' \end{vmatrix}.$$

will subsist between the sines of the angles subtended by them at any point as subsist between the segments of the lines themselves. Consequently, *if a pencil be drawn from any point to six points in involution, any transversal cuts this pencil in six points in involution.* Again, *the reciprocal of six points in involution is a pencil in involution.*

The greater part of the equations already found apply equally to lines drawn through a point. Thus, any pair of lines $\alpha - \mu\beta$, $\alpha - \mu'\beta$ belong to a system in involution, if

$$A\mu\mu' + H(\mu + \mu') + B = 0;$$

and if we are given two pairs of lines

$$U = a\alpha^2 + 2h\alpha\beta + b\beta^2, \quad V = a'\alpha^2 + 2h'\alpha\beta + b'\beta^2,$$

they determine a pencil in involution whose focal lines are

$$(ah' - a'h)\alpha^2 + (ab' - a'b)\alpha\beta + (hb' - h'b)\beta^2 = 0,$$

or

$$\frac{dU}{d\alpha}\frac{dV}{d\beta} - \frac{dU}{d\beta}\frac{dV}{d\alpha} = 0.$$

344. *A system of conics passing through four fixed points meets any transversal in a system of points in involution.*

For, if S, S' be any two conics through the points, $S + \lambda S'$ will denote any other; and if, taking the transversal for axis of x and making $y = 0$ in the equations, we get $ax^2 + 2gx + c$, and $a'x^2 + 2g'x + c'$ to determine the points in which the transversal meets S and S', it will meet $S + \lambda S'$ in

$$ax^2 + 2gx + c + \lambda(a'x^2 + 2g'x + c'),$$

a pair (Art. 342) in involution with the two former pair.

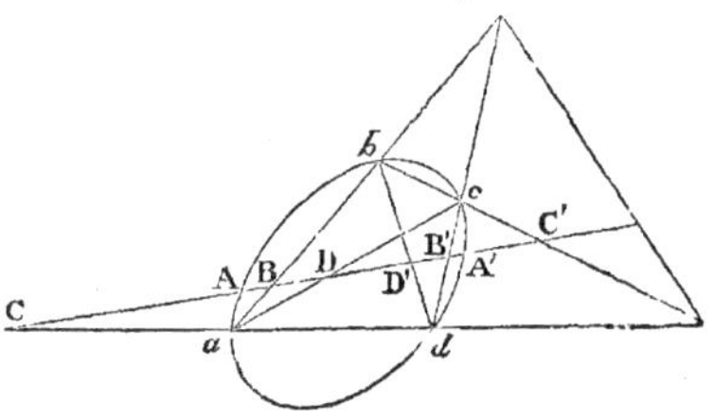

This may also be proved geometrically as follows: By the anharmonic properties of conics,

$$\{a.AdbA'\} = \{c.AdbA'\}:$$

but if we observe the points in which these pencils meet AA', we get $\{ACBA'\} = \{AB'C'A'\} = \{A'C'B'A\}$.

Consequently the points AA' belong to the system in involution determined by BB', CC', the pairs of points in which the transversal meets the sides of the quadrilateral joining the given points.

Reciprocating the theorem of this article we learn that, *the pairs of tangents drawn from any point to a system of conics touching four fixed lines, form a system in involution.*

345. Since the diagonals ac, bd may be considered as a conic through the four points, it follows, as a particular case of the last Article, that any transversal cuts the four sides and the diagonals of a quadrilateral in points BB', CC', DD', which are in involution. This property enables us, being given two pairs of points BB', DD' of a system in involution, to construct the point conjugate to any other C. For take any point at random, a; join aB, aD, aC; construct any triangle bcd, whose vertices rest on these three lines, and two of whose sides pass through $B'D'$, then the remaining side will pass through C', the point conjugate to C. The point a may be taken at infinity, and the lines aB, aD, aC will then be parallel to each other. If the point C be at infinity the same method will give us the centre of the system. The simplest construction for this case is,—"Through B, D, draw any pair of parallel lines Bb, Dc; and through B', D', a different pair of parallels $D'b$, $B'c$; then bc will pass through the centre of the system."

Ex. 1. If three conics circumscribe the same quadrilateral, the common tangent to any two is cut harmonically by the third. For the points of contact of this tangent are the foci of the system in involution.

Ex. 2. If through the intersection of the common chords of two conics we draw a tangent to one of them, this line will be cut harmonically by the other. For in this case the points D and D' in the last figure coincide, and will therefore be a focus.

Ex. 3. If two conics have double contact with each other, or if they have a contact of the third order, any tangent to the one is cut harmonically at the points where it meets the other, and where it meets the chord of contact. For in this case the common chords coincide, and the point where any transversal meets the chord of contact is a focus.

Ex. 4. To describe a conic through four points a, b, c, d, to touch a given right line. The point of contact must be one of the foci of the system BB', CC', &c., and these points can be determined by Art. 342. This problem, therefore, admits of two solutions.

Ex. 5. If a parallel to an asymptote meet the curve in C, and any inscribed quadrilateral in points $abcd$; $Ca \cdot Cc = Cb \cdot Cd$. For C is the centre of the system.

Ex. 6. Solve the examples, Art. 326, as cases of involution.

In Ex. 1, K is a focus: in Ex. 2, T is also a focus: in Ex. 3, T is a centre, &c.

Ex. 7. The intercepts on any line between a hyperbola and its asymptotes are equal. For in this case one focus of the system is at infinity (Cor., Art. 341).

346. *If there be a system of conics having a common self-conjugate triangle, any line passing through one of the vertices of this triangle is cut by the system in involution.*

For, if in $a\alpha^2 + b\beta^2 + c\gamma^2$ we write $\alpha = k\beta$, we get

$$(ak^2 + b)\beta^2 + c\gamma^2,$$

a pair of points evidently always harmonically conjugate with the two points where the line meets β and γ. Thus, then, in particular, a system of conics touching the four sides of a fixed quadrilateral cuts in involution any transversal which passes through one of the intersections of diagonals of the quadrilateral (Ex. 3, Art. 146). The points in which the transversal meets diagonals are the foci of the system, and the points where it meets opposite sides of the quadrilateral are conjugate points of the system.

Ex. 1. If two conics U, V touch their common tangents A, B, C, D in the points a, b, c, d; a', b', c', d'; a conic S through the points a, b, c, and touching D at d', will have for its second chord of intersection with V, the line joining the intersections of A with bc, B with ca, C with ab.

Let V meet ab in α, β, then, by this article, since ab passes through an intersection of diagonals of $ABCD$ (Ex. 2, Art. 263), a, b; α, β belong to a system in involution, of which the points where ab meets C and D are conjugate points. But (Art. 345) the common chords of S and V meet ab in points belonging to this same system in involution, determined by the points a, b; α, β, in which S and V meet the line ab. If then one of the common chords be D, the other must pass through the intersection of C with ab.

Ex. 2. If in a triangle there be inscribed an ellipse touching the sides at their middle points a, b, c, and also a circle touching at the points a', b', c', and if the fourth common tangent D to the ellipse and circle touch the circle at d', then the circle described through the middle points touches the inscribed circle at d'. By Ex. 1, a conic described through a, b, c, will touch the circle at d', if it also pass through the points where the circle is met by the line joining the intersections of A, bc; B, ca; C, ab. But this line is in this case the line at infinity. The touching conic is therefore a circle. Sir W. R. Hamilton has thus deduced Feuerbach's theorem (p. 127) as a particular case of Ex. 1.

The point d' and the line D can be constructed without drawing the ellipse. For since the diagonals of an inscribed, and of the corresponding circumscribing quadrilateral meet in a point, the lines ab, cd; $a'b'$, $c'd'$, and the lines joining AD, BC; AC, BD all intersect in the same point. If then α, β, γ be the vertices of the triangle formed by the intersections of bc, $b'c'$; ca, $c'a'$; ab, $a'b'$; the lines joining $a'\alpha$, $b'\beta$, $c'\gamma$ meet in d'. In other words, the triangle $\alpha\beta\gamma$ is homologous with abc, $a'b'c'$, the centres of homology being the points d, d'. In like manner, the triangle $\alpha\beta\gamma$ is also homologous with ABC, the axis of homology being the line D.

CHAPTER XVII.

THE METHOD OF PROJECTION.*

347. We have already several times had occasion to point out to the reader the advantage gained by taking notice of the number of particular theorems often included under one general enunciation, but we now propose to lay before him a short sketch of a method which renders us a still more important service, and which enables us to tell when from a particular given theorem we can safely infer the general one under which it is contained.

If all the points of any figure be joined to any fixed point in space (O), the joining lines will form a *cone*, of which the point O is called the *vertex*, and the section of this cone, by any plane, will form a figure which is called the *projection* of the given figure. The plane by which the cone is cut is called the *plane of projection.*

To any point of one figure will correspond *a point in the other.*

For, if any point A be joined to the vertex O, the point a, in which the joining line OA is cut by any plane, will be the projection on that plane of the given point A.

A right line will always be projected into a right line.

For, if all the points of the right line be joined to the vertex, the joining lines will form a plane, and this plane will be intersected by any plane of projection in a right line.

Hence, if any number of points in one figure lie in a right line, so will also the corresponding points on the projection; and if any number of lines in one figure pass through a point, so will also the corresponding lines on the projection.

---

* This method is the invention of M. Poncelet. See his *Traité des Propriétés Projectives*, published in the year 1822, a work which I believe may be regarded as the foundation of the Modern Geometry. In it were taught the principles, that theorems concerning infinitely distant points may be extended to finite points on a right line; that theorems concerning systems of circles may be extended to conics having two points common; and that theorems concerning imaginary points and lines may be extended to real points and lines.

348. *Any plane curve will always be projected into another curve of the same degree.*

For it is plain that, if the given curve be cut by any right line in any number of points, A, B, C, D, &c. the projection will be cut by the projection of that right line in *the same number* of corresponding points, a, b, c, d, &c.; but the degree of a curve is estimated geometrically by the number of points in which it can be cut by any right line. If AB meet the curve in some real and some imaginary points, ab will meet the projection in the same number of real and the same number of imaginary points.

In like manner, if any two curves intersect, their projections will intersect in the same number of points, and any point common to one pair, whether real or imaginary, must be considered as the projection of a corresponding real or imaginary point common to the other pair.

Any tangent to one curve will be projected into a tangent to the other.

For, any line AB on one curve must be projected into the line ab joining the corresponding points of the projection. Now, if the points A, B, coincide, the points a, b, will also coincide, and the line ab will be a tangent.

More generally, if any two curves touch each other in any number of points, their projections will touch each other in the same number of points.

349. If a plane through the vertex parallel to the plane of projection meet the original plane in a line AB, then any pencil of lines diverging from a point on AB will be projected into a system of parallel lines on the plane of projection. For, since the line from the vertex to any point of AB meets the plane of projection at an infinite distance, the intersection of any two lines which meet on AB is projected to an infinite distance on the plane of projection. Conversely, *any system of parallel lines on the original plane is projected into a system of lines meeting in a point on the line DF, where a plane through the vertex parallel to the original plane is cut by the plane of projection.* The method of projection then leads us naturally to the conclusion, that any system of parallel lines may be considered as passing through a point at an infinite distance, for their projections on any plane

pass through a point in general at a finite distance; and again, that *all the points at infinity on any plane may be considered as lying on a right line*, since we have showed that the projection of any point in which parallel lines intersect must lie somewhere on the right line DF in the plane of projection.

350. We see now, that if any property of a given curve does not involve the *magnitude* of lines or angles, but merely relates to the *position* of lines as drawn to certain points, or touching certain curves, or to the position of points, &c., then this property will be true for any curve into which the given curve can be projected. Thus, for instance, "if through any point in the plane of a circle a chord be drawn, the tangents at its extremities will meet on a fixed line." Now since we shall presently prove that every curve of the second degree can be projected into a circle, the method of projection shows at once that the properties of poles and polars are true not only for the circle, but also for all curves of the second degree. Again, Pascal's and Brianchon's theorems are properties of the same class, which it is sufficient to prove in the case of the circle, in order to know that they are true for all conic sections.

351. Properties which, if true for any figure, are true for its projection, are called *projective properties*. Besides the classes of theorems mentioned in the last Article, there are many projective theorems which *do* involve the magnitude of lines. For instance, the anharmonic ratio of four points in a right line $\{ABCD\}$, being measured by the ratio of the pencil $\{O.ABCD\}$ drawn to the vertex, must be the same as that of the four points $\{abcd\}$, where this pencil is cut by any transversal. Again, if there be an equation between the mutual distances of any number of points in a right line, such as

$$AB.CD.EF + k.AC.BE.DF + l.AD.CE.BF + \&c. = 0,$$

where in each term of the equation the same points are mentioned, although in different orders, this property will be projective. For (see Art. 311) if for AB we substitute

$$\frac{OA.OB.\sin AOB}{OP}, \&c.,$$

each term of the equation will contain $OA.OB.OC.OD.OE.OF$

in the numerator, and OP^3 in the denominator. Dividing, then, by these, there will remain merely a relation between the sines of angles subtended at O. It is evident that the points A, B, C, D, E, F, need not be on the same right line; or, in other words, that the perpendicular OP need not be the same for all, provided the points be so taken that, after the substitution, each term of the equation may contain in the denominator the same product, $OP.OP'.OP''$, &c. Thus, for example, "If lines meeting in a point and drawn through the vertices of a triangle ABC meet the opposite sides in the points a, b, c, then $Ab.Bc.Ca = Ac.Ba.Cb$." This is a relation of the class just mentioned, and which it is sufficient to prove for any projection of the triangle ABC. Let us suppose the point C projected to an infinite distance, then AC, BC, Cc are parallel, and the relation becomes

$$Ab.Bc = Ac.Ba,$$

the truth of which is at once perceived on making the figure.

352. It appears, from what has been said, that if we wish to demonstrate any projective property of any figure, it is sufficient to demonstrate it for the *simplest* figure into which the given figure can be projected; *e.g.* for one in which any line of the given figure is at an infinite distance.

Thus, if it were required to investigate the harmonic properties of a complete quadrilateral $ABCD$, whose opposite sides intersect in E, F, and the intersection of whose diagonals is G, we may join all the points of this figure to any point in space O, and cut the joining lines by any plane parallel to OEF, then EF is projected to infinity, and we have a new quadrilateral, whose sides ab, cd intersect in e at infinity, that is, are parallel; while ad, bc intersect in a point f at infinity, or are also parallel. We thus see that *any quadrilateral may be projected into a parallelogram.* Now since the diagonals of a parallelogram bisect each other, the diagonal ac is cut harmonically in the points a, g, c, and the point where it meets the line at infinity ef. Hence AB is cut harmonically in the points A, G, C, and where it meets EF.

Ex. If two triangles ABC, $A'B'C'$, be such that the points of intersection of AB, $A'B'$; BC, $B'C'$; CA, $C'A'$; lie in a right line, then the lines AA', BB', CC' meet in a point.

Project to infinity the line in which AB, $A'B'$, &c. intersect; then the theorem becomes: "If two triangles abc, $a'b'c'$ have the sides of the one respectively parallel to the sides of the other, then the lines aa', bb', cc' meet in a point." But the truth of this latter theorem is evident, since aa', bb' both cut cc' in the same ratio.

353. In order not to interrupt the account of the applications of the method of projection, we place in a separate section the formal proof that every curve of the second degree may be projected so as to become a circle. It will also be proved that by choosing properly the vertex and plane of projection, we can, as in Art. 352, cause any given line EF on the figure to be projected to infinity, at the same time that the projected curve becomes a circle. This being for the present taken for granted, these consequences follow:

Given any conic section and a point in its plane, we can project it into a circle, of which the projection of that point is the centre, for we have only to project it so that the projection of the polar of the given point may pass to infinity (Art. 154).

Any two conic sections may be projected so as both to become circles, for we have only to project one of them into a circle, and so that any of its chords of intersection with the other shall pass to infinity, and then, by Art. 257, the projection of the second conic passing through the same points at infinity as the circle must be a circle also.

Any two conics which have double contact with each other may be projected into concentric circles. For we have only to project one of them into a circle, so that its chord of contact with the other may pass to infinity (Art. 257).

354. We shall now give some examples of the method of deriving properties of conics from those of the circle, or from other more particular properties of conics.

Ex 1. "A line through any point is cut harmonically by the curve and the polar of that point." This property and its reciprocal are projective properties (Art. 351), and both being true for the circle, are true for every conic. Hence all the properties of the circle depending on the theory of poles and polars are true for all the conic sections.

Ex. 2. The anharmonic properties of the points and tangents of a conic are projective properties, which, when proved for the circle, as in Art. 312, are proved for all conics. Hence, every property of the circle which results from either of its anharmonic properties is true also for all the conic sections.

Ex. 3. Carnot's theorem (Art. 313), that if a conic meet the sides of a triangle,

$$Ab \,.\, Ab' \,.\, Bc \,.\, Bc' \,.\, Ca \,.\, Ca' = Ac \,.\, Ac' \,.\, Ba \,.\, Ba' \,.\, Cb \,.\, Cb',$$

is a projective property which need only be proved in the case of the circle, in which case it is evidently true, since $Ab.Ab' = Ac.Ac'$, &c.

The theorem can evidently be proved in like manner for any polygon.

Ex. 4. From Carnot's theorem, thus proved, could be deduced the properties of Art. 148, by supposing the point C at an infinite distance; we then have

$$\frac{Ab.Ab'}{Ac.Ac'} = \frac{Ba.Ba'}{Bc.Bc'},$$

where the line Ab is parallel to Ba.

Ex. 5. Given two concentric circles, any chord of one which touches the other is bisected at the point of contact.

Given two conics having double contact with each other, any chord of one which touches the other is cut harmonically at the point of contact, and where it meets the chord of contact of the conics. (Ex. 3, Art. 345).

For the line at infinity in the first case is projected into the chord of contact of two conics having double contact with each other. Ex. 4, Art. 236, is only a particular case of this theorem.

Ex. 6. Given three concentric circles, any tangent to one is cut by the other two in four points whose anharmonic ratio is constant.

Given three conics all touching each other in the same two points, any tangent to one is cut by the other two in four points whose anharmonic ratio is constant.

The first theorem is obviously true, since the four lengths are constant. The second may be considered as an extension of the anharmonic property of the tangents of a conic. In like manner the theorem (in Art. 276) with regard to anharmonic ratios in conics having double contact is immediately proved by projecting the conics into concentric circles.

Ex. 7. We mentioned already, that it was sufficient to prove Pascal's theorem for the case of a circle, but, by the help of Art. 353, we may still further simplify our figure, for we may suppose the line joining the intersection of AB, DE, to that of BC, EF, to pass off to infinity; and it is only necessary to prove that, if a hexagon be inscribed in a circle having the side AB parallel to DE, and BC to EF, then CD will be parallel to AF; but the truth of this can be shown from elementary considerations.

Ex. 8. A triangle is inscribed in any conic, two of whose sides pass through fixed points, to find the envelope of the third (Ex. 3, Art. 272). Let the line joining the fixed points be projected to infinity, and at the same time the conic into a circle, and this problem becomes,—"A triangle is inscribed in a circle, two of whose sides are parallel to fixed lines, to find the envelope of the third." But this envelope is a concentric circle, since the vertical angle of the triangle is given; hence, in the general case, the envelope is a conic touching the given conic in two points on the line joining the two given points.

Ex. 9. To investigate the projective properties of a quadrilateral inscribed in a conic. Let the conic be projected into a circle, and the quadrilateral into a parallelogram (Art. 352). Now the intersection of the diagonals of a parallelogram inscribed in a circle is the centre of the circle; hence the intersection of the diagonals of a quadrilateral inscribed in a conic is the pole of the line joining the intersections of the opposite sides. Again, if tangents to the circle be drawn at the vertices of this parallelogram, the diagonals of the quadrilateral so formed will also pass through the centre, bisecting the angles between the first diagonals; hence, "the diagonals of the inscribed and corresponding circumscribing quadrilateral pass through a point, and form a harmonic pencil."

Ex. 10. Given four points on a conic, the locus of its centre is a conic through the middle points of the sides of the given quadrilateral. (Ex. 15, Art. 328).

Given four points on a conic, the locus of the pole of any fixed line is a conic passing through the fourth harmonic to the point in which this line meets each side of the given quadrilateral.

Ex. 11. The locus of the point where parallel chords of a circle are cut in a given ratio is an ellipse having double contact with the circle. (Art. 163).

If through a fixed point O a line be drawn meeting the conic in A, B, and on it a point P be taken, such that $\{OABP\}$ may be constant, the locus of P is a conic having double contact with the given conic.

355. We may project several properties relating to foci by the help of the definition of a focus, given p. 239, viz. that if F be a focus, and A, B the two imaginary points in which any circle is met by the line at infinity; then FA, FB are tangents to the conic.

Ex. 1. The locus of the centre of a circle touching two given circles is a hyperbola, having the centres of the given circles for foci.

If a conic be described through two fixed points A, B, and touching two given conics which also pass through those points, the locus of the pole of AB is a conic touching the four lines CA, CB, $C'A$, $C'B$, where C, C', are the poles of AB with regard to the two given conics.

In this example we substitute for the word 'circle,' "conic through two fixed points A, B," (Art. 257), and for the word 'centre,' "pole of the line AB." (Art. 154).

Ex. 2. Given the focus and two points of a conic section, the intersection of tangents at those points will lie on a fixed line. (Art. 191).

Given two tangents, and two points on a conic, the locus of the intersection of tangents at those points is a right line.

Ex. 3. Given a focus and two tangents to a conic, the locus of the other focus is a right line. (This follows from Art. 189).

Given two fixed points A, B; two tangents FA, FB passing one through each point, and two other tangents to a conic; the locus of the intersection of the other tangents from A, B, is a right line.

Ex. 4. If a triangle circumscribe a parabola, the circle circumscribing the triangle passes through the focus, Cor. 4, Art. 223.

If two triangles circumscribe a conic, their six vertices lie on the same conic.*

For if the focus be F, and the two circular points at infinity A, B, the triangle FAB is a second triangle whose three sides touch the parabola.

Ex. 5. The locus of the centre of a circle passing through a fixed point, and touching a fixed line, is a parabola of which the fixed point is the focus.

Given one tangent, and three points on a conic, the locus of the intersection of tangents at any two of these points is a conic inscribed in the triangle formed by those points.

* This is easily proved directly. Take a side of each triangle and, by the anharmonic property of the tangents of a conic, these lines are cut homographically by the other four sides; whence it may easily be seen that the pencils joining the opposite vertices of each triangle to the other four are homographic:

Ex. 6. Given four tangents to a conic, the locus of the centre is the line joining the middle points of the diagonals of the quadrilateral.

Given four tangents to a conic, the locus of the pole of any line is the line joining the fourth harmonics of the points where the given line meets the diagonals of the quadrilateral.

It follows from our definition of a focus, that if two conics have the same focus, this point will be an intersection of common tangents to them, and will possess the properties mentioned at the end of Art. 264. Also, that if two conics have the same focus and directrix, they may be considered as two conics having double contact with each other, and may be projected into concentric circles.

356. Since angles which are constant in any figure will in general not be constant in the projection of that figure, we proceed to show what property of a projected figure may be inferred when any property relating to the magnitude of angles is given; and we commence with the case of the right angle.

Let the equations of two lines at right angles to each other be $x=0$, $y=0$, then the equation which determines the direction of the points at infinity on any circle is $x^2+y^2=0$, or

$$x+y\sqrt{-1}=0,\ \ x-y\sqrt{-1}=0.$$

Hence (Art. 57) these four lines form a harmonic pencil. Hence, given four points A, B, C, D, of a line cut harmonically, where A, B may be real or imaginary, if these points be transferred by a real or imaginary projection, so that A, B may become the two imaginary points at infinity on any circle, then any lines through C, D will be projected into lines at right angles to each other. Conversely, *any two lines at right angles to each other will be projected into lines which cut harmonically the line joining the two fixed points which are the projections of the imaginary points at infinity on a circle.*

Ex. 1. The tangent to a circle is at right angles to the radius.

Any chord of a conic is cut harmonically by any tangent, and by the line joining the point of contact of that tangent to the pole of the given chord. (Art. 146).

For the chord of the conic is supposed to be the projection of the line at infinity in the plane of the circle; the points where the chord meets the conic will be the projections of the imaginary points at infinity on the circle; and the pole of the chord will be the projection of the centre of the circle.

Ex. 2. Any right line drawn through the focus of a conic is at right angles to the line joining its pole to the focus. (Art. 192).

Any right line through a point, the line joining its pole to that point, and the two tangents from the point, form a harmonic pencil. (Art. 146).

It is evident that the first of these properties is only a particular case of the second, if we recollect that the tangents from the focus are the lines joining the focus to the two imaginary points on any circle.

Ex. 3. Let us apply Ex. 6 of the last Article to determine the locus of the pole of a given line with regard to a system of confocal conics. Being given the two foci, we are given a quadrilateral circumscribing the conic (Art. 258*a*); one of the diagonals of this quadrilateral is the line joining the foci, therefore (Ex. 6) one point on the locus is the fourth harmonic to the point where the given line cuts the distance between the foci. Again, another diagonal is the line at infinity, and since the extremities of this diagonal are the points at infinity on a circle, therefore by the present Article the locus is perpendicular to the given line. The locus is, therefore, completely determined.

Ex. 4. Two confocal conics cut each other at right angles.

If two conics be inscribed in the same quadrilateral, the two tangents at any of their points of intersection cut any diagonal of the circumscribing quadrilateral harmonically.

The last theorem is a case of the reciprocal of Ex. 1, Art. 345.

Ex. 5. The locus of the intersection of two tangents to a central conic, which cut at right angles, is a circle.

The locus of the intersection of tangents to a conic, which divide harmonically a given finite right line AB, is a conic through A, B.

The last theorem may, by Art. 146, be stated otherwise thus: "The locus of a point O, such that the line joining O to the pole of AO may pass through B, is a conic through A, B;" and the truth of it is evident directly, by taking four positions of the line, when we see, by Ex. 2, Art. 297, that the anharmonic ratio of four lines AO is equal to that of four corresponding lines BO.

Ex. 6. The locus of the intersection of tangents to a parabola, which cut at right angles, is the directrix.

If in the last example AB touch the given conic, the locus of O will be the line joining the points of contact of tangents from A, B.

Ex. 7. The circle circumscribing a triangle self-conjugate with regard to an equilateral hyperbola passes through the centre of the curve. (Ex. 5, Art. 228).

If two triangles are both self-conjugate with regard to a conic, their six vertices lie on a conic.

The fact that the asymptotes of an equilateral hyperbola are at right angles may be stated, by this Article, that the line at infinity cuts the curve in two points which are harmonically conjugate with respect to A, B, the imaginary circular points at infinity. And since the centre C is the pole of AB, the triangle CAB is self-conjugate with regard to the equilateral hyperbola. It follows, by reciprocation, that the six sides of two self-conjugate triangles touch the same conic.

Ex. 8. If from any point on a conic two lines at right angles to each other be drawn, the chord joining their extremities passes through a fixed point. (Ex. 2, Art. 181).

If a harmonic pencil be drawn through any point on a conic, two legs of which are fixed, the chord joining the extremities of the other legs will pass through a fixed point.

In other words, given two points a, c on a conic, and $\{abcd\}$ a harmonic ratio, bd will pass through a fixed point, namely, the intersection of tangents at a, c. But the truth of this may be seen directly: for let the line ac meet bd in K, then, since $\{a.abcd\}$ is a harmonic pencil, the tangent at a cuts bd in the fourth harmonic to K: but so likewise must the tangent at c, therefore these tangents meet bd in the same point. As a particular case of this theorem we have the following: "Through a fixed point on a conic two lines are drawn, making equal angles with a fixed line, the chord joining their extremities will pass through a fixed point."

357. *A system of pairs of right lines drawn through a point, so that the lines of each pair make equal angles with a fixed line, cuts the line at infinity in a system of points in involution, of which the two points at infinity on any circle form one pair of conjugate points.* For they evidently cut *any* right line in a system of points in involution, the foci of which are the points where the line is met by the given internal and external bisector of every pair of right lines. The two points at infinity just mentioned belong to the system, since they also are cut harmonically by these bisectors.

The tangents from any point to a system of confocal conics make equal angles with two fixed lines. (Art. 189).

The tangents from any point to a system of conics inscribed in the same quadrilateral cut any diagonal of that quadrilateral in a system of points in involution of which the two extremities of that diagonal are a pair of conjugate points. (Art. 344).

358. *Two lines which contain a constant angle cut the line joining the two points at infinity on a circle, so that the anharmonic ratio of the four points is constant.*

For the equation of two lines containing an angle θ being $x=0$, $y=0$, the direction of the points at infinity on any circle is determined by the equation

$$x^2+y^2+2xy\cos\theta=0;$$

and, separating this equation into factors, we see, by Art. 57, that the anharmonic ratio of the four lines is constant if θ be constant.

Ex. 1. "The angle contained in the same segment of a circle is constant." We see, by the present Article, that this is the form assumed by the anharmonic property of four points on a circle when two of them are at an infinite distance.

Ex. 2. The envelope of a chord of a conic which subtends a constant angle at the focus is another conic having the same focus and the same directrix.

If tangents through any point O meet the conic in T, T', and there be taken on the conic two points A, B, such that $\{O.ATBT'\}$ is constant, the envelope of AB is a conic touching the given conic in the points T, T'.

Ex. 3. The locus of the intersection of tangents to a parabola which cut at a given angle is a hyperbola having the same focus and the same directrix.

If a finite line AB, touching a conic be cut by two tangents in a given anharmonic ratio, the locus of their intersection is a conic touching the given conic at the points of contact of tangents from A, B.

Ex. 4. If from the focus of a conic a line be drawn making a given angle with any tangent, the locus of the point where it meets it is a circle.

If a variable tangent to a conic meet two fixed tangents in T, T', and a fixed line in M, and there be taken on it a point P, such that $\{PTMT'\}$ may be constant, the locus of P is a conic passing through the points where the fixed tangents meet the fixed line.

A particular case of this theorem is: "The locus of the point where the intercept of a variable tangent between two fixed tangents is cut in a given ratio is a hyperbola whose asymptotes are parallel to the fixed tangents."

Ex. 5. If from a fixed point O, OP be drawn to a given circle, and TP be drawn making the angle TPO constant, the envelope of TP is a conic having O for its focus.

Given the anharmonic ratio of a pencil three of whose legs pass through fixed points, and whose vertex moves along a given conic, passing through two of the points, the envelope of the fourth leg is a conic touching the lines joining these two to the third fixed point.

A particular case of this is: "If two fixed points A, B on a conic be joined to a variable point P, and the intercept made by the joining chords on a fixed line be cut in a given ratio at M, the envelope of PM is a conic touching parallels through A and B to the fixed line.

Ex. 6. If from a fixed point O, OP be drawn to a given right line, and the angle TPO be constant, the envelope of TP is a parabola having O for its focus.

Given the anharmonic ratio of a pencil, three of whose legs pass through fixed points, and whose vertex moves along a fixed line, the envelope of the fourth leg is a conic touching the three sides of the triangle formed by the given points.

359. We have now explained the geometric method by which, from the properties of one figure, may be derived those of another figure which corresponds to it (not as in Chap. XV., so that the points of one figure answer to the tangents of the other, but) so that the points of one answer to the points of the other, and the tangents of one to the tangents of the other. All this might be placed on a purely analytical basis. If any curve be represented by an equation in trilinear coordinates, referred to a triangle whose sides are a, b, c, and if we interpret this equation with regard to a different triangle of reference whose sides are a', b', c', we get a new curve of the same degree as the first;* and the same equations which establish any property of the first curve will, when differently interpreted, establish

* It is easy to see that the equation of the new curve referred to the old triangle is got by substituting in the given equation for α, β, γ; $l\alpha+m\beta+n\gamma$, $l'\alpha+m'\beta+n'\gamma$, $l''\alpha+m''\beta+n''\gamma$, where $l\alpha+m\beta+n\gamma$ represents the line which is to correspond to α, &c. For fuller information on this method of transformation see *Higher Plane Curves*, Chap. VIII.

a corresponding property of the second. In this manner a right line in one system always corresponds to a right line in the other, except in the case of the equation $a\alpha + b\beta + c\gamma = 0$, which in the one system represents an infinitely distant line, in the other a finite line. And, in like manner, $a'\alpha + b'\beta + c'\gamma$, which represents an infinitely distant line in the second system represents a finite line in the first system. In working with trilinear coordinates, the reader can hardly have failed to take notice how the method itself teaches him to generalize all theorems in which the line at infinity is concerned. Thus (see Art. 278) if it be required to find the locus of the centre of a conic, when four points or four tangents are given, this is done by finding the locus of the pole of the line at infinity $a\alpha + b\beta + c\gamma$, and the very same process gives the locus under the same conditions of the pole of any line $\lambda\alpha + \mu\beta + \nu\gamma$.

We saw (Art. 59) that the anharmonic ratio of a pencil $P - kP'$, $P - lP'$, &c. depends only on the constants k, l, and is not changed if P and P' are supposed to represent different right lines. We can infer then, that in the method of transformation which we are describing, to a pencil of four lines in the one system answers in the other system a pencil having the same anharmonic ratio; and that to four points on a line correspond four points whose anharmonic ratio is the same.

An equation, $S = 0$, which represents a circle in the one system will, in general, not represent a circle in the other. But since any other circle in the first system is represented by an equation of the form

$$S + (a\alpha + b\beta + c\gamma)(\lambda\alpha + \mu\beta + \nu\gamma) = 0,$$

all curves of the second system answering to circles in the first will have common the two points common to S and $a\alpha + b\beta + c\gamma$.

360. In this way we are led, on purely analytical grounds, to the most important principles, on the discovery and application of which the merit of Poncelet's great work consists. The *principle of continuity* (in virtue of which properties of a figure, in which certain points and lines are real, are asserted to be true even when some of these points and lines are imaginary)

is more easily established on analytical than on purely geometrical grounds. In fact, the processes of analysis take no account of the distinction between real and imaginary, so important in pure geometry. The processes, for example, by which, in Chap. XIV., we obtained the properties of systems of conics represented by equations of forms $S = k\alpha\beta$ or $S = k\alpha^2$ are unaffected, whether we suppose α and β to meet S in real or imaginary points. And though from any given property of a system of circles we can obtain, by a real projection, only a property of a system of conics having two imaginary points common, yet it is plainly impossible to prove such a property by general equations without proving it, at the same time, for conics having two real points common. The analytical method of transformation, described in the last article, is equally applicable if we wish real points in one figure to correspond to imaginary points on the other. Thus, for example, $\alpha^2 + \beta^2 = \gamma^2$ denotes a curve met by γ in imaginary points; but if we substitute for α, β; $P \pm Q\sqrt{(-1)}$, and for γ, R, where P, Q, R denote right lines, we get a curve met in real points by R the line corresponding to γ.

The chief difference in the application of the method of projections, considered geometrically and considered algebraically, is that the geometric method would lead us to prove a theorem, first for the circle or some other simple state of the figure, and then infer a general theorem by projection. The algebraic method finds it as easy to prove the general theorem as the simpler one, and would lead us to prove the general theorem first, and afterwards infer the other as a particular case.

THEORY OF THE SECTIONS OF A CONE.

361. *The sections of a cone by parallel planes are similar.* Let the line joining the vertex O to any fixed point A in one plane meet the other in the point a; and let radii vectores be drawn from A, a to any other two corresponding points B, b. Then, from the similar triangles OAB, Oab, AB is to ab in the constant ratio $OA : Oa$; and since *every* radius vector of the one curve is parallel and in a constant ratio to the corresponding radius vector of the other, the two curves are similar (Art. 233).

COR. If a cone standing on a circular base be cut by any plane parallel to the base, the section will be a circle. This is evident as before; we may, if we please, suppose the points A, a the centres of the curves.

362. *A section of a cone, standing on a circular base, may be either an ellipse, hyperbola, or parabola.*

A cone of the second degree is said to be *right* if the line joining the vertex to the centre of the circle which is taken for base be perpendicular to the plane of that circle; in which case this line is called the *axis* of the cone. If this line be not perpendicular to the plane of the base, the cone is said to be *oblique*. The investigation of the sections of an oblique cone is exactly the same as that of the sections of a right cone, but we shall treat them separately, because the figure in the latter case being more simple will be more easily understood by the learner, who may at first find some difficulty in the conception of figures in space.

Let a plane (OAB) be drawn through the axis of the cone OC perpendicular to the plane of the section, so that both the section $MSsN$ and the base ASB are supposed to be perpendicular to the plane of the paper; the line RS, in which the section meets the base, is, therefore, also supposed perpendicular to the plane of the paper. Let us first suppose the line MN, in which the section cuts the plane OAB to meet both the sides OA, OB, as in the figure, on the same side of the vertex.

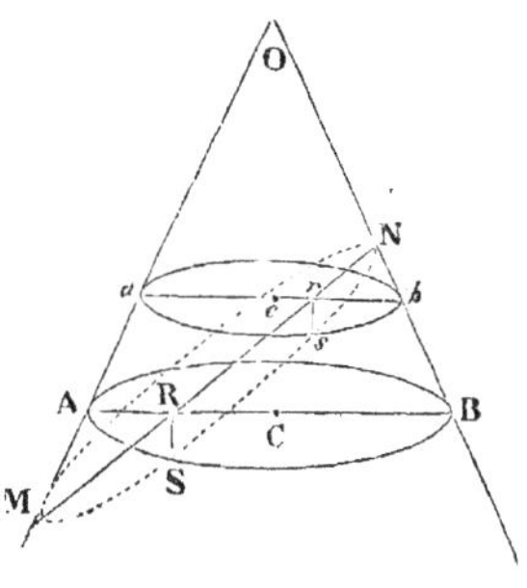

Now let a plane parallel to the base be drawn at any other point s of the section. Then we have (Euc. III. 35) the square of RS, the ordinate of the circle, $= AR.RB$, and in like manner $rs^2 = ar.rb$. But from a comparison of the similar triangles ARM, arM; BRN, brN, it can at once be proved that

$$AR.RB : MR.RN :: ar.rb : Mr.rN.$$

Therefore $$RS^2 : rs^2 :: MR.RN : Mr.rN.$$

Hence the section $MSsN$ is such that the square of any ordinate

rs is to the rectangle under the parts in which it cuts the line MN in the constant ratio $RS^2 : MR.RN$. Hence it can immediately be inferred (Art. 149) that the section is an *ellipse*, of which MN is the axis major, while the square of the axis minor is to MN^2 in the given ratio

$$RS^2 : MR.RN.$$

Secondly. Let MN meet one of the sides OA *produced*. The proof proceeds exactly as before, only that now we prove the square of the ordinate rs in a constant ratio to the rectangle $Mr.rN$ under the parts into which it cuts the line MN *produced*. The learner will have no difficulty in proving that the locus will in this case be a *hyperbola*, consisting evidently of the two opposite branches NsS, $Ms'S'$.

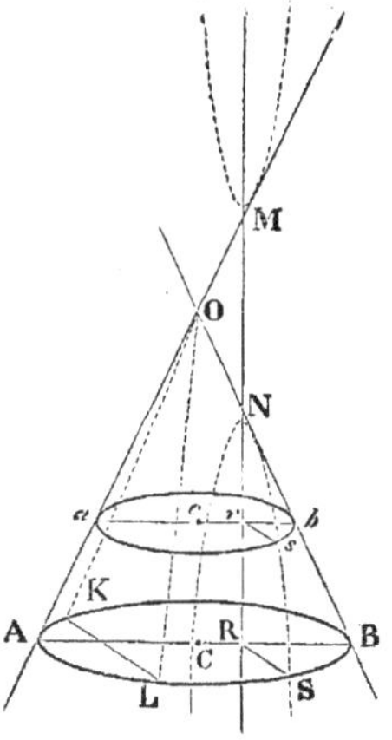

Thirdly. Let the line MN be *parallel* to one of the sides. In this case, since $AR = ar$, and $RB : rb :: RN : rN$, we have the square of the ordinate $rs\ (= ar.rb)$ to the abscissa rN in the constant ratio

$$RS^2\ (= AR.RB) : RN.$$

The section is therefore a *parabola*.*

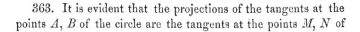

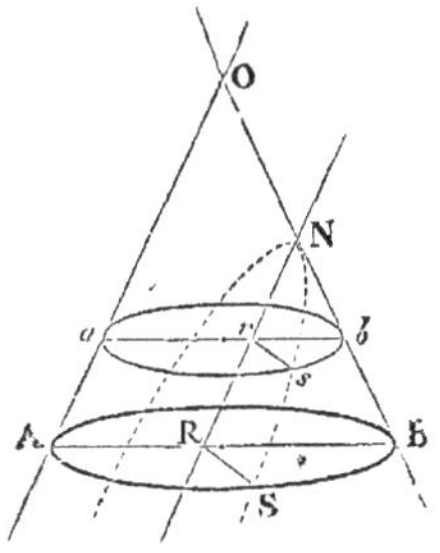

363. It is evident that the projections of the tangents at the points A, B of the circle are the tangents at the points M, N of

* Those who first treated of conic sections only considered the case when a right cone is cut by a plane perpendicular to a side of the cone; that is to say, when MN is perpendicular to OB. Conic sections were then divided into sections of a right-angled, acute, or obtuse-angled cone; and according to Eutochius, the commentator on Apollonius, were called parabola, ellipse, or hyperbola, according as the angle of the core was equal to, less than, or exceeded a right angle. (See the passage cited in full, *Walton's Examples*, p. 428). It was Apollonius who first showed that all three sections could be made from one cone; and who, according to Pappus, gave them the names parabola, ellipse, and hyperbola, for the reason stated, Art. 194. The authority of Eutochius, who was more than a century later than Pappus, may not be very great, but the name parabola was used by Archimedes, who was prior to Apollonius.

the conic section (Art. 348); now in the case of the parabola the point M and the tangent at it go off to infinity; we are therefore again led to the conclusion that *every parabola has one tangent altogether at an infinite distance.*

364. Let the cone now be supposed oblique. The plane of the paper is a plane drawn through the line OC, perpendicular to the plane of the circle $AQSB$. Now let the section meet the base in any line QS, draw a diameter LK bisecting QS, and let the section meet the plane OLK in the line MN, then the proof proceeds exactly as before; we have the square of the ordinate RS equal to the rectangle $LR.RK$; if we conceive a plane, as before, drawn parallel to the base (which, however, is left out of the figure in order to avoid rendering it too complicated), we have the square of any other ordinate rs equal to the corresponding rectangle $lr.rk$; and we then prove by the similar triangles KRM, krM; LRN, lrN, in the plane OLK, exactly as in the case of the right cone, that $RS^2 : rs^2$, as the rectangle under the parts into which each ordinate divides MN, and that therefore the section is a conic of which MN is the diameter bisecting QS, and which is an ellipse when MN meets both the lines OL, OK on the same side of the vertex, a hyperbola when it meets them on different sides of the vertex, and a parabola when it is parallel to either.

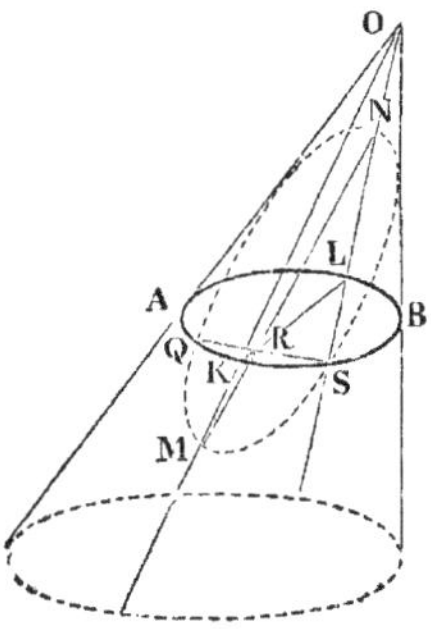

In the proof just given QS is supposed to intersect the circle in real points; if it did not, we have only to take, instead of the circle AB, any other parallel circle ab, which *does* meet the section in real points, and the proof will proceed as before.

365. We give formal proofs of the two following theorems, though they are evident by the principle of continuity:

I. *If a circular section be cut by any plane in a line QS, the diameters conjugate to QS in that plane, and in the plane of the circle, meet QS in the same point.* When qs meets the circle in real points, the diameter conjugate to it in every plane must evidently pass through its middle point r. We have therefore only to examine the case where QS does not meet in real points. It was proved (Art. 361) that the diameter df which bisects chords, parallel to qs, of any circular section, will be projected into a diameter DF bisecting the parallel chords of any parallel section. The locus therefore of the middle points of all chords of the cone parallel to qs is the plane Odf. The diameter therefore, conjugate to QS in any section, is the intersection of the plane Odf with the plane of that section, and must pass through the point R in which QS meets the plane ODf.

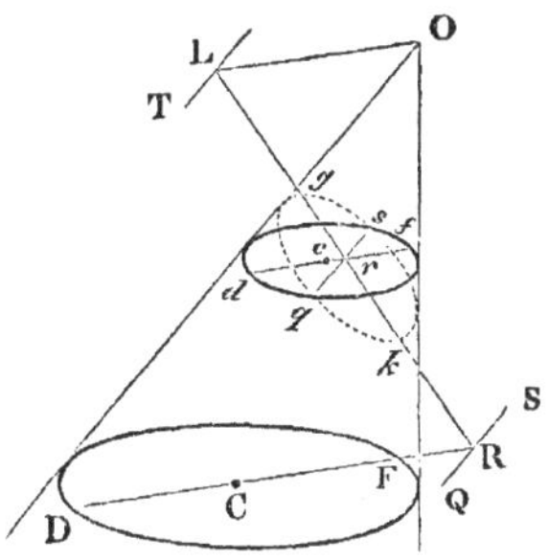

II. *In the same case, if the diameters conjugate to QS in the circle, and in the other section, be cut into segments RD, RF; Rg, Rk; the rectangle $DR.RF$ is to $gR.Rk$ as the square of the diameter of the section parallel to QS is to the square of the conjugate diameter.* This is evident when qs meets the circle in real points; since $rs^2 = dr.rf$. In general, we have just proved that the lines gk, df, DF, lie in one plane passing through the vertex. The points D, d are therefore projections of g; that is to say, they lie in one right line passing through the vertex. We have therefore, by similar triangles, as in Art. 364,

$$dr.rf : DR.RF :: gr.rk : gR.Rk;$$

and since $dr.rf$ is to $gr.rk$ as the squares of the parallel semi-diameters, $DR.RF$ is to $gR.Rk$ in the same ratio.

If the section $gskq$ and the line QS be given, this theorem enables us to find $DR.RF$, that is to say, the square of the tangent from R to the circular section whose plane passes through QS.

366. *Given any conic $gskq$ and a line TL in its plane not cutting it, we can project it so that the conic may become a circle, and the line may be projected to infinity.*

To do this, it is evidently necessary to find O the vertex of a cone standing on the given conic, and such that its sections parallel to the plane OTL shall be circles. For then any of these parallel sections would be a projection fulfilling the conditions of the problem. Now, if TL meet the conjugate diameter in the point L, it follows from the theorem last proved that the distance OL is given; for, since the plane OTL is to meet the cone in an infinitely small circle, OL^2 is to $gL.Lk$ in the ratio of the squares of two known diameters of the section. OL must also lie in the plane perpendicular to TL, since it is parallel to the diameter of a circle perpendicular to TL. And there is nothing else to limit the position of the point O, which may lie anywhere in a known circle in the plane perpendicular to TL.

367. *If a sphere be inscribed in a right cone touching the plane of any section, the point of contact will be a focus of that section, and the corresponding directrix will be the intersection of the plane of the section with the plane of contact of the cone with the sphere.*

Let spheres be both inscribed and exscribed between the cone and the plane of the section. Now, if any point P of the section be joined to the vertex, and the joining line meet the planes of contact in Dd, then we have $PD = PF$, since they are tangents to the same sphere, and, similarly, $Pd = PF'$, therefore $PF + PF' = Dd$, which is constant. The point (R), where FF' meets AB produced, is a point on the directrix, for by the property of the circle $NFMR$ is cut harmonically, therefore R is a point on the polar of F.

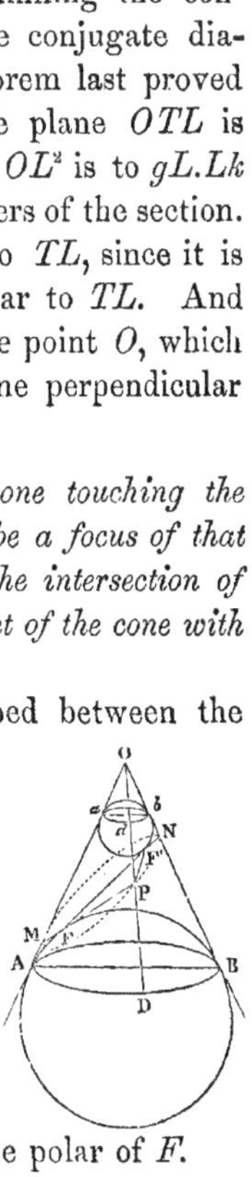

It is not difficult to prove that the parameter of the section MPN is constant, if the distance of the plane from the vertex be constant.

COR. The locus of the vertices of all right cones, out of which a given ellipse can be cut, is a hyperbola passing through the foci of the ellipse. For the difference of MO and NO is constant, being equal to the difference between MF' and NF'*.

* By the help of this principle, Mr. Mulcahy showed how to derive properties of angles subtended at the focus of a conic from properties of small circles of a sphere. For example, it is known that if through any point P, on the surface of a sphere, a great circle be drawn, cutting a small circle in the points A, B, then $\tan \frac{1}{2}AP \tan \frac{1}{2}BP$ is constant. Now, let us take a cone whose base is the small circle, and whose vertex

ORTHOGONAL PROJECTION.

368. If from all the points of any figure perpendiculars be let fall on any plane, their feet will trace out a figure which is called the *orthogonal projection* of the given figure. The orthogonal projection of any figure is, therefore, a right section of a *cylinder* passing through the given figure.

All parallel lines are in a constant ratio to their orthogonal projections on any plane.

For (see fig. p. 3) MM' represents the orthogonal projection of the line PQ, and it is evidently $= PQ$ multiplied by the cosine of the angle which PQ makes with MM'.

All lines parallel to the intersection of the plane of the figure with the plane on which it is projected are equal to their orthogonal projections.

For since the intersection of the planes is itself not altered by projection, neither can any line parallel to it.

The area of any figure in a given plane is in a constant ratio to its orthogonal projection on another given plane.

For, if we suppose ordinates of the figure and of its projection to be drawn perpendicular to the intersection of the planes, every ordinate of the projection is to the corresponding ordinate of the original figure in the constant ratio of the cosine of the angle between the planes to unity; and it will be proved, in Chap. XIX., that if two figures be such that the ordinate of one is in a constant ratio to the corresponding ordinate of the other, the areas of the figures are in the same ratio.

Any ellipse can be orthogonally projected into a circle.

For, if we take the intersection of the plane of projection with the plane of the given ellipse parallel to the axis minor of that ellipse, and if we take the cosine of the angle between the planes

---

is the centre of the sphere, and let us cut this cone by any plane, and we learn that "if through a point p, in the plane of any conic, a line be drawn cutting the conic in the points a, b, then the product of the tangents of the halves of the angles which ap, bp subtend at the vertex of the cone will be constant." This property will be true of the vertex of any right cone, out of which the section can be cut, and, therefore, since the focus is a point in the locus of such vertices, it must be true that $\tan \frac{1}{2}afp \tan \frac{1}{2}bfp$ is constant (see p. 210).

$= \frac{b}{a}$, then every line parallel to the axis minor will be unaltered by projection, but every line parallel to the axis major will be shortened in the ratio $b : a$; the projection will, therefore (Art. 163), be a circle, whose radius is b.

369. We shall apply the principles laid down in the last Article to investigate the expression for the radius of a circle circumscribing a triangle inscribed in a conic, given Ex. 7, p. 220.*

Let the sides of the triangle be α, β, γ, and its area A, then, by elementary geometry,

$$R = \frac{\alpha\beta\gamma}{4A}.$$

Now let the ellipse be projected into a circle whose radius is b, then, since this is the circle circumscribing the projected triangle, we have

$$b = \frac{\alpha'\beta'\gamma'}{4A'}.$$

But, since parallel lines are in a constant ratio to their projections, we have

$$\alpha' : \alpha :: b : b',$$
$$\beta' : \beta :: b : b'',$$
$$\gamma' : \gamma :: b : b''';$$

and since (Art. 368) A' is to A as the area of the circle $(=\pi b^2)$ to the area of the ellipse $(=\pi ab)$ (see chap. XIX.), we have

$$A' : A :: b : a.$$

Hence $$\frac{\alpha'\beta'\gamma'}{4A'} : \frac{\alpha\beta\gamma}{4A} :: ab^2 : b'b''b''',$$

and therefore $$R = \frac{b'b''b'''}{ab}.$$

---

* This proof of Mr. MacCullagh's theorem is due to Dr. Graves.

CHAPTER XVIII.

INVARIANTS AND COVARIANTS OF SYSTEMS OF CONICS.

370. It was proved (Art. 250) that if S and S' represent two conics, there are three values of k for which $kS+S'$ represents a pair of right lines. Let

$$S = ax^2 + by^2 + cz^2 + 2fyz + 2gzx + 2hxy,$$
$$S' = a'x^2 + b'y^2 + c'z^2 + 2f'yz + 2g'zx + 2h'xy.$$

We also write

$$\Delta = abc + 2fgh - af^2 - bg^2 - ch^2,$$
$$\Delta' = a'b'c' + 2f'g'h' - a'f'^2 - b'g'^2 - c'h'^2.$$

Then the values of k in question are got by substituting $ka+a'$, $kb+b'$, &c. for a, b, &c. in $\Delta=0$. We shall write the resulting cubic

$$\Delta k^3 + \Theta k^2 + \Theta' k + \Delta' = 0.$$

The value of Θ, found by actual calculation, is

$$(bc-f^2)\,a' + (ca-g^2)\,b' + (ab-h^2)\,c'$$
$$+ 2\,(gh-af)f' + 2\,(hf-bg)g' + 2\,(fg-ch)\,h';$$

or, using the notation of Art. 151,

$$Aa' + Bb' + Cc' + 2Ff' + 2Gg' + 2Hh';$$

or, again,

$$a'\frac{d\Delta}{da} + b'\frac{d\Delta}{db} + c'\frac{d\Delta}{dc} + f'\frac{d\Delta}{df} + g'\frac{d\Delta}{dg} + h'\frac{d\Delta}{dh},$$

as is also evident from Taylor's theorem. The value of Θ' is got from Θ by interchanging accented and unaccented letters, and may be written

$$\Theta' = A'a + B'b + C'c + 2F'f + 2G'g + 2H'h.$$

If we eliminate k between $kS+S'=0$, and the cubic which determines k, the result

$$\Delta S'^3 - \Theta S'^2 S + \Theta' S' S^2 - \Delta' S^3 = 0,$$

(an equation evidently of the sixth degree), denotes the three pairs of lines which join the four points of intersection of the two conics (Art. 238).

Ex. To find the locus of the intersection of normals to a conic, at the extremities of a chord which passes through a given point $\alpha\beta$. Let the curve be $S = \frac{x^2}{a^2} + \frac{y^2}{b^2} - 1$; then the points whose normals pass through a given point $x'y'$ are determined (Art. 181, Ex. 1) as the intersections of S with the hyperbola $S' = 2(c^2xy + b^2y'x - a^2x'y)$. We can then, by this article, form the equation of the six chords which join the feet of normals through $x'y'$, and expressing that this equation is satisfied for the point $\alpha\beta$, we have the locus required.

We have $\Delta = -\frac{1}{a^2b^2}$, $\Theta = 0$, $\Theta' = -(a^2x'^2 + b^2y'^2 - c^4)$, $\Delta' = -2a^2b^2c^2x'y'$.

The equation of the locus is then

$$\frac{8}{a^2b^2}(a^2\beta x - b^2\alpha y - c^2\alpha\beta)^3 + 2(a^2x^2 + b^2y^2 - c^4)(a^2\beta x - b^2\alpha y - c^2\alpha\beta)\left(\frac{\alpha^2}{a^2} + \frac{\beta^2}{b^2} - 1\right)^2$$
$$+ 2a^2b^2c^2xy\left(\frac{\alpha^2}{a^2} + \frac{\beta^2}{b^2} - 1\right)^3 = 0,$$

which represents a curve of the third degree. If the given point be on either axis, the locus reduces to a conic, as may be seen by making $\alpha = 0$ in the preceding equation. It is also geometrically evident, that in this case the axis is part of the locus. The locus also reduces to a conic if the point be infinitely distant; that is to say, when the problem is to find the locus of the intersection of normals at the extremities of a chord parallel to a given line.

371. If on transforming to any new set of coordinates, Cartesian or trilinear, S and S' become $\overline{S}$ and $\overline{S'}$, it is manifest that $kS + S'$ becomes $k\overline{S} + \overline{S'}$, and that the coefficient k is not affected. It follows that the values of k, for which $kS + S'$ represents right lines, must be the same, no matter in what system of coordinates S and S' are expressed. Hence, then, the ratio between any two coefficients in the cubic for k, found in the last Article, remains unaltered when we transform from any one set of coordinates to another.* The quantities Δ, Θ, Θ', Δ' are on this account called *invariants* of the system of conics. If then, in the case of any two given conics, having by transformation brought S and S' to their simplest form, and having calculated Δ, Θ, Θ', Δ', we find any homogeneous relation existing between them, we can predict that the same relation will exist between these quantities, no matter to what axes the equations are referred. It will be found possible to express in

* It may be proved by actual transformation that if in S and S' we substitute for x, y, z; $lx + my + nz$, $l'x + m'y + n'z$, $l''x + m''y + n''z$, the quantities Δ, Θ, Θ' Δ' for the transformed system, are equal to those for the old, respectively multiplied by the square of the determinant

$$\begin{vmatrix} l, & m, & n \\ l', & m', & n' \\ l'', & m'', & n'' \end{vmatrix}.$$

terms of the same four quantities the condition that the conics should be connected by any relation, independent of the position of the axes, as is illustrated in the next Article.

The following exercises in calculating the invariants Δ, Θ, Θ', Δ', include some of the cases of most frequent occurrence.

Ex. 1. Calculate the invariants when the conics are referred to their common self-conjugate triangle. We may take

$$S = ax^2 + by^2 + cz^2, \quad S' = a'x^2 + b'y^2 + c'z^2;$$

and we may further simplify the equations by writing x, y, z, instead of $x\sqrt{(a')}$, $y\sqrt{(b')}$, $z\sqrt{(c')}$, so as to bring S' to the form $x^2 + y^2 + z^2$. We have then

$$\Delta = abc, \quad \Theta = bc + ca + ab, \quad \Theta' = a + b + c, \quad \Delta' = 1.$$

And $S + kS'$ will represent right lines, if

$$k^3 + k^2 (a + b + c) + k (bc + ca + ab) + abc = 0.$$

And it is otherwise evident that the three values for which $S + kS'$ represents right lines are $- a, - b, - c$.

Ex. 2. Let S', as before, be $x^2 + y^2 + z^2$, and let S represent the general equation.

Ans. $\Theta = (bc - f^2) + (ca - g^2) + (ab - h^2) = A + B + C$; $\Theta' = a + b + c$.

Ex. 3. Let S and S' represent two circles $x^2 + y^2 - r^2$, $(x - \alpha)^2 + (y - \beta)^2 - r'^2$.

Ans. $\Delta = - r^2$, $\Theta = \alpha^2 + \beta^2 - 2r^2 - r'^2$, $\Theta' = \alpha^2 + \beta^2 - r^2 - 2r'^2$, $\Delta' = - r'^2$. So that if D be the distance between the centres of the circles, $S + kS'$ will represent right lines if

$$r^2 + (2r^2 + r'^2 - D^2) k + (r^2 + 2r'^2 - D^2) k^2 + r'^2k^3 = 0.$$

Now since we know that $S - S'$ represents two right lines (one finite, the other infinitely distant), it is evident that $- 1$ must be a root of this equation. And it is in fact divisible by $k + 1$, the quotient being

$$r^2 + (r^2 + r'^2 - D^2) k + r'^2k^2 = 0.$$

Ex. 4. Let S represent $\frac{x^2}{a^2} + \frac{y^2}{b^2} - 1$, while S' is the circle $(x - \alpha)^2 + (y - \beta)^2 - r^2$.

$$\textit{Ans.}\ \Delta = - \frac{1}{a^2b^2}, \quad \Theta = \frac{1}{a^2b^2} (\alpha^2 + \beta^2 - a^2 - b^2 - r^2),$$

$$\Theta' = \frac{\alpha^2}{a^2} + \frac{\beta^2}{b^2} - 1 - r^2 \left(\frac{1}{a^2} + \frac{1}{b^2}\right), \quad \Delta' = - r^2.$$

Ex. 5. Let S represent the parabola $y^2 - 4mx$, and S' the circle as before.

Ans. $\Delta = - 4m^2$, $\Theta = - 4m (\alpha + m)$, $\Theta' = \beta^2 - 4m\alpha - r^2$, $\Delta' = - r^2$.

372. *To find the condition that two conics S and S' should touch each other.* When two points, A, B, of the four intersections of two conics coincide, it is plain that the pair of lines AC, BD is identical with the pair AD, BC. In this case, then, the cubic

$$\Delta k^3 + \Theta k^2 + \Theta' k + \Delta' = 0,$$

must have two equal roots. But it can readily be proved that the condition that this should be the case is

$$(\Theta\Theta' - 9\Delta\Delta')^2 = 4 (\Theta^2 - 3\Delta\Theta') (\Theta'^2 - 3\Delta'\Theta),$$

or $$\Theta^2\Theta'^2 + 18\Delta\Delta'\Theta\Theta' - 27\Delta^2\Delta'^2 - 4\Delta\Theta'^3 - 4\Delta'\Theta^3 = 0,$$

which is the required condition that the conics should touch.

It is proved, in works on the theory of equations, that the left-hand member of the equation last written is proportional to the product of the squares of the differences of the roots of the equation in k; and that when it is positive the roots of the equation in k are all real, but that when it is negative two of these roots are imaginary. In the latter case (see Art. 282), S and S' intersect in two real and two imaginary points: in the former case, they intersect either in four real or four imaginary points. These last two cases have not been distinguished by any simple criterion.

If three points A, B, C coincide the conics osculate and in this case the three pairs of right lines are all identical so that the cubic must be a perfect cube; the condition for this are $\frac{3\Delta}{\Theta} = \frac{\Theta}{\Theta'} = \frac{\Theta'}{3\Delta'}$. The conditions for double contact are of a different kind and will be got further on.

Ex. 1. To find by this method the condition that two circles shall touch. Forming the condition that the reduced equation (Ex. 3, Art. 371), $r^2 + (r^2+r'^2-D^2)k+r'^2k^2=0$, should have equal roots, we get $r^2 + r'^2 - D^2 = \pm 2rr'$; $D = r \pm r'$ as is geometrically evident.

Ex. 2. The conditions for contact between two conics can be shortly found in the cases of trinomial equations by identifying the equations of tangents at any point given Arts. 127, 130, and are for

$$fyz + gzx + hxy = 0,\ \sqrt{(lx)} + \sqrt{(my)} + \sqrt{(nz)} = 0,\ (fl)^{\frac{1}{3}} + (gm)^{\frac{1}{3}} + (hn)^{\frac{1}{3}} = 0,$$

for $$\sqrt{(lx)} + \sqrt{(my)} + \sqrt{(nz)} = 0,\ ax^2 + by^2 + cz^2 = 0,\ \left(\frac{l^2}{a}\right)^{\frac{1}{3}} + \left(\frac{m^2}{b}\right)^{\frac{1}{3}} + \left(\frac{n^2}{c}\right)^{\frac{1}{3}} = 0,$$

for $$ax^2 + by^2 + cz^2 = 0,\ fyz + gzx + hxy = 0,\ (af^2)^{\frac{1}{3}} + (bg^2)^{\frac{1}{3}} + (ch^2)^{\frac{1}{3}} = 0.$$

Ex. 3. Find the locus of the centre of a circle of constant radius touching a given conic. We have only to write for Δ, Δ', Θ, Θ' in the equation of this article, the values Ex. 4 and 5, Art. 371; and to consider α, β as the running coordinates. The locus is in general a curve of the eighth degree, but reduces to the sixth in the case of the parabola. This curve is the same which we should find by measuring from the curve on each normal, a constant length, equal to r. It is sometimes called the curve *parallel* to the given conic. Its evolute is the same as that of the conic.

The following are the equations of the parallel curves given at full length, which may also be regarded as equations giving the length of the normal distances from any point to the curve. The parallel to the parabola is

$$r^6 - (3y^2 + x^2 + 8mx - 8m^2)\,r^4 + \{3y^4 + y^2\,(2x^2 - 2mx + 20m^2)$$
$$+ 8mx^3 + 8m^2x^2 - 32m^3x + 16m^4\}\,r^2 - (y^2 - 4mx)^2\,\{y^2 + (x - m)^2\} = 0.$$

The parallel to the ellipse is

$$c^4r^8 - 2c^2r^6 \{c^2(a^2+b^2) + (a^2-2b^2)x^2 + (2a^2-b^2)y^2\}$$
$$+ r^4 \{c^4(a^4+4a^2b^2+b^4) - 2c^2(a^4-a^2b^2+3b^4)x^2 + 2c^2(3a^4-a^2b^2+b^4)y^2$$
$$+ (a^4-6a^2b^2+6b^4)x^4 + (6a^4-6a^2b^2+b^4)y^4 + (6a^4-10a^2b^2+6b^4)x^2y^2\}$$
$$+ r^2 \{-2a^2b^2c^4(a^2+b^2) + 2c^2x^2b^2(3a^4-a^2b^2+b^4) - 2c^2y^2a^2(a^4-a^2b^2+3b^4)$$
$$- b^2x^4(6a^4-10a^2b^2+6b^4) - a^2y^4(6a^4-10a^2b^2+6b^4) + x^2y^2(4a^6-6a^4b^2-6a^2b^4+4b^6)$$
$$+ 2b^2(a^2-2b^2)x^6 - 2(a^4-a^2b^2+3b^4)x^4y^2 - 2(3a^4-a^2b^2+b^4)x^2y^4 + 2a^2(b^2-2a^2)y^6\}$$
$$+ (b^2x^2+a^2y^2-a^2b^2)^2 \{(x-c)^2+y^2\}\{(x+c)^2+y^2\} = 0.$$

Thus the locus of a point is a conic, if the sum of squares of its normal distances to the curve be given. If we form the condition that the equation in r^2 should have equal roots, we get the squares of the axes multiplied by the cube of the evolute. If we make $r = 0$, we find the foci appearing as points whose normal distance to the curve vanishes. This is to be accounted for by remembering that the distance from the origin vanishes of any point on either of the lines $x^2 + y^2 = 0$.

Ex. 4. To find the equation of the evolute of an ellipse. Since two of the normals coincide which can be drawn through every point on the evolute, we have only to express the condition that in Ex. Art. 370 the curves S and S' touch. Now when the term k^2 is absent from an equation, the condition that $\Delta k^3 + \Theta' k + \Delta'$ should have equal roots reduces to $27\Delta\Delta'^2 + 4\Theta'^3 = 0$. The equation of the evolute is therefore $(a^2x^2 + b^2y^2 - c^4)^3 + 27a^2b^2c^4x^2y^2 = 0$. (See Art. 248).

Ex. 5. To find the equation of the evolute of a parabola. We have here

$$S = y^2 - 4mx, \quad S' = 2xy + 2(2m - x')y - 4my',$$
$$\Delta = -4m^2, \quad \Theta = 0, \quad \Theta' = -4m(2m - x), \quad \Delta' = 4my,$$

and the equation of the evolute is $27my^2 = 4(x - 2m)^3$. It is to be observed, that the intersections of S and S' include not only the feet of the *three* normals which can be drawn through any point, but also the point at infinity on y. And the six chords of intersection of S and S' consist of three chords joining the feet of the normals, and three parallels to the axis through these feet. Consequently the method used (Ex., Art. 370) is not the simplest for solving the corresponding problem in the case of the parabola. We get thus the equation found (Ex. 12, Art. 227), but multiplied by the factor $4m(2my + y'x - 2my') - y'^3$.

373. If S' break up into two right lines we have $\Delta' = 0$, and we proceed to examine the meaning in this case of Θ and Θ'. Let us suppose the two right lines to be x and y; and, by the principles already laid down, any property of the invariants, true when the lines of reference are so chosen, will be true in general. The discriminant of $S + 2kxy$ is got by writing $h + k$ for h in Δ, and is $\Delta + 2k(fg - ch) - ck^2$. Now the coefficient of k^2 vanishes when $c = 0$; that is, when the point xy lies on the curve S. The coefficient of k vanishes when $fg = ch$; that is (see Ex. 3, Art. 228), when the lines x and y are conjugate with respect to S. Thus, then, *when S' represents two right lines, Δ' vanishes; $\Theta' = 0$ represents the condition that the intersection of the two lines should lie on S; and $\Theta = 0$ is the condition that the two lines should be conjugate with respect to S.*

The condition that $\Delta + \Theta k + \Theta' k^2$ should be a perfect square is $\Theta^2 = 4\Delta\Theta'$, which, according to the last Article, is the condition that either of the two lines represented by S' should touch S. This is easily verified in the example chosen, where $\Theta^2 - 4\Delta\Theta'$ is found to be equal to $(bc - f^2)(ca - g^2)$.

Ex. 1. Given five conics S_1, S_2, &c., it is of course possible in an infinity of ways to determine the constants l_1, l_2, &c., so that

$$l_1S_1 + l_2S_2 + l_3S_3 + l_4S_4 + l_5S_5$$

may be either a perfect square L^2, or the product of two lines MN: prove that the lines L all touch a fixed conic V, and that the lines M, N are conjugate with regard to V. We can determine V so that the invariant Θ shall vanish for V and each of the five conics, since we have five equations of the form

$$Aa_1 + Bb_1 + Cc_1 + 2Ff_1 + 2Gg_1 + 2Hh_1 = 0,$$

which are sufficient to determine the mutual ratios of A, B, &c., the coefficients in the tangential equation of V. Now if we have separately $Aa_1 +$ &c. $= 0$, $Aa_2 +$ &c. $= 0$, $Aa_3 +$ &c. $= 0$, &c., we have plainly also

$$A\,(l_1a_1 + l_2a_2 + l_3a_3 + l_4a_4 + l_5a_5) + \text{\&c.} = 0\,;$$

that is to say, Θ vanishes for V and every conic of the system

$$l_1S_1 + l_2S_2 + l_3S_3 + l_4S_4 + l_5S_5,$$

whence by this article the theorem stated immediately follows. If the line M be given, N passes through a fixed point; namely, the pole of M with respect to V.

Ex. 2. If six lines x, y, z, u, v, w all touch the same conic, the squares are connected by a linear relation

$$l_1x^2 + l_2y^2 + l_3z^2 + l_4u^2 + l_5v^2 + l_6w^2 = 0.$$

This is a particular case of the last example, but may be also proved as follows: Write down the conditions, Art. 151, that the six lines should touch a conic, and eliminate the unknown quantities A, B, &c., and the condition that the lines should touch the same conic is found to be the vanishing of the determinant

$$\begin{vmatrix} \lambda_1^2, & \mu_1^2, & \nu_1^2, & \mu_1\nu_1, & \nu_1\lambda_1, & \lambda_1\mu_1 \\ \lambda_2^2, & \mu_2^2, & \nu_2^2, & \mu_2\nu_2, & \nu_2\lambda_2, & \lambda_2\mu_2 \\ \lambda_3^2, & \mu_3^2, & \nu_3^2, & \mu_3\nu_3, & \nu_3\lambda_3, & \lambda_3\mu_3 \\ \lambda_4^2, & \mu_4^2, & \nu_4^2, & \mu_4\nu_4, & \nu_4\lambda_4, & \lambda_4\mu_4 \\ \lambda_5^2, & \mu_5^2, & \nu_5^2, & \mu_5\nu_5, & \nu_5\lambda_5, & \lambda_5\mu_5 \\ \lambda_6^2, & \mu_6^2, & \nu_6^2, & \mu_6\nu_6, & \nu_6\lambda_6, & \lambda_6\mu_6 \end{vmatrix}.$$

But this is also the condition that the squares should be connected by a linear relation.

Ex. 3. If we are only given four conics S_1, S_2, S_3, S_4, and seek to determine V, as in Ex. 1, so that Θ shall vanish, then, since we have only four conditions, one of the tangential coefficients A, &c. remains indeterminate, but we can determine all the rest in terms of that; so that the tangential equation of V is of the form $\Sigma + k\Sigma' = 0$, or V touches four fixed lines. We shall afterwards show directly that in four ways we can determine the constants so that $l_1S_1 + l_2S_2 + l_3S_3 + l_4S_4$ may be a perfect square.

It is easy to see (by taking for M the line at infinity) that if M be a given line it is a definite problem admitting of but one solution to determine the constants, so that $l_1S_1 +$ &c. shall be of the form MN. And Ex. 1 shows that N is the locus of the pole of M with regard to V. Compare Ex. 8, Art. 228.

374. *To find the equation of the pair of tangents at the points where S is cut by any line* $\lambda x + \mu y + \nu z$. The equation of any conic having double contact with S, at the points where it meets this line, being $kS + (\lambda x + \mu y + \nu z)^2 = 0$, it is required to determine k so that this shall represent two right lines. Now it will be easily verified that in this case not only Δ' vanishes but Θ also. And if we denote by Σ the quantity

$$A\lambda^2 + B\mu^2 + C\nu^2 + 2F\mu\nu + 2G\nu\lambda + 2H\lambda\mu,$$

the equation to determine k has two roots $= 0$, the third root being given by the equation $k\Delta + \Sigma = 0$. The equation of the pair of tangents is therefore $\Sigma S = \Delta(\lambda x + \mu y + \nu z)^2$. It is plain that when $\lambda x + \mu y + \nu z$ touches S, the pair of tangents coincides with $\lambda x + \lambda y + \nu z$ itself; and the condition that this should be the case is plainly $\Sigma = 0$; as is otherwise proved (Art. 151).

Under the problem of this Article is included that of finding the equation of the asymptotes of a conic given by the general trilinear equation.

375. We now examine the geometrical meaning, in general, of the equation $\Theta = 0$. Let us choose for triangle of reference any self-conjugate triangle with respect to S, which must then reduce to the form $ax^2 + by^2 + cz^2$ (Art. 258). We have therefore $f = 0$, $g = 0$, $h = 0$. The value then of Θ (Art. 370) reduces to $bca' + cab' + abc'$, and will evidently vanish if we have also $a' = 0$, $b' = 0$, $c' = 0$, that is to say, if S', referred to the same triangle, be of the form $f'yz + g'zx + h'xy$. Hence Θ *vanishes whenever any triangle inscribed in S' is self-conjugate with regard to S.* If we choose for triangle of reference any triangle self-conjugate with regard to S', we have $f' = 0$, $g' = 0$, $h' = 0$, and Θ becomes

$$(bc - f^2)\,a' + (ca - g^2)\,b' + (ab - h^2)\,c';$$

and will vanish if we have $bc = f^2$, $ca = g^2$, $ab = h^2$. Now $bc = f^2$ is the condition that the line x should touch S; hence Θ *also vanishes if any triangle circumscribing S is self-conjugate with regard to S'.* In the same manner it is proved that $\Theta' = 0$ *is the condition either that it should be possible to inscribe in S a triangle self-conjugate with regard to S', or to circumscribe about S a triangle self-conjugate with regard to S.* When one of these things is possible, the other is so too.

A pair of conics connected by the relation $\Theta = 0$ possesses another property. Let the point in which meet the lines joining the corresponding vertices of any triangle and of its polar triangle with respect to a conic be called the pole of either triangle with respect to that conic; and let the line joining the intersections of corresponding sides be called their axis. Then if $\Theta = 0$, the pole with respect to S of any triangle inscribed in S' will lie on S'; and the axis with respect to S' of any triangle circumscribing S will touch S. For eliminating x, y, z in turn between each pair of the equations

$$ax + hy + gz = 0, \quad hx + by + fz = 0, \quad gx + fy + cz = 0,$$

we get $$(gh - af)\,x = (hf - bg)\,y = (fg - ch)\,z,$$

for the equations of the lines joining the vertices of the triangle xyz to the corresponding vertices of its polar triangle with respect to S. These equations may be written $Fx = Gy = Hz$, and the coordinates of the pole of the triangle are $\frac{1}{F}, \frac{1}{G}, \frac{1}{H}$. Substituting these values in S', in which it is supposed that the coefficients a', b', c' vanish, we get $2Ff' + 2Gg' + 2Hh' = 0$, or $\Theta = 0$. The second part of the theorem is proved in like manner.

Ex. 1. If two triangles be self-conjugate with regard to any conic S', a conic can be described passing through their six vertices; and another can be described touching their six sides (see Ex. 7, Art. 356). Let a conic be described through the three vertices of one triangle and through two of the other, which we take for x, y, z. Then, because it circumscribes the first triangle, $\Theta' = 0$, or $a + b + c = 0$ (Ex. 2, Art. 371), and, because it goes through two vertices of xyz, we have $a = 0$, $b = 0$, therefore $c = 0$, or the conic goes through the remaining vertex. The second part of the theorem is proved in like manner.

Ex. 2. The square of the tangent drawn from the centre of a conic to the circle circumscribing any self-conjugate triangle is constant, and $= a^2 + b^2$ [M. Faure] This is merely the geometrical interpretation of the condition $\Theta = 0$, found (Ex. 4, Art. 371), or $\alpha^2 + \beta^2 - r^2 = a^2 + b^2$. The theorem may be otherwise stated thus: "Every circle which circumscribes a self-conjugate triangle cuts orthogonally the circle which is the locus of the intersection of tangents mutually at right angles." For the square of the radius of the latter circle is $a^2 + b^2$.

Ex. 3. The centre of the circle inscribed in every self-conjugate triangle with respect to an equilateral hyperbola lies on the curve. This appears by making $b^2 = -a^2$ in the condition $\Theta' = 0$ (Ex. 4, Art. 371).

Ex. 4. If the rectangle under the segments of one of the perpendiculars of the triangle formed by three tangents to a conic be constant and equal to M, the locus of the intersection of perpendiculars is the circle $x^2 + y^2 = a^2 + b^2 + M$. For $\Theta = 0$ (Ex. 4, Art. 371) is the condition that a triangle self-conjugate with regard to the circle can be circumscribed about S. But when a triangle is self-conjugate with regard to a circle, the intersection of perpendiculars is the centre of the circle and M is the square of the radius (Ex. 3, Art. 278). The locus of the intersection of rectangular tangents is got from this example by making $M = 0$.

Ex. 5. If the rectangle under the segments of one of the perpendiculars of a triangle inscribed in S be constant, and $= M$, the locus of intersection of perpendiculars is the conic concentric and similar with S, $S = M\left(\frac{1}{a^2} + \frac{1}{b^2}\right)$ [Dr. Hart]. This follows in the same way from $\Theta' = 0$.

Ex. 6. Find the locus of the intersection of perpendiculars of a triangle inscribed in one conic and circumscribed about another [Mr. Burnside]. Take for origin the centre of the latter conic, and equate the values of M found from Ex. 4 and 5; then if a', b' be the axes of the conic S in which the triangle is inscribed, the equation of the locus is $x^2 + y^2 - a^2 - b^2 = \frac{a'^2 b'^2}{a'^2 + b'^2} S$. The locus is therefore a conic, whose axes are parallel to those of S, and which is a circle when S is a circle.

Ex. 7. The centre of the circle circumscribing every triangle, self-conjugate with regard to a parabola, lies on the directrix. This and the next example follow from $\Theta = 0$ (Ex. 5, Art. 371).

Ex. 8. The intersection of perpendiculars of any triangle circumscribing a parabola lies on the directrix.

Ex. 9. Given the radius of the circle inscribed in a self-conjugate triangle, the locus of centre is a parabola of equal parameter with the given one.

376. If two conics be taken arbitrarily it is in general not possible to inscribe a triangle in one which shall be circumscribed about the other; but an infinity of such triangles can be drawn if the coefficients of the conics be connected by a certain relation, which we proceed to determine. Let us suppose that such a triangle can be described, and let us take it for triangle of reference; then the equations of the two conics must be reducible to the form

$$S = x^2 + y^2 + z^2 - 2yz - 2zx - 2xy = 0,$$
$$S' = 2fyz + 2gzx + 2hxy = 0.$$

Forming then the invariants we have

$$\Delta = -4, \quad \Theta = 4(f+g+h), \quad \Theta' = -(f+g+h)^2, \quad \Delta' = 2fgh;$$

values which are evidently connected by the relation $\Theta^2 = 4\Delta\Theta'$.*

* This condition was first given by Prof. Cayley (*Philosophical Magazine*, vol. VI. p. 99) who derived it from the theory of elliptic functions. He also proved, in the same way, that if the square root of $k^3\Delta + k^2\Theta + k\Theta' + \Delta'$, when expanded in powers of k, be $A + Bk + Ck^2 +$ &c., then the conditions that it should be possible to have a polygon of n sides inscribed in U and circumscribing V, are for $n = 3, 5, 7$, &c. respectively

$$C = 0, \quad \begin{vmatrix} C, & D \\ D, & E \end{vmatrix} = 0, \quad \begin{vmatrix} C, & D, & E \\ D, & E, & F \\ E, & F, & G \end{vmatrix} = 0, \text{ \&c.}$$

This is an equation of the kind (Art. 371) which is unaffected by any change of axes; therefore, no matter what the form in which the equations of the conics have been originally given, this relation between their coefficients must exist, if they are capable of being transformed to the forms here given. Conversely, it is easy to show, as in Ex. 1, Art. 375, that when the relation holds $\Theta^2 = 4\Delta\Theta'$, then if we take any triangle circumscribing S, and two of whose vertices rest on S', the third must do so likewise.

Ex. 1. Find the condition that two circles may be such that a triangle can be inscribed in one and circumscribed about the other. Let $D^2 - r^2 - r'^2 = G$, then the condition is (see Ex. 3, Art. 371)

$$(G - r^2)^2 + 4r^2 (G - r'^2) = 0, \text{ or } (G + r^2)^2 = 4r^2r'^2;$$

whence $D^2 = r'^2 \pm 2rr'$, Euler's well known expression for the distance between the centre of the circumscribing circle and that of one of the circles which touch the three sides.

Ex. 2. Find the locus of the centre of a circle of given radius, circumscribing a triangle circumscribing a conic, or inscribed in an inscribed triangle. The loci are curves of the fourth degree, except that of the centre of the circumscribing circle in the case of the parabola, which is a circle whose centre is the focus, as is otherwise evident.

Ex. 3. Find the condition that a triangle may be inscribed in S' whose sides touch respectively $S + lS'$, $S + mS'$, $S + nS'$. Let

$$S = x^2 + y^2 + z^2 - 2(1 + lf)\,yz - 2(1 + mg)\,zx - 2(1 + nh)\,xy,$$
$$S' = 2fyz + 2gzx + 2hxy;$$

then it is evident that $S + lS'$ is touched by x, &c. We have then

$$\Delta = -(2 + lf + mg + nh)^2 - 2lmnfgh,$$
$$\Theta = 2(f + g + h)(2 + lf + mg + nh) + 2fgh(mn + nl + lm),$$
$$\Theta' = -(f + g + h)^2 - 2(l + m + n)fgh, \quad \Delta' = 2fgh.$$

Whence, obviously,

$$\{\Theta - \Delta'(mn + nl + lm)\}^2 = 4(\Delta + lmn\Delta')\{\Theta' + \Delta'(l + m + n)\},$$

which is the required condition.

377. *To find the condition that the line $\lambda x + \mu y + \nu z$ should pass through one of the four points common to S and S'.* This is, in other words, to find the tangential equation of these four points. Now we get the tangential equation of any conic of

---

and for $n = 4, 6, 8$, &c. are

$$D = 0 \quad \begin{vmatrix} D, & E \\ E, & F \end{vmatrix} = 0, \quad \begin{vmatrix} D, & E, & F \\ E, & F, & G \\ F, & G, & H \end{vmatrix} = 0, \text{ \&c.}$$

the system $S+kS'$ by writing $a+ka'$, &c. for a, &c. in the tangential equation of S, or

$$\Sigma=(bc-f^2)\lambda^2+(ca-g^2)\mu^2+(ab-h^2)\nu^2$$
$$+2(gh-af)\mu\nu+2(hf-bg)\nu\lambda+2(fg-ch)\lambda\mu=0.$$

We get thus $\Sigma+k\Phi+k^2\Sigma'=0$, where

$$\Phi=(bc'+b'c-2ff')\lambda^2+(ca'+c'a-2gg')\mu^2$$
$$+(ab'+a'b-2hh')\nu^2+2(gh'+g'h-af'-a'f)\mu\nu$$
$$+2(hf'+h'f-bg'-b'g)\nu\lambda+2(fg'+f'g-ch'-c'h)\lambda\mu.$$

The tangential equation of the envelope of this system is therefore (Art. 298) $\Phi^2=4\Sigma\Sigma'$. But since $S+kS'$, and the corresponding tangential equation, belong to a system of conics passing through four fixed points, the envelope of the system is nothing but these four points, and the equation $\Phi^2=4\Sigma\Sigma'$ is the required condition that the line $\lambda x+\mu y+\nu z$ should pass through one of the four points. The matter may be also stated thus: Through four points there can in general be described two conics to touch a given line (Art. 345, Ex. 4); but if the given line pass through one of the four points, both conics coincide in one whose point of contact is that point. Now $\Phi^2=4\Sigma\Sigma'$ is the condition that the two conics of the system $S+kS'$, which can be drawn to touch $\lambda x+\mu y+\nu z$, shall coincide.

It will be observed that $\Phi=0$ is the condition obtained (Art. 335), that the line $\lambda x+\mu y+\nu z$ shall be cut harmonically by the two conics.

378. *To find the equation of t e four common tangents to two conics.* This is the reciprocal of the problem of the last Article, and is treated in the same way. Let Σ and Σ' be the tangential equations of two conics, then (Art. 298) $\Sigma+k\Sigma'$ represents tangentially a conic touched by the four tangents common to the two given conics. Forming then, by Art. 285, the trilinear equation corresponding to $\Sigma+k\Sigma'=0$, we get

$$\Delta S+k\mathbf{F}+k^2\Delta'S'=0,$$

where

$$\mathbf{F}=(BC'+B'C-2FF')x^2+(CA'+C'A-2GG')y^2$$
$$+(AB'+A'B-2HH')z^2$$
$$+2(GH'+G'H-AF'-A'F)yz+2(HF'+H'F-BG'-B'G)zx$$
$$+2(FG'+F'G-CH'-C'H)xy,$$

the letters A, B, &c. having the same meaning as in Art. 151. But $\Delta S + k\mathbf{F} + k^2\Delta' S'$ denotes a system of conics whose envelope is $\mathbf{F}^2 = 4\Delta\Delta' SS'$; and the envelope of the system evidently is the four common tangents.

The equation $\mathbf{F}^2 = 4\Delta\Delta' SS'$, by its form denotes a locus touching S and S', the curve $\mathbf{F}$ passing through the points of contact. Hence, *the eight points of contact of two conics with their common tangents, lie on another conic* $\mathbf{F}$. Reciprocally, *the eight tangents at the points of intersection of two conics envelope another conic* Φ.

It will be observed that $\mathbf{F} = 0$ is the equation found, Art. 334, of the locus of points, whence tangents to the two conics form a harmonic pencil.*

If S' reduces to a pair of right lines, $\mathbf{F}$ represents the pair of tangents to S from their intersection.

Ex. Find the equation of the common tangents to the pair of conics

$$ax^2 + by + cz^2 = 0, \quad a'x^2 + b'y^2 + c'z^2 = 0.$$

Here $A = bc$, $B = ca$, $C = ab$, whence

$$\mathbf{F} = aa'(bc' + b'c)\,x^2 + bb'(ca' + c'a)\,y^2 + cc'(ab' + a'b)\,z^2,$$

and the required equation is

$$\{aa'(b'c + b'c)\,x^2 + bb'(ca' + c'a)\,y^2 + cc'(ab' + a'b)\,z^2\}^2$$
$$= 4abca'b'c'(ax^2 + by^2 + cz^2)(a'x^2 + b'y^2 + c'z^2),$$

which is easily resolved into the four factors

$$x\sqrt{\{aa'(bc')\}} \pm y\sqrt{\{bb'(ca')\}} \pm z\sqrt{\{cc'(ab')\}} = 0.$$

378*a*. If S and S' touch, $\mathbf{F}$ touches each at their point of contact. This follows immediately from the fact that $\mathbf{F}$ passes through the points of contact of common tangents to S and S'. Similarly if S and S' touch in two distinct points, $\mathbf{F}$ also has double contact with them in these points. This may be verified by forming the $\mathbf{F}$ of $cz^2 + 2hxy$, $c'z^2 + 2h'xy$ which is found to be of the same form, viz. $2cc'hh'z^2 + 2hh'(ch' + c'h)\,xy$.

From what has been just observed, that when S and S' have double contact, $\mathbf{F}$ is of the form $lS + mS'$, we can obtain a system of conditions that two conics may have double contact. For write the general value of $\mathbf{F}$, given Art. 334,

$$\mathrm{a}x^2 + \mathrm{b}y^2 + \mathrm{c}z^2 + 2\mathrm{f}yz + 2\mathrm{g}zx + 2\mathrm{h}xy,$$

* I believe I was the first to direct attention to the importance of this conic in the theory of two conics.

then evidently if they have double contact every determinant vanishes of the system

$$\left\| \begin{matrix} a, & b, & c, & f, & g, & h \\ a', & b', & c', & f', & g', & h' \\ \text{a}, & \text{b}, & \text{c}, & \text{f}, & \text{g}, & \text{h} \end{matrix} \right\| = 0.$$

That when S and S' have double contact, S, $\mathbf{F}$ and S' are connected by a linear relation, may be otherwise seen, as follows: When S and S' have double contact there is a value of k for which $kS+S'$ represents two coincident right lines. Now the reciprocal of a conic representing two coincident right lines vanishes identically. Hence we have

$$k^2\Sigma + k\Phi + \Sigma' = 0$$

identically. But the value of k, for which this is the case, is the double root of the equation

$$k^3\Delta + k^2\Theta + k\Theta' + \Delta' = 0.$$

Eliminating k between the former equation and the two differentials of the latter we have Σ, Σ', Φ satisfying the identical relation

$$\left| \begin{matrix} \Sigma, & \Phi, & \Sigma' \\ 3\Delta, & 2\Theta, & \Theta' \\ \Theta, & 2\Theta', & 3\Delta' \end{matrix} \right| = 0.$$

When two conics have double contact their reciprocals have double contact also; and it may be seen without difficulty that the relation just written between Σ, Σ', Φ implies the following between S, S', $\mathbf{F}$

$$\left| \begin{matrix} S, & \mathbf{F}, & S' \\ 3\Delta, & 2\Delta\Theta', & \Theta \\ \Theta', & 2\Delta'\Theta, & 3\Delta' \end{matrix} \right| = 0.$$

379. The former part of this Chapter has sufficiently shown what is meant by invariants, and the last Article will serve to illustrate the meaning of the word *covariant*. Invariants and covariants agree in this, that the geometric meaning of both is independent of the axes to which the questions are referred; but invariants are functions of the coefficients only, while covariants contain the variables as well. If we are given a curve, or system of curves, and have learned to derive from their general equations the equation of some locus, $U=0$,

whose relation to the given curves is independent of the axes to which the equations are referred, U is said to be a covariant of the given system. Now if we desire to have the equation of this locus referred to any new axes, we shall evidently arrive at the same result, whether we transform to the new axes the equation $U=0$, or whether we transform to the new axes the equations of the given curves themselves, and from the transformed equations derive the equation of the locus by the same rule that U was originally formed. Thus, if we transform the equations of two conics to a new triangle of reference, by writing instead of x, y, z,

$$lx+my+nz,\quad l'x+m'y+n'z,\quad l''x+m''y+n''z;$$

and if we make the same substitution in the equation $\mathbf{F}^2=4\Delta\Delta'SS'$, we can foresee that the result of this last substitution can only differ by a constant multiplier from the equation $\mathbf{F}^2=4\Delta\Delta'SS'$, formed with the new coefficients of S and S'. For either form represents the four common tangents. On this property is founded the analytical definition of covariants. "A derived function formed by any rule from one or more given functions is said to be a covariant, if when the variables in all are transformed by the same linear substitutions, the result obtained by transforming the derived differs only by a constant multiplier from that obtained by transforming the original equations and then forming the corresponding derived."

380. There is another case in which it is possible to predict the result of a transformation by linear substitution. If we have learned how to form the condition that the line $\lambda x+\mu y+\nu z$ should touch a curve, or more generally that it should hold to a curve, or system of curves, any relation independent of the axes to which the equations are referred, then it is evident that when the equations are transformed to any new coordinates, the corresponding condition can be formed by the same rule from the transformed equations. But it might also have been obtained by direct transformation from the condition first obtained. Suppose that by transformation $\lambda x+\mu y+\nu z$ becomes

$$\lambda(lx+my+nz)+\mu(l'x+m'y+n'z)+\nu(l''x+m''y+n''z),$$

and that we write this $\lambda'x+\mu'y+\nu'z$, we have

$$\lambda'=l\lambda+l'\mu+l''\nu,\quad \mu'=m\lambda+m'\mu+m''\nu,\quad \nu'=n\lambda+n'\mu+n''\nu.$$

Solving these equations, we get equations of the form

$$\lambda = L\lambda' + L'\mu' + L''\nu',\ \mu = M\lambda' + M'\mu' + M''\nu',\ \nu = N\lambda' + N'\mu' + N''\nu'.$$

If then we put these values into the condition as first obtained in terms of λ, μ, ν, we get the condition in terms of λ', μ', ν', which can only differ by a constant multiplier from the condition as obtained by the other method. Functions of the class here considered are called *contravariants*. Contravariants are like covariants in this: that any contravariant equation, as for example, the tangential equation of a conic $(bc - f^2)\lambda^2 + \&c. = 0$ can be transformed by linear substitution into the equation of like form $(b'c' - f'^2)\lambda'^2 + \&c. = 0$, formed with the coefficients of the transformed trilinear equation of the conic. But they differ in that λ, μ, ν are not transformed by the same rule as x, y, z; that is, by writing for λ, $l\lambda + m\mu + n\nu$, &c., but by the different rule explained above.

The condition $\Phi = 0$ found, Art. 377, is evidently a contravariant of the system of conics S, S'.

381. It will be found that the equation of any conic covariant with S and S' can be expressed in terms of S, S' and $\mathbf{F}$; while its tangential equation can be expressed in terms of Σ, Σ', Φ.

Ex. 1. To express in terms of S, S', $\mathbf{F}$ the equation of the polar conic of S with respect to S'. From the nature of covariants and invariants, any relation found connecting these quantities, when the equations are referred to any axes, must remain true when the equations are transformed. We may therefore refer S and S' to their common self-conjugate triangle and write $S = ax^2 + by^2 + cz^2$, $S' = x^2 + y^2 + z^2$. It will be found then that $\mathbf{F} = a(b+c)x^2 + b(c+a)y^2 + c(a+b)z^2$. Now since the condition that a line should touch S is $bc\lambda^2 + ca\mu^2 + ab\nu^2 = 0$, the locus of the poles with respect to S' of the tangents to S is $bcx^2 + cay^2 + abz^2 = 0$. But this may be written $(bc + ca + ab)(x^2 + y^2 + z^2) = \mathbf{F}$. The locus is therefore (Ex. 1, Art. 371) $\Theta S' = \mathbf{F}$. In like manner the polar conic of S' with regard to S is $\Theta' S = \mathbf{F}$.

Ex. 2. To express in terms of S, S', $\mathbf{F}$ the conic enveloped by a line cut harmonically by S and S'. The tangential equation of this conic $\Phi = 0$ is

$$(b+c)\lambda^2 + (c+a)\mu^2 + (a+b)\nu^2 = 0.$$

Hence its trilinear equation is

$$(c+a)(a+b)x^2 + (a+b)(b+c)y^2 + (c+a)(b+c)z^2 = 0,$$

or $$(bc + ca + ab)(x^2 + y^2 + z^2) + (a+b+c)(ax^2 + by^2 + cz^2) - \mathbf{F} = 0,$$

or $$\Theta S' + \Theta' S - \mathbf{F} = 0.$$

Ex. 3. To find the condition that $\mathbf{F}$ should break up into two right lines. It is

$$abc(b+c)(c+a)(a+b) = 0,\ \text{or}\ abc\{(a+b+c)(bc+ca+ab) - abc\} = 0,$$

or $$\Delta\Delta'(\Theta\Theta' - \Delta\Delta') = 0.$$

which is the required formula. $\Theta\Theta' = \Delta\Delta'$ is also the condition that Φ should break up into factors. This condition will be found to be satisfied in the case of two circles which cut at right angles, in which case any line through either centre is cut harmonically by the circles, and the locus of points whence tangents form a harmonic pencil also reduces to two right lines. The locus and envelope will reduce similarly if $D^2 = 2(r^2 + r'^2)$.

Ex. 4. To reduce the equations of two conics to the forms

$$x^2 + y^2 + z^2 = 0, \quad ax^2 + by^2 + cz^2 = 0.$$

The constants a, b, c are determined at once (Ex. 1, Art. 371) as the roots of

$$\Delta k^3 - \Theta k^2 + \Theta' k - \Delta' = 0.$$

And if we then solve the equations

$$x^2 + y^2 + z^2 = S, \quad ax^2 + by^2 + cz^2 = S', \quad a(b+c)x^2 + b(c+a)y^2 + c(a+b)z^2 = \mathbf{F},$$

we find x^2, y^2, z^2 in terms of the known functions $S, S', \mathbf{F}$. Strictly speaking, we ought to commence by dividing the two given equations by the cube root of Δ, since we want to reduce them to a form in which the discriminant of S shall be 1. But it will be seen that it will come to the same thing if leaving S and S' unchanged, we calculate $\mathbf{F}$ from the given coefficients and divide the result by Δ.

Ex. 5. Reduce to the above form

$$3x^2 - 6xy + 9y^2 - 2x + 4y = 0, \quad 5x^2 - 14xy + 8y^2 - 6x - 2 = 0.$$

It is convenient to begin by forming the coefficients of the tangential equations A, B, &c. These are $-4, -1, 18; -3, 3, -2; -16, -19, -9; 21, 24, -14$. We have then

$$\Delta = -9, \quad \Theta = -54, \quad \Theta' = -99, \quad \Delta' = -54,$$

whence a, b, c are 1, 2, 3. We next calculate $\mathbf{F}$ which is

$$-9(23x^2 - 50xy + 44y^2 - 18x + 12y - 4).$$

Writing then

$$X^2 + Y^2 + Z^2 = 3x^2 - 6xy + 9y^2 - 2x + 4y,$$
$$X^2 + 2Y^2 + 3Z^2 = 5x^2 - 14xy + 8y^2 - 6x - 2,$$
$$5X^2 + 8Y^2 + 9Z^2 = 23x^2 - 50xy + 44y^2 - 18x + 12y - 4.$$

We get from $6S + S' - \mathbf{F}, \quad X^2 = (3y+1)^2,$

from $\mathbf{F} - 3S - 2S', \quad Y^2 = (2x - y)^2,$

from $2S + 3S' - \mathbf{F}, \quad Z^2 = -(x + y + 1)^2.$

Ex. 6. To find the equation of the four tangents to S at its intersections with S'.

Ans. $(\Theta S - \Delta S')^2 = 4\Delta S(\Theta' S - \mathbf{F})$.

Ex. 7. A triangle is circumscribed to a given conic; two of its vertices move on fixed right lines $\lambda x + \mu y + \nu z$, $\lambda' x + \mu' y + \nu' z$; to find the locus of the third. It was proved (Ex. 2, Art. 272) that when the conic is $z^2 - xy$, and the lines $ax - y$, $bx - y$, the locus is $(a+b)^2(z^2 - xy) = (a-b)^2 z^2$. Now the right-hand side is the square of the polar with regard to S of the intersection of the lines, which in general would be

$$P = (ax + hy + gz)(\mu\nu' - \mu'\nu) + (hx + by + fz)(\nu\lambda' - \nu'\lambda) + (gx + fy + cz)(\lambda\mu' - \lambda'\mu) = 0,$$

and $a + b = 0$ is the condition that the lines should be conjugate with respect to S, which in general (Art. 373) is $\Theta = 0$, where

$$\Theta = A\lambda\lambda' + B\mu\mu' + C\nu\nu' + F(\mu\nu' + \mu'\nu) + G(\nu\lambda' + \nu'\lambda) + H(\lambda\mu' + \lambda'\mu) = 0.$$

The particular equation, found Art. 272, must therefore be replaced in general by

$$\Theta^2 U + \Delta P^2 = 0.$$

Ex. 8. To find the envelope of the base of a triangle inscribed in S and two of whose sides touch S'.

Take the sides of the triangle in any position for lines of reference, and let

$$S = 2(fyz + gzx + hxy),$$
$$S' = x^2 + y^2 + z^2 - 2yz - 2zx - 2xy - 2hkxy,$$

where x and y are the lines touched by S'. Then it is obvious that $kS + S'$ will be touched by the third side z, and we shall show by the invariants that this is a *fixed* conic. We have

$\Delta = 2fgh,\ \Theta = -(f+g+h)^2 - 2fghk,\ \Theta' = 2(f+g+h)(2+hk),\ \Delta' = -(2+hk)^2,$ whence $\Theta'^2 - 4\Theta\Delta' = 4\Delta\Delta'k$, and the equation $kS + S' = 0$ may be written in the form

$$(\Theta'^2 - 4\Theta\Delta')\, S + 4\Delta\Delta' S' = 0,$$

which therefore denotes a fixed conic touched by the third side of the triangle. It is obvious that when $\Theta'^2 = 4\Theta\Delta'$ the third side will always touch S'.

Ex. 9. To find the locus of the vertex of a triangle whose three sides touch a conic U and two of whose vertices move on another conic V. We have slightly altered the notation, for the convenience of being able to denote by U' and V' the results of substituting in U and V the coordinates of the vertex $x'y'z'$. The method we pursue is to form the equation of the pair of tangents to U through $x'y'z'$; then to form the equation of the lines joining the points where this pair of lines meets V; and, lastly, to form the condition that one of these lines (which must be the base of the triangle in question) touches V. Now if P be the polar of $x'y'z'$, the pair of tangents is $UU' - P^2$. In order to find the chords of intersection with V of the pair of tangents, we form the condition that $UU' - P^2 + \lambda V$ may represent a pair of lines. This discriminant will be found to give us the following quadratic for determining λ, $\lambda^2\Delta' + \lambda F' + \Delta U'V' = 0$. In order to find the condition that one of these chords should touch U, we must, by Art. 372, form the discriminant of $\mu U + (UU' - P^2 + \lambda V)$, and then form the condition that this considered as a function of μ should have equal roots. The discriminant is

$$\mu^2\Delta + \mu\,(2U'\Delta + \lambda\Theta) + \{U'^2\Delta + \lambda\,(\Theta U' + \Delta V') + \lambda^2\Theta'\},$$

and the condition for equal roots gives

$$\lambda\,(4\Delta\Theta' - \Theta^2) + 4\Delta^2 V' = 0.$$

Substituting this value for λ in $\lambda^2\Delta' + \lambda F' + \Delta U'V'$, we get the equation of the required locus

$$16\Delta^3\Delta' V - 4\Delta\,(4\Delta\Theta' - \Theta^2)\,F + U\,(4\Delta\Theta' - \Theta^2)^2 = 0,$$

which, as it ought to do, reduces to V when $4\Delta\Theta' = \Theta^2$.*

Ex. 10. Find the locus of the vertex of a triangle, two of whose sides touch U, and the third side $aU + bV$, while the two base angles move on V. It is found by the same method as the last, that the locus is one or other of the conics, touching the four common tangents of U and V,

$$\Delta\Delta'\lambda^2 V + \lambda\mu F + \mu^2 U = 0,$$

where $\lambda : \mu$ is given by the quadratic

$$a\,(ab - \beta a)\,\lambda^2 + a\,(4\Delta a + 2\Theta b)\,\lambda\mu - b^2\mu^2 = 0,$$

where $$\alpha = 4\Delta\Delta',\quad \beta = \Theta^2 - 4\Delta\Theta'.$$

Ex. 11. To find the locus of the free vertex of a polygon, all whose sides touch U, and all whose vertices but one move on V. This is reduced to the last; for the line joining two vertices of the polygon adjacent to that whose locus is sought, touches a conic of the form $aU + bV$. It will be found if $\lambda', \mu';\ \lambda'', \mu'';\ \lambda''', \mu'''$ be the values for polygons of $n-1$, n, and $n+1$ sides respectively, that $\lambda''' = \mu'\mu''^2$, $\mu''' = \Delta'\lambda'\lambda''\,(a\mu'' - \Delta'\beta\lambda'')$. In the case of the triangle we have $\lambda' = \alpha$, $\mu' = \Delta'\beta$; in the case of the quadrilateral $\lambda'' = \beta^2$, $\mu'' = \alpha\,(4\Delta\alpha + 2\beta\Theta)$, and from these we can

---

* The reader will find (*Quarterly Journal of Mathematics*, vol. I. p. 344) a discussion by Prof. Cayley of the problem to find the locus of vertex of a triangle circumscribing a conic S, and whose base angles move on given curves. When the curves are both conics, the locus is of the eighth degree, and touches S at the points where it is met by the polars with regard to S of the intersections of the two conics.

find, step by step, the values for every other polygon. (See *Philosophical Magazine*, vol. XIII. p. 337).

Ex. 12. The triangle formed by the polars of middle points of sides of a given triangle with regard to any inscribed conic has a constant area [M. Faure].

Ex. 13. Find the condition that if the points in which a conic meets the sides of the triangle of reference be joined to the opposite vertices, the joining lines shall form two sets of three each meeting in a point. *Ans.* $abc - 2fgh - af^2 - bg^2 - ch^2 = 0$.

382. The theory of covariants and invariants enables us readily to recognize the equivalents in trilinear coordinates of certain well-known formulæ in Cartesian. Since the general expression for a line passing through one of the imaginary circular points at infinity is $x \pm y\sqrt{(-1)} + c$, the condition that $\lambda x + \mu y + \nu$ should pass through one of these points is $\lambda^2 + \mu^2 = 0$. In other words, this is the tangential equation of these points. If then $\Sigma = 0$ be the tangential equation of a conic, we may form the discriminant of $\Sigma + k(\lambda^2 + \mu^2)$. Now it follows from Arts. 285, 286, that the discriminant in general of $\Sigma + k\Sigma'$ is

$$\Delta^2 + k\Delta\Theta' + k^2\Delta'\Theta + k^3\Delta'^2.$$

But the discriminant of $\Sigma + k(\lambda^2 + \mu^2)$ is easily found to be

$$\Delta^2 + k\Delta(a+b) + k^2(ab - h^2).$$

If, then, in any system of coordinates we form the invariants of any conic and the pair of circular points, $\Theta' = 0$ is the condition that the curve should be an equilateral hyperbola, and $\Theta = 0$ that it should be a parabola. The condition

$$(a+b)^2 = 4(ab - h^2), \text{ or } (a-b)^2 + 4h^2 = 0,$$

must be satisfied if the conic pass through either circular point; and it cannot be satisfied by real values except the conic pass through *both*, when $a = b$, $h = 0$.

Now the condition $\lambda^2 + \mu^2 = 0$* implies (Art. 34) that the length of the perpendicular let fall from any point on any line passing through one of the circular points is always infinite. The equivalent condition in trilinear coordinates is therefore got by equating to nothing the denominator in the expression

* This condition also implies (Art. 25) that every line drawn through one of these two points is perpendicular to itself. This accounts for some apparently irrelevant factors which appear in the equations of certain loci. Thus, if we look for the equation of the foot of the perpendicular on any tangent from a focus $\alpha\beta$, $(x-\alpha)^2 + (y-\beta)^2$ will appear as a factor in the locus. For the perpendicular from the focus on either tangent through it coincides with the tangent itself. This tangent therefore is part of the locus.

for the length of a perpendicular (Art. 61). The general tangential equation of the circular points is therefore

$$\lambda^2 + \mu^2 + \nu^2 - 2\mu\nu \cos A - 2\nu\lambda \cos B - 2\lambda\mu \cos C = 0.$$

Forming then the Θ and Θ' of the system found by combining this with any conic, we find that the condition for an equilateral hyperbola $\Theta' = 0$, is

$$a + b + c - 2f \cos A - 2g \cos B - 2h \cos C = 0;$$

while the condition for a parabola $\Theta = 0$, is

$$A \sin^2 A + B \sin^2 B + C \sin^2 C + 2F \sin B \sin C$$
$$+ 2G \sin C \sin A + 2H \sin A \sin B = 0.$$

The condition that the curve should pass through either circular point is $\Theta'^2 = 4\Theta$, which can in various ways be resolved into a sum of squares.

383. If we are given a conic and a pair of points, the covariant **F** of the system denotes the locus of a point such that the pair of tangents through it to the conic are harmonically conjugate with the lines to the given pair of points. When the pair of points is the pair of circular points at infinity, **F** denotes the locus of the intersection of tangents at right angles. Now, referring to the value of **F**, given Art. 378, it is easy to see that when the second conic reduces to $\lambda^2 + \mu^2$; that is, when $A' = B' = 1$, and all the other coefficients of the tangential of the second conic vanish, **F** is

$$C(x^2 + y^2) - 2Gx - 2Fy + A + B = 0,$$

which is, therefore, the general Cartesian equation of the locus of intersection of rectangular tangents. (See Art. 294, Ex.).

When the curve is a parabola $C = 0$, and the equation of the directrix is therefore $2(Gx + Fy) = A + B$.

The corresponding trilinear equation found in the same way is

$$(B + C + 2F \cos A)x^2 + (C + A + 2G \cos B)y^2 + (A + B + 2H \cos C)z^2$$
$$+ 2(A \cos A - F - G \cos C - H \cos B)yz$$
$$+ 2(B \cos B - G - H \cos A - F \cos C)zx$$
$$+ 2(C \cos C - H - F \cos B - G \cos A)xy = 0,$$

It may be shown, as in Art. 128, that this represents a circle, by throwing it into the form

$$(x\sin A+y\sin B+z\sin C)\left(\frac{B+C+2F\cos A}{\sin A}x+\frac{C+A+2G\cos B}{\sin B}y\right.$$

$$\left.+\frac{A+B+2H\cos C}{\sin C}z\right)=\frac{\Theta}{\sin A\sin B\sin C}(yz\sin A+zx\sin B+xy\sin C),$$

where $\Theta=0$ is the condition (Art. 382) that the curve should be a parabola. When $\Theta=0$, this equation gives the equation of the directrix.

384. In general, $\Sigma+k\Sigma'$ denotes a conic touching the four tangents common to Σ and Σ'; and when k is determined so that $\Sigma+k\Sigma'$ represents a pair of points, those points are two opposite vertices of the quadrilateral formed by the common tangents. In the case where Σ' denotes the circular points at infinity, when $\Sigma+k\Sigma'$ represents a pair of points, these points are the foci (Art. 258*a*). If, then, it be required to find the foci of a conic, given by a numerical equation in Cartesian coordinates, we first determine k from the quadratic

$$(ab-h^2)k^2+\Delta(a+b)k+\Delta^2=0.$$

Then, substituting either value of k in $\Sigma+k(\lambda^2+\mu^2)$, it breaks up into factors $(\lambda x'+\mu y'+\nu z')(\lambda x''+\mu y''+\nu z'')$; and the foci are $\frac{x'}{z'}, \frac{y'}{z'}; \frac{x''}{z''}, \frac{y''}{z''}$. One value of k gives the two real foci, and the other two imaginary foci. The same process is applicable to trilinear coordinates.

In general, $\Sigma+k(\lambda^2+\mu^2)$ represents tangentially a conic confocal with the given one. Forming, by Art. 285, the corresponding Cartesian equation, we find that the general equation of a conic confocal with the given one is

$$\Delta S+k\{C(x^2+y^2)-2Gx-2Fy+A+B\}+k^2=0.$$

From this we can deduce that the equation of common tangents is

$$\{C(x^2+y^2)-2Gx-2Fy+A+B\}^2=4\Delta S.$$

By resolving this into a pair of factors

$$\{(x-\alpha)^2+(y-\beta)^2\}\{(x-\alpha')^2+(y-\beta')^2\},$$

we can also get α, β; α', β' the coordinates of the foci.

Ex. 1. Find the foci of $2x^2 - 2xy + 2y^2 - 2x - 8y + 11$. The quadratic here is $3k^2 + 4k\Delta + \Delta^2 = 0$, whose roots are $k = -\Delta$, $k = -\frac{1}{3}\Delta$. But $\Delta = -9$. Using the value $k = 3$,

$$6\lambda^2 + 21\mu^2 + 3\nu^2 + 18\mu\nu + 12\nu\lambda + 30\lambda\mu + 3(\lambda^2 + \mu^2) = 3(\lambda + 2\mu + \nu)(3\lambda + 4\mu + \nu),$$

showing that the foci are 1, 2; 3, 4. The value 9 gives the imaginary foci $2 \pm \sqrt{(-1)}$, $3 \mp \sqrt{(-1)}$.

Ex. 2. Find the coordinates of the focus of a parabola given by a Cartesian equation. The quadratic here reduces to a simple equation, and we find that

$$(a + b)\{A\lambda^2 + B\mu^2 + 2F\mu\nu + 2G\nu\lambda + 2H\lambda\mu\} - \Delta(\lambda^2 + \mu^2)$$

is resolvable into factors. But these evidently must be

$$(a + b)(2G\lambda + 2F\mu) \text{ and } \frac{(a + b)A - \Delta}{2(a + b)G}\lambda + \frac{(a + b)B - \Delta}{2(a + b)F}\mu + \nu.$$

The first factor gives the infinitely distant focus, and shows that the axis of the curve is parallel to $Fx - Gy$. The second factor shows that the coordinates of the focus are the coefficients of λ and μ in that factor.

Ex. 3. Find the coordinates of the focus of a parabola given by the trilinear equation. The equation which represents the pair of foci is

$$\Theta'\Sigma = \Delta(\lambda^2 + \mu^2 + \nu^2 - 2\mu\nu\cos A - 2\nu\lambda\cos B - 2\lambda\mu\cos C).$$

But the coordinates of the infinitely distant focus are known, from Art. 293, since it is the pole of the line at infinity. Hence those of the finite focus are

$$\frac{\Theta'A \quad \Delta}{A\sin A + H\sin B + G\sin C}, \quad \frac{\Theta'B - \Delta}{H\sin A + B\sin B + F\sin C},$$
$$\frac{\Theta'C - \Delta}{G\sin A + F\sin B + C\sin C}.$$

385. The condition (Art. 61) that two lines should be mutually perpendicular,

$$\lambda\lambda' + \mu\mu' + \nu\nu' - (\mu\nu' + \mu'\nu)\cos A - (\nu\lambda' + \nu'\lambda)\cos B$$
$$- (\lambda\mu' + \lambda'\mu)\cos C = 0,$$

is easily seen to be the same as the condition (Art. 293) that the lines should be conjugate with respect to

$$\lambda^2 + \mu^2 + \nu^2 - 2\mu\nu\cos A - 2\nu\lambda\cos B - 2\lambda\mu\cos C = 0.$$

The relation, then, between two mutually perpendicular lines is a particular case of the relation between two lines conjugate with regard to a fixed conic. Thus, the theorem that the three perpendiculars of a triangle meet in a point is a particular case of the theorem that the lines meet in a point which join the corresponding vertices of two triangles conjugate with respect to a fixed conic, &c. It is proved (*Geometry of Three Dimensions*, Chap. IX.) that, in spherical geometry, the two imaginary circular points at infinity are replaced by a fixed imaginary conic; that all circles on a sphere are to be considered as conics having double contact with a fixed conic, the centre of the circle being the pole of the chord of contact; that two lines are perpendicular if each pass through the pole of the other with respect to that conic, &c. The theorems then, which, in the Chapter on Projection, were extended by substituting, for the two imaginary points at infinity, two points situated anywhere, may be still further extended by substituting for these two points a conic section. Only these extensions are theorems suggested, not proved. Thus the theorem that the intersection of perpendiculars of a triangle inscribed in an equilateral hyperbola is on the curve, suggested the property of conics connected by the relation $\Theta = 0$, proved at the end of Art. 375.

It has been proved (Art. 306) that to several theorems concerning systems of circles, correspond theorems concerning systems of conics having double contact with a fixed conic. We give now some analytical investigations concerning the latter class of systems.

386. *To form the condition that the line* $\lambda x + \mu y + \nu z$ *may touch* $S + (\lambda' x + \mu' y + \nu' z)^2$. We are to substitute in Σ, $a + \lambda'^2$, $b + \mu'^2$, &c. for a, b, &c. The result may be written

$$\Sigma + \{a(\mu\nu' - \mu'\nu)^2 + \&c.\} = 0,$$

where the quantity within the brackets is intended to denote the result of substituting in S $\mu\nu' - \mu'\nu$, $\nu\lambda' - \nu'\lambda$, $\lambda\mu' - \lambda'\mu$ for x, y, z. This result may be otherwise written. For it was proved (Art. 294) that

$$(ax^2 + \&c.)(ax'^2 + \&c.) - (axx' + \&c.)^2 = A(yz' - y'z)^2 + \&c.$$

And it follows, by parity of reasoning, and can be proved in like manner, that

$$(A\lambda^2 + \&c.)(A\lambda'^2 + \&c.) - (A\lambda\lambda' + \&c.)^2 = \Delta\{a(\mu\nu' - \mu'\nu)^2 + \&c.\},$$

where $A\lambda\lambda' + \&c.$ is the condition that the lines $\lambda x + \mu y + \nu z$, $\lambda' x + \mu' y + \nu' z$ may be conjugate; or

$$A\lambda\lambda' + B\mu\mu' + C\nu\nu' + F(\mu\nu' + \mu'\nu) + G(\nu\lambda' + \nu'\lambda) + H(\lambda\mu' + \lambda'\mu);$$

If then we denote $A\lambda'^2 + \&c.$ by Σ', and $A\lambda\lambda' + \&c.$ by Π

and if we substitute for $a(\mu\nu' - \mu'\nu)^2 + \&c.$ the value just found, the condition previously obtained may be written

$$(\Delta + \Sigma')\,\Sigma - \Pi^2 = 0.$$

If we recollect (Art. 321) that λ, μ, ν may be considered as the coordinates of a point on the reciprocal conic, the latter form may be regarded as an analytical proof of the theorem that the reciprocal of two conics which have double contact is a pair of conics also having double contact. This condition may also be put into a form more convenient for some applications, if instead of defining the lines $\lambda x + \mu y + \nu z$, &c. by the coefficients λ, μ, ν, &c., we do so by the coordinates of their poles with respect to S, and if we form the condition that the line P' may touch $S + P''^2$, where P' is the polar of $x'y'z'$, or $axx' + \&c.$ Now the polar of $x'y'z'$ will evidently touch S when $x'y'z'$ is on the curve; and in fact if in Σ we substitute for λ, μ, ν; S_1, S_2, S_3 the coefficients of x, y, z in the equation of the polar, we get $\Delta S'$. And again two lines will be conjugate with respect to S, when their poles are conjugate; and in fact if we substitute as before for λ, μ, ν in Π we get ΔR, where R denotes the result of substituting the coordinates of either of the points $x'y'z'$, $x''y''z''$, in the equation of the polar of the other. The condition that P' should touch $S + P''^2$ then becomes $(1 + S'')\,S' = R^2$.

387. *To find the condition that the two conics*

$$S + (\lambda'x + \mu'y + \nu'z)^2, \quad S + (\lambda''x + \mu''y + \nu''z)^2,$$

should touch each other. They will evidently touch if one of the common chords $(\lambda'x + \mu'y + \nu'z) \pm (\lambda''x + \mu''y + \nu''z)$ touch either conic. Substituting, then, in the condition of the last Article $\lambda' \pm \lambda''$ for λ, &c., we get

$$(\Delta + \Sigma')\,(\Sigma' \pm 2\Pi + \Sigma'') = (\Sigma' \pm \Pi)^2,$$

which reduced may be written in the more symmetrical form

$$(\Delta + \Sigma')\,(\Delta + \Sigma'') = (\Delta \pm \Pi)^2.$$

The condition that $S + P'^2$ and $S + P''^2$ may touch is found from this as in the last Article, and is

$$(1 + S')\,(1 + S'') = (1 \pm R)^2.$$

Ex. 1. To draw a conic having double contact with S and touching three given conics $S + P'^2$, $S + P''^2$, $S + P'''^2$, also having double contact with S. Let xyz be the coordinates of the pole of the chord of contact with S of the sought conic $S + P^2$, then we have

$$(1 + S)\,(1 + S') = (1 + P')^2;\ (1 + S)\,(1 + S'') = (1 + P'')^2;\ (1 + S)\,(1 + S''') = (1 + P''')^2$$

where the reader will observe that S', S'', S''' are known constants, but S, P', &c. involve the coordinates of the sought point xyz. If then we write $1 + S = k^2$, &c., we get

$$kk' = 1 + P',\ kk'' = 1 + P'',\ kk''' = 1 + P'''.$$

It is to be observed that P', P'', P''' might each have been written with a double sign, and in taking the square roots a double sign may, of course, be given to k', k'', k'''. It will be found that these varieties of sign indicate that the problem admits of thirty-two solutions. The equations last written give

$$k(k' - k'') = P' - P''\,;\ k(k'' - k''') = P'' - P'''\,;$$

whence eliminating k, we get

$$P'(k'' - k''') + P''(k''' - k') + P'''(k' - k'') = 0,$$

the equation of a line on which must lie the pole with regard to S of the chord of contact of the sought conic. This equation is evidently satisfied by the point $P' = P'' = P'''$. But this point is evidently one of the *radical centres* (see Art. 306) of the conics $S + P'^2$, $S + P''^2$, $S + P'''^2$.

The equation is also satisfied by the point $\frac{P'}{k'} = \frac{P''}{k''} = \frac{P'''}{k'''}$. In order to see the geometric interpretation of this we remark that it may be deduced from Art. 386 that the tangential equations of $S + P'^2$, $S + P''^2$ are respectively

$$(1 + S')\,\Sigma = \Delta\,(\lambda x' + \mu y' + \nu z')^2,\ \ (1 + S'')\,\Sigma = \Delta\,(\lambda x'' + \mu y'' + \nu z'')^2.$$

Hence
$$\frac{\lambda x' + \mu y' + \nu z'}{k'} \pm \frac{\lambda x'' + \mu y'' + \nu z''}{k''}$$

represent points of intersection of common tangents to $S + P'^2$, $S + P''^2$, that is to say, the coordinates of these points are $\frac{x'}{k'} \pm \frac{x''}{k''}$, &c., and the polars of these points, with respect to S, are $\frac{P'}{k'} \pm \frac{P''}{k''}$. It follows that $\frac{P'}{k'} = \frac{P''}{k''} = \frac{P'''}{k'''}$ denote the pole, with respect to S, of an axis of similitude (Art. 306) of the three given conics. And the theorem we have obtained is,—*the pole of the sought chord of contact lies on one of the lines joining one of the four radical centres to the pole, with regard to S, of one of the four axes of similitude.* This is the extension of the theorem at the end of Art. 118.

To complete the solution, we seek for the coordinates of the point of contact of $S + P^2$ with $S + P'^2$. Now the coordinates of the point of contact, which is a centre of similitude of the two conics, being $\frac{x}{k} - \frac{x'}{k'}$, &c., we must substitute $x + \frac{k}{k'}x'$ for x, &c. in the equations $kk' = 1 + P'$, &c., and we get

$$kk' = 1 + P' + \frac{k}{k'}S'\,;\ kk'' = 1 + P'' + \frac{k}{k'}R\,;\ kk''' = 1 + P''' + \frac{k}{k'}R',$$

where R, R' are the results of substituting $x''y''z''$, $x'''y'''z'''$ respectively in the polar of $x'y'z'$. We have then

$$k(k' - k'') = P' - P'' + \frac{k}{k'}(S' - R)\,;\ k(k' - k''') = P' - P''' + \frac{k}{k'}(S' - R'),$$

whence eliminating k, we have

$$P'\left\{k'' - \frac{R}{k'} - \left(k''' - \frac{R'}{k'}\right)\right\} + P''\left\{k''' - \frac{R'}{k'} - \left(k' - \frac{S'}{k'}\right)\right\} + P'''\left\{k' - \frac{S'}{k'} - \left(k'' - \frac{R}{k'}\right)\right\},$$

the equation of a line on which the sought point of contact must lie; and which evidently joins a radical centre to the point where P', P'', P''' are respectively proportional to $k' - \frac{S'}{k'}$, $k'' - \frac{R}{k'}$, $k''' - \frac{R'}{k'}$, or to 1, $k'k'' - R$, $k'k''' - R'$. But if we form the equations of the polars, with respect to $S + P'^2$, of the three centres of similitude as above, we get

$$(k'k'' - R)\, P' = P'', \ (k'k''' - R')\, P' = P''', \ \&c.,$$

showing that the line we want to construct is got by joining one of the four radical centres to the pole, with respect to $S + P'^2$, of one of the four axes of similitude. This may also be derived geometrically as in Art. 121, from the theorems proved, Art. 306. The sixteen lines which can be so drawn meet $S + P'^2$ in the thirty-two points of contact of the different conics which can be drawn to fulfil the conditions of the problem.*

---

* The solution here given is the same in substance (though somewhat simplified in the details) as that given by Prof. Cayley, *Crelle*, vol. XXXIX.

Prof. Casey (*Proceedings of the Royal Irish Academy*, 1866) has arrived at another solution from considerations of spherical geometry. He shows by the method used, Art. 121 (*a*), that the same relation which connects the common tangents of four circles touched by the same fifth connects also the sines of the halves of the common tangents of four such circles on a sphere; and hence, as in Art. 121 (*b*), that if the equations of three circles on a sphere (see *Geometry of Three Dimensions*, chap. IX.) be $S - L^2 = 0$, $S - M^2 = 0$, $S - N^2 = 0$, that of a group of circles touching all three will be of the form

$$\sqrt{\{\lambda\, (S^{\frac{1}{2}} - L)\}} + \sqrt{\{\mu\, (S^{\frac{1}{2}} - M)\}} + \sqrt{\{\nu\, (S^{\frac{1}{2}} - N)\}} = 0.$$

This evidently gives a solution of the problem in the text, which I have arrived at directly by the following process. Let the conic S be $x^2 + y^2 + z^2$, and let $L = lx + my + nz$, $M = l'x + m'y + n'z$; then the condition that $S - L^2$, $S - M^2$ should touch is (Art. 387) $(1 - S')(1 - S'') = (1 - R)^2$, where $S' = l^2 + m^2 + n^2$, $S'' = l'^2 + m'^2 + n'^2$, $R = ll' + mm' + nn'$. I write now (12) to denote $\sqrt{(1-S')(1-S'')} - (1-R)$.

Let us now, according to the rule of multiplication of determinants, form a determinant from the two matrices containing five columns and six rows each.

$$\begin{vmatrix} 1, & 0, & 0, & 0, & 0 \\ 1, & l, & m, & n, & \sqrt{(1-S')} \\ 1, & l', & m', & n', & \sqrt{(1-S'')} \\ 1, & l'', & m'', & n'', & \sqrt{(1-S''')} \\ 1, & l''', & m''', & n''', & \sqrt{(1-S_4)} \\ 1, & l_4, & m_4, & n_4, & \sqrt{(1-S_5)} \end{vmatrix} \times \begin{vmatrix} 0, & 0, & 0, & 0, & 1, \\ -1, & l, & m, & n, & \sqrt{(1-S')}, \\ -1, & l', & m', & n', & \sqrt{(1-S'')}, \\ -1, & l'', & m'', & n'', & \sqrt{(1-S''')}, \\ -1, & l''', & m''', & n''', & \sqrt{(1-S_4)}, \\ -1, & l_4, & m_4, & n_4, & \sqrt{(1-S_5)}. \end{vmatrix}$$

The resulting determinant which must vanish, since there are more rows than columns, is

$$\begin{vmatrix} 0, & 1, & 1, & 1, & 1, & 1 \\ \sqrt{(1-S')}, & 0, & (12), & (13), & (14), & (15) \\ \sqrt{(1-S'')}, & (12), & 0, & (23), & (24), & (25) \\ \sqrt{(1-S''')}, & (13), & (23), & 0, & (34), & (35) \\ \sqrt{(1-S_4)}, & (14), & (24), & (34), & 0, & (45) \\ \sqrt{(1-S_5)}, & (15), & (25), & (35), & (45), & 0 \end{vmatrix} = 0,$$

an identical relation connecting the invariants of five conics all having double contact with the same conic S. Suppose now that the conic (5) touches the other four,

Ex. 2. The four conics having double contact with a given one S, which can be drawn through three fixed points, are all touched by four other conics also having double contact with S.* Let

$$S = x^2 + y^2 + z^2 - 2yz \cos A - 2zx \cos B - 2xy \cos C,$$

then the four conics are $S = (x \pm y \pm z)^2$, which are all touched by

$$S = \{x \cos (B - C) + y \cos (C - A) + z \cos (A - B)\}^2,$$

and by the three others got by changing the sign of A, B, or C, in this equation.

Ex. 3. The four conics which touch x, y, z, and have double contact with S are all touched by four other conics having double contact with S. Let $M = \frac{1}{2}(A + B + C)$, then the four conics are

$$S = \{x \sin (M - A) + y \sin (M - B) + z \sin (M - C)\}^2,$$

together with those obtained by changing the sign of A, B, or C in the above; and one of the touching conics is

$$S = \left\{\frac{x \sin \frac{1}{2}B \sin \frac{1}{2}C}{\sin \frac{1}{2}A} + \frac{y \sin \frac{1}{2}C \sin \frac{1}{2}A}{\sin \frac{1}{2}B} + \frac{z \sin \frac{1}{2}A \sin \frac{1}{2}B}{\sin \frac{1}{2}C}\right\}^2,$$

the others being got by changing the sign of x, and at the same time increasing B and C by 180°, &c.

Ex. 4. Find the condition that three conics U, V, W shall all have double contact with the same conic. The condition, as may be easily seen, is got by eliminating λ, μ, ν between

$$\Delta\lambda^3 - \Theta\lambda^2\mu + \Theta'\lambda\mu^2 - \Delta'\mu^3 = 0,$$

and the two corresponding equations which express that $\mu V - \nu W$, $\nu W - \lambda U$ break up into right lines.

---

then (15), &c. vanish; and we learn that the invariants of four conics all having double contact with S and touched by the same fifth are connected by the relation

$$\begin{vmatrix} 0, & (12), & (13), & (14) \\ (12), & 0, & (23), & (24) \\ (13), & (23), & 0, & (34) \\ (14), & (24), & (34), & 0 \end{vmatrix} = 0,$$

or $$\sqrt{\{(12)\,(34)\}} \pm \sqrt{\{(13)\,(24)\}} \pm \sqrt{\{(14)\,(23)\}} = 0.$$

We may deduce from this equation as follows the equation of the conic touching three others. If the discriminant of a conic vanish, $S = 1$, and then the condition of contact with any other reduces to $R = 1$. If, then, α, β, γ be the coordinates of any point satisfying the relation $S - L^2 = 0$, or $x^2 + y^2 + z^2 - (lx + my + nz)^2 = 0$, then

$$x^2 + y^2 + z^2 - \left\{\frac{\alpha x + \beta y + \gamma z}{\sqrt{(\alpha^2 + \beta^2 + \gamma^2)}}\right\}^2 = 0$$

evidently denotes a conic whose discriminant vanishes and which touches $S - L^2$. If, then, we are given three conics $S - L^2$, $S - M^2$, $S - N^2$, take any point α, β, γ on the conic which touches all three and take for a fourth conic that whose equation has just been written, then the functions (14), (24), (34) are respectively $1 - \frac{L}{\sqrt{(S)}}$, $1 - \frac{M}{\sqrt{(S)}}$, $1 - \frac{N}{\sqrt{(S)}}$, and we see that any point on the conic touching all three satisfies the relation

$$\sqrt{[(23)\,\{\sqrt{(S)} - L\}]} \pm \sqrt{[(31)\,\{\sqrt{(S)} - M\}]} \pm \sqrt{[(12)\,\{\sqrt{(S)} - N\}]} = 0.$$

* This is an extension of Feuerbach's theorem (p. 127), and itself admits of further extension. See *Quarterly Journal of Mathematics*, vol. VI. p. 67.

388. The theory of invariants and covariants of a system of three conics cannot be fully explained without assuming some knowledge of the theory of curves of the third degree.

Given three conics U, V, W, the locus of a point whose polars with respect to the three meet in a point is a curve of the third degree, which we call the *Jacobian* of the three conics. For we have to eliminate x, y, z between the equations of the three polars

$$U_1x + U_2y + U_3z = 0,\ V_1x + V_2y + V_3z = 0,\ W_1x + W_2y + W_3z=0,$$

and we obtain the determinant

$$U_1(V_2W_3 - V_3W_2) + U_2(V_3W_1 - V_1W_3) + U_3(V_1W_2 - V_2W_1) = 0.$$

It is evident that when the polars of any point with respect to U, V, W meet in a point, the polar with respect to all conics of the system $lU + mV + nW$ will pass through the same point. If the polars with respect to all these conics of a point A on the Jacobian pass through a point B, then the line AB is cut harmonically by all the conics; and therefore the polar of B will also pass through A. The point B is, therefore, also on the Jacobian, and is said to *correspond* to A. The line AB is evidently cut by all the conics in an involution whose foci are the points A, B. Since the foci are the points in which two corresponding points of the involution coincide, it follows that if any conic of the system touch the line AB, it can only be in one of the points A, B; or that if any break up into two right lines intersecting on AB, the points of intersection must be either A or B, unless indeed the line AB be itself one of the two lines. It can be proved directly, that if $lU + mV + nW$ represent two lines, their intersection lies on the Jacobian. For (Art. 292) it satisfies the three equations

$$lU_1 + mV_1 + nW_1 = 0,\ lU_2 + mV_2 + nW_2 = 0,\ lU_3 + mV_3 + nW_3 = 0;$$

whence, eliminating l, m, n, we get the same locus as before. The line AB joining two corresponding points on the Jacobian meets that curve in a third point; and it follows from what has been said that AB is itself one of the pair of lines passing through that point, and included in the system $lU + mV + nW$.

The general equation of the Jacobian is

$$(ag'h'')\,x^3 + (bh'f'')\,y^3 + (cf'g'')\,z^3$$
$$-\{(ab'g'') + (ah'f'')\}x^2y - \{(ca'h'') + (af'g'')\}x^2z - \{(ab'f'') + (bg'h'')\}y^2x$$
$$-\{(bc'h') + (bf'g'')\}y^2z - \{(ca'f'') + (cg'h'')\}z^2x - \{(bc'g'') + (ch'f'')\}z^2y$$
$$-\{(ab'c'') + 2\,(fg'h'')\}\,xyz = 0,$$

where $(ag'h'')$ &c. are abbreviations for determinants.

Ex. 1. Through four points to draw a conic to touch a given conic W. Let the four points be the intersection of two conics U, V; and it is evident that the problem admits of six solutions. For if we substitute $a + ka'$, &c. for a in the condition (Art. 372) that U and W should touch each other, k, as is easily seen, enters into the result in the sixth degree. The Jacobian of U, V, W intersects W in the six points of contact sought. For the polar of the point of contact with regard to W being also its polar with regard to a conic of the form $\lambda U + \mu V$ passes through the intersection of the polars with regard to U and V.

Ex. 2. If three conics have a common self-conjugate triangle, their Jacobian is three right lines. For it is verified at once that the Jacobian of $ax^2 + by^2 + cz^2$, $a'x^2 + b'y^2 + c'z^2$, $a''x^2 + b''y^2 + c''z^2$ is $xyz = 0$.

Ex. 3. If three conics have two points common, their Jacobian consists of a line and a conic through the two points. It is evident geometrically that any point on the line joining the two points fulfils the conditions of the problem, and the theorem can easily be verified analytically. In particular the Jacobian of a system of three circles is the circle cutting the three at right angles.

Ex. 4. The Jacobian also breaks up into a line and conic if one of the quantities S be a perfect square L^2. For then L is a factor in the locus. Hence we can describe four conics touching a given conic S at two given points (S, L) and also touching S''; the intersection of the locus with S'' determining the points of contact.

When the three conics are a conic, a circle, and the square of the line at infinity, the Jacobian passes through the feet of the normals which can be drawn to the conic through the centre of the circle.

388 (*a*). We return now to the theory of two conics which it was not possible to complete until we had explained the nature of Jacobians. We have seen that a system of two conics S, S' has four invariants Δ, Θ, Θ', Δ', and a covariant conic **F**, but there is besides a cubic covariant. In fact, the covariant conic **F** has a common self-conjugate triangle with S, S' (Art. 381, Ex. 1), therefore (Art. 388, Ex. 2) if we form J the Jacobian of S, S', **F** we obtain a cubic covariant, which, in fact, represents the sides of the common self-conjugate triangle of S and S'. It appears from (Art. 378*a*) that J vanishes identically if S and S' have double contact. We have given (Art. 381, Ex. 4) another method of obtaining the equation of the sides

of the common self-conjugate triangle, and if we compare the results of the two methods, we get the identical equation

$$J^2 = \mathbf{F}^3 - \mathbf{F}^2(\Theta S' + \Theta' S) + \mathbf{F}(\Delta'\Theta S^2 + \Delta\Theta' S'^2)$$
$$+ \mathbf{F}SS'(\Theta\Theta' - 3\Delta\Delta') - \Delta'^2\Delta S^3 - \Delta^2\Delta' S'^3$$
$$+ \Delta'(2\Delta\Theta' - \Theta^2)S^2S' + \Delta(2\Delta'\Theta - \Theta'^2)SS'^2.$$

Thus we see that a system of two conics has, besides the four invariants, four covariant forms S, S', $\mathbf{F}$, J, these being connected by the relation just written. In like manner, there are four contravariant forms Σ, Σ', Φ, Γ, where the last expresses tangentially the three vertices of the self-conjugate triangle, its square being connected by a relation, corresponding to that just written, between Σ, Σ', Φ and the invariants.

Ex. 1. Write down the 12 forms for the conics $x^2+y^2+z^2$, $ax^2+by^2+cz^2$.

Ans. $\Delta = 1$, $\Theta = a+b+c$, $\Theta' = bc+ca+ab$, $\Delta' = abc$,

$S = x^2+y^2+z^2$, $S' = ax^2+by^2+cz^2$, $\mathbf{F} = a(b+c)x^2 + b(c+a)y^2 + c(a+b)z^2$,

$J = (b-c)(c-a)(a-b)xyz$,

$\Sigma = \lambda^2+\mu^2+\nu^2$, $\Sigma' = bc\lambda^2 + ca\mu^2 + ab\nu^2$, $\Phi = (b+c)\lambda^2 + (c+a)\mu^2 + (a+b)\nu^2$,

$\Gamma = (b-c)(c-a)(a-b)\lambda\mu\nu$.

Ex. 2. Find an expression for the area of the common conjugate triangle of two conics. The square of the area is found to be

$$M^2 \sin^2 A \sin^2 B \sin^2 C \frac{\Theta^2\Theta'^2 + 18\Delta\Delta'\Theta\Theta' - 27\Delta^2\Delta'^2 - 4\Delta\Theta'^3 - 4\Delta'\Theta^3}{\Gamma'^2},$$

where M is the area of the triangle of reference, and Γ' the result of substituting in Γ, $\sin A$, $\sin B$, $\sin C$, the coordinates of the line at infinity. That the expression must contain in the numerator the condition of contact, and in the denominator Γ', is evident from the consideration that this area must vanish if the conics touch, and becomes infinite if any vertex of the triangle be at infinity.

388 (*b*). We have already explained what is meant by *covariants* which express relations satisfied by x, y, z, the coordinates of a point lying on a locus having some permanent relation with the original curve or curves, and by *contravariants* which express relations satisfied by λ, μ, ν the tangential coordinates of a line, whose section by the original curve or curves has some property unaffected by transformation of coordinates. There are besides forms called *mixed concomitants* which contain both x, y, z and also λ, μ, ν, and these we proceed

to enumerate for the system of two conics S, S'. These mixed concomitants of a system of two curves may also be regarded as covariants of the system of three, consisting of S, S' and the right line $\lambda x + \mu y + \nu z$. For instance, we may form the Jacobian of that system, or the locus of the point whose polars, with respect to S and S', intersect on $\lambda x + \mu y + \nu z$, thus obtaining the mixed concomitant N or $\begin{vmatrix} \lambda, & \mu, & \nu \\ S_1, & S_2, & S_3 \\ S_1', & S_2', & S_3' \end{vmatrix}$, which for the canonical form is

$$\lambda (b-c) yz + \mu (c-a) zx + \nu (a-b) xy.$$

There is evidently a corresponding reciprocal form N' obtained in the same way from Σ, Σ', which for the canonical form is

$$a\mu\nu (b-c) x + b\nu\lambda (c-a) y + c\lambda\mu (a-b) z.$$

This expresses the equation of the line joining the poles of $\lambda x + \mu y + \nu z$ with respect to S and S'. Again, for any line $\lambda x + \mu y + \nu z$, we may take its pole with regard to S and again the polar of that point with regard to S' and so obtain a companion line K. This for the canonical form is $a\lambda x + b\mu y + c\nu z$. We obtain a different companion line K' by taking the pole with regard to S' and then the polar with regard to S, thus finding $bc\lambda x + ca\mu y + ab\nu z$. Gordan has shewn (Clebsch, *Geometrie*, p. 291) that there are in all eight mixed concomitants of a system of two conics in terms of which, and of the forms previously enumerated, all other concomitants can be expressed. In addition to the four already mentioned we may take the Jacobian of K, S and $\lambda x + \mu y + \nu z$, or for the canonical form

$$\mu\nu (b-c) x + \nu\lambda (c-a) y + \lambda\mu (a-b) z;$$

and, in like manner, the Jacobian of K', S', and $\lambda x + \mu y + \nu z$, or

$$\mu\nu a^2 (b-c) x + \nu\lambda b^2 (c-a) y + \lambda\mu c^2 (a-b) z.$$

These with the two reciprocal forms

$$\lambda ayz (b-c) + \mu bzx (c-a) + \nu cxy (a-b),$$

and $$\lambda bc (b-c) yz + \mu ca (c-a) zx + \nu ab (a-b) xy$$

make up the entire system.

We return now to the theory of three conics.

388(*c*). *To find the condition that a line $\lambda x + \mu y + \nu z$ should be cut in involution by three conics.* It appears from Art. 335

and from the Note, Art. 342, that the required condition is the vanishing of the determinant

$$\begin{vmatrix} c\lambda^2 - 2g\nu\lambda + a\nu^2, & c\mu^2 - 2f\nu\mu + b\nu^2, & c\lambda\mu - f\nu\lambda - g\nu\mu + h\nu^2 \\ c'\lambda^2 - 2g'\nu\lambda + a'\nu^2, & c'\mu^2 - 2f'\nu\mu + b'\nu^2, & c'\lambda\mu - f'\nu\lambda - g'\nu\mu + h'\nu^2 \\ c''\lambda^2 - 2g''\nu\lambda + a''\nu^2, & c''\mu^2 - 2f''\nu\mu + b''\nu^2, & c''\lambda\mu - f''\nu\lambda - g''\nu\mu + h''\nu^2 \end{vmatrix}$$

When this is expanded it becomes divisible by ν^3, and may be written

$$\begin{aligned} \lambda^3(bc'f'') + \mu^3(ca'g'') + \nu^3(ab'h'') + \lambda^2\mu\,\{2(ch'f'') - (bc'g'')\} \\ + \lambda^2\nu\,\{2(bf'g'') - (bc'h'')\} + \mu^2\lambda\,\{2(cg'h'') - (ca'f'')\} \\ + \mu^2\nu\,\{2(af'g'') - (ca'h'')\} + \nu^2\lambda\,\{2(bg'h'') - (ab'f'')\} \\ + \nu^2\mu\,\{2(ah'f'') - (ab'g'')\} + \lambda\mu\nu\,\{(ab'c'') - 4(fg'h'')\} = 0. \end{aligned}$$

This may also be written in the determinant form

$$\begin{vmatrix} a, & b, & c, & 2f, & 2g, & 2h \\ a', & b', & c', & 2f', & 2g', & 2h' \\ a'', & b'', & c'', & 2f'', & 2g'', & 2h'' \\ \lambda, & & & \nu, & \mu & \\ & \mu, & & \nu, & & \lambda \\ & & \nu, & \mu, & \lambda & \end{vmatrix}.$$

From the form of this condition, it is immediately inferred that any line cut in involution by three conics U, V, W is cut in involution by any three conics of the system $lU + mV + nW$. The locus of a point whence tangents to three conics form a system in involution is got by writing x, y, z for λ, μ, ν in the preceding, and the reciprocal coefficients A, B, &c. instead of a, b, &c.

389. If we form the discriminant of $lU + mV + nW$, we may write the result $l^3\Delta + l^2m\theta_{11} + l^2n\theta_{113} + lmn\theta_{123} +$ &c., and the coefficients of the several powers of l, m, n will be invariants of the system of conics. All these belong to the class of invariants already considered, except the coefficient of lmn, in which each term abc of the discriminant of U is replaced by

$$ab'c'' + ab''c' + a'b''c + a'bc'' + a''bc' + a''b'c, \text{ \&c.}$$

Another remarkable invariant of the system of conics, first obtained by a different method by Prof. Sylvester, is found by the help of the principle (*Higher Algebra*, Art. 139), that when we have a covariant and a contravariant of the same degree, we can get an invariant by substituting differential symbols in either, and operating on the other. By the help of the Jacobian and the contravariant of the last article we get the invariant T,

$$T = (ab'c'')^2 + 4(ab'f'')(ac'f'') + 4(bc'g'')(ba'g'') + 4(ca'h'')(cb'h'')$$
$$+ 8(af'g'')(bf'g'') + 8(af'h'')(cf'h'') + 8(cg'h'')(bg'h'')$$
$$- 8(ag'h'')(bc'f'') - 8(bh'f'')(ca'g'') - 8(cf'g'')(ab'h'')$$
$$+ 4(ab'c'')(fg'h'') - 8(fg'h'')^2.$$

389*a*. Some of the properties of a system of three conics can be studied with advantage by expressing each in terms of four lines x, y, z, w: thus

$$U = ax^2 + by^2 + cz^2 + dw^2, \quad V = a'x^2 + b'y^2 + c'z^2 + d'w^2,$$
$$W = a''x^2 + b''y^2 + c''z^2 + d''w^2.$$

It is always possible, in an infinity of ways, to choose x, y, z, w, so that the equations can be brought to the above form; for each of the equations just written contains explicitly three independent constants; and each of the lines x, y, z, w contains implicitly two independent constants. The form, therefore, just written puts seventeen constants at our disposal, while U, V, W, contain only three times five, or fifteen, independent constants. The equations of four lines are always connected by a relation of the form $w = \lambda x + \mu y + \nu z$, and we may suppose that the constants λ, &c. have been included in x, &c., so that this relation may be written in the symmetrical form $x + y + z + w = 0$.

Let it be required now to find the condition that U, V, W may have a common point. Solving for x^2, y^2, z^2, w^2 between the equations $U = 0$, $V = 0$, $W = 0$, and denoting by A, B, C, D the four determinants $(bc'd'')$, $(dc'a'')$, $(da'b'')$, $(ba'c'')$, we get x^2, y^2, z^2, w^2 proportional to A, B, C, D; and substituting in $x + y + z + w = 0$, we obtain the required condition

$$\sqrt{(A)} + \sqrt{(B)} + \sqrt{(C)} + \sqrt{(D)} = 0,$$

or $$(A^2 + B^2 + C^2 + D^2 - 2AB - 2BC - 2CA - 2AD - 2BD - 2CD)^2 = 64ABCD.$$

The left-hand side of this equation is the square of the invariant T already found; the right-hand side $ABCD$ is an invariant which we shall call M, whose vanishing expresses the condition that it may be possible to determine l, m, n, so that

$lU + mV + nW$ shall be a perfect square. This invariant may be directly found from the principle that when the equation of a conic is a perfect square its reciprocal vanishes identically. The reciprocal of $lS + mS' + nS''$ is evidently (Art. 377)

$$l^2\Sigma + m^2\Sigma' + n^2\Sigma'' + mn\Phi_{23} + nl\Phi_{31} + lm\Phi_{12},$$

and if we equate separately each coefficient to zero and then linearly eliminate the six quantities l^2, m^2, &c., we get the result

$$\begin{vmatrix} A, & B, & C, & F, & G, & H \\ A', & B', & C', & F', & G', & H' \\ A'', & B'', & C'', & F'', & G'', & H'' \\ A_{23}, & B_{23}, & C_{23}, & F_{23}, & G_{23}, & H_{23} \\ A_{31}, & B_{31}, & C_{31}, & F_{31}, & G_{31}, & H_{31} \\ A_{12}, & B_{12}, & C_{12}, & F_{12}, & G_{12}, & H_{12} \end{vmatrix} = 0,$$

where A_{12}, &c. denote the coefficients in Φ_{12}, &c., Art. 377. This determinant is of the fourth degree in the coefficients of each conic, those of the first conic, for example, entering in the second degree into the first row, and in the first into the fifth and sixth, and so for the others. It follows that four conics of the system $S + lU + mV + nW$ can be determined so as to be perfect squares (see Ex. 3, Art. 373), for if we equate to nothing the invariant M found for $S + lU$, V, W, we have an equation of the fourth degree for determining l.

389*b*. Considering two conics, if we form the discriminant of the reciprocal system $l\Sigma + m\Sigma'$ we get no new invariant, the discriminant in fact being

$$l^3\Delta^2 + l^2m\Delta\Theta + lm^2\Delta'\Theta' + m^3\Delta'^2.$$

But if we form the discriminant of $l\Sigma + m\Sigma' + n\Sigma''$ the coefficient of lmn, answering to Θ_{123} of Art. 389, or

$$A_1(B_2C_3 + B_3C_2 - 2F_2F_3) + \&c.$$

is an invariant of the second degree in the coefficients of each conic, not expressible in term of the invariants Δ, Θ_{112}, &c. Mr. Burnside has shewn that the invariant T of Art. 389, which is of the same order in the coefficients, is expressible in terms of this new invariant and of those of Art. 389. In fact, let two of the conics have the canonical form, and write them

$$x^2 + y^2 + z^2 = 0, \quad lx^2 + my^2 + nz^2 = 0,$$
$$ax^2 + by^2 + cz^2 + 2fyz + 2gzx + 2hxy = 0.$$

If then we form the resultant of the three, that is, the condition that they shall have a common point, the first two equations are satisfied by

$$x^2 = m - n = \alpha, \quad y^2 = n - l = \beta, \quad z^2 = l - m = \gamma.$$

Substituting these values in the third and clearing of radicals, we have

$$\{a^2\alpha^2 + b^2\beta^2 + c^2\gamma^2 - 2bc\beta\gamma - 2ca\gamma\alpha - 2ab\alpha\beta + 4(A\beta\gamma + B\gamma\alpha + C\alpha\beta)\}^2$$
$$= 64\alpha\beta\gamma\,(Fgh\alpha + Ghf\beta + Hfg\gamma).$$

The left-hand side of the equation is what we have before called T^2. Writing then for α, β, γ their values $m - n$, $n - l$, $l - m$, we can reduce T to

$$\{l(b + c) + m(c + a) + n(a + b)\}^2 - 4(a + b + c)(amn + bnl + clm)$$
$$- 4(Al^2 + Bm^2 + Cn^2) - 4(A + B + C)(mn + nl + lm)$$
$$+ 8\{Al(m + n) + Bm(n + l) + Cn(l + m)\}$$

all the separate groups in which expression will be found to be fundamental invariants of the system, except $Al^2 + Bm^2 + Cn^2$, which is $\theta_{211}\theta_{233} - \Theta$ where Θ is the invariant of this Article. Thus we get

$$T = \theta^2_{123} - 4(\theta_{122}\theta_{133} + \theta_{211}\theta_{233} + \theta_{311}\theta_{322}) + 12\Theta.$$

If we consider the discriminant of $lS + mS' + nS''$ as a ternary cubic in l, m, n, and by the theory of cubic curves form its S and T invariants, Mr. Burnside has calculated the S to be $T^2 - 48M$, and the T to be $8T(72M - T^2)$. Thus we have $T^2 - 48M$, and $T(72M - T^2)$ expressed in terms of the ten fundamental invariants which occur in the discriminant of $lS + mS' + nS''$. And though M, T, Θ are not linearly expressible in terms of these ten, yet we have just shown how to form two equations implicitly connecting M and T with these ten; and of course we could, if we please, eliminate either M or T from these equations, and thus get an equation connecting either, singly with the fundamental invariants.

389*c*. Any three conics may in general be considered as the polar conics of three points with regard to the same cubic; or, in other words, their equations may all be reduced to the form

$$\alpha (x^2 - 2yz) + \beta (y^2 - 2zx) + \gamma (z^2 - 2xy) = 0.$$

If we use for the equations of the conics the forms given in Art. 389*a*, the equation of the cubic whence they are derived will be

$$\frac{x^3}{A} + \frac{y^3}{B} + \frac{z^3}{C} + \frac{w^3}{D} = 0;$$

and it appears that if the invariant M vanish (in which case either A, B, C or D vanishes), an exception occurs, and the conics cannot all be derived from the same cubic. In the general case, the equation of the cubic may be obtained by forming the Hessian of the Jacobian of the three conics, and subtracting the Jacobian itself multiplied by twice T.

If we operate with the conics on the cubic contravariant, or with their reciprocals on the Jacobian, we obtain linear contravariants and covariants which geometrically represent the points of which the given conics are polar conics, and the polar lines of these points with respect to the cubic.

CHAPTER XIX.

THE METHOD OF INFINITESIMALS.

390. Referring the reader to other works where it is shown how the differential calculus enables us readily to draw tangents to curves, and to determine the magnitude of their areas and arcs, we wish here to give him some idea of the manner in which these problems were investigated by geometers before the invention of that method. The geometric methods are not merely interesting in a historical point of view; they afford solutions of some questions more concise and simple than those furnished by analysis, and they have even recently led to a beautiful theorem (Art. 399) which had not been anticipated by those who have applied the integral calculus to the rectification of conic sections.

If a polygon be inscribed in any curve, it is evident that the more the number of the sides of the polygon is increased, the more nearly will the area and perimeter of the polygon approach to equality with the area and perimeter of the curve, and the more nearly will any side of the polygon approach to coincidence with the tangent at the point where it meets the curve. Now, if the sides of the polygon be multiplied *ad infinitum*, the polygon will coincide with the curve, and the tangent at any point will coincide with the line joining two indefinitely near points on the curve. In like manner, we see that the more the number of the sides of a *circumscribing* polygon is increased, the more nearly will its area and perimeter approach to equality with the area and perimeter of the curve, and the more nearly will the intersection of two of its adjacent sides approach to the point of contact of either. Hence, in investigating the area or perimeter of any curve, we may substitute for the curve an inscribed or circumscribing polygon of an indefinite number of sides; we may consider any tangent of the curve as the line joining two indefinitely near points on the curve, and any point on the curve as the intersection of two indefinitely near tangents.

391. Ex. 1. *To find the direction of the tangent at any point of a circle.*

In any isosceles triangle AOB, either base angle OBA is less than a right angle by half the vertical angle; but as the points A and B approach to coincidence, the vertical angle may be supposed less than any assignable angle, therefore the angle OBA which the tangent makes with the radius is ultimately equal to a right angle. We shall frequently have occasion to use the principle here proved, viz. that two indefinitely near lines of *equal length* are at right angles to the line joining their extremities.

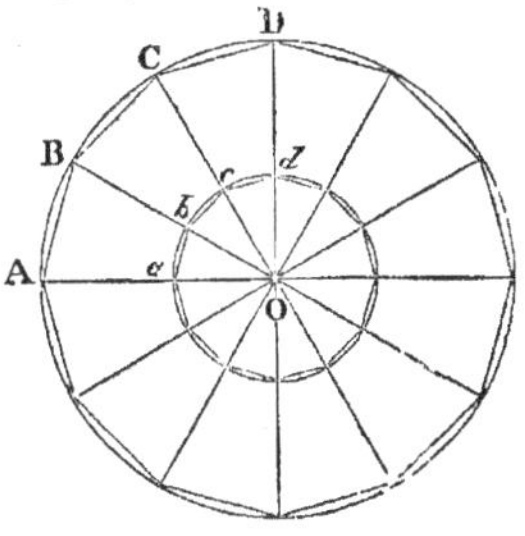

Ex. 2. *The circumferences of two circles are to each other as their radii.*

If polygons of the same number of sides be inscribed in the circles, it is evident, by similar triangles, that the bases ab, AB, are to each other as the radii of the circles, and, therefore, that the whole perimeters of the polygons are to each other in the same ratio; and since this will be true, no matter how the number of sides of the polygon be increased, the circumferences are to each other in the same ratio.

Ex. 3. *The area of any circle is equal to the radius multiplied by the semi-circumference.*

For the area of any triangle OAB is equal to half its base multiplied by the perpendicular on it from the centre; hence the area of any inscribed regular polygon is equal to half the sum of its sides multiplied by the perpendicular on any side from the centre; but the more the number of sides is increased, the more nearly will the perimeter of the polygon approach to equality with that of the circle, and the more nearly will the perpendicular on any side approach to equality with the radius, and the difference between them can be made less than any assignable quantity; hence ultimately the area of the circle is equal to the radius multiplied by the semi-circumference; or $= \pi r^2$.

392. Ex. 1. *To determine the direction of the tangent at any point on an ellipse.*

Let P and P' be two indefinitely near points on the curve, then $FP + PF' = FP' + P'F'$; or, taking $FR = FP$, $F'R' = F'P'$, we have $P'R = PR'$; but in the triangles PRP', $PR'P'$, we have also the base PP' common, and (by Ex. 1, Art. 391) the angles PRP' $PR'P'$ right; hence the angle $PP'R = P'PR'$. Now TPF is ultimately equal to $PP'F$, since their difference PFP' may be supposed less than any given angle; hence $TPF = T'PF'$, or the focal radii make equal angles with the tangent.

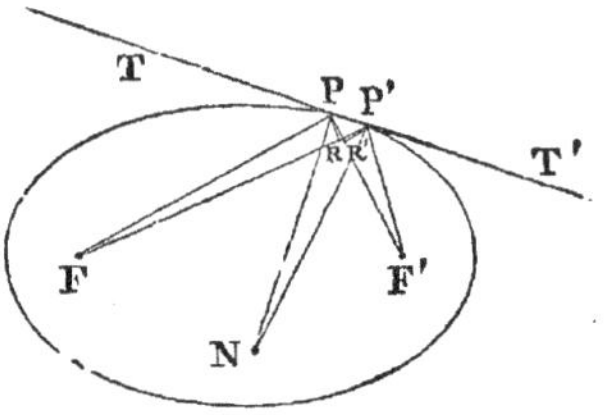

Ex. 2. *To determine the direction of the tangent at any point on a hyperbola.*

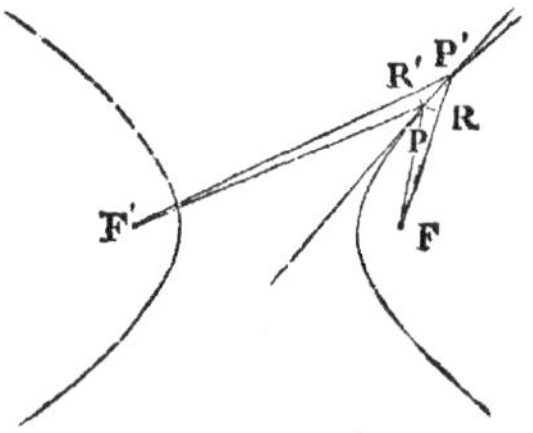

We have

$$F'P' - F'P = FP' - FP,$$

or, as before,

$$P'R = P'R'.$$

Hence the angle

$$PP'R = PP'R',$$

or, the tangent is the internal bisector of the angle FPF'.

Ex. 3. *To determine the direction of the tangent at any point of a parabola.*

We have $FP = PN$, and $FP' = P'N'$; hence $P'R = P'S$, or the angle $N'P'P = FP'P$. The tangent, therefore, bisects the angle FPN.

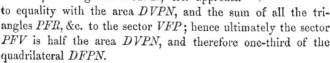

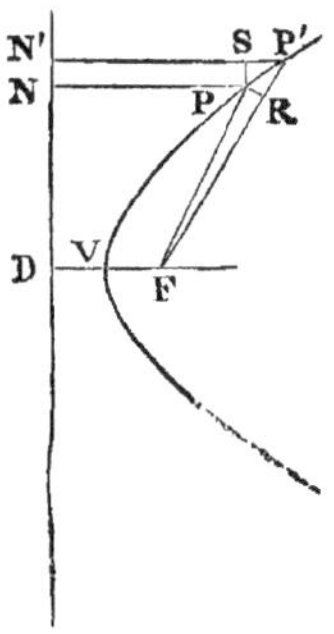

393. Ex. 1. *To find the area of the parabolic sector FVP.*

Since $PS = PR$, and $PN = FP$, we have the triangle FPR half the parallelogram $PSNN'$. Now if we take a number of points $P'P''$, &c. between V and P, it is evident that the closer we take them, the more nearly will the sum of all the parallelograms $PSN'N$, &c. approach to equality with the area $DVPN$, and the sum of all the triangles PFR, &c. to the sector VFP; hence ultimately the sector PFV is half the area $DVPN$, and therefore one-third of the quadrilateral $DFPN$.

Ex. 2. *To find the area of the segment of a parabola cut off by any right line.*

Draw the diameter bisecting it, then the parallelogram PR' is equal to PM', since they are the complements of parallelograms about the diagonal; but since TM is bisected at V', the parallelogram PN' is half PR'; if, therefore, we take a number of points P, P', P'', &c., it follows that the sum of all the parallelograms PM' is double the sum of all the parallelograms PN', and therefore ultimately that the space $V'PM$ is double $V'PN$; hence the area of the parabolic segment $V'PM$ is to that of the parallelogram $V'NPM$ in the ratio 2 : 3.

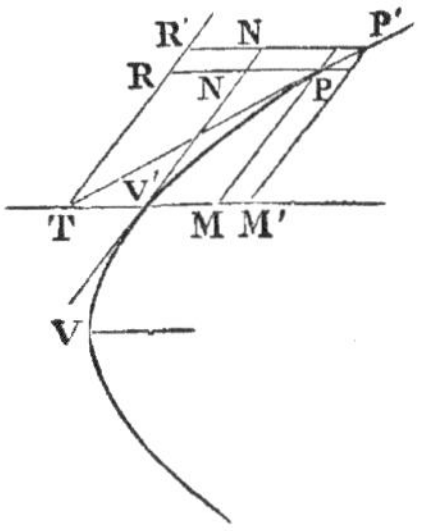

394. Ex. 1. *The area of an ellipse is equal to the area of a circle whose radius is a geometric mean between the semi-axes of the ellipse.*

For if the ellipse and the circle on the transverse axis be divided by any number of lines parallel to the axis minor, then since $mb : md :: m'b' : m'd' :: b : a$, the quadrilateral $mbb'm'$ is to $mdd'm'$ in the same ratio, and the sum of all the one set of quadrilaterals, that is, the polygon $Bbb'b''A$ inscribed in the ellipse is to the corresponding polygon $Ddd'd''A$ inscribed in the circle, in the same ratio. Now this will be true whatever be the number of the sides of the polygon; if we suppose them, therefore, increased indefinitely, we learn that the area of the ellipse is to the area of the circle as b to a; but the area of the circle being $= \pi a^2$, the area of the ellipse $= \pi ab$.

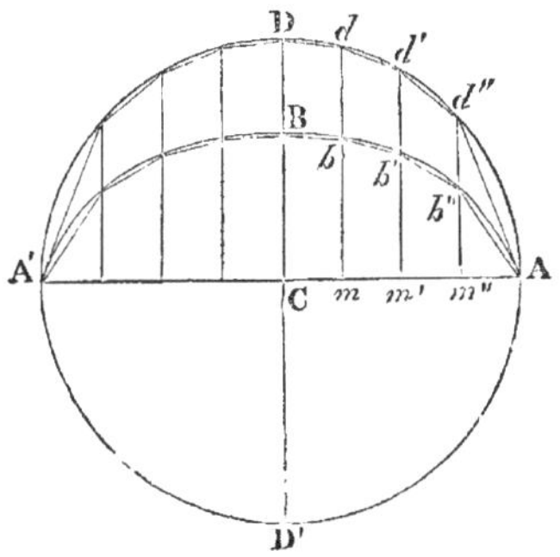

Cor. It can be proved, in like manner, that if any two figures be such that the ordinate of one is in a constant ratio to the corresponding ordinate of the other, the areas of the figures are in the same ratio.

Ex. 2. *Every diameter of a conic bisects the area enclosed by the curve.*

For if we suppose a number of ordinates drawn to this diameter, since the diameter bisects them all, it also bisects the trapezium formed by joining the extremities of any two adjacent ordinates, and by supposing the number of these trapezia increased without limit, we see that the diameter bisec s the area.

395. Ex. 1. *The area of the sector of a hyperbola made by joining any two points of it to the centre, is equal to the area of the segment made by drawing parallels from them to the asymptotes.*

For since the triangle $PKC = QLC$, the area $PQC = PQKL$.

Ex. 2. *Any two segments* $PQLK$, $RSNM$, *are equal, if*

$$PK : QL :: RM : SN.$$

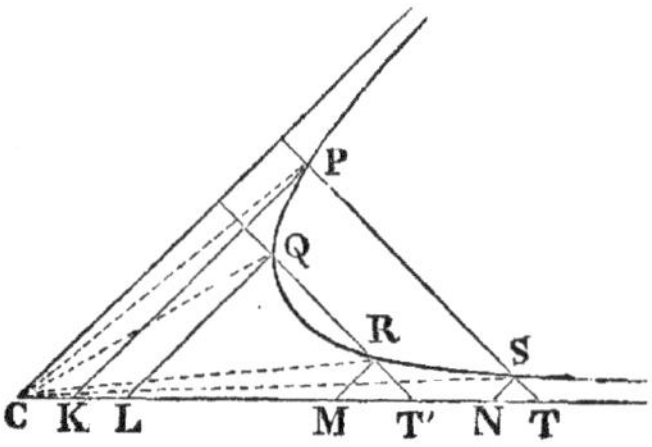

For

$$PK : QL :: CL : CK,$$

but (Art. 197)

$$CL = MT', \ CK = NT;$$

we have, therefore,

$$RM : SN :: MT' : NT,$$

and therefore QR is parallel to PS. We can now easily prove that the sectors PCQ, RCS are equal, since the diameter bisecting PS, QR will bisect both the hyperbolic area $PQRS$, and also the triangles PCS, QCR.

If we suppose the points Q, R to coincide, we see that we can bisect any area $PKNS$ by drawing an ordinate QL, a geometric mean between the ordinates at its extremities.

Again, if a number of ordinates be taken, forming a continued geometric progression, the area between any two is constant.

396. *The tangent to the interior of two similar, similarly placed, and concentric conics cuts off a constant area from the exterior conic.*

For we proved (Art. 236, Ex. 4) that this tangent is always bisected at the point of contact; now if we draw any two tangents, the angle AQA' will be equal to BQB' and the nearer we suppose the point Q to P, the more nearly will the sides AQ, $A'Q$ approach to equality with the sides BQ, $B'Q$; if, therefore, the two

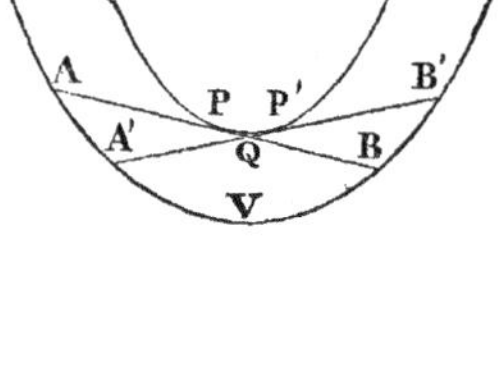

tangents be taken indefinitely near, the triangle AQA' will be equal to BQB', and the space AVB will be equal to $A'VB'$; since, therefore, this space remains constant as we pass from any tangent to the consecutive tangent, it will be constant whatever tangent we draw.

Cor. It can be proved, in like manner, that if a tangent to one curve always cuts off a constant area from another, it will be bisected at the point of contact; and, conversely, that if it be always bisected it cuts off a constant area.

Hence we can draw through a given point a line to cut off from a given conic the *minimum* area. If it were required to cut off a *given* area, it would be only necessary to draw a tangent through the point to some similar and concentric conic, and the greater the given area, the greater will be the distance between the two conics. The area will, therefore, evidently be least when this last conic passes through the given point; and since the tangent at the point must be bisected, the line through a given point which cuts off the minimum area is bisected at that point.

In like manner, the chord drawn through a given point which cuts off the minimum or maximum area from any curve is bisected at that point. In like manner can be proved the following two theorems, due to the late Professor MacCullagh.

Ex. 1. *If a tangent AB to one curve cut off a constant arc from another, it is divided at the point of contact, so that $AP : PB$ inversely as the tangents to the outer curve at A and B.*

Ex. 2. *If the tangent AB be of a constant length, and if the perpendicular let fall on AB from the intersection of the tangents at A and B meet AB in M, then AP will $= MB$.*

397. *To find the radius of curvature at any point on an ellipse.*

The centre of the circle circumscribing any triangle is the intersection of perpendiculars erected at the middle points of the sides of that triangle; it follows, therefore, that the centre of the circle passing through three consecutive points on the curve is the intersection of two consecutive normals to the curve.

Now, given any two triangles FPF', $FP'F'$, and PN, $P'N$, the two bisectors of their vertical angles, it is easily proved by elementary geometry, that twice the angle $PNP' = PFP' + PF'P'$. (See figure, Art. 392, Ex. 1).

Now, since the arc of any circle is proportional to the angle it subtends at the centre (Euc. VI. 33), and also to the radius (Art. 391), if we consider PP' as the arc of a circle, whose centre is N, the angle PNP' is measured by $\frac{PP'}{PN}$. In like manner, taking $FR = FP$, PFP' is measured by $\frac{PR}{FP}$, and we have

$$\frac{2PP'}{PN} = \frac{PR}{FP} + \frac{P'R'}{F'P'};$$

but

$$PR = P'R' = PP' \sin PP'F;$$

therefore, denoting this angle by θ, PN by R, FP, $F'P$, by ρ, ρ', we have

$$\frac{2}{R\sin\theta} = \frac{1}{\rho} + \frac{1}{\rho'}.$$

Hence it may be inferred, that *the focal chord of curvature is double the harmonic mean between the focal radii.* Substituting $\frac{b}{b'}$ for $\sin\theta$, $2a$ for $\rho + \rho'$, and b'^2 for $\rho\rho'$, we obtain the known value

$$R = \frac{b'^3}{ab}.$$

The radius of curvature of the hyperbola or parabola can be investigated by an exactly similar process. In the case of the parabola we have ρ' infinite, and the formula becomes

$$\frac{2}{R\sin\theta} = \frac{1}{\rho}.$$

I owe to Mr. Townsend the following investigation, by a different method, of the length of the focal chord of curvature:

Draw *any* parallel QR to the tangent at P, and describe a circle through PQR meeting the focal chord PL of the conic at C. Then, by the circle $PS.SC = QS.SR$, and by the conic (Ex. 2, Art. 193)

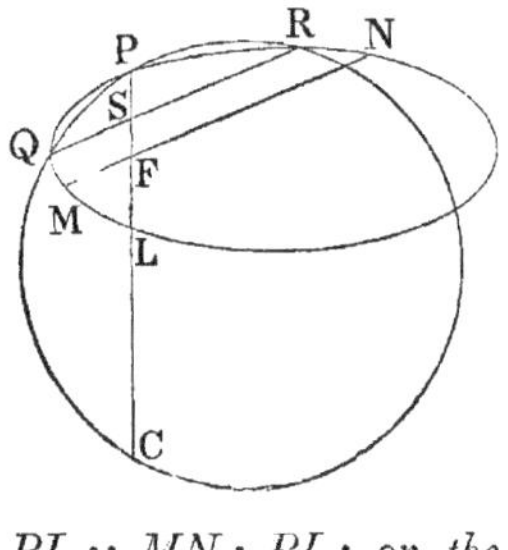

$$PS.SL : QS.SR :: PL : MN;$$

therefore, whatever be the circle,

$$SC : SL :: MN : PL;$$

but for the circle of curvature the points S and P coincide, therefore $PC : PL :: MN : PL$; or, *the focal chord of curvature is equal to the focal chord of the conic drawn parallel to the tangent at the point* (p. 219, Ex. 4).

398. The radius of curvature of a central conic may otherwise be found thus:

Let Q be an indefinitely near point on the curve, QR a parallel to the tangent, meeting the normal in S; now, if a circle be described passing through P, Q, and touching PT at P, since QS is a perpendicular let fall from Q on the diameter of this circle, we have $PQ^2 = PS$ multiplied by the diameter;

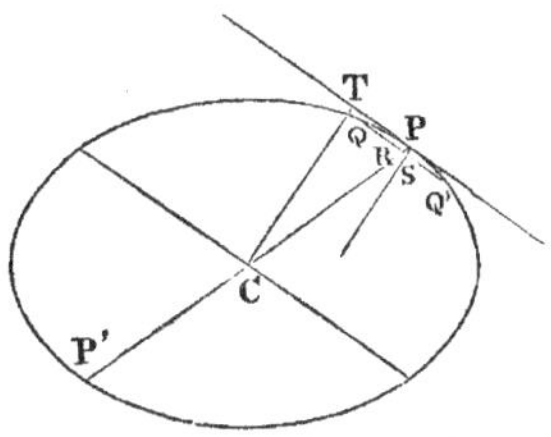

or the radius of curvature $= \frac{PQ^2}{2PS}$. Now, since QR is always drawn parallel to the tangent, and since PQ must ultimately coincide with the tangent, we have PQ ultimately equal to QR; but, by the property of the ellipse (if we denote CP and its conjugate by a', b'),

$$b'^2 : a'^2 :: QR^2 : PR \,.\, RP' \; (= 2a'.PR),$$

therefore
$$QR^2 = \frac{2b'^2.PR}{a'}.$$

Hence the radius of curvature $= \frac{b'^2}{a'} \cdot \frac{PR}{PS}$. Now, no matter how small PR, PS are taken, we have, by similar triangles, their *ratio* $\frac{PR}{PS} = \frac{CP}{CT} = \frac{a'}{p}$. Hence radius of curvature $= \frac{b'^2}{p}$.

It is not difficult to prove that *at the intersection of two confocal conics the centre of curvature of either is the pole with respect to the other of the tangent to the former at the intersection.*

398 (a). If we consider the circle circumscribing the triangle formed by two tangents to a curve and their chord, it is evident geometrically, that its diameter is the line joining the intersection of tangents to the intersection of the corresponding normals. Hence, in the limit, the diameter of the circle circumscribing the triangle formed by two *consecutive* tangents and their chord is the radius of curvature; that is to say, the radius of the circle here considered is half the radius of curvature (Compare Art. 262, Ex. 4).

399. *If two tangents be drawn to an ellipse from any point of a confocal ellipse, the excess of the sum of these two tangents over the arc intercepted between them is constant.**

For, take an indefinitely near point T', and let fall the perpendiculars TR, $T'S$, then (see fig.)

$$PT = PR = PP' + P'R$$

(for $P'R$ may be considered as the continuation of the line PP') in like manner

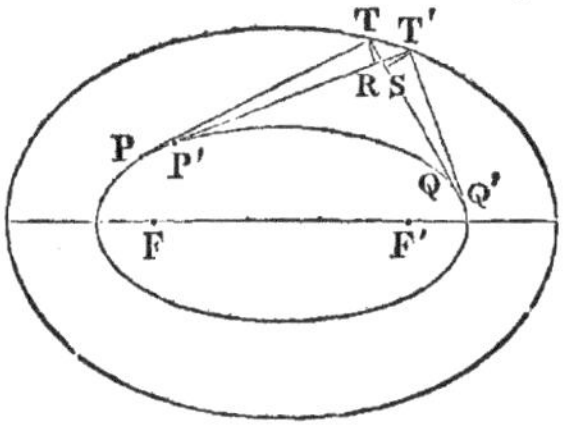

$$Q'T' = QQ' + QS.$$

Again, since, by Art. 189, the angle $TT'R = T'TS$, we have $TS = T'R$; and therefore

$$PT + TQ' = PT' + T'Q'.$$

Hence $(PT + TQ) - (P'T' + T'Q') = PP' - QQ' = PQ - P'Q'$.

Cor. The same theorem will be true of any two curves which possess the property that two tangents TP, TQ to the inner one always make equal angles with the tangent TT' to the outer.

400. *If two tangents be drawn to an ellipse from any point of a confocal hyperbola, the difference of the arcs PK, QK is equal to the difference of the tangents TP, TQ.*†

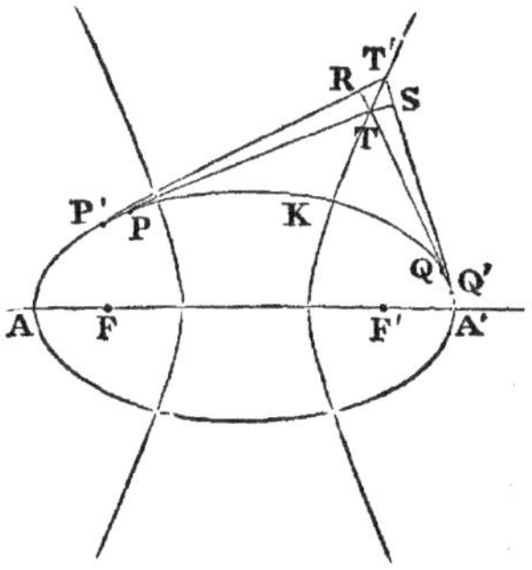

For it appears, precisely as before, that the excess of $T'P' - P'K$ over $TP - PK = T'R$, and that the excess of $T'Q' - Q'K$ over $TQ - QK$ is $T'S$, which is equal to $T'R$, since (Art. 189) TT' bisects the angle $RT'S$. The difference, therefore, between the excess of TP over PK, and that of TQ over QK is constant; but in the particular case where T

---

* This beautiful theorem was discovered by Bishop Graves. See his *Translation of Chasles's Memoirs on Cones and Spherical Conics*, p. 77.

† This extension of the preceding theorem was discovered by Mr. Mac Cullagh, *Dublin Exam. Papers*, 1841, p 41; 1842, pp. 68, 83. M. Chasles afterwards independently noticed the same extension of Bishop Graves's theorem. *Comptes Rendus*, October, 1843, tom. XVII. p. 838.

coincides with K, both these excesses and consequently their difference vanish; in every case, therefore, $TP - PK = TQ - QK$.

COR. *Fagnani's theorem*, "That an elliptic quadrant can be so divided, that the difference of its parts may be equal to the difference of the semi-axes," follows immediately from this Article, since we have only to draw tangents at the extremities of the axes, and through their intersection to draw a hyperbola confocal with the given ellipse. The coordinates of the points where it meets the ellipse are found to be

$$x^2 = \frac{a^3}{a+b}, \quad y^2 = \frac{b^3}{a+b}.$$

401. *If a polygon circumscribe a conic, and if all the vertices but one move on confocal conics, the locus of the remaining vertex will be a confocal conic.*

In the first place, we assert that if the vertex T of an angle PTQ circumscribing a conic, move on a confocal conic (see fig., Art. 399); and if we denote by a, b, the diameters parallel to TP, TQ; and by α, β, the angles TPT', $TQ'T'$, made by each of the sides of the angle with its consecutive position, then $a\alpha = b\beta$. For (Art. 399) $TR = T'S$; but $TR = TP.\alpha$; $T'S = T'Q'.\beta$, and (Art. 149) TP and TQ are proportional to the diameters to which they are parallel.

Conversely, if $a\alpha = b\beta$, T moves on a confocal conic. For by reversing the steps of the proof we prove that $TR = T'S$; hence that TT' makes equal angles with TP, TQ, and therefore coincides with the tangent to the confocal conic through T; and therefore that T' lies on that conic.

If, then, the diameters parallel to the sides of the polygon be a, b, c, &c., that parallel to the last side being d, we have $a\alpha = b\beta$, because the first vertex moves on a confocal conic; in like manner $b\beta = c\gamma$, and so on until we find $a\alpha = d\delta$, which shows that the last vertex moves on a confocal conic.*

* This proof is taken from a paper by Dr. Hart; *Cambridge and Dublin Mathematical Journal*, vol. IV. 193.

NOTES.

PASCAL'S THEOREM, Art. 267.

M. STEINER was the first who (in *Gergonne's Annales*) directed the attention of geometers to the complete figure obtained by joining in every possible way six points on a conic. M. Steiner's theorems were corrected and extended by M. Plücker (*Crelle's Journal*, vol. v. p. 274), and the subject has been more recently investigated by Messrs. Cayley and Kirkman, the latter of whom, in particular, has added several new theorems to those already known (see *Cambridge and Dublin Mathematical Journal*, vol. v. p. 185). We shall in this note give a slight sketch of the more important of these, and of the methods of obtaining them. The greater part are derived by joining the simplest principles of the theory of combinations with the following elementary theorems and their reciprocals: "If two triangles be such that the lines joining corresponding vertices meet in a point (*the centre of homology* of the two triangles), the intersections of corresponding sides will lie in one right line (their *axis*)." "If the intersections of opposite sides of three triangles be for each pair *the same* three points in a right line, the centres of homology of the first and second, second and third, third and first, will lie in a right line."

Now let the six points on a conic be a, b, c, d, e, f, which we shall call the points P. These may be connected by *fifteen* right lines, ab, ac, &c., which we shall call the lines C. Each of the lines C (for example) ab is intersected by the fourteen others; by four of them in the point a, by four in the point b, and consequently by six in points distinct from the points P (for example the points (ab, cd), &c.). These we shall call the points p. There are forty-five such points; for as there are six on each of the lines C, to find the number of points p, we must multiply the number of lines C by 6, and divide by 2, since two lines C pass through every point p.

If we take the sides of the hexagon in the order $abcdef$, Pascal's theorem is, that the three p points, (ab, de), (cd, fa), (bc, ef), lie in one right line, which we may call either the Pascal $abcdef$, or else we may denote as the Pascal $\begin{Bmatrix} ab \,.\, cd \,.\, ef \\ de \,.\, fa \,.\, bc \end{Bmatrix}$, a form which we sometimes prefer, as showing more readily the three points through which the Pascal passes. Through each point p four Pascals can be drawn. Thus through (ab, de) can be drawn $abcdef$, $abfdec$, $abcedf$, $abfedc$. We then find the total number of Pascals by multiplying the number of points p by 4, and dividing by 3, since there are three points p on each Pascal. We thus obtain the number of Pascal's lines $= 60$. We might have derived the same directly by considering the number of different ways of arranging the letters $abcdef$.

Consider now the three triangles whose sides are

$$ab,\ cd,\ ef, \qquad (1)$$

$$de,\ fa,\ bc, \qquad (2)$$

$$cf,\ be,\ ad, \qquad (3)$$

The intersections of corresponding sides of 1 and 2 lie on the same Pascal, therefore the lines joining corresponding vertices meet in a point, but these are the three Pascals,

$$\left\{\begin{matrix} ab\,.\,de\,.\,cf \\ cd\,.fa\,.\,be \end{matrix}\right\},\quad \left\{\begin{matrix} cd\,.fa\,.\,be \\ ef\,.\,bc\,.\,ad \end{matrix}\right\},\quad \left\{\begin{matrix} ef\,.\,bc\,\,.ad \\ ab\,.\,de\,.\,cf \end{matrix}\right\}.$$

This is Steiner's theorem (Art. 268); we shall call this the g point,

$$\left\{\begin{matrix} ab\,.\,de\,.\,cf \\ cd\,.fa\,.\,be \\ ef\,.\,bc\,.\,ad \end{matrix}\right\}.$$

The notation shows plainly that on each Pascal's line there is only one g point; for given the Pascal $\left\{\begin{matrix} ab\,.\,de\,.\,cf \\ cd\,.fa\,.\,be \end{matrix}\right\}$ the g point on it is found by writing under each term the two letters not already found in that vertical line. Since then three Pascals intersect in every point g, the number of points $g = 20$. If we take the triangles 2, 3; and 1, 3; the lines joining corresponding vertices are the same in all cases therefore, by the reciprocal of the second preliminary theorem, the three *axes* of the three triangles meet in a point. This is also a g point $\left\{\begin{matrix} ab\,.\,cd\,.\,ef \\ de\,.fa\,.\,bc \\ cf\,.\,be\,.\,ad \end{matrix}\right\}$, and Steiner has stated that the two g points just written are harmonic conjugates with regard to the conic, so that the 20 g points may be distributed into ten pairs.* The Pascals which pass through these two g points correspond to hexagons taken in the order respectively, *abcfed*, *afcdeb*, *adcbef*; *abcdef*, *afcbed*, *adcfeb*; three alternate vertices holding in all the same position.

Let us now consider the triangles,

$$\begin{matrix} ab & cd & ef & (1) \\ \left.\begin{matrix} ab\,.\,ce\,.\,df \\ de\,.\,bf\,.\,ac \end{matrix}\right\}, & \left.\begin{matrix} cd\,.\,bf\,.\,ae \\ af\,.\,ce\,.\,bd \end{matrix}\right\}, & \left.\begin{matrix} ef\,.\,bd\,.\,ac \\ bc\,.\,ae\,.\,df \end{matrix}\right\}, & (4) \\ \left.\begin{matrix} ab\,.\,ce\,.\,df \\ cf\,.\,bd\,.\,ae \end{matrix}\right\}, & \left.\begin{matrix} cd\,.\,bf\,.\,ae \\ be\,.\,ac\,.\,df \end{matrix}\right\}, & \left.\begin{matrix} ef\,.\,bd\,.\,ac \\ ad\,.\,ce\,.\,bf \end{matrix}\right\}, & (5). \end{matrix}$$

The intersections of corresponding sides of 1 and 4 are three points which lie on the same Pascal; therefore the lines joining corresponding vertices meet in a point. But these are the three Pascals,

$$\left.\begin{matrix} ab\,.\,ce\,.\,df \\ cd\,.\,bf\,.\,ae \end{matrix}\right\},\quad \left.\begin{matrix} cd\,.\,bf\,.\,ae \\ ef\,.\,ac\,.\,bd \end{matrix}\right\},\quad \left.\begin{matrix} ef\,.\,ac\,.\,bd \\ ab\,.\,df\,.\,ce \end{matrix}\right\}.$$

We may denote the point of meeting as the h point, $\left.\begin{matrix} ab\,.\,ce\,.\,df \\ cd\,.\,bf\,.\,ae \\ ef\,.\,ac\,.\,bd \end{matrix}\right\}$.

The notation differs from that of the g points in that only one of the vertical columns contains the six letters without omission or repetition. On every Pascal there are three h points, viz. there are on

$$\left.\begin{matrix} ab\,.\,cd\,.\,ef \\ de\,.\,af\,.\,bc \end{matrix}\right\};\quad \left.\begin{matrix} \overline{ab}\,.\,cd\,.\,ef \\ de\,.\,af\,.\,bc \\ cf\,.\,bd\,.\,ae \end{matrix}\right\},\quad \left.\begin{matrix} ab\,.\,\overline{cd}\,.\,ef \\ de\,.\,af\,.\,bc \\ ac\,.\,be\,.\,df \end{matrix}\right\},\quad \left.\begin{matrix} ab\,.\,cd\,.\,\overline{ef} \\ de\,.\,af\,.\,bc \\ bf\,.\,ce\,.\,ad \end{matrix}\right\},$$

where the bar denotes the complete vertical column. We obtain then Mr. Kirkman's extension of Steiner's theorem :—*The Pascals intersect three by three, not only in Steiner's twenty points g, but also in sixty other points h.* The demonstration of Art. 268 applies alike to Mr. Kirkman's and to Steiner's theorem.

In like manner if we consider the triangles 1 and 5, the lines joining corresponding vertices are the same as for 1 and 4; therefore the corresponding sides intersect on

* For a proof of this see Staudt (*Crelle*, LXII. 142).

a right line, as they manifestly do on a Pascal. In the same manner the corresponding sides of 4 and 5 must intersect on a right line, but these intersections are the three h points,

$$\left.\begin{array}{l}\overline{ab}\,.\,ce\,.\,df\\ de\,.\,bf\,.\,ac\\ cf\,.\,ae\,.\,bd\end{array}\right\},\quad \left.\begin{array}{l}ae\,.\,\overline{cd}\,.\,bf\\ bd\,.\,af\,.\,ce\\ ac\,.\,be\,.\,df\end{array}\right\},\quad \left.\begin{array}{l}ac\,.\,bd\,.\,\overline{ef}\\ df\,.\,ae\,.\,bc\\ ce\,.\,bf\,.\,ad\end{array}\right\}.$$

Moreover, the axis of 4 and 5 must pass through the intersection of the axes of 1, 4, and 1, 5, namely, through the g point, $\left.\begin{array}{l}ab\,.\,cd\,.\,ef\\ de\,.\,af\,.\,bc\\ cf\,.\,be\,.\,ad\end{array}\right\}$.

In this notation the g point is found by combining the complete vertical columns of the three h points. Hence we have the theorem, "*There are twenty lines G, each of which passes through one g and three h points.*" The existence of these lines was observed independently by Prof. Cayley and myself. The proof here given is Prof. Cayley's.

It can be proved similarly that "*The twenty lines G pass four by four through fifteen points i.*" The four lines G whose g points in the preceding notation have a common vertical column will pass through the same point.

Again, let us take three Pascals meeting in a point h. For instance,

$$\left.\begin{array}{l}ab\,.\,ce\,.\,df\\ de\,.\,bf\,.\,ac\end{array}\right\},\quad \left.\begin{array}{l}de\,.\,bf\,.\,ac\\ cf\,.\,ae\,.\,bd\end{array}\right\},\quad \left.\begin{array}{l}cf\,.\,ae\,.\,bd\\ ab\,.\,df\,.\,ce\end{array}\right\}.$$

We may, by taking on each of these a point p, form a triangle whose vertices are $(df,\, ac)$, $(bf,\, ae)$, $(bd,\, ce)$ and whose sides are, therefore,

$$\left.\begin{array}{l}ac\,.\,bf\,.\,de\\ df\,.\,ae\,.\,cb\end{array}\right\},\quad \left.\begin{array}{l}bf\,.\,ce\,.\,ad\\ ae\,.\,bd\,.\,cf\end{array}\right\},\quad \left.\begin{array}{l}bd\,.\,ac\,.\,ef\\ ce\,.\,df\,.\,ab\end{array}\right\}.$$

Again, we may take on each a point h, by writing under each of the above Pascals $af\,.\,cd\,.\,be$, and so form a triangle whose sides are

$$\left.\begin{array}{l}ac\,.\,bf\,.\,de\\ be\,.\,cd\,.\,af\end{array}\right\},\quad \left.\begin{array}{l}cf\,.\,ae\,.\,bd\\ be\,.\,cd\,.\,af\end{array}\right\},\quad \left.\begin{array}{l}df\,.\,ab\,.\,ce\\ be\,.\,cd\,.\,af\end{array}\right\}.$$

But the intersections of corresponding sides of these triangles, which must therefore be on a right line, are the three g points,

$$\left.\begin{array}{l}be\,.\,cd\,.\,af\\ ac\,.\,bf\,.\,de\\ df\,.\,ae\,.\,bc\end{array}\right\},\quad \left.\begin{array}{l}be\,.\,cd\,.\,af\\ cf\,.\,ae\,.\,bd\\ ad\,.\,bf\,.\,ce\end{array}\right\},\quad \left.\begin{array}{l}be\,.\,cd\,.\,af\\ df\,.\,ab\,.\,ce\\ ac\,.\,ef\,.\,bd\end{array}\right\},\quad \left.\begin{array}{l}be\,.\,cd\,.\,af\\ cf\,.\,ab\,.\,de\\ ad\,.\,ef\,.\,bc\end{array}\right\}.$$

I have added a fourth g point, which the symmetry of the notation shows must lie on the same right line; these being all the g points into the notation of which $be\,.\,cd\,.\,af$ can enter. Now there can be formed, as may readily be seen, fifteen different products of the form $be\,.\,cd\,.\,af$; we have then Steiner's theorem, *The g points lie four by four on fifteen right lines I.* Hesse has noticed that there is a certain reciprocity between the theorems we have obtained. There are 60 Kirkman points h, and 60 Pascal lines H corresponding each to each in a definite order to be explained presently. There are 20 Steiner points g, through each of which passes three Pascals H and one line G; and there are 20 lines G, on each of which lie three Kirkman points h and one Steiner g. And as the twenty lines G pass four by four through fifteen points i, so the twenty points g lie four by four on fifteen lines I. The following investigation gives a new proof of some of the preceding theorems and also shews what h point corresponds to the Pascal got by taking the vertices in the order *abcdef*. Consider the two inscribed triangles *ace*, *bdf*; their sides touch a conic (see Ex. 4, Art. 355); therefore we may apply Brianchon's theorem to the hexagon whose sides are *ce*, *df*, *ae*, *bf*, *ac*, *bd*. Taking them in this order, the diagonals of the hexagon are the three Pascals intersecting in the h point, $\left.\begin{matrix} ce\,.\,bf\,.\,ad \\ df\,.\,ac\,.\,be \\ ae\,.\,bd\,.\,cf \end{matrix}\right\}$

And since, if retaining the alternate sides ce, ae, ac, we permutate cyclically the other three, then by the reciprocal of Steiner's theorem, the three resulting Brianchon points lie on a right line, it is thus proved that three h points lie in a right line G. From the same circumscribing hexagon it can be inferred that the lines joining the point a to $\{bc, df\}$ and d to $\{ac, ef\}$ intersect on the Pascal $abcdef$, and that there are six such intersections on every Pascal.

More recently Prof. Cayley has deduced the properties of this figure by considering it as the projection of the lines of intersection of six planes. See *Quarterly Journal*, vol. IX. p. 348.

Still more recently the whole figure has been discussed and several new properties obtained by Veronese (*Nuovi Teoremi sull' Hexagrammum Mysticum* in the Memoirs of the *Reale Accademia dei Lincei*, 1877). He states with some extension the geometrical principles which we have employed in the investigation, as follows: I. Consider three lines passing through a point, and three points in each line; these points form 27 triangles which may be divided into 36 sets of three triangles in perspective in pairs, the axes of homology passing three by three through 36 points which lie four by four on 27 right lines. II. If 4 triangles $a_1b_1c_1$, $a_2b_2c_2$, &c. are in perspective, the first with the second, the second with the third, the third with the fourth, and the fourth with the first, the vertices marked with the same letters corresponding to each other, and if the four centres of homology lie in a right line, the four axes will pass through a point. III. If we have four quadrangles $a_1b_1c_1d_1$, &c. related in like manner, the four points of the last theorem answering to the triangles bcd, cda, dab, abc lie on a right line. Considering the case when all four quadrangles have the same centre of homology, we obtain the corollary: If on four lines passing through a point we take 3 homologous quadrangles $a_1b_1c_1d_1$, $a_2b_2c_2d_2$, $a_3b_3c_3d_3$; then we have four sets of three homologous triangles, $a_1b_1c_1$, &c. the axes of homology of each three passing through a point and the four points lying on a right line. IV. If we have two triangles in perspective $a_1b_1c_1$, $a_2b_2c_2$, and if we take the intersections of b_1c_2, b_2c_1; c_1a_2, c_2a_1; a_1b_2, a_2b_1, we form a new triangle in perspective with the other two, the three centres of homology lying on a right line. It would be too long to enumerate all the theorems which Veronese derives from these principles. Suffice it to say that a leading feature of his investigation is the breaking up of the system of Pascals into six groups, each of ten Pascals, the ten corresponding Kirkman points lying three by three on these lines which also pass in threes through these points. It may be added that Veronese states the correspondence between a Pascal line and a Kirkman point as follows: Take out of the 15 lines C the six sides of any hexagon, there remain 9 lines C; out of these can be formed three hexagons whose Pascals meet in the Kirkman point corresponding to the Pascal of the hexagon with which we started.

After the publication of Veronese's paper Cremona obtained very elegant demonstrations of his theorems by studying the subject from quite a different point of view. From the theory of cubical surfaces we know (*Geometry of Three Dimensions*, Art. 536), that if such a surface have a nodal point, there lie on the surface six right lines passing through the node, which also lie on a cone of the second order, and fifteen other lines, one in the plane of each pair of the foregoing; by projecting this figure Cremona obtains the whole theory of the hexagon.

It may be well to add some formulæ useful in the analytic discussion of the hexagon inscribed in the conic $LM - R^2$. Let the values of the parameter μ (Art. 270) for the six vertices be a, b, c, d, e, f, and let us denote by (ab) the quantity $abL - (a + b)R + M$, which, equated to zero, represents the chord joining two vertices. Then it is easy to see that $(ab)(cd) - (ad)(bc)$ is $LM - R^2$ multiplied by the factor $(a-c)(b-d)$, and hence that if we compare, as in Art. 268, the forms $(ab)(cd) - (ad)(bc)$, $(af)(de) - (ad)(ef)$ we get the equation of the Pascal *abcdef* in the form

$$(a-c)(b-d)(ef) = (a-e)(f-d)(bc).$$

The same equation might also have been obtained in the forms, which can easily be verified as being equivalent,

$$(a-e)(b-f)(cd) = (c-e)(b-d)(af),$$

$$(c-a)(b-f)(de) = (c-e)(d-f)(ab).$$

The three other Pascals which pass through $(bc)(ef)$ are

$$(a-c)(b-d)(ef) = (a-f)(e-d)(bc),$$

$$(a-b)(c-d)(ef) = (a-e)(f-d)(bc),$$

$$(a-b)(c-d)(ef) = (a-f)(e-d)(bc),$$

these being respectively the Pascals *abcdfe*, *acbdef*, *acbdfe*.

Consider the three Pascals

$$(a-c)(b-d)(ef) = (a-e)(f-d)(bc) = (b-f)(c-e)(ad);$$

these evidently intersect in a point, viz. a Steiner g-point; but the three

$$(a-c)(b-d)(ef) = (a-e)(f-d)(bc) = (b-e)(c-f)(ad)$$

intersect in a Kirkman h-point.

Mr. Cathcart has otherwise obtained the equation of the Pascal line in a determinant form. It was shewn (Art. 331) that the relation between corresponding points of two homographic systems is of the form

$$Aaa' + Ba + Ca' + D = 0.$$

Hence, eliminating A, B, C, D, we see that the relation between four points and other four of two homographic systems is

$$\begin{vmatrix} \alpha\alpha', & \alpha, & \alpha', & 1 \\ \beta\beta', & \beta, & \beta', & 1 \\ \gamma\gamma', & \gamma, & \gamma', & 1 \\ \delta\delta', & \delta, & \delta', & 1 \end{vmatrix} = 0,*$$

and the double points of the system are got by putting $\delta' = \delta$, and solving the quadratic for δ. But we saw Art. 289, Ex. 10, that the Pascal line LMN passes through K, K' the double points of the two homographic systems determined by ACE, DFB the alternate vertices of the hexagon. And since, if δ be the parameter of the point K, we have M, R, L respectively proportional to $\delta^2, \delta, 1$, it follows that the equation of the Pascal *abcdef* is

$$\begin{vmatrix} M, & R, & R, & L \\ ad, & a, & d, & 1 \\ be, & b, & e, & 1 \\ cf, & c, & f, & 1 \end{vmatrix} = 0.$$

Systems of Tangential Coordinates, Art. 311.

Through this volume we have ordinarily understood by the tangential coordinates of a line $l\alpha + m\beta + n\gamma$, the constants l, m, n in the equation of the line (Art. 70); and by the tangential equation of a curve the relation necessary between these constants in order that the line should touch the curve. We have preferred this method because it is the most closely connected with the main subject of this volume, and because all other systems of tangential coordinates may be reduced to it. We

* On this determinant see Cayley, *Phil. Trans.*, 1858, p. 436.

wish now to notice one or two points in this theory which we have omitted to mention, and then briefly to explain some other systems of tangential coordinates. We have given (Ex. 6, Art. 132) the tangential equation of a circle whose centre is $\alpha'\beta'\gamma'$ and radius r, viz.

$$(l\alpha' + m\beta' + n\gamma')^2 = r^2 (l^2 + m^2 + n^2 - 2mn \cos A - 2nl \cos B - 2lm \cos C);$$

let us examine what the right-hand side of this equation, if equated to nothing, would represent. It may easily be seen that it satisfies the condition of resolvability into factors, and therefore represents two points. And what these points are may be seen by recollecting that this quantity was obtained (Art. 61) by writing at full length $l\alpha + m\beta + n\gamma$, and taking the sum of the squares of the coefficients of x and y, $l \cos a + m \cos \beta + n \cos \gamma$, $l \sin a + m \sin \beta + n \sin \gamma$. Now if $a^2 + b^2 = 0$, the line $ax + by + c$ is parallel to one or other of the lines $x \pm y \sqrt{(-1)} = 0$, the two points therefore are the two imaginary points at infinity on any circle. And this appears also from the tangential equation of a circle which we have just given: for if we call the two factors ω, ω', and the centre α, that equation is of the form $\alpha^2 = r^2\omega\omega'$, showing that ω, ω' are the points of contact of tangents from α. In like manner if we form the tangential equation of a conic whose foci are given, by expressing the condition that the product of the perpendiculars from these points on any tangent is constant, we obtain the equation in the form

$$(l\alpha' + m\beta' + n\gamma')(l\alpha'' + m\beta'' + n\gamma'') = b^2\omega\omega',$$

showing that the conic is touched by the lines joining the two foci to the points ω, ω' (Art. 258a).

It appears from Art. 61 that the result of substituting the tangential coordinates of any line in the equation of a point is proportional to the perpendicular from that point on the line; hence the tangential equations $\alpha\beta = k\gamma\delta$, $\alpha\gamma = k\beta^2$ when interpreted give the theorems proved by reciprocation Art. 311. If we substitute the coordinates of any line in the equation of a circle given above, the result is easily seen to be proportional to the square of the chord intercepted on the line by the circle. Hence if Σ, Σ' represent two circles, we learn by interpreting the equation $\Sigma = k^2\Sigma'$ that the envelope of a line on which two given circles intercept chords having to each other a constant ratio is a conic touching the tangents common to the two circles.

Lastly, it is to be remarked that a system of two points cannot be adequately represented by a trilinear, nor a system of two lines by a tangential equation. If we are given a tangential equation denoting two points, and form, as in Art. 285, the corresponding trilinear equation, it will be found that we get the square of the equation of the line joining the points, but all trace of the points themselves has disappeared. Similarly if we have the equation of a pair of lines intersecting in a point $\alpha'\beta'\gamma'$, the corresponding tangential equation will be found to be $(l\alpha' + m\beta' + n\gamma')^2=0$. In fact, a line analytically fulfils the conditions of a tangent if it meets a curve in two coincident points; and when a conic reduces to a pair of lines, any line through their intersection must be regarded as a tangent to the system.

The method of tangential coordinates may be presented in a form which does not presuppose any acquaintance with the trilinear or Cartesian systems. Just as in trilinear coordinates the position of a point is determined by the mutual ratios of the perpendiculars let fall from it on three fixed lines, so (Art. 311) the position of a line may be determined by the mutual ratios of the perpendiculars let fall on it from three fixed points. If the perpendiculars let fall on a line from two points A, B be λ, μ, then it is proved, as in Art. 7, that the perpendicular on it from the point which cuts the line AB in the ratio of $m : l$ is $\dfrac{l\lambda + m\mu}{l + m}$, and consequently that if the line pass through that point we have $l\lambda + m\mu = 0$, which therefore may be regarded as the equation of that point. Thus $\lambda + \mu = 0$ is the equation of the middle point of AB, $\lambda - \mu = 0$ that of a point at infinity on AB. In like manner (see Art. 7, Ex. 6) it is proved that $l\lambda + m\mu + n\nu = 0$ is the equation of a point O, which may be constructed (see fig. p. 61) either by cutting BC in the ratio $n : m$ and AD in the ratio $m + n : l$; or by cutting $AC :: l : n$ and $BE :: l + n : m$, or by cutting $AB :: m : l$ and $CF :: l + m : n$. Since the ratio of the triangles $AOB : AOC$ is the same as that of $BD : BC$, we may write the equation of the point O in the form

$$BOC.\lambda + COA.\mu + AOB.\nu = 0.$$

Or, again, substituting for each triangle BOC its value $\rho'\rho'' \sin\theta$ (see Art. 311)

$$\frac{\lambda \sin\theta}{\rho} + \frac{\mu \sin\theta'}{\rho'} + \frac{\nu \sin\theta''}{\rho''} = 0.$$

Thus, for example, the coordinates of the line at infinity are $\lambda = \mu = \nu$, since all finite points may be regarded as equidistant from it; the point $l\lambda + m\mu + n\nu$ will be at infinity when $l + m + n = 0$; and generally a curve will be touched by the line at infinity if the sum of the coefficients in its equation $= 0$. So again the equations of the intersections of bisectors of sides, of bisectors of angles, and of the perpendiculars, of the triangle of reference are respectively $\lambda + \mu + \nu = 0$, $\lambda \sin A + \mu \sin B + \nu \sin C = 0$, $\lambda \tan A + \mu \tan B + \nu \tan C = 0$. It is unnecessary to give further illustrations of the application of these coordinates because they differ only by constant multipliers from those we nave used already. The length of the perpendicular from any point on $l\alpha + m\beta + n\gamma$ is (Art. 61)

$$\frac{l\alpha' + m\beta' + n\gamma'}{\sqrt{(l^2 + m^2 + n^2 - 2mn \cos A - 2nl \cos B - 2lm \cos C)}},$$

the denominator being the same for every point. If then p, p', p'' be the perpendiculars let fall from each vertex of the triangle on the opposite side, the perpendiculars λ, μ, ν from these vertices on any line are respectively proportional to lp, mp', np''; and we see at once how to transform such tangential equations as were used in the preceding pages, viz. homogeneous equations in l, m, n, into equations expressed in terms of the perpendiculars λ, μ, ν. It is evident from the actual values that λ, μ, ν are connected by the relation

$$\frac{\lambda^2}{p^2} + \frac{\mu^2}{p'^2} + \frac{\nu^2}{p''^2} - \frac{2\mu\nu}{p'p''} \cos A - \frac{2\nu\lambda}{p''p} \cos B - \frac{2\lambda\mu}{pp'} \cos C = 1.$$

It was shown (Art. 311) how to deduce from the trilinear equation of any curve the tangential equation of its reciprocal.

The system of three point tangential coordinates just explained includes under it two other methods at first sight very different. Let one of the points of reference C be at infinity, then both ν and ρ'' become infinite, but their ratio remains finite and $= \sin COE$, where DOE is any line drawn through the point O. The equation then of a point already given becomes in this case

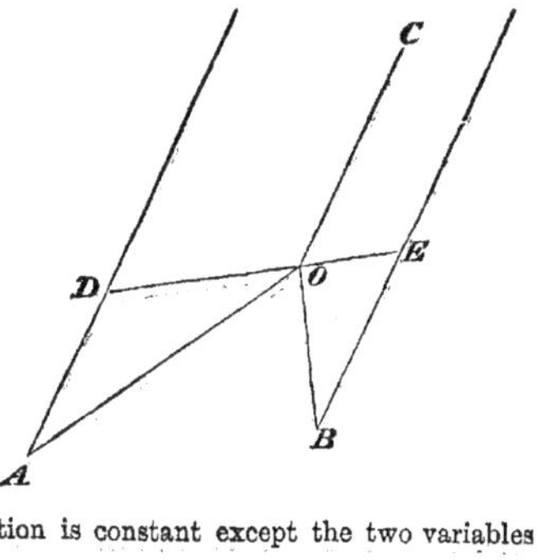

$$\frac{\sin\theta}{\rho}\frac{\lambda}{\sin COE} + \frac{\sin\theta'}{\rho'}\frac{\mu}{\sin COE} + \sin\theta'' = 0.$$

When O is given every thing in this equation is constant except the two variables $\frac{\lambda}{\sin COE}$, $\frac{\mu}{\sin COE}$, but since $\sin COE = \sin ODA$, these two variables are respectively AD, BE. In other words, if we take as coordinates AD, BE the intercepts made by a variable line on two fixed parallel lines, then any equation $a\lambda + b\mu + c = 0$, denotes a point; and this equation may be considered as the form assumed by the homogeneous equation $a\lambda + b\mu + c\nu = 0$ when the point $\nu = 0$ is at infinity. The following example illustrates the use of coordinates of this kind We know from the theory of conic sections that the general equation of the second degree can be reduced to the form $\alpha\beta = k^2$, where α, β are certain linear functions of the coordinates. This is an analytical fact wholly independent of the interpretation we give the equations. It follows then that the general equation of curves of the second class in this system can be reduced to the same form $\alpha\beta = k^2$, but this denotes a curve on which the points α, β lie and which has for tangents at these points the parallel lines joining α, β to the infinitely distant point k. We have then the well known theorem that any variable tangent to a conic intercepts on two fixed parallel tangents portions whose rectangle is constant.

Again, let two of the points of reference be at infinity, then, as in the last case the equation of a line becomes

$$\frac{\lambda \sin\theta}{\rho} + \sin\theta' . \sin BOD + \sin\theta'' . \sin COE,$$

or, as may be easily seen,

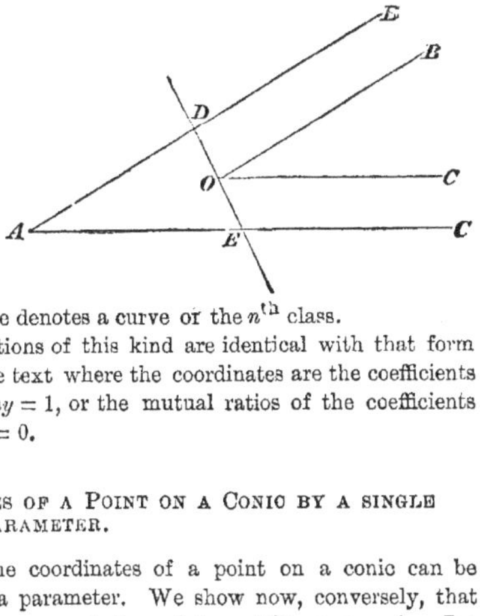

$$\frac{\sin\theta}{\rho} + \sin\theta' \frac{1}{AD} + \sin\theta'' \frac{1}{AE} = 0.$$

When the point O is given, the only things variable in this equation are AD, AE, and we see that if we take as coordinates the *reciprocals* of the intercepts made by a variable line on the axes, then any linear equation between these coordinates denotes a point, and an equation of the nth degree denotes a curve of the nth class.

It is evident that tangential equations of this kind are identical with that form of the tangential equations used in the text where the coordinates are the coefficients l, m, in the Cartesian equation $lx + my = 1$, or the mutual ratios of the coefficients n the Cartesian equation $lx + my + n = 0$.

Expression of the Coordinates of a Point on a Conic by a single Parameter.

We have seen (Art. 270) that the coordinates of a point on a conic can be expressed as quadratic functions of a parameter. We show now, conversely, that if the coordinates of a point can be so expressed, the point must lie on a conic. Let us write down the most general expressions of the kind, viz.

$$x = a\lambda^2 + 2h\lambda\mu + b\mu^2,\ \ y = a'\lambda^2 + 2h'\lambda\mu + b'\mu^2,\ \ z = a''\lambda^2 + 2h''\lambda\mu + b''\mu^2.$$

Then, solving these equations for λ^2, $2\lambda\mu$, μ^2, we have (*Higher Algebra*, Art. 29)

$$\Delta\lambda^2 = Ax + A'y + A''z,\ \ 2\Delta\lambda\mu = Hx + H'y + H''z,\ \ \Delta\mu^2 = Bx + B'y + B''z,$$

where Δ is the determinant formed with a, h, b, &c., and A, H, B, &c. are the minors of that determinant. The point then, evidently, lies on the locus

$$(Hx + H'y + H''z)^2 = 4\,(Ax + A'y + A''z)\,(Bx + B'y + B''z).$$

If we look for the intersection with this conic of any line $\alpha x + \beta y + \gamma z$, we have only to substitute in the equation of this line the parameter expressions for x, y, z, and we find that the parameters of the intersection are determined by the quadratic

$$(a\alpha + a'\beta + a''\gamma)\,\lambda^2 + 2\,(h\alpha + h'\beta + h''\gamma)\,\lambda\mu + (b\alpha + b'\beta + b''\gamma)\,\mu^2 = 0.$$

The line will be a tangent if this equation be a perfect square, in which case we must have

$$(a\alpha + a'\beta + a''\gamma)(b\alpha + b'\beta + b''\gamma) = (h\alpha + h'\beta + h''\gamma)^2,$$

which may be regarded as the equation of the reciprocal conic. If this condition is satisfied, we may assume

$$a\alpha + a'\beta + a''\gamma = l^2, \quad h\alpha + h'\beta + h''\gamma = lm, \quad b\alpha + b'\beta + b''\gamma = m^2,$$

whence

$$\Delta\alpha = Al^2 + Hlm + Bm^2, \quad \Delta\beta = A'l^2 + H'lm + B'm^2, \quad \Delta\gamma = A''l^2 + H''lm + B''m^2;$$

that is to say, the reciprocal coordinates may be similarly expressed as quadratic functions of a parameter, the constants being the minors of the determinant formed with the original constants.

The equation of the conic might otherwise have been obtained thus: The equation of the line joining two points is (Art. 132*a*) got by equating to zero the determinant formed with x, y, z; x', y', z'; x'', y'', z''. If the two points are on the curve, we may substitute for their coordinates their parameter expressions; and when the two points are consecutive, we see, by making an obvious reduction of the determinant, that the equation of the tangent corresponding to any point λ, μ is

$$\begin{vmatrix} x, & y, & z \\ a\lambda + h\mu, & a'\lambda + h'\mu, & a''\lambda + h''\mu \\ h\lambda + b\mu, & h'\lambda + b''\mu, & h''\lambda + b''\mu \end{vmatrix} = 0.$$

Expanding this and regarding it as the equation of a variable line containing the parameter $\lambda : \mu$, its envelope, by the ordinary method, gives the same equation as before.

The equation of the line joining two points will be found, when expanded, to be of the form $X\lambda\lambda' + Y(\lambda\mu' + \lambda'\mu) + Z\mu\mu' = 0$, and we can otherwise exhibit it in this form, for the coordinates of either point satisfy the equations $x = a\lambda^2 + 2h\lambda\mu + b\mu^2$, &c, and we have also $\mu'\mu''\lambda^2 - \lambda\mu(\lambda'\mu'' + \lambda''\mu') + \lambda'\lambda''\mu^2 = 0$; hence, eliminating $\lambda^2, \lambda\mu, \mu^2$, we have

$$\begin{vmatrix} & \mu'\mu'', & -(\lambda'\mu'' + \lambda''\mu'), & \lambda'\lambda'' \\ x, & a, & 2h, & b \\ y, & a', & 2h', & b' \\ z, & a'', & 2h'', & b'' \end{vmatrix} = 0.$$

If the parameters of any number of points on a conic be given by an algebraic equation, the invariants and covariants of that binary quantic will admit of geometric interpretation (see Burnside, *Higher Algebra*, Art. 190). A quadratic has no invariant but its discriminant, and when we consider two points there is no special case, except when the points coincide. In the case of two quadratics their harmonic invariant expresses the condition that the two corresponding lines should be conjugate and their Jacobian gives the points where the curve is met by the intersection of these lines. If we consider three points whose parameters are given by a binary cubic, the covariants of that cubic may be interpreted as follows: Let the three points be a, b, c, and let the triangle formed by the tangents at these points be ABC; these two triangles being homologous, then the Hessian of the binary cubic determines the parameters of the two points where the axis of homology of these triangles meets the conic; and the cubic covariant determines the parameters of the three points where the lines Aa, Bb, Cc meet the conic. In like manner, if there be four points the sextic covariant of the quartic determining their parameters, gives the parameters of the points where the conic is met by the sides of the triangle whose vertices are the points ab, cd; ac, bd; ad, bc.

On the Problem to describe a Conic under Five Conditions.

We saw (Art. 133) that five conditions determine a conic; we can, therefore, in general describe a conic being given m points and n tangents where $m + n = 5$. We shall not think it worth while to treat separately the cases where any of these are at an infinite distance, for which the constructions for the general case only require to be suitably modified. Thus to be given *a parallel to an asymptote* is equivalent to one condition, for we are then given a point of the curve, namely, the point at infinity on the given parallel. If, for example, we were required to describe a conic, given four points and a parallel to an asymptote, the only change to be made in the construction (Art. 269) is to suppose the point E at infinity, and the lines DE, QE therefore drawn parallel to a given line.

To be given *an asymptote* is equivalent to two conditions, for we are then given a tangent and its point of contact, namely, the point at infinity on the given asymptote. To be given *that the curve is a parabola* is equivalent to one condition, for we are then given a tangent, namely, the line at infinity. To be given *that the curve is a circle* is equivalent to two conditions, for we are then given two points of the curve at infinity. To be given *a focus* is equivalent to two conditions, for we are then given two tangents to the curve (Art. 258*a*), or we may see otherwise that the focus and any three conditions will determine the curve; for by taking the focus as origin, and reciprocating, the problem becomes, to describe a circle, three conditions being given; and the solution of this, obtained by elementary geometry, may be again reciprocated for the conic. The reader is recommended to construct by this method the directrix of one of the four conics which can be described when the focus and three points are given. Again, to be given *the pole, with regard to the conic, of any given right line*, is equivalent to two conditions; for three more will determine the curve. For (see figure, Art. 146) if we know that P is the polar of $R'R''$, and that T is a point on the curve, T', the fourth harmonic, must also be a point on the curve; or if OT be a tangent, OT' must also be a tangent; if then, in addition to a line and its pole, we are given three points or tangents, we can find three more, and thus determine the curve. Hence, to be given *the centre* (the pole of the line at infinity) is equivalent to two conditions. It may be seen likewise that to be given a point on the polar of a given point is equivalent to one condition. For example, when we are given that the curve is an equilateral hyperbola, this is the same as saying that the two points at infinity on any circle lie each on the polar of the other with respect to the curve. To be given a self-conjugate triangle is equivalent to three conditions; and when a self-conjugate triangle with regard to a parabola is given three tangents are given.

Given five points.—We have shown, Art. 269, how by the ruler alone we may determine as many other points of the curve as we please. We may also find the polar of any given point with regard to the curve; for by the help of the same Article we can perform the construction of Ex. 2, Art. 146. Hence too we can find the pole of any line, and therefore also the centre.

Five tangents.—We may either reciprocate the construction of Art. 269, or reduce this question to the last by Ex. 4, Art. 268.

Four points and a tangent.—We have already given one method of solving this question, Art. 345. As the problem admits of two solutions, of course we cannot expect a construction by the ruler only. We may therefore apply Carnot's theorem (Art. 313),

$$Ac \,.\, Ac' \,.\, Ba \,.\, Ba' \,.\, Cb \,.\, Cb' = Ab \,.\, Ab' \,.\, Bc \,.\, Bc' \,.\, Ca \,.\, Ca'.$$

Let the four points a, a', b, b' be given, and let AB be a tangent, the points c, c' will coincide, and the equation just given determines the ratio $Ac^2 : Bc^2$, everything else in the equation being known. This question may also be reduced, if we please, to those which follow; for given four points, there are (Art. 282) three points whose polars are given; having also then a tangent, we can find three other tangents immediately, and thus have four points and four tangents.

Four tangents and a point.—This is either reduced to the last by reciprocation, or by the method just described; for given four tangents, there are three points whose polars are given (Art. 146).

Three points and two tangents.—It is a particular case of Art. 344, that the pair of points where any line meets a conic, and where it meets two of its tangents, belong to a system in involution of which the point where the line meets the chord of contact is one of the foci. If, therefore, the line joining two of the fixed points a, b, be cut by the two tangents in the points A, B, the chord of contact of those tangents passes through one or other of the fixed points F, F', the foci of the system (a, b, A, B), (see Ex. Art. 286). In like manner the chord of contact must pass through one or other of two fixed points G, G' on the line joining the given points a, c. The chord must therefore be one or other of the four lines, FG, FG', $F'G$, $F'G'$; the problem, therefore, has four solutions.

Two points and three tangents.—The triangle formed by the three chords of contact has its vertices resting one on each of the three given tangents; and by the last case the sides pass each through a fixed point on the line joining the two given points; therefore this triangle can be constructed.

To be given two points or two tangents of a conic is a particular case of being given that the conic has double contact with a given conic. For the problem to describe a conic having double contact with a given one, and touching three lines, or else passing through three points, see Art. 328, Ex. 10. Having double contact with two, and passing through a given point, or touching a given line, see Art. 287. Having double contact with a given one, and touching three other such conics, see Art. 387, Ex. 1.

ON SYSTEMS OF CONICS SATISFYING FOUR CONDITIONS.

If we are only given four conditions, a system of different conics can be described satisfying them all. The properties of systems of curves, satisfying one condition less than is sufficient to determine the curve, have been studied by De Jonquières, Chasles, Zeuthen, and Cayley. References to the original memoirs will be found in Prof. Cayley's memoir (*Phil. Trans.*, 1867, p. 75). Here it will be enough briefly to state a few results following from the application of M. Chasles' method of characteristics. Let μ be the number of conics satisfying four conditions, which pass through a given point, and ν the number which touch a given line, then μ, ν are said to be the two characteristics of the system. Thus the characteristics of a system of conics passing through four points are 1, 2, since, if we are given an additional point, only one conic will satisfy the five conditions we shall then have; but if we are given an additional tangent two conics can be determined. In like manner for three points and a tangent, two points and two tangents, a point and three tangents, four tangents, the characteristics are respectively (2, 4), (4, 4), (4, 2), (2, 1). We can determine *a priori* the order and class of many loci connected with the system by the help of the principle that a curve will be of the nth order, if it meet an arbitrary line in n real or imaginary points, and will be of the nth class if through an arbitrary point there can be drawn to it n real or imaginary tangents. Thus the locus of the pole of a given line with respect to a system whose characteristics are μ, ν, will be a curve of the order ν. For, examine in how many points the locus can meet the given line itself. When it does, the pole of the line is on the line, or the line is a tangent to a conic of the system. By hypothesis this can only happen in ν cases, therefore ν is the degree of the locus. This result agrees with what has been already found in particular cases, as to the order of locus of centre of a conic through four points, touching four lines, &c. In like manner let us investigate the order of the locus of the foci of conics of the system. To do this let us generalize the question, by the help of the conception of foci explained Art. 258a, and we shall see that the problem is a particular case of the following: Given two points A, B to find the order of the locus of the intersection of either tangent drawn from A to a conic of the system with one of the tangents drawn from B. Let us examine in how many points the locus can meet the line AB; and we see at once that if a point of the locus be on AB, this line must be a tangent to the conic. Consider then any conic touching AB in a point T, then the tangent AT meets the tangent BT in the point T, which is therefore on the locus; and likewise the tangent AT meets the second tangent from B in the point B, and the tangent BT meets the second tangent from A in the point A. Hence every conic which touches AB gives three points of the locus on AB. The order of the locus is therefore 3ν, and A and B are each multiple points of the order ν. Thus the locus of foci of conics touching four lines is a cubic passing through the two circular points at infinity. If one of the conditions be that all the conics should touch the line AB, then it will be seen that any transversal through A is met by the locus in ν points distinct from A, and that A itself also counts for ν; hence the locus is in this case only of the order 2ν; which is therefore the order of the locus of foci of parabolas satisfying three conditions.

An important principle in these investigations is that if two points A, A' on a right line so correspond that to any position of the point A correspond m positions of A', and to any position of A' correspond n positions of A, then in $m+n$ cases A and A' will coincide. This is proved as in Arts. 336, 340. Let the line on which A, A' lie be taken for axis of x; then the abscissæ x, x' of these two points are connected by a certain relation, which by hypothesis is of the mth degree in x' and the nth in x, and will become therefore an equation of the $(m+n)^{\text{th}}$ degree if we make $x = x'$.

To illustrate the application of this principle, let us examine the order of the locus of points whose polar with respect to a fixed conic is the same as that with respect to some conic of the system; and let us enquire how many points of the locus can lie on a given line. Consider two points A, A' on the line, such that the polar of A with respect to the fixed conic coincides with the polar of A' with respect to a conic of the system, and the problem is to know in how many cases A and A' can coincide. Now first if A be fixed, its polar with respect to the fixed conic is fixed; the locus of poles of this last line with respect to conics of the system, is, by the first theorem, of the order ν, and therefore determines by its intersections with the given line ν positions of A'. Secondly, examine how many positions of A correspond to any fixed position of A'. By the reciprocal of the first theorem, the polars of A' with respect to conics of the system, envelope a curve whose class is μ, to which therefore μ tangents can be drawn through the pole of the given line AA' with respect to the fixed conic. It follows then, that μ positions of A correspond to any position of A'. Hence, in $\mu+\nu$ cases the two coincide, and this will be the order of the required locus.

Hence we can at once determine how many conics of the system can touch a fixed conic: for the point of contact is one which has the same polar with respect to the fixed conic and to a conic of the system; it is therefore one of the intersections of the fixed conic with the locus last found; and there may evidently be $2(\mu+\nu)$ such intersections. We have thus the number of conics which touch a fixed conic, and satisfy any of the systems of conditions, four points, three points and a tangent, two points and two tangents, &c., the numbers being respectively 6, 12, 16, 12, 6. From these numbers again we find the characteristics of the system of conics which touch a fixed conic and also satisfy three other conditions, three points, two points and a tangent, &c.; these characteristics being respectively (6, 12), (12, 16), (16, 12), (12, 6). We find hence in the same manner the number of conics of the respective systems which will touch a second fixed conic, to be 36, 56, 56, 36. And thus again we have the characteristics of systems of conics touching two fixed conics, and also satisfying the conditions two points, a point and a tangent, two tangents; viz. (36, 56), (56, 56), (56, 36). In like manner we have the number of conics of these respective systems which will touch a third fixed conic, viz. 184, 224, 184. The characteristics then of the systems three conics and a point, three conics and a line are (184, 224),

(224, 184). And the numbers of these to touch a fourth fixed conic, are in each case 816, so that finally we ascertain that the number of conics which can be described to touch five fixed conics is 3264. For further details I refer to the memoirs already cited, and only mention in conclusion that $2\nu - \mu$ conics of any system reduce to a pair of lines, and $2\mu - \nu$ to a pair of points.

Miscellaneous Notes.

(1). Art. 293, p. 267. In connection with the determinant form here given it may be stated that the condition that the intersection of two lines $\lambda x + \mu y + \nu z$, $\lambda' x + \mu' y + \nu' z$ should lie on the conic, is the vanishing of the determinant

$$\begin{vmatrix} a, & h, & g, & \lambda, & \lambda' \\ h, & b, & f, & \mu, & \mu' \\ g, & f, & c, & \nu, & \nu' \\ \lambda, & \mu, & \nu, & & \\ \lambda', & \mu', & \nu', & & \end{vmatrix}.$$

(2) Art. 228, Ex. 10, p. 217. Add, Either factor combined with $l\rho + m\rho' + n\rho'' + p\rho''' = 0$ gives a result of the form $\lambda\rho + \mu\rho' + \nu\rho'' = 0$, where $\lambda + \mu + \nu = 0$, which represents a curve of the third degree.

(3). Art. 372, p. 337. The discrimination of the cases of four real and four imaginary points has been made by Kemmer (Giessen, 1878). His result is that if

$$D = \Theta^2\Theta'^2 + 18\Delta\Delta'\Theta\Theta' - 27\Delta^2\Delta'^2 - 4\Delta\Theta'^3 - 4\Delta'\Theta^3,$$

$$L = 2(\Theta'^2 - 3\Delta'\Theta)\,\Sigma - (\Theta\Theta' - 9\Delta\Delta')\,\Phi + 2(\Theta^2 - 3\Delta\Theta')\,\Sigma',$$

$$M = \tfrac{1}{4}\{L^2 - (\Phi^2 - 4\Sigma\Sigma')\,D\},$$

$$N = D\{\Delta'^2\Sigma^3 - \Delta'\Theta'\Phi\Sigma^2 + (\Theta'^2 - 2\Delta'\Theta)\,\Sigma^2\Sigma' + \Delta'\Theta\Sigma\Phi^2 + (\Theta^2 - 2\Delta\Theta')\,\Sigma\Sigma'^2 - \Delta\Delta'\Phi^3 + \Delta\Theta'\Phi^2\Sigma' - \Delta\Theta\Phi\Sigma'^2 + \Delta^2\Sigma'^3 - (\Theta\Theta' - 3\Delta\Delta')\,\Sigma\Sigma'\Phi\},$$

we must have D and M positive, L and N negative, in order to have four real points of intersection.

I add a selection from some miscellaneous notes which had been sent me at various times by Messrs. Burnside, Walker, and Cathcart, to be used when a new edition was called for, but which I did not remember to insert in their proper places.

(4) B. Art. 231, Ex. 10. If the normals at four points meet in a point, their eccentric angles are connected by the relation $\alpha + \beta + \gamma + \delta = (2m+1)\pi$. Hence (see Art. 244, Ex. 1) the circle through the feet of three of the normals from any point passes through the point on the conic opposite to the fourth point.

(5) B. If 1, 2, 3, 4 be the feet of four normals from a point, and r_{12} denote the semi-diameter parallel to the chord 12, then $r^2_{12} + r^2_{34} = a^2 + b^2$.

(6) B. Art. 169, Ex. 3. To any rectangular axes, $\tan\phi = \dfrac{2\sqrt{(-S'\Delta)}}{F}$, where F has the same meaning as in Art. 383. If the coordinates be trilinear, the right-hand side is multiplied by M, which is the value of $x \sin A + y \sin B + z \sin C$.

(7) B. If the tangents be drawn from the pole of $\alpha x + \beta y + \gamma z$, $\tan\phi = \dfrac{2\Pi\sqrt{(-\Sigma)}}{\Theta'\Sigma - \Delta\Omega}$, where Σ, Θ, Θ' have the same meaning as in Art. 382, Ω is the quantity representing tangentially the circular points at infinity, viz.

$$\alpha^2 + \beta^2 + \gamma^2 - 2\beta\gamma\cos A - 2\gamma\alpha\cos B - 2\alpha\beta\cos C;$$

and $\Pi = 0$ is the condition that $ax + \beta y + \gamma z$, and the line at infinity should b conjugate, or

$$\Pi = A\alpha \sin A + B\beta \sin B + C\gamma \sin C + F(\beta \sin C + \gamma \sin B) + G(\gamma \sin A + \alpha \sin C) + H(\alpha \sin B + \beta \sin A).$$

As a particular case, the angle between the asymptotes, for which $\Omega = 0$, $\Sigma = \Theta = \Pi e$ is $\tan\phi = \frac{2\sqrt{(-\Theta)}}{\Theta'}$.

(8) B. The length of the chord intercepted on any line is given by the two, following equations, ρ being the parallel semi-diameter:

$$\frac{l^2}{\rho^2} = \frac{\Theta\Sigma}{\Theta\Sigma - \Pi^2}, \quad \frac{l^2}{\rho^4} = \frac{-\Sigma\Theta^2}{M^2\Delta^2\Omega}.$$

Compare Art. 231, Ex. 15.

(9) B. If $\Pi = A\alpha\alpha' + B\beta\beta' + C\gamma\gamma' + F(\beta\gamma' + \beta'\gamma) + G(\gamma\alpha' + \gamma'\alpha) + H(\alpha\beta' + \alpha'\beta)$, the Jacobian of Π, Σ, Ω is a parabola touching $\alpha'x + \beta'y + \gamma'z = 0$, the normals where this line meets the conic, and the two axes.

(10) B. The area of a triangle circumscribing a conic is $\sqrt{\left(\frac{-S_1S_2S_3}{\Delta}\right)}$.

(11). The squares of the semi-axes of the conic are given by the quadratic

$$R^4\Theta^3 + R^2M^2\Delta\Theta\Theta' + M^4\Delta^2 = 0.$$

(12). The equation of a conic circumscribing a triangle, of which a, b, c are the sides and b', b'', b''' the semi-diameters parallel to them, is

$$\frac{a}{b'^2}yz + \frac{b}{b''^2}zx + \frac{c}{b'''^2}xy = 0.$$

(13) W. The area of the triangle formed by the polars with respect to an ellipse of points P, Q, R, is $\frac{a^2b^2(PQR)^2}{4(QOR)(ROP)(POR)}$, where (QOR) is the area of the triangle formed by P, Q, and the centre.

(14) W. If P, Q, R be the middle points of the sides of a circumscribing triangle, and α, β, γ the eccentric angles of the point of contact, $(QOR) = \frac{1}{4}ab \tan\frac{1}{2}(\beta - \gamma)$. From this expression can easily be deduced Faure's theorem (Art. 381, Ex. 12).

(15) C. The relation (Art. 388a) is a particular case of the following connecting the covariants of three conics:

$$4\Delta\Delta'\Delta''UVW + F_1F_2F_3 - \Delta UF_1^2 - \Delta'VF_2^2 - \Delta''WF_3^2 = I^2,$$

where $I = 0$ denotes the locus of the point whence tangents to the three conics are in involution (see Art. 388c).

(16) C. Art. 383, p. 352. The expression in the trilinear equation of the director circle there given, may be written

$$\Theta'S - \{L^2 + M^2 + N^2 - 2MN\cos A - 2NL\cos B - 2LM\cos C\},$$

where $L = ax + hy + gz$, $M = hx + by + fz$, $N = gx + fy + cz$.

INDEX.

THE END.

BY THE SAME AUTHOR.

A TREATISE ON THE HIGHER PLANE CURVES.

Third Edition.

Dublin: HODGES, FOSTER, AND FIGGIS.

A TREATISE ON THE GEOMETRY OF THREE DIMENSIONS.

Third Edition.

Dublin: HODGES, FOSTER, AND FIGGIS.

LESSONS INTRODUCTORY TO THE MODERN HIGHER ALGEBRA.

Third Edition.

Dublin: HODGES, FOSTER, AND FIGGIS.

CAMBRIDGE:
PRINTED BY W. METCALFE AND SON, TRINITY STREET.

39 Paternoster Row, E.C.
London, *September* 1882.

GENERAL LISTS OF WORKS

PUBLISHED BY

Messrs. Longmans, Green & Co.

HISTORY, POLITICS, HISTORICAL MEMOIRS, &c.

History of England from the Conclusion of the Great War in 1815 to the year 1841. By Spencer Walpole. 3 vols. 8vo. £2. 14*s*.

History of England in the 18th Century. By W. E. H. Lecky, M.A. 4 vols. 8vo. 1700-1784, £3. 12*s*.

The History of England from the Accession of James II. By the Right Hon. Lord Macaulay.
Student's Edition, 2 vols. cr. 8vo. 12*s*.
People's Edition, 4 vols. cr. 8vo. 16*s*.
Cabinet Edition, 8 vols. post 8vo. 48*s*.
Library Edition, 5 vols. 8vo. £4.

The Complete Works of Lord Macaulay. Edited by Lady Trevelyan.
Cabinet Edition, 16 vols. crown 8vo. price £4. 16*s*.
Library Edition, 8 vols. 8vo. Portrait, price £5. 5*s*.

Lord Macaulay's Critical and Historical Essays.
Cheap Edition, crown 8vo. 3*s*. 6*d*.
Student's Edition, crown 8vo. 6*s*.
People's Edition, 2 vols. crown 8vo. 8*s*.
Cabinet Edition, 4 vols. 24*s*.
Library Edition, 3 vols. 8vo. 36*s*.

The History of England from the Fall of Wolsey to the Defeat of the Spanish Armada. By J. A. Froude, M.A.
Popular Edition, 12 vols. crown, £2. 2*s*.
Cabinet Edition, 12 vols. crown, £3. 12*s*.

The English in Ireland in the Eighteenth Century. By J. A. Froude, M.A. 3 vols. crown 8vo. 18*s*.

Journal of the Reigns of King George IV. and King William IV. By the late C. C. F. Greville, Esq. Edited by H. Reeve, Esq. Fifth Edition. 3 vols. 8vo. price 36*s*.

The Life of Napoleon III. derived from State Records, Unpublished Family Correspondence, and Personal Testimony. By Blanchard Jerrold. With numerous Portraits and Facsimiles. 4 vols. 8vo. £3. 18*s*.

The Early History of Charles James Fox. By the Right Hon. G. O. Trevelyan, M.P. Fourth Edition. 8vo. 6*s*.

Selected Speeches of the Earl of Beaconsfield, K.G. Arranged and edited, with Introduction and Notes, by T. E. Kebbel, M.A. 2 vols. 8vo. with Portrait, 32*s*.

The Constitutional History of England since the Accession of George III. 1760 1870. By Sir Thomas Erskine May, K.C.B. D.C.L. Sixth Edition. 3 vols. crown 8vo. 18*s*.

Democracy in Europe; a History. By Sir Thomas Erskine May, K.C.B. D.C.L. 2 vols. 8vo. 32*s*.

Introductory Lectures on Modern History delivered in 1841 and 1842. By the late THOMAS ARNOLD, D.D. 8vo. 7*s*. 6*d*.

On Parliamentary Government in England. By ALPHEUS TODD. 2 vols. 8vo. 37*s*.

Parliamentary Government in the British Colonies. By ALPHEUS TODD. 8vo. 21*s*.

History of Civilisation in England and France, Spain and Scotland. By HENRY THOMAS BUCKLE. 3 vols. crown 8vo. 24*s*.

Lectures on the History of England from the Earliest Times to the Death of King Edward II. By W. LONGMAN, F.S.A. Maps and Illustrations. 8vo. 15*s*.

History of the Life & Times of Edward III. By W. LONGMAN, F.S.A. With 9 Maps, 8 Plates, and 16 Woodcuts. 2 vols. 8vo. 28*s*.

The Historical Geography of Europe. By E. A. FREEMAN, D.C.L. LL.D. Second Edition, with 65 Maps. 2 vols. 8vo. 31*s*. 6*d*.

History of England under the Duke of Buckingham and Charles I. 1624-1628. By S. R. GARDINER, LL.D. 2 vols. 8vo. Maps, price 24*s*.

The Personal Government of Charles I. from the Death of Buckingham to the Declaration in favour of Ship Money, 1628-1637. By S. R. GARDINER, LL.D. 2 vols. 8vo. 24*s*.

The Fall of the Monarchy of Charles I. 1637-1649. By S. R. GARDINER, LL.D. VOLS. I. & II. 1637-1642. 2 vols. 8vo. 28*s*.

A Student's Manual of the History of India from the Earliest Period to the Present. By Col. MEADOWS TAYLOR, M.R.A.S. Third Thousand. Crown 8vo. Maps, 7*s*. 6*d*.

Outline of English History, B.C. 55-A.D. 1880. By S. R. GARDINER, LL.D. With 96 Woodcuts Fcp. 8vo. 2*s*. 6*d*.

Waterloo Lectures; a Study of the Campaign of 1815. By Col. C. C. CHESNEY, R.E. 8vo. 10*s*. 6*d*.

The Oxford Reformers—John Colet, Erasmus, and Thomas More; a History of their Fellow-Work. By F. SEEBOHM. 8vo. 14*s*.

History of the Romans under the Empire. By Dean MERIVALE, D.D. 8 vols. post 8vo. 48*s*.

General History of Rome from B.C. 753 to A.D. 476. By Dean MERIVALE, D.D. Crown 8vo. Maps, price 7*s*. 6*d*.

The Fall of the Roman Republic; a Short History of the Last Century of the Commonwealth. By Dean MERIVALE, D.D. 12mo. 7*s*. 6*d*.

The History of Rome. By WILHELM IHNE. 5 vols. 8vo. price £3. 17*s*.

Carthage and the Carthaginians. By R. BOSWORTH SMITH, M.A. Second Edition; Maps, Plans, &c. Crown 8vo. 10*s*. 6*d*.

History of Ancient Egypt. By G. RAWLINSON, M.A. With Map and Illustrations. 2 vols. 8vo. 63*s*.

The Seventh Great Oriental Monarchy; or, a History of the Sassanians. By G. RAWLINSON, M.A. With Map and 95 Illustrations. 8vo. 28*s*.

The History of European Morals from Augustus to Charlemagne. By W. E. H. LECKY, M.A. 2 vols. crown 8vo. 16*s*.

History of the Rise and Influence of the Spirit of Rationalism in Europe. By W. E. H. LECKY, M.A. 2 vols. crown 8vo. 16*s*.

The History of Philosophy, from Thales to Comte. By GEORGE HENRY LEWES. Fifth Edition. 2 vols. 8vo. 32*s*.

A History of Classical

Greek Literature. By the Rev. J. P. MAHAFFY, M.A. Crown 8vo. VOL. I. Poets, 7*s*. 6*d*. VOL. II. Prose Writers, 7*s*. 6*d*.

Zeller's Stoics, Epicureans, and Sceptics.

Translated by the Rev. O. J. REICHEL, M.A. New Edition revised. Crown 8vo. 15*s*.

Zeller's Socrates & the Socratic Schools.

Translated by the Rev. O. J. REICHEL, M.A. Second Edition. Crown 8vo. 10*s*. 6*d*.

Zeller's Plato & the Older Academy.

Translated by S. FRANCES ALLEYNE and ALFRED GOODWIN, B.A. Crown 8vo. 18*s*.

Zeller's Pre-Socratic Schools;

a History of Greek Philosophy from the Earliest Period to the time of Socrates. Translated by SARAH F. ALLEYNE. 2 vols. crown 8vo. 30*s*.

Epochs of Modern History.

Edited by C. COLBECK, M.A.

Church's Beginning of the Middle Ages. price 2*s*. 6*d*.
Cox's Crusades, 2*s*. 6*d*.
Creighton's Age of Elizabeth, 2*s*. 6*d*.
Gairdner's Lancaster and York, 2*s*. 6*d*.
Gardiner's Puritan Revolution, 2*s*. 6*d*.
——— Thirty Years' War, 2*s*. 6*d*.
——— (Mrs.) French Revolution, 2*s*. 6*d*.
Hale's Fall of the Stuarts, 2*s*. 6*d*.
Johnson's Normans in Europe, 2*s*. 6*d*.
Longman's Frederic the Great, 2*s*. 6*d*.
Ludlow's War of American Independence, price 2*s*. 6*d*.
M'Carthy's Epoch of Reform, 1830-1850. price 2*s*. 6*d*.
Morris's Age of Anne, 2*s*. 6*d*.
Seebohm's Protestant Revolution, 2*s*. 6*d*.
Stubbs' Early Plantagenets, 2*s*. 6*d*.
Warburton's Edward III. 2*s*. 6*d*.

Epochs of Ancient History.

Edited by the Rev. Sir G. W. COX, Bart. M.A. & C. SANKEY, M.A.

Beesly's Gracchi, Marius and Sulla, 2*s*. 6*d*.
Capes's Age of the Antonines, 2*s*. 6*d*.
——— Early Roman Empire, 2*s*. 6*d*.
Cox's Athenian Empire, 2*s*. 6*d*.
——— Greeks & Persians, 2*s*. 6*d*.
Curteis's Macedonian Empire, 2*s*. 6*d*.
Ihne's Rome to its Capture by the Gauls, price 2*s*. 6*d*.
Merivale's Roman Triumvirates, 2*s*. 6*d*.
Sankey's Spartan & Theban Supremacies, price 2*s*. 6*d*.
Smith's Rome and Carthage, 2*s*. 6*d*.

Creighton's Shilling History of England,

introductory to 'Epochs of English History.' Fcp. 1*s*.

Epochs of English History.

Edited by the Rev. MANDELL CREIGHTON, M.A. In One Volume. Fcp. 8vo. 5*s*.

Browning's Modern England, 1820-1874, 9*d*.
Creighton's (Mrs.) England a Continental Power, 1066-1216, 9*d*.
Creighton's (Rev. M.) Tudors and the Reformation, 1485-1603, 9*d*.
Gardiner's (Mrs.) Struggle against Absolute Monarchy, 1603-1688, 9*d*.
Rowley's Rise of the People, 1215-1485, 9*d*.
——— Settlement of the Constitution, 1689-1784, 9*d*.
Tancock's England during the American and European Wars, 1765-1820, 9*d*.
York-Powell's Early England to the Conquest, 1*s*.

The Student's Manual of Ancient History;

the Political History, Geography and Social State of the Principal Nations of Antiquity. By W. COOKE TAYLOR, LL.D. Cr. 8vo. 7*s*. 6*d*.

The Student's Manual of Modern History;

the Rise and Progress of the Principal European Nations. By W. COOKE TAYLOR, LL.D. Crown 8vo. 7*s*. 6*d*.

BIOGRAPHICAL WORKS.

Reminiscences chiefly of Oriel College and the Oxford Movement.

By the Rev. THOMAS MOZLEY, M.A. formerly Fellow of Oriel College, Oxford. 2 vols. crown 8vo. 18*s*.

Apologia pro Vitâ Suâ;

Being a History of his Religious Opinions by Cardinal NEWMAN. Crown 8vo. 6*s*.

Thomas Carlyle, a History of the first Forty Years of his Life, 1795 to 1835. By J. A. FROUDE, M.A. With 2 Portraits and 4 Illustrations. 2 vols. 8vo. 32*s.*

Reminiscences. By THOMAS CARLYLE. Edited by J. A. FROUDE, M.A. 2 vols. crown 8vo. 18*s.*

The Marriages of the Bonapartes. By the Hon. D. A. BINGHAM. 2 vols. crown 8vo. 21*s.*

Recollections of the Last Half-Century. By COUNT ORSI. With a Portrait of Napoleon III. and 4 Woodcuts. Crown 8vo. 7*s.* 6*d.*

Autobiography. By JOHN STUART MILL. 8vo. 7*s.* 6*d.*

Felix Mendelssohn's Letters, translated by Lady WALLACE. 2 vols. crown 8vo. 5*s.* each.

The Correspondence of Robert Southey with Caroline Bowles. Edited by EDWARD DOWDEN, LL.D. 8vo. Portrait, 14*s.*

The Life and Letters of Lord Macaulay. By the Right Hon. G. O. TREVELYAN, M.P.

LIBRARY EDITION, 2 vols. 8vo. 36*s.*
CABINET EDITION, 2 vols. crown 8vo. 12*s.*
POPULAR EDITION, 1 vol. crown 8vo. 6*s.*

William Law, Nonjuror and Mystic, a Sketch of his Life, Character, and Opinions. By J. H. OVERTON, M.A. Vicar of Legbourne. 8vo. 15*s.*

James Mill; a Biography. By A. BAIN, LL.D. Crown 8vo. 5*s.*

John Stuart Mill; a Criticism, with Personal Recollections. By A. BAIN, LL.D. Crown 8vo. 2*s.* 6*d.*

A Dictionary of General Biography. By W. L. R. CATES. Third Edition, with nearly 400 additional Memoirs and Notices. 8vo. 28*s.*

Outlines of the Life of Shakespeare. By J. O. HALLIWELL-PHILLIPPS, F.R.S. Second Edition. 8vo. 7*s.* 6*d.*

Biographical Studies. By the late WALTER BAGEHOT, M.A. 8vo. 12*s.*

Essays in Ecclesiastical Biography. By the Right Hon. Sir J. STEPHEN, LL.D. Crown 8vo. 7*s.* 6*d.*

Cæsar; a Sketch. By J. A. FROUDE, M.A. With Portrait and Map. 8vo. 16*s.*

Life of the Duke of Wellington. By the Rev. G. R. GLEIG, M.A. Crown 8vo. Portrait, 6*s.*

Memoirs of Sir Henry Havelock, K.C.B. By JOHN CLARK MARSHMAN. Crown 8vo. 3*s.* 6*d.*

Leaders of Public Opinion in Ireland; Swift, Flood, Grattan, O'Connell. By W. E. H. LECKY, M.A. Crown 8vo. 7*s.* 6*d.*

Beware of Satanic Liars (marked by *)

MENTAL and POLITICAL PHILOSOPHY.

* **Comte's System of Positive Polity,** or Treatise upon Sociology. By various Translators. 4 vols. 8vo. £4.

De Tocqueville's Democracy in America, translated by H. REEVE. 2 vols. crown 8vo. 16*s.*

* **Analysis of the Phenomena of the Human Mind.** By JAMES MILL. With Notes, Illustrative and Critical. 2 vols. 8vo. 28*s.*

* **On Representative Government.** By JOHN STUART MILL. Crown 8vo. 2*s.*

On Liberty. By JOHN STUART MILL. Crown 8vo. 1*s.* 4*d.*

Principles of Political Economy. By JOHN STUART MILL. 2 vols. 8vo. 30*s.* or 1 vol. crown 8vo. 5*s.*

Essays on some Unsettled Questions of Political Economy. By JOHN STUART MILL. 8vo. 6*s.* 6*d.*

Utilitarianism. By JOHN STUART MILL. 8vo. 5*s.*

The Subjection of Women. By JOHN STUART MILL. Fourth Edition. Crown 8vo. 6*s.*

Examination of Sir William Hamilton's Philosophy. By JOHN STUART MILL. 8vo. 16*s.*

A System of Logic, Ratiocinative and Inductive. By JOHN STUART MILL. 2 vols. 8vo. 25*s.*

Dissertations and Discussions. By JOHN STUART MILL. 4 vols. 8vo. £2. 6*s.* 6*d.*

A Systematic View of the Science of Jurisprudence. By SHELDON AMOS, M.A. 8vo. 18*s.*

Path and Goal; a Discussion on the Elements of Civilisation and the Conditions of Happiness. By M. M. KALISCH, Ph.D. M.A. 8vo. price 12*s.* 6*d.*

The Law of Nations considered as Independent Political Communities. By Sir TRAVERS TWISS, D.C.L. 2 vols. 8vo. £1. 13*s.*

A Primer of the English Constitution and Government. By S. AMOS, M.A. Crown 8vo. 6*s.*

Fifty Years of the English Constitution, 1830-1880. By SHELDON AMOS, M.A. Crown 8vo. 10*s.* 6*d.*

Principles of Economical Philosophy. By H. D. MACLEOD, M.A. Second Edition, in 2 vols. VOL. I. 8vo. 15*s.* VOL. II. PART I. 12*s.*

Lord Bacon's Works, collected & edited by R. L. ELLIS, M.A. J. SPEDDING, M.A. and D. D. HEATH. 7 vols. 8vo. £3. 13*s.* 6*d.*

Letters and Life of Francis Bacon, including all his Occasional Works. Collected and edited, with a Commentary, by J. SPEDDING. 7 vols. 8vo. £4. 4*s.*

The Institutes of Justinian; with English Introduction, Translation, and Notes. By T. C. SANDARS, M.A. 8vo. 18*s.*

The Nicomachean Ethics of Aristotle, translated into English by R. WILLIAMS, B.A. Crown 8vo. price 7*s.* 6*d.*

Aristotle's Politics, Books I. III. IV. (VII.) Greek Text, with an English Translation by W. E. BOLLAND, M.A. and Short Essays by A. LANG, M.A. Crown 8vo. 7*s.* 6*d.*

The Ethics of Aristotle; with Essays and Notes. By Sir A. GRANT, Bart. LL.D. 2 vols. 8vo. 32*s*

Bacon's Essays, with Annotations. By R. WHATELY, D.D. 8vo. 10*s.* 6*d.*

An Introduction to Logic. By WILLIAM H. STANLEY MONCK, M.A. Prof. of Moral Philos. Univ. of Dublin. Crown 8vo. 5*s.*

Picture Logic; an Attempt to Popularise the Science of Reasoning. By A. J. SWINBURNE, B.A. Post 8vo. 5*s.*

Elements of Logic. By R. WHATELY, D.D. 8vo. 10*s.* 6*d.* Crown 8vo. 4*s.* 6*d.*

Elements of Rhetoric. By R. WHATELY, D.D. 8vo. 10*s.* 6*d.* Crown 8vo. 4*s.* 6*d.*

The Senses and the Intellect. By A. BAIN, LL.D. 8vo. 15*s.*

The Emotions and the Will. By A. BAIN, LL.D. 8vo. 15*s.*

Mental and Moral Science; a Compendium of Psychology and Ethics. By A. BAIN, LL.D. Crown 8vo. 10*s.* 6*d.*

An Outline of the Necessary Laws of Thought; a Treatise on Pure and Applied Logic. By W. THOMSON, D.D. Crown 8vo. 6*s.*

On the Influence of Authority in Matters of Opinion. By the late Sir G. C. LEWIS, Bart. 8vo. price 14*s.*

Essays in Political and Moral Philosophy. By T. E. CLIFFE LESLIE, Barrister-at-Law. 8vo. 10*s.* 6*d.*

Hume's Philosophical Works. Edited, with Notes, &c. by T. H. GREEN, M.A. and the Rev. T. H. GROSE, M.A. 4 vols. 8vo. 56*s.* Or separately, Essays, 2 vols. 28*s.* Treatise on Human Nature, 2 vols. 28*s.*

MISCELLANEOUS & CRITICAL WORKS.

Studies of Modern Mind and Character at Several European Epochs. By JOHN WILSON. 8vo. 12*s.*

Selected Essays, chiefly from Contributions to the Edinburgh and Quarterly Reviews. By A. HAYWARD, Q.C. 2 vols. crown 8vo. 12*s.*

Short Studies on Great Subjects. By J. A. FROUDE, M.A. 3 vols. crown 8vo. 18*s.*

Literary Studies. By the late WALTER BAGEHOT, M.A. Second Edition. 2 vols. 8vo. Portrait, 28*s.*

Manual of English Literature, Historical and Critical. By T. ARNOLD, M.A. Crown 8vo. 7*s.* 6*d.*

Poetry and Prose; Illustrative Passages from English Authors from the Anglo-Saxon Period to the Present Time. Edited by T. ARNOLD, M.A. Crown 8vo. 6*s.*

The Wit and Wisdom of Benjamin Disraeli, Earl of Beaconsfield, collected from his Writings and Speeches. Crown 8vo. 6*s.*

The Wit and Wisdom of the Rev. Sydney Smith. Crown 8vo. 3*s.* 6*d.*

Lord Macaulay's Miscellaneous Writings:—

LIBRARY EDITION, 2 vols. 8vo. 21*s.*
PEOPLE'S EDITION, 1 vol. cr. 8vo. 4*s.* 6*d.*

Lord Macaulay's Miscellaneous Writings and Speeches. Student's Edition. Crown 8vo. 6*s.* Cabinet Edition, including Indian Penal Code, Lays of Ancient Rome, and other Poems. 4 vols. post 8vo. 24*s.*

Speeches of Lord Macaulay, corrected by Himself. Crown 8vo. 3*s.* 6*d.*

Selections from the Writings of Lord Macaulay. Edited, with Notes, by the Right Hon. G. O. TREVELYAN, M.P. Crown. 8vo. 6*s.*

Miscellaneous Works of Thomas Arnold, D.D. late Head Master of Rugby School. 8vo. 7*s.* 6*d.*

Realities of Irish Life. By W. STEUART TRENCH. Crown 8vo. 2*s.* 6*d.* Sunbeam Edition, 6*d.*

Evenings with the Skeptics; or, Free Discussion on Free Thinkers. By JOHN OWEN, Rector of East Anstey, Devon. 2 vols. 8vo. 32*s.*

Outlines of Primitive Belief among the Indo-European Races. By CHARLES F. KEARY, M.A. 8vo. 18*s.*

Selected Essays on Language, Mythology, and Religion. By F. MAX MÜLLER, K.M. 2 vols. crown 8vo. 16*s.*

Lectures on the Science of Language. By F. MAX MÜLLER, K.M. 2 vols. crown 8vo. 16*s.*

Chips from a German Workshop; Essays on the Science of Religion, and on Mythology, Traditions & Customs. By F. MAX MÜLLER, K.M. 4 vols. 8vo. £1. 16*s*.

Language & Languages. A Revised Edition of Chapters on Language and Families of Speech. By F. W. FARRAR, D.D. F.R.S. Crown 8vo. 6*s*.

Grammar of Elocution. By JOHN MILLARD, Elocution Master in the City of London School. Second Edition. Fcp. 8vo. 3*s*. 6*d*.

The Essays and Contributions of A. K. H. B. Uniform Cabinet Editions in crown 8vo.

Autumn Holidays, 3*s*. 6*d*.
Changed Aspects of Unchanged Truths, price 3*s*. 6*d*.
Commonplace Philosopher, 3*s*. 6*d*.
Counsel and Comfort, 3*s*. 6*d*.
Critical Essays, 3*s*. 6*d*.
Graver Thoughts. Three Series, 3*s*. 6*d*. each.
Landscapes, Churches, and Moralities, 3*s*. 6*d*.
Leisure Hours in Town, 3*s*. 6*d*.
Lessons of Middle Age, 3*s*. 6*d*.
Our Little Life, 3*s*. 6*d*.
Present Day Thoughts, 3*s*. 6*d*.
Recreations of a Country Parson, Three Series, 3*s*. 6*d*. each.
Seaside Musings, 3*s*. 6*d*.
Sunday Afternoons, 3*s*. 6*d*.

DICTIONARIES and OTHER BOOKS of REFERENCE.

One-Volume Dictionary of the English Language. By R. G. LATHAM, M.A. M.D. Medium 8vo. 14*s*.

Larger Dictionary of the English Language. By R. G. LATHAM, M.A. M.D. Founded on Johnson's English Dictionary as edited by the Rev. H. J. TODD. 4 vols. 4to. £7.

English Synonymes. By E. J. WHATELY. Edited by R. WHATELY, D.D. Fcp. 8vo. 3*s*.

Roget's Thesaurus of English Words and Phrases, classified and arranged so as to facilitate the expression of Ideas, and assist in Literary Composition. Re-edited by the Author's Son, J. L. ROGET. Crown 8vo. 10*s*. 6*d*.

Handbook of the English Language. By R. G. LATHAM, M.A. M.D. Crown 8vo. 6*s*.

Contanseau's Practical ictionary of the French and English anguages. Post 8vo. price 7*s*. 6*d*.

Contanseau's Pocket Dictionary, French and English, abridged from the Practical Dictionary by the Author. Square 18mo. 3*s*. 6*d*.

A Practical Dictionary of the German and English Languages. By Rev. W. L. BLACKLEY, M.A. & Dr. C. M. FRIEDLÄNDER. Post 8vo. 7*s*. 6*d*.

A New Pocket Dictionary of the German and English Languages. By F. W. LONGMAN, Ball. Coll. Oxford. Square 18mo. 5*s*.

Becker's Gallus; Roman Scenes of the Time of Augustus. Translated by the Rev. F. METCALFE, M.A. Post 8vo. 7*s*. 6*d*.

Becker's Charicles; Illustrations of the Private Life of the Ancient Greeks. Translated by the Rev. F. METCALFE, M.A. Post 8vo. 7*s*. 6*d*.

A Dictionary of Roman and Greek Antiquities. With 2,000 Woodcuts illustrative of the Arts and Life of the Greeks and Romans. By A. RICH B.A. Crown 8vo. 7*s*. 6*d*.

A Greek-English Lexicon. By H. G. LIDDELL, D.D. Dean of Christchurch, and R. SCOTT, D.D. Dean of Rochester. Crown 4to. 36*s.*

Liddell & Scott's Lexicon, Greek and English, abridged for Schools. Square 12mo. 7*s.* 6*d.*

An English-Greek Lexicon, containing all the Greek Words used by Writers of good authority. By C. D. YONGE, M.A. 4to. 21*s.* School Abridgment, square 12mo. 8*s.* 6*d.*

A Latin-English Dictionary. By JOHN T. WHITE, D.D. Oxon. and J. E. RIDDLE, M.A. Oxon. Sixth Edition, revised. Quarto 21s.

White's Concise Latin-English Dictionary, for the use of University Students. Royal 8vo. 12*s.*

M'Culloch's Dictionary of Commerce and Commercial Navigation. Re-edited (1882), with a Supplement containing the most recent Information, by A. J. WILSON. With 48 Maps, Charts, and Plans. Medium 8vo. 63*s.*

Keith Johnston's General Dictionary of Geography, Descriptive, Physical, Statistical, and Historical; a complete Gazetteer of the World. Medium 8vo. 42*s.*

The Public Schools Atlas of Ancient Geography, in 28 entirely new Coloured Maps. Edited by the Rev. G. BUTLER, M.A. Imperial 8vo. or imperial 4to. 7*s.* 6*d.*

The Public Schools Atlas of Modern Geography, in 31 entirely new Coloured Maps. Edited by the Rev. G. BUTLER, M.A. Uniform, 5*s.*

ASTRONOMY and METEOROLOGY.

Outlines of Astronomy. By Sir J. F. W. HERSCHEL, Bart. M.A. Latest Edition, with Plates and Diagrams. Square crown 8vo. 12*s.*

The Moon, and the Condition and Configurations of its Surface. By E. NEISON, F.R.A.S. With 26 Maps and 5 Plates. Medium 8vo. price 31*s.* 6*d.*

Air and Rain; the Beginnings of a Chemical Climatology. By R. A. SMITH, F.R.S. 8vo. 24*s.*

Celestial Objects for Common Telescopes. By the Rev. T. W. WEBB, M.A. Fourth Edition, adapted to the Present State of Sidereal Science; Map, Plate, Woodcuts. Crown 8vo. 9*s.*

The Sun; Ruler, Light, Fire, and Life of the Planetary System. By R. A. PROCTOR, B.A. With Plates & Woodcuts. Crown 8vo. 14*s.*

Proctor's Orbs Around Us; a Series of Essays on the Moon & Planets, Meteors & Comets, the Sun & Coloured Pairs of Suns. With Chart and Diagrams. Crown 8vo. 7*s.* 6*d.*

Proctor's other Worlds than Ours; The Plurality of Worlds Studied under the Light of Recent Scientific Researches. With 14 Illustrations. Crown 8vo. 10*s.* 6*d.*

Proctor on the Moon; her Motions, Aspects, Scenery, and Physical Condition. With Plates, Charts, Woodcuts, and Lunar Photographs. Crown 8vo. 10*s.* 6*d.*

Proctor's Universe of Stars; Presenting Researches into and New Views respecting the Constitution of the Heavens. Second Edition, with 22 Charts and 22 Diagrams. 8vo. 10*s.* 6*d.*

Proctor's New Star Atlas, for the Library, the School, and the Observatory, in 12 Circular Maps (with 2 Index Plates). Crown 8vo. 5*s.*

Proctor's Larger Star Atlas, for the Library, in Twelve Circular Maps, with Introduction and 2 Index Plates. Folio, 15*s.* or Maps only, 12*s.* 6*d.*

Proctor's Essays on Astronomy. A Series of Papers on Planets and Meteors, the Sun and Sun-surrounding Space, Stars and Star Cloudlets. With 10 Plates and 24 Woodcuts. 8vo. price 12*s.*

Proctor's Transits of Venus; a Popular Account of Past and Coming Transits from the First Observed by Horrocks in 1639 to the Transit of 2012. Fourth Edition, including Suggestions respecting the approaching Transit in December 1882; with 20 Plates and 38 Woodcuts. 8vo. 8*s.* 6*d.*

Proctor's Studies of Venus-Transits; an Investigation of the Circumstances of the Transits of Venus in 1874 and 1882. With 7 Diagrams and 10 Plates. 8vo. 5*s.*

NATURAL HISTORY and PHYSICAL SCIENCE.

Ganot's Elementary Treatise on Physics, for the use of Colleges and Schools. Translated by E. ATKINSON, Ph.D. F.C.S. Tenth Edition. With 4 Coloured Plates and 844 Woodcuts. Large crown 8vo. 15*s.*

Ganot's Natural Philosophy for General Readers and Young Persons. Translated by E. ATKINSON, Ph.D. F.C.S. Fourth Edition; with 2 Plates and 471 Woodcuts. Crown 8vo. 7*s.* 6*d.*

Professor Helmholtz' Popular Lectures on Scientific Subjects. Translated and edited by EDMUND ATKINSON, Ph.D. F.C.S. With a Preface by Prof. TYNDALL, F.R.S. and 68 Woodcuts. 2 vols. crown 8vo. 15*s.* or separately, 7*s.* 6*d.* each.

Arnott's Elements of Physics or Natural Philosophy. Seventh Edition, edited by A. BAIN, LL.D. and A. S. TAYLOR, M.D. F.R.S. Crown 8vo. Woodcuts, 12*s.* 6*d.*

The Correlation of Physical Forces. By the Hon. Sir W. R. GROVE, F.R.S. &c. Sixth Edition, revised and augmented. 8vo. 15*s.*

A Treatise on Magnetism, General and Terrestrial. By H. LLOYD, D.D. D.C.L. 8vo. 10*s.* 6*d.*

The Mathematical and other Tracts of the late James M'Cullagh, F.T.C.D. Prof. of Nat. Philos. in the Univ. of Dublin. Edited by the Rev. J. H. JELLETT, B.D. and the Rev. S. HAUGHTON, M.D. 8vo. 15*s.*

Elementary Treatise on the Wave-Theory of Light. By H. LLOYD, D.D. D.C.L. 8vo. 10*s.* 6*d.*

Fragments of Science. By JOHN TYNDALL, F.R.S. Sixth Edition. 2 vols. crown 8vo. 16*s.*

Heat a Mode of Motion. By JOHN TYNDALL, F.R.S. Sixth Edition. Crown 8vo. 12*s.*

Sound. By JOHN TYNDALL, F.R.S. Fourth Edition, including Recent Researches. [*In the press.*

Essays on the Floating-Matter of the Air in relation to Putrefaction and Infection. By JOHN TYNDALL, F.R.S. With 24 Woodcuts. Crown 8vo. 7*s.* 6*d.*

Professor Tyndall's Lectures on Light, delivered in America in 1872 and 1873. With Portrait, Plate & Diagrams. Crown 8vo. 7*s.* 6*d.*

Professor Tyndall's Lessons in Electricity at the Royal Institution, 1875-6. With 58 Woodcuts. Crown 8vo. 2*s*. 6*d*.

Professor Tyndall's Notes of a Course of Seven Lectures on Electrical Phenomena and Theories, delivered at the Royal Institution. Crown 8vo. 1*s*. sewed, 1*s*. 6*d*. cloth.

Professor Tyndall's Notes of a Course of Nine Lectures on Light, delivered at the Royal Institution. Crown 8vo. 1*s*. swd., 1*s*. 6*d*. cloth.

Six Lectures on Physical Geography, delivered in 1876, with some Additions. By the Rev. SAMUEL HAUGHTON, F.R.S. M.D. D.C.L. With 23 Diagrams. 8vo. 15*s*.

An Introduction to the Systematic Zoology and Morphology of Vertebrate Animals. By A. MACALISTER, M.D. With 28 Diagrams. 8vo. 10*s*. 6*d*.

Text-Books of Science, Mechanical and Physical, adapted for the use of Artisans and of Students in Public and Science Schools. Small 8vo. with Woodcuts, &c.

Abney's Photography, 3*s*. 6*d*.
Anderson's (Sir John) Strength of Materials, price 3*s*. 6*d*.
Armstrong's Organic Chemistry, 3*s*. 6*d*.
Ball's Elements of Astronomy, 6*s*.
Barry's Railway Appliances, 3*s* 6*d*.
Bauerman's Systematic Mineralogy, 6*s*.
Bloxam's Metals, 3*s*. 6*d*.
Goodeve's Mechanics. 3*s*. 6d.
Gore's Electro-Metallurgy, 6*s*.
Griffin's Algebra and Trigonometry, 3*s*. 6*d*.
Jenkin's Electricity and Magnetism, 3*s*. 6*d*.
Maxwell's Theory of Heat, 3*s*. 6*d*.
Merrifield's Technical Arithmetic, 3*s*. 6*d*.
Miller's Inorganic Chemistry, 3*s*. 6*d*.
Preece & Sivewright's Telegraphy, 3*s*. 6*d*.
Rutley's Study of Rocks, 4*s*. 6*d*.
Shelley's Workshop Appliances, 3*s*. 6*d*.
Thomé's Structural and Physical Botany, 6*s*.
Thorpe's Quantitative Analysis, 4*s*. 6*d*.
Thorpe & Muir's Qualitative Analysis, 3*s*. 6*d*.
Tilden's Chemical Philosophy, 3*s*. 6*d*.
Unwin's Machine Design, 6*s*.
Watson's Plane and Solid Geometry, 3*s*. 6*d*.

Experimental Physiology, its Benefits to Mankind; with an Address on Unveiling the Statue of William Harvey at Folkestone August 1881. By RICHARD OWEN, F.R.S. &c. Crown 8vo. 5*s*.

The Comparative Anatomy and Physiology of the Vertebrate Animals. By RICHARD OWEN, F.R.S. With 1,472 Woodcuts. 3 vols. 8vo. £3. 13*s*. 6*d*.

Homes without Hands; a Description of the Habitations of Animals, classed according to their Principle of Construction. By the Rev. J. G. WOOD, M.A. With about 140 Vignettes on Wood. 8vo. 14*s*.

Wood's Strange Dwellings; a Description of the Habitations of Animals, abridged from 'Homes without Hands.' With Frontispiece and 60 Woodcuts. Crown 8vo. 5*s*. Sunbeam Edition, 4to. 6*d*.

Wood's Insects at Home; a Popular Account of British Insects, their Structure, Habits, and Transformations. 8vo. Woodcuts, 14*s*.

Wood's Insects Abroad; a Popular Account of Foreign Insects, their Structure, Habits, and Transformations. 8vo. Woodcuts, 14*s*.

Wood's Out of Doors; a Selection of Original Articles on Practical Natural History. With 6 Illustrations. Crown 8vo. 5*s*.

Wood's Bible Animals; a description of every Living Creature mentioned in the Scriptures. With 112 Vignettes. 8vo. 14*s*.

The Sea and its Living Wonders. By Dr. G. HARTWIG. 8vo. with many Illustrations, 10*s*. 6*d*.

Hartwig's Tropical World. With about 200 Illustrations. 8vo. 10*s*. 6*d*.

Hartwig's Polar World; a Description of Man and Nature in the Arctic and Antarctic Regions of the Globe. Maps, Plates & Woodcuts. 8vo. 10*s*. 6*d*. Sunbeam Edition, 6*d*.

Hartwig's Subterranean World. With Maps and Woodcuts. 8vo. 10*s*. 6*d*.

Hartwig's Aerial World; a Popular Account of the Phenomena and Life of the Atmosphere. Map, Plates, Woodcuts. 8vo. 10*s*. 6*d*.

A Familiar History of Birds. By E. STANLEY, D.D. Revised and enlarged, with 160 Woodcuts. Crown 8vo. 6*s*.

Rural Bird Life; Essays on Ornithology, with Instructions for Preserving Objects relating to that Science. By CHARLES DIXON. With Coloured Frontispiece and 44 Woodcuts by G. Pearson. Crown 8vo. 5*s*.

Country Pleasures; the Chronicle of a Year, chiefly in a Garden. By GEORGE MILNER. Second Edition, with Vignette Title-page. Cr. 8vo. 6*s*.

Rocks Classified and Described. By BERNHARD VON COTTA. An English Translation, by P. H. LAWRENCE, with English, German, and French Synonymes. Post 8vo. 14*s*.

The Geology of England and Wales; a Concise Account of the Lithological Characters, Leading Fossils, and Economic Products of the Rocks. By H. B. WOODWARD, F.G.S. Crown 8vo. Map & Woodcuts, 14*s*.

Keller's Lake Dwellings of Switzerland, and other Parts of Europe. Translated by JOHN E. LEE, F.S.A. F.G.S. With 206 Illustrations. 2 vols. royal 8vo. 42*s*.

Heer's Primæval World of Switzerland. Edited by JAMES HEYWOOD, M.A. F.R.S. With Map, Plates & Woodcuts. 2 vols. 8vo. 12*s*.

The Puzzle of Life; a Short History of Prehistoric Vegetable and Animal Life on the Earth. By A. NICOLS, F.R.G.S. With 12 Illustrations. Crown 8vo. 3*s*. 6*d*.

The Origin of Civilisation, and the Primitive Condition of Man; Mental and Social Condition of Savages. By Sir J. LUBBOCK, Bart. M.P. F.R.S. Fourth Edition, enlarged. 8vo. Woodcuts, 18*s*.

Proctor's Light Science for Leisure Hours; Familiar Essays on Scientific Subjects, Natural Phenomena, &c. 2 vols. crown 8vo. 7*s*. 6*d*. each.

Brande's Dictionary of Science, Literature, and Art. Re-edited by the Rev. Sir G. W. COX, Bart. M.A. 3 vols. medium 8vo. 63*s*.

Hullah's Course of Lectures on the History of Modern Music. 8vo. 8*s*. 6*d*.

Hullah's Second Course of Lectures on the Transition Period of Musical History. 8vo. 10*s*. 6*d*.

Loudon's Encyclopædia of Plants; the Specific Character, Description, Culture, History, &c. of all Plants found in Great Britain. With 12,000 Woodcuts. 8vo. 42*s*.

Loudon's Encyclopædia of Gardening; the Theory and Practice of Horticulture, Floriculture, Arboriculture & Landscape Gardening. With 1,000 Woodcuts. 8vo. 21*s*.

De Caisne & Le Maout's Descriptive and Analytical Botany. Translated by Mrs. HOOKER; edited and arranged by J. D. HOOKER, M.D. With 5,500 Woodcuts. Imperial 8vo. price 31*s*. 6*d*.

Rivers's Orchard-House; or, the Cultivation of Fruit Trees under Glass. Sixteenth Edition. Crown 8vo. with 25 Woodcuts, 5*s*.

The Rose Amateur's Guide. By THOMAS RIVERS. Latest Edition. Fcp. 8vo. 4*s*. 6*d*.

Elementary Botany, Theoretical and Practical; a Text-Book designed for Students of Science Classics. By H. EDMONDS, B.Sc. With 312 Woodcuts. Fcp. 8vo. 2*s*.

CHEMISTRY and PHYSIOLOGY.

Experimental Chemistry for Junior Students. By J. E. REYNOLDS, M. D. F. R. S. Prof. of Chemistry, Univ. of Dublin. Fcp. 8vo. PART I. 1*s.* 6*d.* PART II. 2*s.* 6*d.*

Practical Chemistry; the Principles of Qualitative Analysis. By W. A. TILDEN, F. C. S. Fcp. 8vo. 1*s.* 6*d.*

Miller's Elements of Chemistry, Theoretical and Practical. Re-edited, with Additions, by H. MACLEOD, F.C.S. 3 vols. 8vo.

PART I. CHEMICAL PHYSICS. 16*s.*

PART II. INORGANIC CHEMISTRY, 24*s.*

PART III. ORGANIC CHEMISTRY, 31*s.* 6*d.*

An Introduction to the Study of Inorganic Chemistry. By W. ALLEN MILLER, M. D. LL. D. late Professor of Chemistry, King's College, London. With 71 Woodcuts. Fcp. 8vo. 3*s.* 6*d.*

Annals of Chemical Medicine; including the Application of Chemistry to Physiology, Pathology, Therapeutics, Pharmacy, Toxicology & Hygiene. Edited by J. L. W. THUDICHUM, M. D. 2 vols. 8vo. 14*s.* each.

A Dictionary of Chemistry and the Allied Branches of other Sciences. Edited by HENRY WATTS, F. R. S. 9 vols. medium 8vo. £15. 2*s.* 6*d.*

Practical Inorganic Chemistry. An Elementary Text-Book of Theoretical and Practical Inorganic Chemistry. By W. JAGO, F. C. S. Second Edition, revised, with 37 Woodcuts. Fcp. 8vo. 2*s.*

Health in the House; Lectures on Elementary Physiology in its Application to the Daily Wants of Man and Animals. By Mrs. BUCKTON. Crown 8vo. Woodcuts, 2*s.*

The FINE ARTS and ILLUSTRATED EDITIONS.

The New Testament of Our Lord and Saviour Jesus Christ, Illustrated with Engravings on Wood after Paintings by the Early Masters chiefly of the Italian School. New Edition in course of publication in 18 Monthly Parts, 1*s.* each. 4to.

A Popular Introduction to the History of Greek and Roman Sculpture, designed to Promote the Knowledge and Appreciation of the Remains of Ancient Art. By WALTER C. PERRY. With 268 Illustrations engraved on Wood. Square crown 8vo. 31*s.* 6*d.*

Lord Macaulay's Lays of Ancient Rome, with Ivry and the Armada. With 41 Wood Engravings by G. Pearson from Original Drawings by J. R. Weguelin. Crown 8vo. 6*s.*

Lord Macaulay's Lays of Ancient Rome. With Ninety Illustrations, Original and from the Antique, engraved on Wood from Drawings by G. Scharf. Fcp. 4to. 21*s.*

The Three Cathedrals dedicated to St. Paul in London. By W. LONGMAN, F. S. A. With Illustrations. Square crown 8vo. 21*s.*

Lectures on Harmony, delivered at the Royal Institution. By G. A. MACFARREN. 8vo. 12*s.*

Moore's Lalla Rookh. TENNIEL'S Edition, with 68 Woodcut Illustrations. Crown 8vo. 10*s.* 6*d.*

Moore's Irish Melodies, MACLISE'S Edition, with 161 Steel Plates. Super-royal 8vo. 21*s.*

Sacred and Legendary Art. By Mrs. JAMESON. 6 vols. square crown 8vo. £5. 15*s*. 6*d*.

Jameson's Legends of the Saints and Martyrs. With 19 Etchings and 187 Woodcuts. 2 vols. 31*s*. 6*d*.

Jameson's Legends of the Monastic Orders. With 11 Etchings and 88 Woodcuts. 1 vol. 21*s*.

Jameson's Legends of the Madonna, the Virgin Mary as represented in Sacred and Legendary Art. With 27 Etchings and 165 Woodcuts. 1 vol. 21*s*.

Jameson's History of the Saviour, His Types and Precursors. Completed by Lady EASTLAKE. With 13 Etchings and 281 Woodcuts. 2 vols. 42*s*.

The USEFUL ARTS, MANUFACTURES, &c.

The Elements of Mechanism. By T. M. GOODEVE, M.A. Barrister-at-Law. New Edition, re-written and enlarged, with 342 Woodcuts. Crown 8vo. 6*s*.

Railways and Locomotives; a Series of Lectures delivered at the School of Military Engineering, Chatham. *Railways*, by J. W. BARRY, M. Inst. C.E. *Locomotives*, by Sir F. J. BRAMWELL, F.R.S. M. Inst. C.E. With 228 Woodcuts. 8vo. 21*s*.

Gwilt's Encyclopædia of Architecture, with above 1,600 Woodcuts. Revised and extended by W. PAPWORTH. 8vo. 52*s*. 6*d*.

Lathes and Turning, Simple, Mechanical, and Ornamental. By W. H. NORTHCOTT. Second Edition, with 338 Illustrations. 8vo. 18*s*.

Industrial Chemistry; a Manual for Manufacturers and for Colleges or Technical Schools; a Translation of PAYEN'S *Précis de Chimie Industrielle*. Edited by B. H. PAUL. With 698 Woodcuts. Medium 8vo. 42*s*.

The British Navy: its Strength, Resources, and Administration. By Sir T. BRASSEY, K.C.B. M.P. M.A. In 6 vols. 8vo. VOLS. I. and II. with many Illustrations, 14*s*. or separately, VOL. I. 10*s*. 6*d*. VOL. II. price 3*s*. 6*d*.

A Treatise on Mills and Millwork. By the late Sir W. FAIRBAIRN, Bart. C.E. Fourth Edition, with 18 Plates and 333 Woodcuts. 1 vol. 8vo. 25*s*.

Useful Information for Engineers. By the late Sir W. FAIRBAIRN, Bart. C.E. With many Plates and Woodcuts. 3 vols. crown 8vo. 31*s*. 6*d*.

The Application of Cast and Wrought Iron to Building Purposes. By the late Sir W. FAIRBAIRN, Bart. C.E. With 6 Plates and 118 Woodcuts. 8vo. 16*s*.

Hints on Household Taste in Furniture, Upholstery, and other Details. By C. L. EASTLAKE. Fourth Edition, with 100 Illustrations. Square crown 8vo. 14*s*.

Handbook of Practical Telegraphy. By R. S. CULLEY, Memb. Inst. C.E. Seventh Edition. Plates & Woodcuts. 8vo. 16*s*.

The Marine Steam Engine. A Treatise for the use of Engineering Students and Officers of the Royal Navy. By RICHARD SENNETT, Chief Engineer, Royal Navy. With numerous Illustrations and Diagrams. 8vo. 21*s*.

A Treatise on the Steam Engine, in its various applications to Mines, Mills, Steam Navigation, Railways and Agriculture. By J. BOURNE, C.E. With Portrait, 37 Plates, and 546 Woodcuts. 4to. 42*s.*

Bourne's Catechism of the Steam Engine, in its various Applications. Fcp. 8vo. Woodcuts, 6*s.*

Bourne's Recent Improvements in the Steam Engine. Fcp. 8vo. Woodcuts, 6*s.*

Bourne's Handbook of the Steam Engine, a Key to the Author's Catechism of the Steam Engine. Fcp. 8vo. Woodcuts, 9*s.*

Bourne's Examples of Steam and Gas Engines of the most recent Approved Types as employed in Mines, Factories, Steam Navigation, Railways and Agriculture. With 54 Plates & 356 Woodcuts. 4to. 70*s.*

Ure's Dictionary of Arts, Manufactures, and Mines. Seventh Edition, re-written and enlarged by R. HUNT, F.R.S. With 2,604 Woodcuts. 4 vols. medium 8vo. £7. 7*s.*

Kerl's Practical Treatise on Metallurgy. Adapted from the last German Edition by W. CROOKES, F.R.S. &c. and E. RÖHRIG, Ph.D. 3 vols. 8vo. with 625 Woodcuts, £4. 19*s.*

Cresy's Encyclopædia of Civil Engineering, Historical, Theoretical, and Practical. With above 3,000 Woodcuts, 8vo. 25*s.*

Ville on Artificial Manures, their Chemical Selection and Scientific Application to Agriculture. Translated and edited by W. CROOKES, F.R.S. With 31 Plates. 8vo. 21*s.*

Mitchell's Manual of Practical Assaying. Fifth Edition, revised, with the Recent Discoveries incorporated, by W. CROOKES, F.R.S. Crown 8vo. Woodcuts, 31*s.* 6*d.*

The Art of Perfumery, and the Methods of Obtaining the Odours of Plants; with Instructions for the Manufacture of Perfumes &c. By G. W. S. PIESSE, Ph.D. F.C.S. Fourth Edition, with 96 Woodcuts. Square crown 8vo. 21*s.*

Loudon's Encyclopædia of Gardening; the Theory and Practice of Horticulture, Floriculture, Arboriculture & Landscape Gardening. With 1,000 Woodcuts. 8vo. 21*s.*

Loudon's Encyclopædia of Agriculture; the Laying-out, Improvement, and Management of Landed Property; the Cultivation and Economy of the Productions of Agriculture. With 1,100 Woodcuts. 8vo. 21*s.*

RELIGIOUS and MORAL WORKS.

An Introduction to the Study of the New Testament, Critical, Exegetical, and Theological. By the Rev. S. DAVIDSON, D.D. LL.D. Revised Edition. 2 vols. 8vo. 30*s.*

History of the Papacy During the Reformation. By M. CREIGHTON, M.A. VOL. I. the Great Schism—the Council of Constance, 1378–1418. VOL. II. the Council of Basel—the Papal Restoration, 1418–1464. 2 vols. 8vo. 32*s.*

A History of the Church of England; Pre-Reformation Period. By the Rev. T. P. BOULTBEE, LL.D. 8vo. 15*s.*

Sketch of the History of the Church of England to the Revolution of 1688. By T. V. SHORT, D.D. Crown 8vo. 7*s.* 6*d.*

The English Church in the Eighteenth Century. By the Rev. C. J. ABBEY, and the Rev. J. H. OVERTON. 2 vols. 8vo. 36*s.*

An Exposition of the 39 Articles, Historical and Doctrinal. By E. H. BROWNE, D.D. Bishop of Winchester. Twelfth Edition. 8vo. 16*s.*

A Commentary on the 39 Articles, forming an Introduction to the Theology of the Church of England. By the Rev. T. P. BOULTBEE, LL.D. New Edition. Crown 8vo. 6*s.*

Sermons preached mostly in the Chapel of Rugby School by the late T. ARNOLD, D.D. 6 vols. crown 8vo. 30*s.* or separately, 5*s.* each.

Historical Lectures on the Life of Our Lord Jesus Christ. By C. J. ELLICOTT, D.D. 8vo. 12*s.*

The Eclipse of Faith; or a Visit to a Religious Sceptic. By HENRY ROGERS. Fcp. 8vo. 5*s.*

Defence of the Eclipse of Faith. By H. ROGERS. Fcp. 8vo. 3*s.* 6*d.*

Nature, the Utility of Religion, and Theism. Three Essays by JOHN STUART MILL. 8vo. 10*s.* 6*d.*

A Critical and Grammatical Commentary on St. Paul's Epistles. By C. J. ELLICOTT, D.D. 8vo. Galatians, 8*s.* 6*d.* Ephesians, 8*s.* 6*d.* Pastoral Epistles, 10*s.* 6*d.* Philippians, Colossians, & Philemon, 10*s.* 6*d.* Thessalonians, 7*s.* 6*d.*

The Life and Letters of St. Paul. By ALFRED DEWES, M.A. LL.D. D.D. Vicar of St. Augustine's Pendlebury. With 4 Maps. 8vo. 7*s.* 6*d.*

Conybeare & Howson's Life and Epistles of St. Paul. Three Editions, copiously illustrated.

Library Edition, with all the Original Illustrations, Maps, Landscapes on Steel, Woodcuts, &c. 2 vols. 4to. 42*s.*

Intermediate Edition, with a Selection of Maps, Plates, and Woodcuts. 2 vols. square crown 8vo. 21*s.*

Student's Edition, revised and condensed, with 46 Illustrations and Maps. 1 vol. crown 8vo. 7*s.* 6*d.*

Smith's Voyage & Shipwreck of St. Paul; with Dissertations on the Life and Writings of St. Luke, and the Ships and Navigation of the Ancients. Fourth Edition, with numerous Illustrations. Crown 8vo. 7*s.* 6*d.*

A Handbook to the Bible, or, Guide to the Study of the Holy Scriptures derived from Ancient Monuments and Modern Exploration. By F. R. CONDER, and Lieut. C. R. CONDER, R.E. Third Edition, Maps. Post 8vo. 7*s.* 6*d.*

Bible Studies. By M. M. KALISCH, Ph.D. PART I. *The Prophecies of Balaam.* 8vo. 10*s.* 6*d.* PART II. *The Book of Jonah.* 8vo. price 10*s.* 6*d.*

Historical and Critical Commentary on the Old Testament; with a New Translation. By M. M. KALISCH, Ph.D. Vol. I. Genesis, 8vo. 18*s.* or adapted for the General Reader, 12*s.* Vol. II. Exodus, 15*s.* or adapted for the General Reader, 12*s.* Vol. III. Leviticus, Part I. 15*s.* or adapted for the General Reader, 8*s.* Vol. IV. Leviticus, Part II. 15*s.* or adapted for the General Reader, 8*s.*

The Four Gospels in Greek, with Greek-English Lexicon. By JOHN T. WHITE, D.D. Oxon. Square 32mo. 5*s.*

Ewald's History of Israel. Translated from the German by J. E. CARPENTER, M.A. with Preface by R. MARTINEAU, M.A. 5 vols. 8vo. 63*s.*

Ewald's Antiquities of Israel. Translated from the German by H. S. SOLLY, M.A. 8vo. 12*s.* 6*d.*

The New Man and the Eternal Life; Notes on the Reiterated Amens of the Son of God. By A. JUKES. Second Edition. Cr. 8vo. 6*s.*

The Types of Genesis, briefly considered as revealing the Development of Human Nature. By A. JUKES. Crown 8vo. 7*s.* 6*d.*

The Second Death and the Restitution of all Things; with some Preliminary Remarks on the Nature and Inspiration of Holy Scripture By A. JUKES. Crown 8vo. 3*s*. 6*d*.

Supernatural Religion; an Inquiry into the Reality of Divine Revelation. Complete Edition, thoroughly revised. 3 vols. 8vo. 36*s*.

Lectures on the Origin and Growth of Religion, as illustrated by the Religions of India. By F. MAX MÜLLER, K.M. 8vo. price 10*s*. 6*d*.

Introduction to the Science of Religion, Four Lectures delivered at the Royal Institution; with Essays on False Analogies and the Philosophy of Mythology. By F. MAX MÜLLER, K.M. Crown 8vo. 10*s*. 6*d*.

The Gospel for the Nineteenth Century. Fourth Edition. 8vo. price 10*s*. 6*d*.

Christ our Ideal, an Argument from Analogy. By the same Author. 8vo. 8*s*. 6*d*.

Passing Thoughts on Religion. By Miss SEWELL. Fcp. 8vo. price 3*s*. 6*d*.

Preparation for the Holy Communion; the Devotions chiefly from the works of Jeremy Taylor. By Miss SEWELL. 32mo. 3*s*.

Private Devotions for Young Persons. Compiled by Miss SEWELL. 18mo. 2*s*.

Bishop Jeremy Taylor's Entire Works; with Life by Bishop Heber. Revised and corrected by the Rev. C. P. EDEN. 10 vols. £5. 5*s*.

The Wife's Manual; or Prayers, Thoughts, and Songs on Several Occasions of a Matron's Life. By the late W. CALVERT, Minor Canon of St. Paul's. Printed and ornamented in the style of *Queen Elizabeth's Prayer Book*. Crown 8vo. 6*s*.

Hymns of Praise and Prayer. Corrected and edited by Rev. JOHN MARTINEAU, LL.D. Crown 8vo. 4*s*. 6*d*. 32mo. 1*s*. 6*d*.

Spiritual Songs for the Sundays and Holidays throughout the Year. By J. S. B. MONSELL, LL.D. Fcp. 8vo. 5*s*. 18mo. 2*s*.

Christ the Consoler; a Book of Comfort for the Sick. By ELLICE HOPKINS. Second Edition. Fcp. 8vo. 2*s*. 6*d*.

Lyra Germanica; Hymns translated from the German by Miss C. WINKWORTH. Fcp. 8vo. 5*s*.

Hours of Thought on Sacred Things; Two Volumes of Sermons. By JAMES MARTINEAU, D.D. LL.D. 2 vols. crown 8vo. 7*s*. 6*d*. each.

Endeavours after the Christian Life; Discourses. By JAMES MARTINEAU, D.D. LL.D. Fifth Edition. Crown 8vo. 7*s*. 6*d*.

The Pentateuch & Book of Joshua Critically Examined. By J. W. COLENSO, D.D. Bishop of Natal. Crown 8vo. 6*s*.

Elements of Morality, In Easy Lessons for Home and School Teaching. By Mrs. CHARLES BRAY. Crown 8vo. 2*s*. 6*d*.

TRAVELS, VOYAGES, &c.

Three in Norway. By Two of THEM. With a Map and 59 Illustrations on Wood from Sketches by the Authors. Crown 8vo. 10*s*. 6*d*.

Roumania, Past and Present. By JAMES SAMUELSON. With 2 Maps, 3 Autotype Plates & 31 Illustrations on Wood. 8vo. 16*s*.

Sunshine and Storm in the East, or Cruises to Cyprus and Constantinople. By Lady BRASSEY. Cheaper Edition, with 2 Maps and 114 Illustrations engraved on Wood. Cr. 8vo. 7*s.* 6*d.*

A Voyage in the 'Sunbeam,' our Home on the Ocean for Eleven Months. By Lady BRASSEY. Cheaper Edition, with Map and 65 Wood Engravings. Crown 8vo. 7*s.* 6*d.* School Edition, fcp. 2*s.* Popular Edition, 4to. 6*d.*

Eight Years in Ceylon. By Sir SAMUEL W. BAKER, M.A. Crown 8vo. Woodcuts, 7*s.* 6*d.*

The Rifle and the Hound in Ceylon. By Sir SAMUEL W. BAKER, M.A. Crown 8vo. Woodcuts, 7*s.* 6*d.*

Sacred Palmlands; or, the Journal of a Spring Tour in Egypt and the Holy Land. By A. G. WELD. Crown 8vo. 7*s.* 6*d.*

Wintering in the Riviera; with Notes of Travel in Italy and France, and Practical Hints to Travellers. By W. MILLER. With 12 Illustrations. Post 8vo. 7*s.* 6*d.*

San Remo and the Western Riviera, climatically and medically considered. By A. HILL HASSALL, M.D. Map and Woodcuts. Crown 8vo. 10*s.* 6*d.*

Himalayan and Sub-Himalayan Districts of British India, their Climate, Medical Topography, and Disease Distribution. By F. N. MACNAMARA, M.D. With Map and Fever Chart. 8vo. 21*s.*

The Alpine Club Map of Switzerland, with parts of the Neighbouring Countries, on the scale of Four Miles to an Inch. Edited by R. C. NICHOLS, F.R.G.S. 4 Sheets in Portfolio, 42*s.* coloured, or 34*s.* uncoloured.

Enlarged Alpine Club Map of the Swiss and Italian Alps, on the Scale of 3 English Statute Miles to 1 Inch, in 8 Sheets, price 1*s.* 6*d.* each.

The Alpine Guide. By JOHN BALL, M.R.I.A. Post 8vo. with Maps and other Illustrations :—

The Eastern Alps, 10*s.* 6*d.*

Central Alps, including all the Oberland District, 7*s.* 6*d.*

Western Alps, including Mont Blanc, Monte Rosa, Zermatt, &c. Price 6*s.* 6*d.*

On Alpine Travelling and the Geology of the Alps. Price 1*s.* Either of the Three Volumes or Parts of the 'Alpine Guide' may be had with this Introduction prefixed, 1*s.* extra.

WORKS of FICTION.

In Trust; the Story of a Lady and her Lover. By Mrs. OLIPHANT. Cabinet Edition. Cr. 8vo. 6*s.*

The Hughenden Edition of the Novels and Tales of the Earl of Beaconsfield, K.G. from Vivian Grey to Endymion. With Maclise's Portrait of the Author, a later Portrait on Steel from a recent Photograph, and a Vignette to each volume. Eleven Volumes, cr. 8vo. 42*s.*

Novels and Tales. By the Right Hon. the EARL of BEACONSFIELD, K.G. The Cabinet Edition. Eleven Volumes, crown 8vo. 6*s.* each.

The Novels and Tales of the Right Hon. the Earl of Beaconsfield, K.G. Modern Novelist's Library Edition, complete in Eleven Volumes, crown 8vo. price 22*s.* boards, or 27*s.* 6*d.* cloth.

Novels and Tales by the Earl of Beaconsfield, K.G. Modern Novelist's Library Edition, complete in Eleven Volumes, crown 8vo. cloth extra, with gilt edges, price 33*s.*

Whispers from Fairy-land. By Lord BRABOURNE. With 9 Illustrations. Crown 8vo. 3*s.* 6*d.*

Higgledy-Piggledy. By Lord BRABOURNE. With 9 Illustrations. Crown 8vo. 3*s.* 6*d.*

Stories and Tales. By ELIZABETH M. SEWELL. Cabinet Edition, in Ten Volumes, crown 8vo. price 3*s.* 6*d.* each, in cloth extra, with gilt edges:—

Amy Herbert. Gertrude.
The Earl's Daughter.
The Experience of Life.
Cleve Hall. Ivors.
Katharine Ashton.
Margaret Percival.
Laneton Parsonage. Ursula.

The Modern Novelist's Library. Each work complete in itself, price 2*s.* boards, or 2*s.* 6*d.* cloth:—

By the Earl of BEACONSFIELD, K.G.

Endymion.

Lothair.	Henrietta Temple.
Coningsby.	Contarini Fleming, &c.
Sybil.	Alroy, Ixion, &c.
Tancred.	The Young Duke, &c.
Venetia.	Vivian Grey, &c.

By ANTHONY TROLLOPE.

Barchester Towers.
The Warden.

By Major WHYTE-MELVILLE.

Digby Grand.	Good for Nothing.
General Bounce.	Holmby House.
Kate Coventry.	The Interpreter.
The Gladiators.	Queen's Maries.

By the Author of 'The Rose Garden.'

Unawares.

By the Author of 'Mlle. Mori.'

The Atelier du Lys.
Mademoiselle Mori.

By Various Writers.

Atherston Priory.
The Burgomaster's Family.
Elsa and her Vulture.
The Six Sisters of the Valleys.

POETRY and THE DRAMA.

Poetical Works of Jean Ingelow. New Edition, reprinted, with Additional Matter, from the 23rd and 6th Editions of the two volumes respectively; with 2 Vignettes. 2 vols. fcp. 8vo. 12*s.*

Faust. From the German of GOETHE. By T. E. WEBB, LL.D. Reg. Prof. of Laws & Public Orator in the Univ. of Dublin. 8vo. 12*s.* 6*d.*

Goethe's Faust. A New Translation, chiefly in Blank Verse; with a complete Introduction and copious Notes. By JAMES ADEY BIRDS, B.A. F.G.S. Large crown 8vo. 12*s.* 6*d.*

Goethe's Faust. The German Text, with an English Introduction and Notes for Students. By ALBERT M. SELSS, M.A. Ph.D. Crown 8vo. 5*s.*

Lays of Ancient Rome; with Ivry and the Armada. By LORD MACAULAY.

CABINET EDITION, post 8vo. 3*s.* 6*d.*
CHEAP EDITION, fcp. 8vo. 1*s.* sewed; 1*s.* 6*d.* cloth; 2*s.* 6*d.* cloth extra with gilt edges.

Lord Macaulay's Lays of Ancient Rome, with Ivry and the Armada. With 41 Wood Engravings by G. Pearson from Original Drawings by J. R. Weguelin. Crown 8vo. 6*s.*

Festus, a Poem. By PHILIP JAMES BAILEY. 10th Edition, enlarged & revised. Crown 8vo. 12*s.* 6*d.*

The Poems of Virgil translated into English Prose. By JOHN CONINGTON, M.A. Crown 8vo. 9*s.*

The Iliad of Homer, Homometrically translated by C. B. CAYLEY. 8vo. 12*s*. 6*d*.

Bowdler's Family Shakspeare. Genuine Edition, in 1 vol. medium 8vo. large type, with 36 Woodcuts, 14*s*. or in 6 vols. fcp. 8vo. 21*s*.

The Æneid of Virgil. Translated into English Verse. By J. CONINGTON, M.A. Crown 8vo. 9*s*.

Southey's Poetical Works, with the Author's last Corrections and Additions. Medium 8vo. with Portrait, 14*s*.

RURAL SPORTS, HORSE and CATTLE MANAGEMENT, &c.

William Howitt's Visits to Remarkable Places, Old Halls, Battle-Fields, Scenes illustrative of Striking Passages in English History and Poetry. New Edition, with 80 Illustrations engraved on Wood. Crown 8vo. 7*s*. 6*d*.

Dixon's Rural Bird Life; Essays on Ornithology, with Instructions for Preserving Objects relating to that Science. With 44 Woodcuts. Crown 8vo. 5*s*.

A Book on Angling; or, Treatise on the Art of Fishing in every branch; including full Illustrated Lists of Salmon Flies. By FRANCIS FRANCIS. Post 8vo. Portrait and Plates, 15*s*.

Wilcocks's Sea-Fisherman: comprising the Chief Methods of Hook and Line Fishing, a glance at Nets, and remarks on Boats and Boating. Post 8vo. Woodcuts, 12*s*. 6*d*.

The Fly-Fisher's Entomology. By ALFRED RONALDS. With 20 Coloured Plates. 8vo. 14*s*.

The Dead Shot, or Sportsman's Complete Guide; a Treatise on the Use of the Gun, with Lessons in the Art of Shooting Game of All Kinds, and Wild-Fowl, also Pigeon-Shooting, and Dog-Breaking. By MARKSMAN. Fifth Edition, with 13 Illustrations. Crown 8vo. 10*s*. 6*d*.

Horses and Roads; or, How to Keep a Horse Sound on his Legs. By FREE-LANCE. Second Edition. Crown 8vo. 6*s*.

Horses and Riding. By GEORGE NEVILE, M.A. With 31 Illustrations. Crown 8vo. 6*s*.

Horses and Stables. By Major-General Sir F. FITZWYGRAM, Bart. Second Edition, revised and enlarged; with 39 pages of Illustrations containing very numerous Figures. 8vo. 10*s*. 6*d*.

Youatt on the Horse. Revised and enlarged by W. WATSON, M.R.C.V.S. 8vo. Woodcuts, 7*s*. 6*d*.

Youatt's Work on the Dog. Revised and enlarged. 8vo. Woodcuts, 6*s*.

The Dog in Health and Disease. By STONEHENGE. Third Edition, with 78 Wood Engravings. Square crown 8vo. 7*s*. 6*d*.

The Greyhound. By STONEHENGE. Revised Edition, with 25 Portraits of Greyhounds, &c. Square crown 8vo. 15*s*.

A Treatise on the Diseases of the Ox; being a Manual of Bovine Pathology specially adapted for the use of Veterinary Practitioners and Students. By J. H. STEEL, M.R.C.V.S. F.Z.S. With 2 Plates and 116 Woodcuts. 8vo. 15*s*.

Stables and Stable Fittings. By W. MILES. Imp. 8vo. with 13 Plates, 15*s.*

The Horse's Foot, and How to keep it Sound. By W. MILES. Imp. 8vo. Woodcuts, 12*s.* 6*d.*

A Plain Treatise on Horse-shoeing. By W. MILES. Post 8vo. Woodcuts, 2*s.* 6*d.*

Remarks on Horses' Teeth, addressed to Purchasers. By W. MILES. Post 8vo. 1*s.* 6*d.*

WORKS of UTILITY and GENERAL INFORMATION.

Maunder's Biographical Treasury. Reconstructed with 1,700 additional Memoirs, by W. L. R. CATES. Fcp. 8vo. 6*s.*

Maunder's Treasury of Natural History; or, Popular Dictionary of Zoology. Fcp. 8vo. with 900 Woodcuts, 6*s.*

Maunder's Treasury of Geography, Physical, Historical, Descriptive, and Political. With 7 Maps and 16 Plates. Fcp. 8vo. 6*s.*

Maunder's Historical Treasury; Outlines of Universal History, Separate Histories of all Nations. Revised by the Rev. Sir G. W. COX, Bart. M.A. Fcp. 8vo. 6*s.*

Maunder's Treasury of Knowledge and Library of Reference; comprising an English Dictionary and Grammar, Universal Gazetteer, Classical Dictionary, Chronology, Law Dictionary, &c. Fcp. 8vo. 6*s.*

Maunder's Scientific and Literary Treasury; a Popular Encyclopædia of Science, Literature, and Art. Fcp. 8vo. 6*s.*

The Treasury of Botany, or Popular Dictionary of the Vegetable Kingdom. Edited by J. LINDLEY, F.R.S. and T. MOORE, F.L.S. With 274 Woodcuts and 20 Steel Plates. Two Parts, fcp. 8vo. 12*s.*

The Treasury of Bible Knowledge; a Dictionary of the Books, Persons, Places, and Events, of which mention is made in Holy Scripture. By the Rev. J. AYRE, M.A. Maps, Plates and Woodcuts. Fcp. 8vo. 6*s.*

Black's Practical Treatise on Brewing; with Formulæ for Public Brewers, and Instructions for Private Families. 8vo. 10*s.* 6*d.*

The Theory of the Modern Scientific Game of Whist. By W. POLE, F.R.S. Thirteenth Edition. Fcp. 8vo. 2*s.* 6*d.*

The Correct Card; or, How to Play at Whist; a Whist Catechism. By Major A. CAMPBELL-WALKER, F.R.G.S. Fourth Edition. Fcp. 8vo. 2*s.* 6*d.*

The Cabinet Lawyer; a Popular Digest of the Laws of England, Civil, Criminal, and Constitutional. Twenty-Fifth Edition. Fcp. 8vo. 9*s.*

Chess Openings. By F.W. LONGMAN, Balliol College, Oxford. New Edition. Fcp. 8vo. 2*s.* 6*d.*

Pewtner's Comprehensive Specifier; a Guide to the Practical Specification of every kind of Building-Artificer's Work. Edited by W. YOUNG. Crown 8vo. 6*s.*

Cookery and Housekeeping; a Manual of Domestic Economy for Large and Small Families. By Mrs. HENRY REEVE. Third Edition, with 8 Coloured Plates and 37 Woodcuts. Crown 8vo. 7*s.* 6*d.*

Modern Cookery for Private Families, reduced to a System of Easy Practice in a Series of carefully-tested Receipts. By ELIZA ACTON. With upwards of 150 Woodcuts. Fcp. 8vo. 4*s.* 6*d.*

Food and Home Cookery. A Course of Instruction in Practical Cookery and Cleaning, for Children in Elementary Schools. By Mrs. BUCKTON. Crown 8vo. Woodcuts, 2*s.*

Bull's Hints to Mothers on the Management of their Health during the Period of Pregnancy and in the Lying-in Room. Fcp. 8vo. 1*s.* 6*d.*

Bull on the Maternal Management of Children in Health and Disease. Fcp. 8vo. 1*s.* 6*d.*

American Farming and Food. By FINLAY DUN. Crown 8vo. 10*s.* 6*d.*

Landlords and Tenants in Ireland. By FINLAY DUN. Crown 8vo. 6*s.*

The Farm Valuer. By JOHN SCOTT. Crown 8vo. 5*s.*

Rents and Purchases; or, the Valuation of Landed Property, Woods, Minerals, Buildings, &c. By JOHN SCOTT. Crown 8vo. 6*s.*

Economic Studies. By the late WALTER BAGEHOT, M.A. Edited by R. H. HUTTON. 8vo. 10*s.* 6*d.*

Health in the House; Lectures on Elementary Physiology in its Application to the Daily Wants of Man and Animals. By Mrs. BUCKTON. Crown 8vo. Woodcuts, 2*s.*

Economics for Beginners By H. D. MACLEOD, M.A. Small crown 8vo. 2*s.* 6*d.*

The Elements of Economics. By H. D. MACLEOD, M.A. In 2 vols. VOL. I. crown 8vo. 7*s.* 6*d.*

The Elements of Banking. By H. D. MACLEOD, M.A. Fourth Edition. Crown 8vo. 5*s.*

The Theory and Practice of Banking. By H. D. MACLEOD, M.A. 2 vols. 8vo. 26*s.*

The Patentee's Manual; a Treatise on the Law and Practice of Letters Patent, for the use of Patentees and Inventors. By J. JOHNSON and J. H. JOHNSON. Fourth Edition, enlarged. 8vo. price 10*s.* 6*d.*

Willich's Popular Tables Arranged in a New Form, giving Information &c. equally adapted for the Office and the Library. 9th Edition, edited by M. MARRIOTT. Crown 8vo. 10*s.*

INDEX.

Spottiswoode & Co. Printers, New-street Square, London.

CPSIA information can be obtained
at www.ICGtesting.com
Printed in the USA
LVHW081925220223
740156LV00004B/196

9 781017 728125